Springer Texts in Statistics

Advisors:
George Casella Stephen Fienberg Ingram Olkin

Springer
New York
Berlin
Heidelberg
Hong Kong
London
Milan
Paris
Tokyo

Springer Texts in Statistics

Alfred: Elements of Statistics for the Life and Social Sciences
Berger: An Introduction to Probability and Stochastic Processes
Bilodeau and Brenner: Theory of Multivariate Statistics
Blom: Probability and Statistics: Theory and Applications
Brockwell and Davis: Introduction to Times Series and Forecasting, Second Edition
Carmona: Statistical Analysis of Financial Data in S-Plus
Chow and Teicher: Probability Theory: Independence, Interchangeability, Martingales, Third Edition
Christensen: Advanced Linear Modeling: Multivariate, Time Series, and Spatial Data; Nonparametric Regression and Response Surface Maximization, Second Edition
Christensen: Log-Linear Models and Logistic Regression, Second Edition
Christensen: Plane Answers to Complex Questions: The Theory of Linear Models, Third Edition
Creighton: A First Course in Probability Models and Statistical Inference
Davis: Statistical Methods for the Analysis of Repeated Measurements
Dean and Voss: Design and Analysis of Experiments
du Toit, Steyn, and Stumpf: Graphical Exploratory Data Analysis
Durrett: Essentials of Stochastic Processes
Edwards: Introduction to Graphical Modelling, Second Edition
Finkelstein and Levin: Statistics for Lawyers
Flury: A First Course in Multivariate Statistics
Jobson: Applied Multivariate Data Analysis, Volume I: Regression and Experimental Design
Jobson: Applied Multivariate Data Analysis, Volume II: Categorical and Multivariate Methods
Kalbfleisch: Probability and Statistical Inference, Volume I: Probability, Second Edition
Kalbfleisch: Probability and Statistical Inference, Volume II: Statistical Inference, Second Edition
Karr: Probability
Keyfitz: Applied Mathematical Demography, Second Edition
Kiefer: Introduction to Statistical Inference
Kokoska and Nevison: Statistical Tables and Formulae
Kulkarni: Modeling, Analysis, Design, and Control of Stochastic Systems
Lange: Applied Probability
Lehmann: Elements of Large-Sample Theory
Lehmann: Testing Statistical Hypotheses, Second Edition
Lehmann and Casella: Theory of Point Estimation, Second Edition
Lindman: Analysis of Variance in Experimental Design
Lindsey: Applying Generalized Linear Models

(continued after index)

René A. Carmona

Statistical Analysis of Financial Data in S-Plus

With 144 Figures

René A. Carmona
Department of Statistics
University of Princeton
Princeton, NJ 08544-5263
USA
rcarmona@princeton.edu

Editorial Board

George Casella
Department of Statistics
University of Florida
Gainesville, FL 32611-8545
USA

Stephen Fienberg
Department of Statistics
Carnegie Mellon University
Pittsburgh, PA 15213-3890
USA

Ingram Olkin
Department of Statistics
Stanford University
Stanford, CA 94305
USA

Library of Congress Cataloging-in-Publication Data
Carmona, R. (René)
Statistical analysis of financial data in S-PLUS / René A. Carmona.
p. cm. — (Springer texts in statistics)
Based on the author's lecture notes for a course at Princeton University.
Includes bibliographical references and index.

1. Finance—Mathematical models. 2. Finance—Econometric models. 3. S-Plus. I. Title.
II. Series.
HG106.C37 2003
332′01′51955—dc22 2003066218

ISBN 978-1-4419-1908-3 Printed on acid-free paper.

© 2004 Springer-Verlag New York, Inc.
Softcover reprint of the hardcover 1st edition 2004
All rights reserved. This work may not be translated or copied in whole or in part without the written permission of the publisher (Springer-Verlag New York, Inc., 175 Fifth Avenue, New York, NY 10010, USA), except for brief excerpts in connection with reviews or scholarly analysis. Use in connection with any form of information storage and retrieval, electronic adaptation, computer software, or by similar or dissimilar methodology now known or hereafter developed is forbidden.
The use in this publication of trade names, trademarks, service marks, and similar terms, even if they are not identified as such, is not to be taken as an expression of opinion as to whether or not they are subject to proprietary rights.

9 8 7 6 5 4 3 2 1 SPIN 10953280

Springer-Verlag is a part of *Springer Science+Business Media*

springeronline.com

To Chanel, Chelsea and Stéphanie

Preface

This book grew out of lectures notes written for a one-semester junior statistics course offered to the undergraduate students majoring in the Department of Operations Research and Financial Engineering at Princeton University. Tidbits of the history of this course will shed light on the nature and spirit of the book.

The purpose of the course is to introduce the students to modern data analysis with an emphasis on a domain of application that is of interest to most of them: financial engineering. The prerequisites for this course are minimal, however it is fair to say that all of the students have already taken a basic introductory statistics course. Thus the elementary notions of random variables, expectation and correlation are taken for granted, and earlier exposure to statistical inference (estimation, tests and confidence intervals) is assumed. It is also expected that the students are familiar with a minimum of linear algebra as well as vector and matrix calculus.

Because of my background, the course is both computational and mathematical in nature. Most problems considered are formulated in a rigorous manner. Mathematical facts are motivated by applications, stated precisely, justified at an intuitive level, but essentially never proven rigorously. The emphasis is more on the relevance of concepts and on the practical use of tools, rather than on their theoretical underpinnings.

I chose to illustrate concepts, manipulate data, build models, and implement estimation and prediction procedures in the `S-Plus` computer environment. For this reason an introduction to `S` and `S-Plus` (reproduced in appendix) is offered at the beginning of the course each semester, and many lectures are sprinkled with the `S` commands needed to perform the analyses discussed in class. The first two incarnations of this course were using `S-Plus` on Unix platforms and not all the students were able to cope with the steep learning curve. Moreover, the two textbooks used for the class did not seem to be of very much help to the students. So I decided to prepare lecture notes focused on the material covered in class, and to switch to Windows in order to work with a friendlier implementation of S.

The present manuscript is a polished version of the class notes. It is divided into three parts. Part I, *Exploratory Data Analysis,* reviews the most commonly used methods of statistical data exploration. Part II, *Regression,* introduces the students to modern regression with an emphasis on robustness and non-parametric techniques. Part III, *Time Series and State Space Models,* is concerned with the theories of time series and of state space models.

Contents

Part I is a patchwork of many exploratory data analysis techniques. It begins with a discussion of various methods of density estimation, including histograms and kernel density estimators. Since the emphasis of the course is on financial applications, the notion of *heavy tail* is immediately showcased with examples. A good part of the first chapter is concerned with the practical estimation of heavy tailed distributions, their detection, their estimation and their simulation. We use the statistical concept of percentile to introduce the notion of value-at-risk so important in the financial industry, and we demonstrate its use on a couple of illustrative examples. The second chapter is concerned with multivariate distributions and the various concepts of dependence. We study the classical correlation coefficients, but we also spend a good amount of time understanding the notion of copula, and the important role it plays when the marginal distributions have heavy tails. As in the univariate case, we learn how to detect unusual dependencies, to estimate them, and to simulate them. We also give a complete discussion of principal component analysis and illustrate its power on two applications to fixed income markets.

Part II is concerned with regression, and it is naturally divided into two chapters: the first devoted to parametric methods, and the second to non-parametric ones. Chapter 3 deals with linear models and their applications. The notion of robustness is introduced and examples are used to illustrate the differences between least squares and least absolute deviations regressions. Applications of linear models include polynomial and more general nonlinear regressions. We use financial examples throughout and we analyze the term structure of interest rates in detail. Chapter 4 is concerned with nonparametric regression. We compare the properties of data smoothers for univariate data, and we analyze in detail the multivariate kernel regression and density estimation for intermediate values of the dimension. For large values of the dimension we consider projection pursuit. To illustrate, we analyze energy forward curves and intra-day tick data on S&P 500 futures contracts. The last part of this chapter is devoted to a demonstration of the use of semi-parametric and nonparametric methods in option pricing. We review the derivation of the classical Black-Scholes pricing formula, we illustrate its shortcomings, and we walk the reader through the implementation of modern regression techniques as pricing alternatives. The actual implementations are done on liquid S&P 500 futures option data.

The first chapter of Part III is devoted to the classical linear models for time series, and to the idiosyncrasies of the `S-Plus` objects and methods needed to fit them. We discuss auto regressive and moving-average models, and we give examples of their use in practice. The main application of the material of this chapter is concerned with the analysis of temperature data. Even if it may not appear to be much of a financial application at first, we recast this analysis in the framework of financial risk management via a thorough discussion of the booming market of weather derivatives. We give practical examples to illustrate the use of the statistical techniques introduced in this chapter to the pricing of these new financial instruments.

In the following two chapters, we turn to the analysis of partially observed state space systems. Chapter 6 deals with linear models and the classical Kalman filter. For illustration purposes, we study two financial applications, one related to an extension of the CAPM model, and a second dealing with the analysis of quarterly company earnings. Chapter 7 is devoted to the analysis of nonlinear time series. We first consider the natural generalizations of the linear time series models and we provide an extensive review of the theory and the practice of the famous ARCH and GARCH models. We also consider models from continuous time finance through their discretized forms. A special section is devoted to the use of scenarios for economic modeling. We concentrate on scenarios for a stock index and the short and long interest rates. These scenarios are of crucial importance in risk management where they are used as input to large stochastic optimization programs. Finally, we revisit the theory presented in the case of partially observed linear systems, and we extend the filtering paradigm to nonlinear systems with the help of recent advances in Monte Carlo techniques. We give several applications of this material, including to the estimation of stochastic volatility and commodity convenience yield.

Each chapter contains a problem section. Most problems are of a financial nature. They are preceded with symbols Ⓔ, Ⓢ, and/or Ⓣ to indicate if they are of an empirical, simulation, and/or theoretical nature. Each chapter ends with a section called Notes & Complements that includes bibliographic references which can be used by readers interested in acquiring a deeper understanding of the topics of that chapter. The book ends with two appendices as a suite of indexes. Appendix A contains the introductory session set up to initiate the students to `S-Plus` at the beginning of the course, and Appendix B gives information on how to download the library `EVANESCE` and the home-grown functions used in the text, as well as the data sets used in the text and in the problems.

The code together with the data used in the text can be downloaded from the author web page at the URL:

`http://www.princeton.edu/~rcarmona/safd/`

This web page will be updated regularly, and corrections, complements, new data sets, updates, etc., will be posted frequently.

Acknowledgments

First and foremost, I want to thank all the students who suffered painfully through early versions of the course, and primitive drafts of the lecture notes. Their patience and their encouragement helped me persevere, and over the years, figure out the recipe for the form and the content of the course. I feel guilty to have used them as guinea pigs, but I am glad that the process finally converged. My Chairman Erhan Çinlar trusted me with this course, and gave me total freedom to reshape it. What seemed like foolishness to some, may have been great insight with what needed to be done. I am grateful for his confidence and his relentless encouragements.

My interest in computational statistics was sparked over twenty years ago by two dear friends: Anestis Antoniadis and Jacques Berruyer. Time and distance pulled us apart, but what their collaboration taught me will remain with me forever. The first part of the book would not have been possible without Julia Morrison's contribution. It was a real pleasure to work with her on the development of the `S-Plus` library `EVANESCE`. I am very grateful for this experience. I also want to thank Yacine Ait-Sahalia for enlightening discussions on nonparametric asset pricing. I am also indebted to Cliona Golden for a superb job proofreading an early version of the manuscript. Finally, I want to thank my wife Debra for tolerating my insane working habits, and my three wonderful daughters Stephanie, Chelsea and Chanel for their limitless patience and unconditional love. I may not deserve it, but I sure am proud of it.

René Carmona
Princeton, N.J.

November 4, 2003

Contents

Part I

DATA EXPLORATION, ESTIMATION AND SIMULATION

1

UNIVARIATE EXPLORATORY DATA ANALYSIS

The goal of this chapter is to present basic tools of univariate data analysis. We take a statistical approach firmly grounded in the calculus of probability. In particular, our interpretation of the data given by a set of observations is to view them as realizations of random objects with the same distribution. In the simplest case, these objects will be real numbers. The analysis of multivariate samples (i.e. observations of finite dimensional vectors with real components) is postponed to the next chapter. In fact, the title of this two-chapter set could as well have been "Distribution Modeling and Estimation". Our emphasis is on the implementation of methods of data analysis, more than on the mathematical underpinnings of these methods. In other words, even though we introduce clear mathematical notation and terminology, we spend more time discussing practical examples than discussing mathematical results. After reviewing standard tools such as histograms, we consider the kernel density estimation method. We also emphasize the importance of quantiles plots, and of random simulations. Data used by financial institutions serve as illustrations of the power, as well as the limitations, of standard nonparametric estimation methods. Thus, when it comes to analyzing heavy-tailed distributions, we introduce semi-parametric estimation tools based on the theory of extreme value distributions.

1.1 DATA, RANDOM VARIABLES AND THEIR DISTRIBUTIONS

The data used in this chapter come in the form of a sample:

$$x_1, x_2, \ldots\ldots, x_n$$

where the x_j are real numbers. They are analyzed with statistical tools based on concepts of statistical data analysis which we introduce in this chapter. Our presentation is sprinkled with numerical illustrations anchored on specific examples. We proceed to the introduction of the first of these examples.

1.1.1 The PCS Data

The PCS Index is the year-to-date aggregate amount of total damage reported in the United States to the insurance industry. PCS stands for Property Claim Services. It is a division of ISO Inc (Insurance Services Office). Regional indexes do also exist, and different regions have unique indexes: e.g. California, Florida, Texas, ... , but we only consider the national index in this book. Each index value represents $100 million worth of damage. For example, a value of 72.4 for the national index in 1966, means that \$ (72.4×100) million (i.e. \$ 7.24 billion) in damage were recorded on that year.

The Chicago Board of Trade began trading options on the PCS Index in 1996. Options and futures contracts on the PCS Index offer a possibility to securitize insurance catastrophe risk in a standardized fashion. These financial products were seen in the mid 90's by the insurance industry as a way to tap into the enormous appetite for risk in the global investor community.

For the purpose of our analysis, we do not use the index values. Instead, we use some of the individual claim reports used to compute the final values of the index. The data we use come in the form of a matrix:

```
 13 4.00
 16 0.07
 46 0.35
 60 0.25
 87 0.36
 95 1.00
115 0.60
124 0.90
126 1.00
135 1.30
... ...
```

where the first column contains a time stamp (i.e. a code for the date of a catastrophic event), and the second column contains the aggregate amount (again in \$ 100 million) of all the claims reported after this event. These data are contained in an `S-Plus` matrix called `PCS`. The plot of the data is given in Figure 1.1. In a first analysis, we do not use the information of the timing of the catastrophes. In other words, we first work with the second column of the data set, i.e. with the dollar amounts only. So, at least in this chapter, the data of interest to us will be:

$$x_1 = 4.00,\ x_2 = 0,07,\ x_3 = 0.35,\ x_4 = 0.25, \ldots\ldots$$

and we encapsulate this data set in an `S-Plus` object `PCS.index` (a vector in this case) with the command:

```
> PCS.index <- PCS[,2]
```

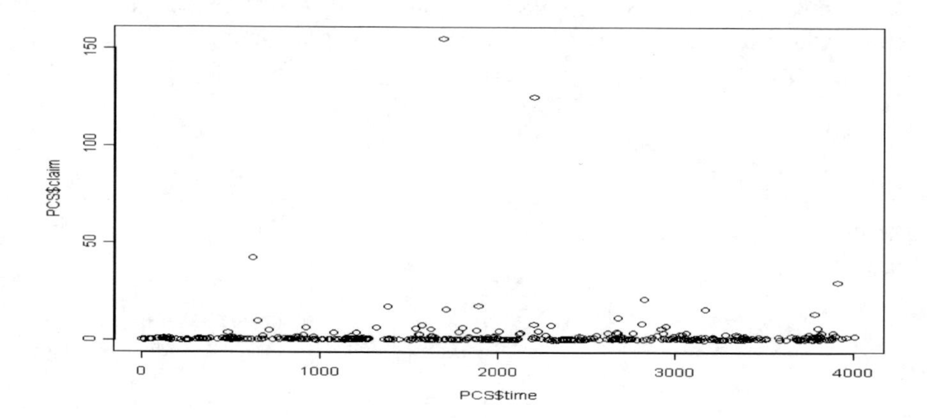

Fig. 1.1. Approximately 10 years worth of individual catastrophe costs used to compute the PCS index. For each catastrophic event included in the data set, the time stamp is reported on the horizontal axis, and the aggregate dollar amount of the claims attributed to this catastrophic event is reported on the vertical axis.

which extracts the second column of the matrix `PCS`, and renames it `PCS.index`. From now on, we work with these values, ignoring the timing of the actual catastrophes. The analysis of the time dependence is the subject of time series analysis which we will tackle in the third part of the book.

We refer to the `S-Plus` Tutorial contained in the Appendix for details about how to start `S-Plus`, and for information on the basic commands used in this chapter.

1.1.2 The S&P 500 Index and Financial Returns

For this second example, the data give the weekly closing values of the S&P 500 index. They come in the form:

```
01/04/1960 59.50
01/11/1960 58.38
01/18/1960 57.38
01/25/1960 55.61
02/01/1960 55.98
02/08/1960 55.46
02/15/1960 56.24
02/23/1960 56.16
   ...     ...
```

These data are contained in an `S-Plus` object `WSP.ts` of class `timeSeries` and plotted in Figure 1.2. `timeSeries` objects will be studied in Chapter 5. In the present chapter, we do not make use of the time stamp appearing in the first entry of each row. We concentrate on the data appearing in the second column. For the sake of convenience, we organized these data in an `S-Plus` numeric column vector which we called `WSP`. We are not really interested in the raw values of the index.

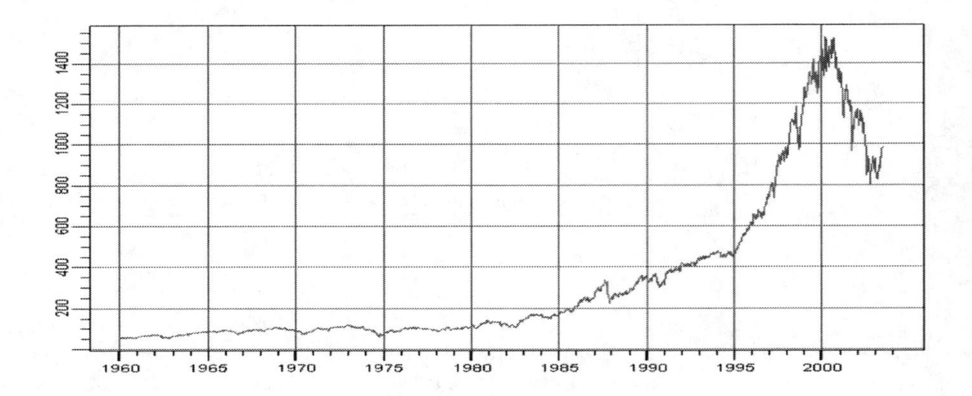

Fig. 1.2. Weekly values of the S&P index

For reasons which will become clear later, we would rather analyze the returns, so we transform the data for the purposes of our analysis. However, before doing so we define the two notions of *return* used in finance. For the sake of simplicity, we ignore the possibility of dividend payments. Given the value of the index at time t, say S_t, and its value after a period of length Δt, say $S_{t+\Delta t}$, the raw-return over that period is defined as:

$$RR_t = \frac{S_{t+\Delta t} - S_t}{S_t} = \frac{S_{t+\Delta t}}{S_t} - 1 \tag{1.1}$$

while the log-return over the period is defined by the formula:

$$LR_t = \log \frac{S_{t+\Delta t}}{S_t}. \tag{1.2}$$

Log-returns are natural in the context of continuous time discounting, while raw-returns are more natural when discounting is done at discrete time intervals. Notice that since $\log(1+x) \sim x$ when x is small, then:

$$LR_t = \log\left(1 + \frac{S_{t+\Delta t} - S_t}{S_t}\right) \sim \frac{S_{t+\Delta t} - S_t}{S_t} = RR_t$$

whenever the ratio $S_{t+\Delta t}/S_t$ is close to 1. So we expect that the two methods of computing returns will give essentially the same results when Δt is small, or when the value of the index does not change much over a period of length Δt. Notice that both notions depend upon the time period over which the returns are computed. Δt is one week in the present situation. It will be one month in some examples, or even one quarter in others. However, it will be one day in most of the examples considered in the book.

The practical computation of the log-returns proceeds as follows. Except for the first value, we divide each closing value by its value the previous week (computing in this way the weekly return), and we then compute the logarithm of this ratio, obtaining in this way the weekly log-return. The `S-Plus` command needed to do this is:

```
> WSPLRet <- diff(log(WSP))
```

When applied to a vector, the `S-Plus` function `log` produces a vector of the same length, whose entries are given by the logarithms of the entries of the original vector. The function `diff` gives a vector that is one element shorter than the original vector, and whose entries are given by the difference of two successive terms of that sequence. The log-returns form the data set on interest in this chapter. Its first entries are:

$$x_1 = -0.019003,\ x_2 = -0.017278,\ x_3 = -0.031332,\ x_4 = 0.006631, \dots\dots$$

As before, we shall restrict ourselves to the analysis of these log-return values, irrespective of the timings of their occurrences. Practically speaking, this means that, were we to shuffle these numbers and change the order in which they appear, the results of our analysis would remain unchanged.

1.1.3 **Random Variables and Their Distributions**

In order to introduce notation and terminology which will be used throughout the book, we review two fundamental concepts from the calculus of probability.

A *random variable* is a numerical quantity whose value cannot be predicted or measured with certainty. We use upper case letters to denote random variables. If X is such a random variable, we use the lower case x to denote possible values or empirical observations of sample realizations of the random variable X. In most cases of interest, the possible values of X are not equally likely, and we use probabilities to quantify the likelihood with which the values of X do occur. The distribution of these probabilities is best encapsulated in a function, say F_X, which we call the *cumulative distribution function* of X (cdf for short.) The defining relation between a random variable X and its cdf F_X is given by:

$$F_X(x) = \mathbb{P}\{X \le x\}, \tag{1.3}$$

which states that the probability that the value of X is not greater than x is exactly the number $F_X(x)$. From this definition, one easily sees that $F_X(x)$ is always a number between 0 and 1 (since it is a probability), and that $F_X(x)$ is increasing from 0 to 1 as the value of x increases. Except for the possibility of jumps, such increasing functions are differentiable, and when they are, their derivatives offer a convenient alternative to the characterization of the probability distributions. Whenever it exists, the derivative $f_X(x) = F'_X(x)$ is called the *density* of the probability distribution. Notice that a density function is a nonnegative function with integral 1.

1.1.4 Examples of Probability Distribution Families

We now review the most frequently used examples of probability distributions.

The Uniform Distribution

The uniform distribution is the distribution of numbers equally likely to fall into different intervals as long as the lengths of those intervals are equal. It is also (hopefully) the distribution of the samples produced by the random number generators provided by most computing environments. The density of the uniform distribution on the interval $[a, b]$ is given by the formula:

$$f_{a,b}(x) = \begin{cases} 0 & \text{if } x < a \text{ or } x > b \\ 1/(b-a) & \text{if } a \le x \le b. \end{cases} \tag{1.4}$$

The corresponding cumulative distribution function is given by:

$$F_{a,b}(x) = \begin{cases} 0 & \text{if } x \le a \\ (x-a)/(b-a) & \text{if } a \le x \le b, \\ 1 & \text{if } x > b \end{cases} \tag{1.5}$$

The probability distribution is denoted by $U(a, b)$. The uniform distribution over the unit interval $[0, 1]$ is most frequently used. It corresponds to the end points a and b given by $a = 0$ and $b = 1$. Figure 1.3 gives a plot of the density and of the cdf of this uniform distribution. Values of $f_{a,b}(x)$ and $F_{a,b}(x)$ can be computed with the `S-Plus` functions `dunif` and `punif`.

Remark. Formulae (1.4) and (1.5) are simple enough so that we should not need special commands for the computations of the values of the density and the cdf of the uniform distribution. Nevertheless, we mention the existence of these commands because their format is the same for all the common distribution families.

The Normal Distribution

The univariate normal distribution, also called the Gaussian distribution, is most often defined by means of its density function. It depends upon two parameters μ and σ^2, and is given by the formula:

$$f_{\mu,\sigma^2}(x) = \frac{1}{\sqrt{2\pi}\sigma} e^{-(x-\mu)^2/2\sigma^2}, \qquad x \in \mathbb{R}. \tag{1.6}$$

The two parameters μ and σ^2 are the mean and the variance of the distribution respectively. Indeed, if X is a random variable with such a distribution (in which case we use the notation $X \sim N(\mu, \sigma^2)$) we have:

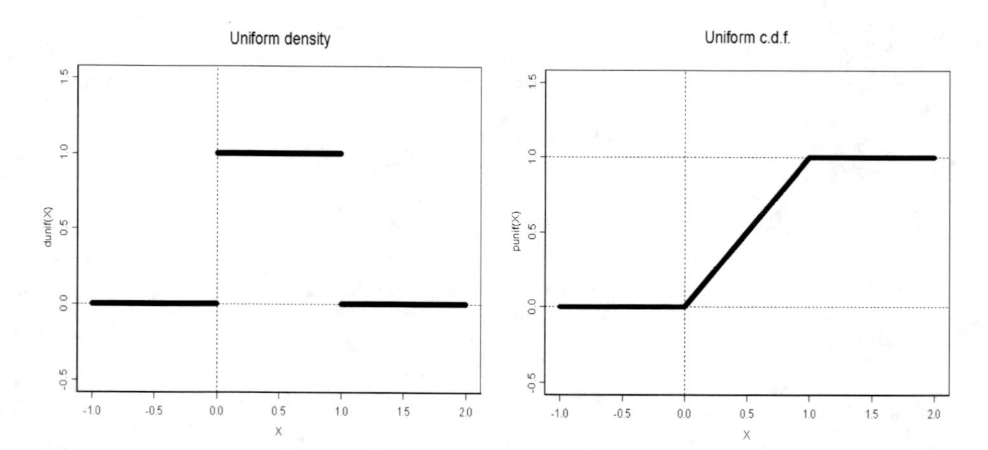

Fig. 1.3. Graph of the density (left) and corresponding cdf (right) of the uniform distribution $U(0,1)$ on the unit interval $[0,1]$.

$$\mathbb{E}\{X\} = \mu, \qquad \text{and} \qquad \text{var}\{X\} = \sigma^2.$$

The corresponding cdf

$$F_{\mu,\sigma^2}(x) = \int_{-\infty}^{x} f_{\mu,\sigma^2}(x')\, dx' \tag{1.7}$$

cannot be given by a formula in closed form involving standard functions. For this reason, we introduce a special notation Φ_{μ,σ^2} for this function. We shall drop the reference to the mean and the variance when $\mu = 0$ and $\sigma^2 = 1$. In this case, we talk about the standard normal distribution, or standard Gaussian distribution, and we use the notation $N(0,1)$. Figure 1.4 gives plots of three Gaussian densities. The density with the smallest variance has the highest central peak and it gets close to zero faster than the other two densities. The density with the largest variance has a flatter central bump, and it goes to zero later than the other ones. By shifting a general normal distribution in order to center it around 0, and rescaling it by its standard deviation, one sees that:

$$X \sim N(\mu,\sigma^2) \quad \Longleftrightarrow \quad \frac{X-\mu}{\sigma} \sim N(0,1) \tag{1.8}$$

Because of this fact, most computations are done with the $N(0,1)$ distribution only.

One can compute values of the cumulative distribution function of a normal distribution using the command `pnorm` with arguments giving, respectively, the list of the values at which the computations are desired, the mean and the standard deviation of the normal distribution in question. For example, the following command computes the probabilities that a standard normal variate is within one, two and three standard deviations of its mean.

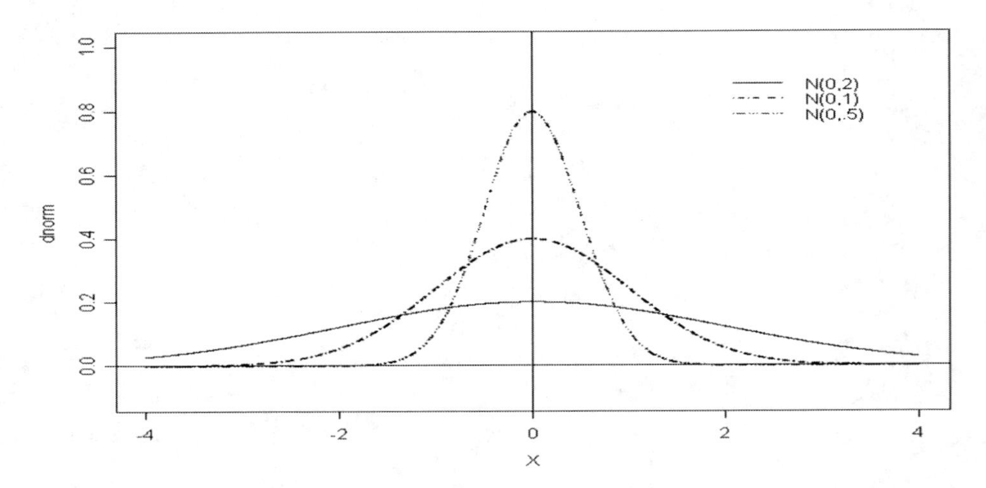

Fig. 1.4. Densities of the mean zero normal distributions $N(0,.5)$, $N(0,1)$ and $N(0,2)$ with variances .5, 1 and 2 respectively.

```
> pnorm(c(1,2,3),mean=0,sd=1)-pnorm(c(-1,-2,-3),mean=0,sd=1)
[1] 0.6826895 0.9544997 0.9973002
```

So, if $X \sim N(\mu, \sigma^2)$, then the scaling property (1.8) implies that $Z = (X - \mu)/\sigma \sim N(0,1)$ and consequently:

$$\mathbb{P}\{-\sigma \le X - \mu \le \sigma\} = \mathbb{P}\{-1 \le Z \le 1\} = \Phi(1) - \Phi(-1) = 0.683$$
$$\mathbb{P}\{-2\sigma \le X - \mu \le 2\sigma\} = \mathbb{P}\{-2 \le Z \le 2\} = \Phi(2) - \Phi(-2) = 0.955$$
$$\mathbb{P}\{-3\sigma \le X - \mu \le 3\sigma\} = \mathbb{P}\{-3 \le Z \le 3\} = \Phi(3) - \Phi(-3) = 0.997.$$

These facts can be restated in words as:

- the probability that a normal r.v. is one standard deviation, or less, away from its mean is 0.683;
- the probability that a normal r.v. is two standard deviations, or less, away from its mean is 0.955;
- the probability that a normal r.v. is three standard deviations, or less, away from its mean is 0.997.

In other words, Gaussian variates are most frequently found within two standard deviations of their mean, essentially always within three standard deviations.

The Exponential Distribution

The exponential distribution is extremely useful in modeling the length of time-intervals separating successive arrivals in a stochastic process. The problems analyzed with these models include internet traffic, physical queues, insurance, catastrophe and rainfall modeling, failure and reliability problems, ...

Random samples from the exponential distribution are positive numbers: the density is non-zero on the positive axis only. In particular, the extreme values are only on the positive side, and the distribution has only one tail at $+\infty$. This distribution depends upon a parameter $\lambda > 0$, and it is denoted by $E(\lambda)$. It can be defined from its density function $f_\lambda(x)$ which is given by the formula:

$$f_\lambda(x) = \begin{cases} 0 & \text{if } x < 0 \\ \lambda e^{-\lambda x} & \text{if } x \geq 0. \end{cases} \tag{1.9}$$

The positive number λ is called the rate of the distribution. It is easy to see that the mean of this distribution is the inverse of the rate. The corresponding cdf is given by the following formula:

$$F_\lambda(x) = \begin{cases} 0 & \text{if } x < 0 \\ 1 - e^{-\lambda x} & \text{if } x \geq 0 \end{cases} \tag{1.10}$$

Figure 1.5 gives plots of three exponential densities. The distribution with the high-

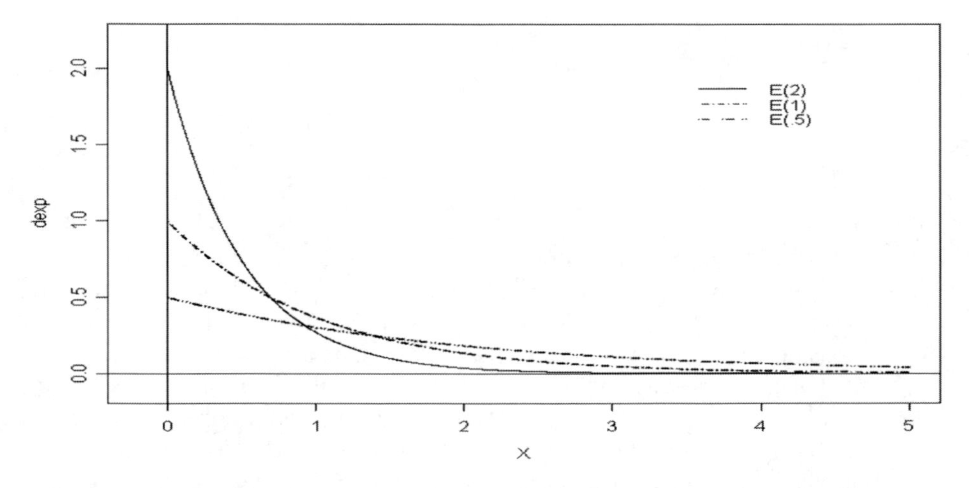

Fig. 1.5. Graphs of the densities of the exponential distributions $E(.5)$, $E(1)$ and $E(2)$ with rates .5, 1 and 2. We plot the graphs only over the positive part of the x-axis because these densities vanish on the negative part of the x-axis.

est rate has the highest starting point on the y-axis, and it tails off faster that the other two densities. The distribution with the lowest rate starts lower on the y-axis, and decays more slowly than the other. Values of the density and the cdf of the exponential distribution can be computed in `S-Plus` with the functions `dexp` and `pexp`, both of which take the rate of the distribution as parameter.

The Cauchy Distribution

Among the probability distributions introduced in this section, the Cauchy distribution is the least known of the four. It is a particular element of the class of Generalized Pareto Distributions (GDP for short) which we will study in detail later in this chapter for the thickness of their tails. It is of great pedagogical (and possibly practical) interest because it is essentially the only distribution from this class for which we can derive explicit closed formulae. Like the Gaussian distribution, it depends upon two parameters: a location parameter, say m, and a scale parameter, say λ. It can be defined from it density function $f_{m,\lambda}(x)$ by the formula:

$$f_{m,\lambda}(x) = \frac{1}{\pi}\frac{\lambda}{\lambda^2 + (x-m)^2}, \qquad x \in \mathbb{R}. \tag{1.11}$$

This distribution is denoted by $C(m, \lambda)$. Because its use is not as widespread, we give the details of the computation of the cdf $F_{m,\lambda}(x)$ of this distribution.

$$\begin{aligned} F_{m,\lambda}(x) &= \int_{-\infty}^{x} f_{m,\lambda}(y)dy \\ &= \frac{1}{\pi}\int_{-\infty}^{x} \frac{1}{1+[(y-m)/\lambda]^2}\frac{dy}{\lambda} \\ &= \frac{1}{\pi}\int_{-\infty}^{(x-m)/\lambda} \frac{1}{1+z^2}dz \\ &= \frac{1}{\pi}[\tan^{-1}\frac{x-m}{\lambda} - \tan^{-1}(-\infty)] \\ &= \frac{1}{\pi}\tan^{-1}\frac{x-m}{\lambda} + \frac{1}{2}, \end{aligned} \tag{1.12}$$

where we used the substitution $z = (y-m)/\lambda$ to compute the indefinite integral. Figure 1.6 gives plots of three Cauchy densities. The distribution with the smallest scale parameter λ has the highest central peak, and it tails off faster that the two other densities. The distribution with the largest scale parameter has a wider central bump, and as a consequence, it goes to zero later than the other ones. This figure seems to be very similar to Figure 1.4 which shows graphs of normal densities. Indeed, both normal and Cauchy distributions are unimodal in the sense that the graph of the density has a unique maximum. This maximum is located at the mean in the case of the normal distribution, and at the value of the location parameter m in the case of the Cauchy distribution. Moreover, if we associate the standard deviation of the normal distribution to the scale parameter of the Cauchy distribution, then the discussion of the qualitative features of the graphs in Figure 1.4 also applies to those in Figure 1.6. Nevertheless, major differences exist between these two families of distributions. Indeed, as we can see from Figure 1.7, where the graphs of densities from both families are superimposed on the same plot, the tails of the normal distribution are much thinner than those of the Cauchy distribution. What we mean here is not so much that the density of the normal distribution approaches zero earlier than the

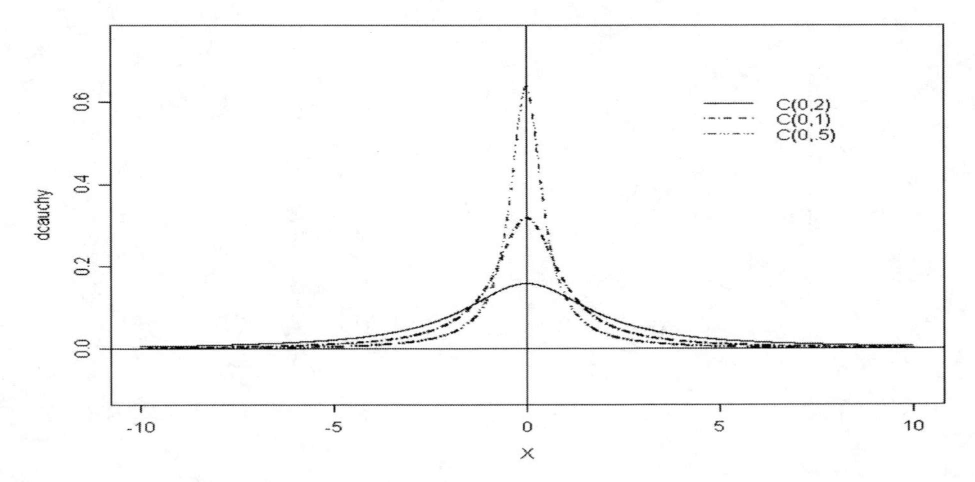

Fig. 1.6. Graphs of the densities of the Cauchy distributions $C(0,.5)$, $C(0,1)$ and $C(0,2)$ located around 0 and with scales .5, 1 and 2 respectively.

density of the Cauchy distribution, but that it does so much faster. This is because the decay toward zero away from the center of the density is exponential in the negative of the square distance to the center, instead of being merely polynomial in this distance. These rates of convergence to zero are very different, and one should not be mislead by the apparent similarities between the two unimodal density graphs.

See Figure 1.14 below for the effects of these tail differences on random samples from the two families of distributions. Values of the density and cumulative distribution functions of the Cauchy distribution can be computed in `S-Plus` with the functions `dcauchy` and `pcauchy` both of which take the location and the rate of the distribution as parameters.

1.2 First Exploratory Data Analysis Tools

As before, we begin the section with a clear statement of some of the conventions adopted in the book.

1.2.1 Random Samples

When considering a data sample, say $x_1, x_2, \ldots, x_n$, we implicitly assume that the observations are realizations of independent identically distributed (i.i.d. for short) random variables $X_1, X_2, \ldots, X_n$ with common distribution function F. Random variables $X_1, X_2, \ldots, X_n$ are said to be independent if

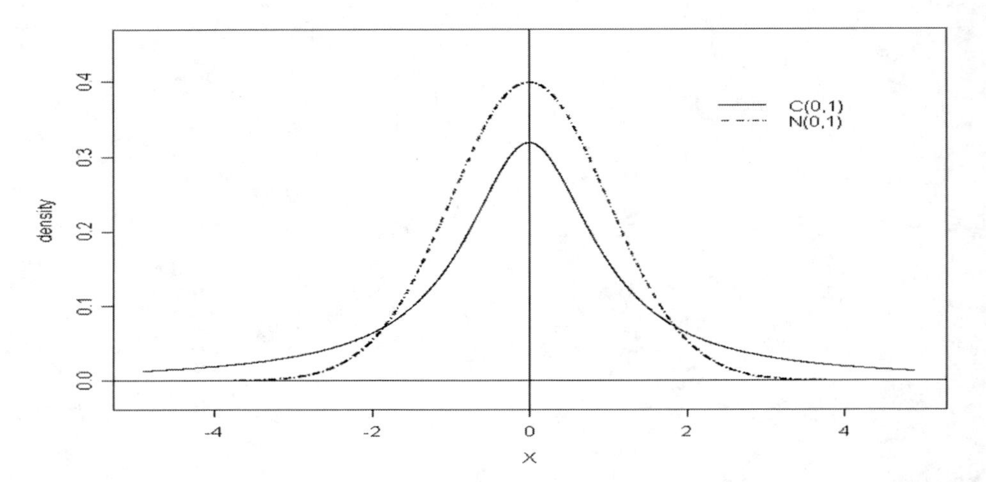

Fig. 1.7. Graphical comparison of the Cauchy distribution $C(0,1)$ and the Gaussian distribution $N(0,1)$.

$$\mathbb{P}\{X_1 \le \alpha_1,\, X_2 \le \alpha_2,\, \dots, X_n \le \alpha_n\} = \mathbb{P}\{X_1 \le \alpha_1\}\mathbb{P}\{X_2 \le \alpha_2\}\mathbb{P}\{X_n \le \alpha_n\}$$

for all possible choices of the real numbers $\alpha_1, \alpha_2, \dots, \alpha_n$. In other words, the random variables are independent if the joint cdf is the product of the marginal cdf's. But since such a definition involves multivariate notions introduced in the next chapter, we shall refrain from emphasizing it at this stage, and we shall rely on the intuitive notion of independence. In any case, the statistical challenge is to estimate F and/or some of its characteristics from the numerical values $x_1, x_2, \dots, x_n$ of the sample.

Nowadays, one cannot open a newspaper without finding a form of graphical data summary, whether it is a histogram, a bar chart, or even a pie chart. We now give a short review of the most commonly used graphical data summaries. Since we are mostly interested in numerical data (as opposed to categorical data), we shall not discuss the graphical representations based on pie charts and bar charts.

1.2.2 **Histograms**

The histogram is presumably the most pervasive of all the graphical data representations. Statistically speaking, a histogram can be viewed as a graphical summary intended to help in the visualization of properties of a distribution of numbers, but it can also be viewed as a (nonparametric) estimator of a density function. Given a set of numerical values $x_1, \dots, x_n$, a histogram is constructed by grouping the data into bins (intervals) and plotting the relative frequencies of the x_i's falling in each interval. Figure 1.8 shows two histograms. They were produced with the commands:

```
> par(mfrow=c(1,2))
```

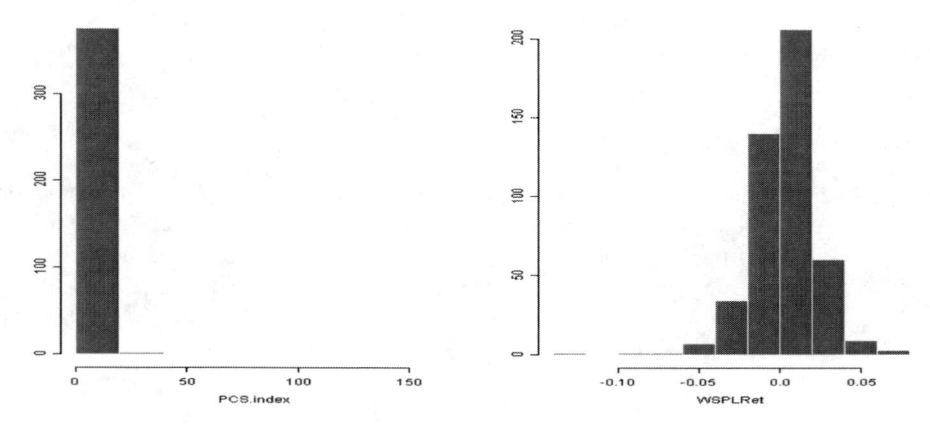

Fig. 1.8. Examples of histograms: PCS index (left) and S&P weekly log-returns (right).

```
> hist(PCS.index)
> hist(WSPLRet)
> par(mfrow=c(1,1))
```

The first command is intended to divide the plotting area into a 1×2 matrix in which plots are inserted in sequence. The next two commands actually produce the histograms. The data used for the computation and the plot of the histograms are accessed from the `S-Plus` objects `PCS.index` and `WSPLRet` whose creation was explained earlier. Histograms are produced in `S-Plus` with the command `hist`. See the help file for details on the options available, including the choice of the number of bins, their size, . . . The fourth command restores the default option of one plot per *screen*. The histogram on the left is for the PCS data which we introduced earlier, and which we will study in detail later in this chapter. Obviously, there are no negative data points (something we could have expected based on the meaning of the data) and the bulk of the data is concentrated near 0. The histogram on the right is for the weekly log-returns on the S&P 500 index. It shows a unimodal distribution of values on both sides of 0, with a lack of symmetry between the right and left ends, which are usually called the tails of the distribution.

Whenever the numerical values $x_1, \ldots, x_n$ can be interpreted as independent observations of random variables with the same distribution, the histogram should be viewed as an estimate of the density of the common probability distribution. One of the main shortcomings of histograms is the weakness of the information they convey because of their strong dependence upon the choice of the positions and sizes of the bins. They can give a pretty good idea of the distribution in regions where the bulk of the data points lie, but they can be very misleading in regions with relatively few data points. Indeed in these other types of regions, the density can be grossly underestimated or overestimated depending on the coverage of the bins.

The look of a histogram can change significantly when the number of bins and the origin of the bins are changed. The reader is encouraged to produce different histograms for the same data sample by setting the value of the parameter `nclass` to different integers.

Remark. The commands given above (as well as most of the commands in this book) can be used both on a Unix/Linux platform and under Windows. There are many other ways to produce plots, especially under Windows. For example, one can select the columns of the variables to be plotted, and then click on the appropriate button of the 2-D plot palette. In fact, some of these alternative methods give plots of better quality. Nevertheless, our approach will remain to provide `S-Plus` commands and function codes which can be used on any platform supported by `S-Plus`, and essentially with any version of the program.

1.3 More Nonparametric Density Estimation

A good part of classical parametric estimation theory can be recast in the framework of density estimation: indeed, estimating the mean and the variance of a normal population is just estimating the density of a normal population. Indeed, a Gaussian distribution is entirely determined by its first two moments, and knowing its mean and variance is enough to determine the entire distribution. Similarly, estimating the mean of an exponential population is the same as estimating the density of the population since the exponential distribution is completely determined by its rate parameter, which in turn is determined by the mean of the distribution. We are not interested in these forms of parametric density estimation in this section. Instead, we concentrate on nonparametric procedures.

Like most nonparametric function estimation procedures, the histogram relies on the choice of some parameters, two to be precise. Indeed, in order to produce a histogram, one has to choose the width of the bins, and the origin from which the bins are defined. The dependence of the histogram upon the choice of the origin is an undesirable artifact of the method. In order to circumvent this shortcoming, the notion of averaged histogram was introduced: one histogram is computed for each of a certain number of choices of the origin, and all these histograms are averaged out to produce a smoother curve expected to be robust to shifts in the origin. This estimate is called the `ASH` estimate of the density of the population, the three initials A,S and H standing for "average shifted histogram". See the Notes & Complements at the end of this chapter for references.

Even though ASH estimates are free of the artificial dependence on the choice of the origin, they are still dependent on the particular choice of the bin width, the latter being responsible for the look of the final product: ragged curves from a choice of small bin widths, and smoother looking blocks from a choice of larger bin widths. The decisive influence of this parameter should be kept in mind as we inch our way toward the introduction of our favorite density estimation procedure.

For the time being, we limit ourselves to the following remark: building a histogram is done by piling up rectangles. Given the choice of the subdivision of the range of the data into intervals of equal lengths (the so-called bins), the contribution of any given observation is a rectangle of height $1/(nb)$ (where n is the population size and b is the bin width), the rectangle being set on top of the bin in which the observation falls. In particular, if many observations fall near the boundary of a bin, the piling up of these rectangles will create an undesirable effect, since a lot of mass is created away from the observation points. Let us give a concrete example of this shortcoming. Let us assume that the bins have been chosen to be the unit intervals $[0,1)$, $[1,2)$, $[2,3)$, ..., $[5,6)$, and let us assume that the data is comprised of $6 \times 8 = 48$ points grouped in 8's around each of the integers $1,2,3,4,5$ and 6, in such a way that for each of these integers, four of the data points are smaller than it (and very close to it), the other four (still very close) being greater. Obviously, the distribution of the points shows a periodic regularity, and one would want the density estimator to account for it: the high concentration of points near the integers should be reflected in the presence of high values for the density, while this same density should vanish in the middle of the inter-integer intervals which are empty of data points. Unfortunately, our histogram will completely miss this pattern. Indeed, the bins having been chosen as they were, the histogram is flat throughout the interval $[0,6)$ leading us (misleading us should I say!) to believe that the distribution is uniform over that interval.

One possible way out of this problem is to center the bins around the observation values themselves. Doing so is just computing the kernel density estimator proposed below with the particular choice of the `box` kernel function !!!

1.3.1 Kernel Density Estimation

Given a sample $x_1, \ldots, x_n$ from a distribution with (unknown) density $f(x)$, the formal kernel density estimator of f is the function $\hat{f}_b$ defined by:

$$\hat{f}_b(x) = \frac{1}{nb} \sum_{i=1}^{n} K\left(\frac{x - x_i}{b}\right) \tag{1.13}$$

where the function K is a given non-negative function which integrates to one (i.e. a probability density function) which we call the kernel, and $b > 0$ is a positive number which we call the bandwidth. The interpretation of formula (1.13) is simple. Over each point x_i of the sample, we center a scaled copy of the kernel function K, and the final density estimate is the superposition of all these "bumps". The division by nb guarantees that the total mass is one (i.e. the integral of $\hat{f}_b(x)$ is one.)

Univariate Kernel Density Estimation in `S-Plus`

`S-Plus` provides two functions for univariate kernel density estimation. They are called `ksmooth` and `density` respectively. Unfortunately, even in the latest version of S-Plus, there are annoying differences between the two, and the terminology discrepancies have not been fixed. For the sake of consistency with our future discussion of kernel regression, we encourage the reader to use the function ksmooth. But for the sake of completeness, we mention some of the most salient differences. Examples of sets of commands are postponed to the discussion of the daily S&P data given in Subsection 1.3.3 below. The kernel functions $K(x)$ used in the kernel density estimation function density are the four kernel functions plotted in Figure 1.9. The choice of the kernel function is set by the parameter window

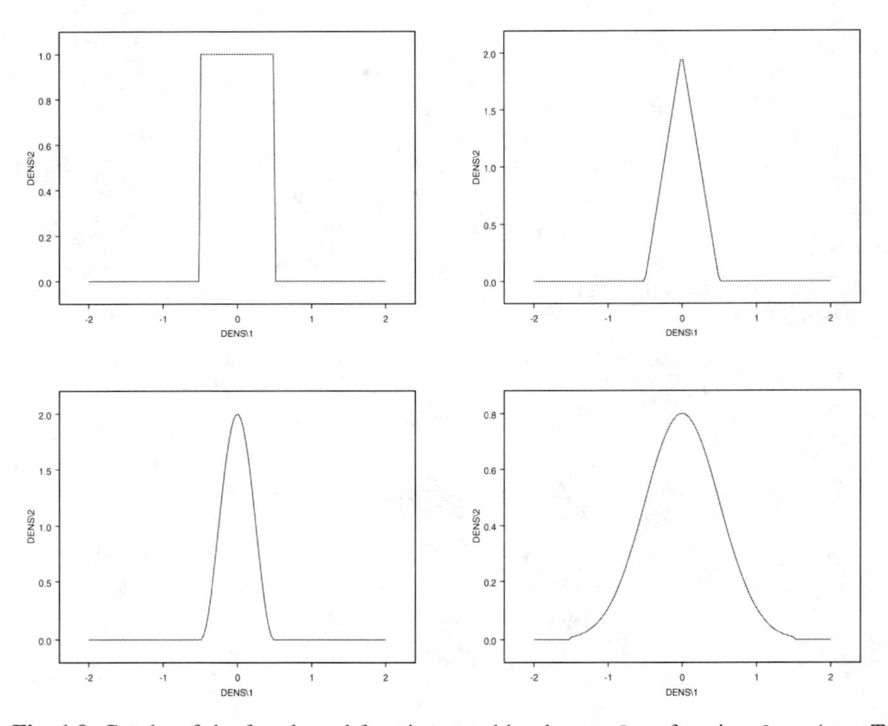

Fig. 1.9. Graphs of the four kernel functions used by the S-Plus function density. Top Left: box kernel. Top Right: triangle kernel. Bottom Left: cosine kernel. Bottom Right: Gaussian kernel.

which can be given one of the four character strings: "cosine", "gaussian", "rectangular", "triangular". This set of kernel functions differs from the four kernel functions given in the Table 4.5 of Subsection 4.3.5. Indeed, the latter comprises "box", "triangle", "parzen" and "normal". To make matters even worse, the kernel function used by density is set by a parameter called window instead of the more natural kernel terminology used in ksmooth. Also, the bandwidth used by density is given by a parameter called width which can be a number from which the actual bandwidth is computed, or a character string giving the name of a method to compute the bandwidth from the data. Finally, the points at which the density kernel estimates are computed are also defined differently. Please check the help files before using any of these functions. After all that, it is not as bad as I make it sound, and since we are aware of these problems, we shall try to use only one of these functions.

1.3.2 **Comparison with the Histogram**

Both histogram and kernel estimates start from basic data in the form of a sample x_1, x_2, ..., x_n. To construct a histogram we choose an origin and n bins, and we define the histogram as the graph of the function:

$$x \hookrightarrow \text{Hist}(x) = \frac{1}{n}\sum_{i=1}^{n}\theta(x, x_i) \qquad (1.14)$$

where $\theta(x, x_i) = 1/b$ if x and x_i belong to the same bin, and 0 otherwise (remember that we use the notation b for the width of the bins.) Notice that definition (1.13) of the kernel density estimator has exactly the same form as the re-formulation (1.14) of the definition of the histogram, provided we re-define the function $\theta(x, x_i)$ by:

$$\theta(x, x_i) = \frac{1}{b}K\left(\frac{x - x_i}{b}\right).$$

The similarity is striking. Nevertheless there are fundamental differences between these two nonparametric density estimation methods.

• *First, at the level of the "bumps":* they are rectangular for the histogram, while their shape is given by the graph of the kernel function $K(x)$ rescaled by b in the case of the kernel estimate:

• *At the level of the locations of the bumps:* these locations are fixed independently of the data for the histogram, while they are centered around the data points in the case of the kernel estimate:

• *At the level of the smoothness of the resulting density estimate:* It is determined by the number (and size) of the bins in the case of the histogram while it is determined by the value of the bandwidth $b > 0$ in the case of the kernel estimate.

1.3.3 **S&P Daily Returns**

The goal of this subsection is to show how to perform kernel density estimation in `S-Plus`, and to illustrate graphically some of the key features of the resulting estimates. We resume our analysis of the S&P index, and for the purposes of illustration, we change the sampling frequency from weekly to daily. The data now comprise the series of daily closing values of the S&P 500 index over the period ranging from January 1, 1960 to June 6, 2003. Figure 1.10 gives a sequential plot of these values,

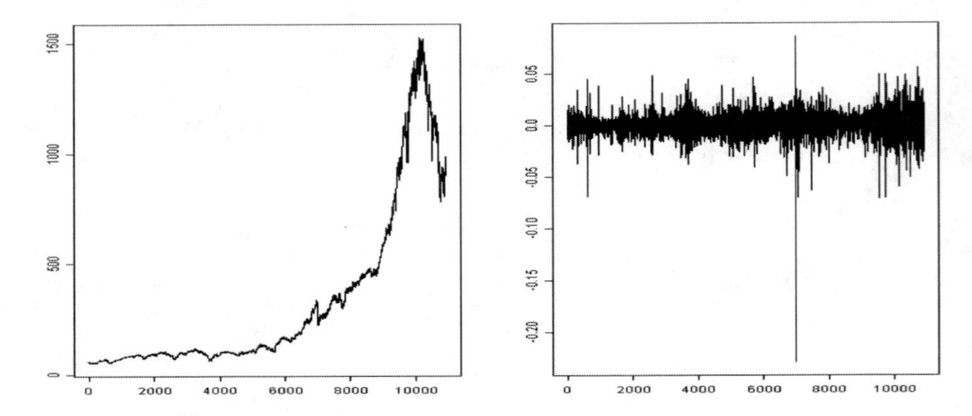

Fig. 1.10. Daily S&P 500 closing prices as imported from the Web (left) and daily log-returns over the same period (right).

together with the series of the daily log-returns. The introductory session to `S-Plus` given in appendix explains how we downloaded the data from the Web, how we imported it into `S-Plus`, and how we produced these plots. Notice that in computing the returns, we ignore the fact that the returns are actually computed over periods of different lengths. Indeed we ignore the presence of weekends (Monday's close follows the preceding Friday's close, and we use the quotes as if they were from consecutive days) and holidays, and more generally the days the market was closed, like the week of September 11, 2001 of the terrorist attack on New York. Also, we produce low tech plots where the dates do not appear on the horizontal axis. We will change all that when we define (mathematical) time series, and when we work with `timeSeries` objects in the third part of the book.

The other important point which needs to be made at this stage is that we are not interested in the time evolution of the values of the series, but instead, in the distribution of these values on a given day. In other words, instead of worrying about how the value on a given day depends upon the values on previous days, we care about statistics which would not change if we were to change the order of the measurements. This seems to be quite a reasonable requirement in the case of the log-returns of the right pane of Figure 1.10. We use the commands:

```
> hist(DSPLR,probability=T,ylim=c(0,30))
> DENS <- density(DSPLR,n=100)
> points(DENS, type="l")
```

to produce Figure 1.11. Notice that we use the option "`probability=T`" to force the area of the histogram to be one, so it will be on the same vertical scale as the kernel density estimate computed next. Also, we set the parameter `ylim` to force the limits on the vertical axis to be 0 and 30 so as to make sure that the kernel density estimate will be plotted inside the plot area. The second command actually produces a matrix with two columns. The first column contains the x-values at which the density is computed, while the second one contains the y-values given by the kernel estimate computed at these x-values. Since we did not specify the parameter `window`, the Gaussian kernel function is used by default. Finally, we use the command "`points`" with the argument `DENS` to add to the existing graph one point for each couple (x-value,y-value) contained in a row of `DENS`. We use the parameter `type="l"` in order to force the plot of line segments joining the various points to produce a continuous curve. Alternatively, we could have used the command `ksmooth` instead of

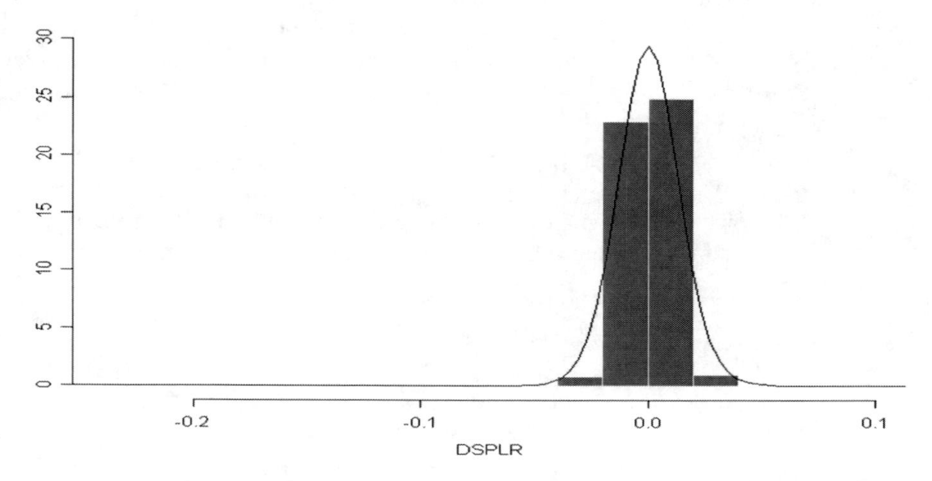

Fig. 1.11. Histogram and kernel density estimates of the daily log-returns of the S&P 500 index during the period 1/1/60 - 06/06/03.

the function `density`, replacing the second command by:

```
> DENS <- ksmooth(DSPLR,kernel="normal",bandwidth=.03)
```

`DENS` would still be a matrix of the same size and structure as before, the main difference resulting in part from the number and the values of the points at which the estimate is computed. By default, the estimate is computed for x-values which are regularly spaced in the range of `DSPLR`. The parameter `n.points` gives the number of these x-values (typically, the is divided into `n.points` -1 intervals of equal lengths). It is set by default to the length of `DSPLR`, but it can also be chosen by the user. Also, the default kernel function is not the Gaussian kernel, so we need to set the parameter `kernel` to `"normal"` if we want this kernel to be used. Finally, we had to set the parameter `bandwidth` to .03 in order to get results qualitatively similar to the results obtained with the function `density`. The use of the function `ksmooth` is desirable because we shall use the same function for nonparametric regression later. But its parameters are set by default to values which are not well suited for density estimation, and finding the right parameters requires experimentation by trial and error. On the other hand, it is often possible to use the function `density` with its default parameters and get reasonable results.

The histogram reproduced in Figure 1.11 is not satisfactory for two related reasons. First it gives very little information about the center of the distribution, the two central bars capturing too much of the information. Second, the extreme values to the left and to the right of the distribution do not appear because they are in small numbers, and the heights of the bars they create are too small compared to the heights of the central bars, for the former to be visible. Because of this shortcoming, we decided to use the kernel method as an alternative to the histogram, but despite a smoother look, the same problem plagues the estimate: the contributions of both ends of the distribution (which we call tails) to the graph are overwhelmed by the central part of the distribution .

1.3.4 **Importance of the Choice of the Bandwidth**

The choice of bandwidth can have drastic consequences on the final look of the density estimate. Figure 1.12 shows two kernel density estimates for the same sample

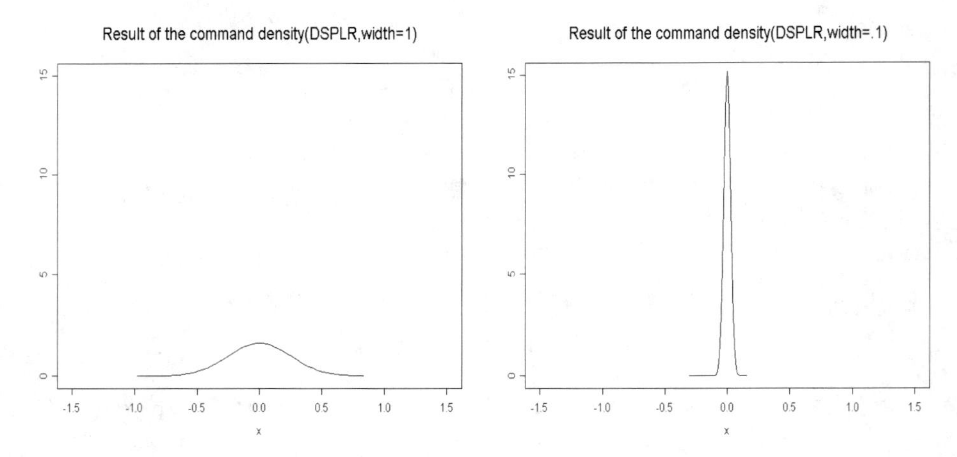

Fig. 1.12. Kernel density estimates of the daily log-returns of the S&P 500 produced by the `S-Plus` function `density` with bandwidths computed by setting the parameter `window` to 1 (left) and .1 (right).

of daily log-returns of the S&P 500. The left pane was obtained with a bandwidth equal to 1 while the right pane was produced with a bandwidth equal to .1. The results look very different. Notice that in order to allow for a meaningful comparison, we forced identical scales on the vertical axes of the two panes. This ensures that the observed differences are not due to an artifact from the choice of axis scales. Indeed, the plots of these two density estimates would look almost identical if we were to let the program adapt the limits of the axes to the actual values of the functions. The strong influence of the value of the bandwidth will also be emphasized in Problem 1.2.

In any case, both histograms and kernel density estimators are unable to give a good account of the nature of the tails of a distribution, and we need to appeal to other graphical tools to exhibit the tail properties in a more suggestive way. However, before defining these graphical tools, we need to introduce more of the standard concepts from the calculus of probability.

1.4 QUANTILES AND Q-Q PLOTS

Given a (theoretical) distribution function F, and a number $q \in [0,1]$, the q-quantile, or the $100q$th percentile of the distribution is the number $\pi_q = \pi_q(F)$ satisfying $F(\pi_q) = q$. If we think of F as the distribution function of a random variable X, then the quantile definition can be restated as:

$$F(\pi_q) = \mathbb{P}\{X \le \pi_q\} = q. \tag{1.15}$$

In words, the $100q$th percentile, is the number π_q such that the probability that X is not greater than π_q is exactly equal to q.

Remark. Even though statement (1.15) is very intuitive, it cannot be a non-ambiguous definition. Indeed, there may not be any real number x satisfying $F(x) = \mathbb{P}\{X \le x\} = q$. For example, this is the case for some value of q when the distribution function F has jumps, i.e. when the random variable X can take discrete values with positive probabilities. When such jumps occur, there may be plenty of possible choices. In fact, all the real numbers x satisfying:

$$\mathbb{P}\{X < x\} \le q \le \mathbb{P}\{X \le x\} = F(x)$$

can be regarded as reasonable candidates. A more precise definition would state that the set of q-quantiles is the closed interval $[x_q^-, x_q^+]$ where:

$$x_q^- = \inf\{x;\ F(x) \ge q\} \qquad \text{and} \qquad x_q^+ = \inf\{x;\ F(x) > q\}.$$

In all cases we get a uniquely defined quantile $\pi_q(F)$ (i.e. we have $x_q^- = x_q^+$) except for at most countably many q's. For the sake of definiteness, for any of these countably many values of q, we shall use the left endpoint x_q^- of the interval as our definition of the percentile.

Most of the cdf's F used in this book are invertible. When it exists, the inverse function F^{-1} is called the quantile function because (1.15) can be rewritten as:

$$\pi_q = F^{-1}(q). \tag{1.16}$$

As an illustration, let us consider the problem of the computation of the quantiles of the classical distributions introduced earlier in the chapter. These computations will be of great use when we discuss random number generators below.

The quantiles of the uniform distribution are very easy to compute. However, they are rarely needed. The quantiles of the normal distribution cannot be given by closed formulas: we cannot compute the normal cdf in close form, neither can we compute its inverse in closed form: we need to rely on numerical approximations. We give examples of these computations in the next subsection below. For the exponential distribution, the percentiles are easily computed from formula (1.10) which gives a simple expression for the cdf $F_\lambda(x)$. One finds:

$$\pi_q = \frac{1}{\lambda} \log \frac{1}{1-q}.$$

Finally, in the case of the Cauchy distribution, the explicit form of the cdf can also be inverted, and from trigonometry we find that the quantile function is given by:

$$\pi_q = F_{m,\lambda}^{-1}(q) = m + \lambda \tan\left(q\pi - \frac{\pi}{2}\right) \tag{1.17}$$

Percentiles are numbers dividing the real line into intervals in which a prescribed proportion of the probability distribution lives. For example, if we compute the quantiles $\pi(q)$ for a sequence of regularly spaced numbers q, patterns in the distribution of this set of percentiles can be interpreted as properties of the probability distribution. This remark is particularly useful when it comes to comparing several distributions (whether these distributions are empirical or theoretical). A Q-Q plot is a plot of the percentiles of one distribution against the same percentiles of another distribution.

1.4.1 Understanding the Meaning of Q-Q Plots

We illustrate the suggestive power of Q-Q plots with a small experiment intended to illustrate the properties of the quantiles of a distribution as they relate to the thickness of its tails.

We already emphasized how thin the tails of the normal distribution are by quantifying the concentration of the probability mass around the mean of the distribution. We now compute quantiles of the normal distribution with the command `qnorm`, and arguments giving respectively the list of quantiles we want, and the mean and standard deviation of the distribution. Because of the symmetry of the distribution, we restrict ourselves to the upper tail.

```
> qnorm(c(.8,.85,.9,.95,.975,.99),mean=0,sd=1)
[1] 0.8416 1.0364 1.2816 1.6449 1.9599 2.3263
```

In words, these numbers tell us that 80% of the probability mass is to the left of $x = 0.8416$, 85% is to the left of $x = 1.0364$, ... The computation of the corresponding quantiles for the Cauchy distribution gives:

	$X \sim N(0,1)$	$X \sim C(0,1)$
$\pi_{.8} = F_X^{-1}(.8)$	0.842	1.376
$\pi_{.85} = F_X^{-1}(.85)$	1.036	1.963
$\pi_{.9} = F_X^{-1}(.9)$	1.282	3.078
$\pi_{.95} = F_X^{-1}(.95)$	1.645	6.314
$\pi_{.975} = F_X^{-1}(.975)$	1.960	12.706
$\pi_{.99} = F_X^{-1}(.99)$	2.326	31.821

Table 1.1. Comparison of the quantiles of the standard normal and Cauchy distributions.

```
> qcauchy(c(.8,.85,.9,.95,.975,.99),location=0,scale=1)
[1]  1.376  1.963  3.078  6.314 12.706 31.821
```

We display these results in tabular form in Table 1.1.

We see that in the case of the Cauchy distribution, in order to have 80% of the probability mass to its left, a quantile candidate has to be as large as $x = 1.376$, which is greater than $x = 0.842$ as found for the normal distribution. Obviously, the same is true for the other quantiles in the above lists. This pattern may be visualized by plotting the quantiles of the Cauchy distribution against the corresponding quantiles of the normal distribution, i.e. by producing a theoretical Q-Q plot for these two distributions. In particular, we would have to plot the points

$$(0.8416212, 1.376382), (1.0364334, 1.962611), (1.2815516, 3.077684),$$
$$(1.6448536, 6.313752), (1.9599640, 12.706205), (2.3263479, 31.820516).$$

Note that all these points are above the diagonal $y = x$, and in fact they drift further and further away above this diagonal. This fact is at the core of the interpretation of a Q-Q plot comparing two distributions: points above the diagonal in the rightmost part of the plot indicate that the upper tail of the first distribution (whose quantiles are on the horizontal axis) is thinner than the tail of the distribution whose quantiles are on the vertical axis. Similarly, points below the diagonal on the left part of the plot indicate that the second distribution has a heavier lower tail. This latter phenomenon, as well as the phenomenon present in the above example, occur in the empirical Q-Q plots of Figures 1.13 and 1.15.

In the following section we explain how to estimate the quantiles of a distribution from an empirical sample.

1.4.2 Value at Risk and Expected Shortfall

For better or worse, Value at Risk (VaR for short) is nowadays a crucial component of most risk analysis/management systems in the financial and insurance industries. Whether this computation is imposed by regulators, or it is done on a voluntary basis by portfolio managers, is irrelevant here. We shall merely attempt to understand the rationale behind this measure of risk.

We introduce the concept at an intuitive level, with the discussion of a simple example of a dynamical model. Let us imagine that we need to track the performance of a portfolio. For the sake of simplicity, we consider a portfolio which comprises a certain number of instruments, we denote by P_t the value of this portfolio at time t, and we choose a specific level of confidence, say $\alpha = 2\%$, and a time horizon Δt. Under these conditions, the required capital RC_t is defined as the capital needed to guarantee that the book will be in the red at time $t + \Delta t$ with probability no greater than α. . In other words, RC_t, is defined by the identity:

$$\mathbb{P}\{P_{t+\Delta t} + RC_t < 0\} = \alpha.$$

The Value at Risk is then defined as the sum of the *current endowment* plus the *required capital*

$$VaR = P_t + RC_t.$$

Notice that $\mathbb{P}\{P_{t+\Delta t} - P_t + VaR_t < 0\} = \alpha$, which says that $-VaR_t$ is the α-quantile of the distribution of the change $P_{t+\Delta t} - P_t$ in the value of the portfolio over the given horizon. It is often more convenient to express VaR_t in units of P_t, i.e. to set $VaR_t = \widetilde{VaR}_t * P_t$. The definition

$$\mathbb{P}\{P_{t+\Delta t} - P_t + VaR_t < 0\} = \alpha$$

of VaR can then be rewritten as:

$$\mathbb{P}\left\{\frac{P_{t+\Delta t} - P_t}{P_t} + \widetilde{VaR}_t < 0\right\} = \alpha$$

which shows that, expressed in units of P_t, the negative of the value at risk is nothing more than the α-quantile of the distribution of the raw return over the period in question. Recall that, since for small relative changes we have:

$$\frac{P_{t+\Delta t} - P_t}{P_t} \sim \log \frac{P_{t+\Delta t}}{P_t}$$

we will often call value at risk at the level α the negative of the α-quantile of the distribution of the log-return !

So if we recast the above discussion in the static framework used so far, for the purposes of this subsection, and in order to get rid of the annoying negative sign which we had to deal with above, it is convenient to think of the distribution function F as modeling the loss, i.e. the down side of the profit and loss (P&L for short) distribution, in the standard financial jargon, associated to a specific financial position. The random variable X could be for example the cumulative catastrophic insurance losses in a reporting period, or the credit losses of a bank or a credit card company, or more generally, the daily negative returns on a financial portfolio. Note that the length Δt of the time horizon can, and will change, from one application to another. We will consider many examples in which the frequency will be different from a day. One can think for example of the minute by minute quotes used by a day trader, or the weekly or monthly quotes used by a fund manager.

For a given level q, the value at risk VaR_q is defined as the $100q$th percentile of the loss distribution. For the sake of definiteness, we shall choose the left hand point of the interval of possible percentiles when the loss distribution is not continuous. Value at Risk is widely used in the financial industry as a risk measure. Indeed, according to the Basel agreement, financial institutions have to control the VaR of their exposures. Nevertheless, this VaR is not a satisfactory measure of risk for several reasons. The first one is quite obvious: it does not involve the actual size of the losses. The second one is more important: it does not encourage diversification. This property is more subtle and more difficult to prove mathematically, and we refer the interested reader to the references given in the Notes and Complements at the end of the chapter. For these reasons we introduce another way to quantify the risk of a financial position modeled by a random variable X. For a given level q, the shortfall distribution is given by the cdf Θ_q defined by:

$$\Theta_q(x) = \mathbb{P}\{X \le x | X > VaR_q\} \tag{1.18}$$

This distribution is just the conditional loss distribution given that the loss exceeds the Value at Risk at the same level. The mean or expected value of this distribution is called the expected shortfall, and is denoted by ES_q. Mathematically, ES_q is given by:

$$ES_q = \mathbb{E}\{X | X > VaR_q\} = \int x \, d\Theta_q(x) = \frac{1}{q} \int_{x > VaR_q} x \, dF(x). \tag{1.19}$$

A good part of risk analysis concentrates on the estimation of the value at risk VaR_q and the expected shortfall ES_q of the various portfolio exposures. The main difficulty with this comes from the fact that the theoretical cdf F is unknown, and its estimation is extremely delicate since it involves the control of rare events. Indeed, by the definition of a tail event, very few observations are available for that purpose. More worrisome is the fact that the computation of the expected shortfall involves integrals which, in most cases, need to be evaluated numerically.

In order to emphasize once more the dramatic effect that the choice of a particular distribution can have, we recast our percentile comparison in the framework of the present discussion. Choosing the level $q = 2.5\%$ for the sake of definiteness, if a portfolio manager assumes that the P&L distribution is normal then she will report that the value at risk is 1.96 (never mind the units, just bear with me), but if she assumes that the P&L distribution is Cauchy, the reported value at risk will jump to 12.71. Quite a difference !!! This shocking example illustrates the crucial importance of the choice of model for the P&L distribution. As we just proved, this choice is open to abuse.

1.5 Estimation from Empirical Data

We now consider the challenging problem of the statistical estimation of the probability concepts discussed in the previous section. As we already explained, statistical inference is based on the analysis of a sample of observations, say:

$$x_1, x_2, \ldots\ldots, x_n.$$

In such a situation, we assume that these observations are realizations of i.i.d. random variables $X_1, X_2, \ldots, X_n$ with common cdf F, and the statistical challenge is to estimate F and/or some of its characteristics from the data.

Mean and standard deviation of the distribution are typical examples of characteristics targeted for estimation. However, in financial applications, VaR_q, ES_q, Θ_q, ... are also often of crucial importance. Because all of the quantities we want to estimate are functions of the cdf F, the general strategy is to use the sample observations to derive an estimate, say $\hat{F}$, of the cdf F, and then to use the characteristics of $\hat{F}$ as estimates of the corresponding characteristics of F. In other words, we will propose the mean and standard deviation of $\hat{F}$ as estimates for the mean and standard deviation of F, and $VaR_q(\hat{F})$, $ES_q(\hat{F})$, ... as estimates for the corresponding $VaR_q(F)$, $ES_q(F)$, ...

If the random mechanism governing the generation of the data is known to be producing samples from a given parametric family, estimating the distribution reduces to estimating the parameters of the distribution, and we can use the classical estimation methods, such as maximum likelihood, method of moments, ... to get estimates for these parameters. We can then plug these values into the formulae defining the parametric family and compute the desired characteristics accordingly. However, if the unknown characteristics of the random mechanism cannot be captured by a small set of parameters, we need to use a nonparametric approach.

1.5.1 The Empirical Distribution Function

Different methods have to be brought to bear when no parametric family gives a reasonable fit to the data. In the absence of any specific information about the type of statistical distribution involved, the only recourse left is the so-called empirical distribution function $\hat{F}_n$ defined by:

$$\hat{F}_n(x) = \frac{1}{n}\#\{i;\ 1 \le i \le n,\ x_i \le x\}. \tag{1.20}$$

So $\hat{F}_n(x)$ represents the proportion of observations not greater than x. The function $\hat{F}_n(x)$ is piecewise constant, since it is equal to j/n for x in between the jth and the $(j+1)$th observations. Its graph is obtained by ordering the observations x_j's, and then plotting the flat plateaus in between the ordered observations.

Once the empirical distribution function has been chosen as the estimate of the (theoretical) unknown cdf, the percentiles can be estimated by the percentiles computed from $\hat{F}_n$. So, according to the strategy spelled out earlier, the $100q$th percentile $\pi_q = \pi_q(F)$ is estimated by the empirical percentile $\pi_q(\hat{F}_n)$. In particular, the value at risk VaR_q can be estimated by $\widehat{VaR_q} = \pi_q(\hat{F}_n)$.

The rationale behind this estimation procedure is the fact that the empirical cdf $\hat{F}_n(x)$ converges uniformly in $x \in \mathbb{R}$ toward the unknown cdf $F(x)$. This is known as the Glivenko-Cantelli theorem (or the fundamental theorem of statistics). Its proof can be found in many elementary texts in statistics. See the Notes and Complements for references. So according to the above discussion, everything depends upon the estimate $\hat{F}_n$ and the computations of its characteristics. Practical details are given in the next subsection.

1.5.2 **Order Statistics**

The actual construction of the empirical cdf given above emphasized the special role of the ordered observations:

$$\min\{x_1, \ldots, x_n\} = x_{(1),n} \le x_{(2),n} \le \cdots \le x_{(n),n} = \max\{x_1, \ldots, x_n\}$$

Warning. The ordered observations $x_{(k),n}$ are not defined uniquely because of possible ties. Ties cannot occur (to be more specific, we should say that the probability that they do occur is zero) when the distribution function F is continuous. The empirical cdf's are never continuous, but most of the cdf's occurring in practice are continuous, so since such an assumption does not restrict the scope of our analysis, we shall subsequently assume that there are no ties in the sample observations.

The ordered observations $x_{(k),n}$ can be viewed as realizations of random variables $X_{(1),n}, X_{(2),n}, \ldots X_{(n),n}$. Note that the latter are neither independent nor identically distributed. The random variable $X_{(k),n}$ is called the k-th order statistic. Our definition of the percentiles of a distribution with jumps, and of the left inverse of a cdf imply that for each level q:

$$x_{(k),n} = \hat{\pi}_q = \pi_q(\hat{F}_n) \qquad \text{for} \qquad \frac{k-1}{n} < q \le \frac{k}{n}.$$

The following argument will have far-reaching consequences in the analysis of the tails of the distribution.

Let us assume that $x_1, x_2, \ldots\ldots, x_n$ form a sample from an unknown cdf F, and let us assume that we are interested in the estimation of a quantile $\pi_q = \pi_q(F)$ which could represent a value at risk, for example. The probability that exactly k of the sample observations x_j fall in between π_q and 1 is the same as the probability that

$$X_{(n-k),n} \le \pi_q < X_{(n-k+1),n}. \tag{1.21}$$

Since the cdf F is monotone increasing, the inequalities (1.21) can be equivalently rewritten as:

$$F(X_{(n-k),n}) \leq q < F(X_{(n-k+1),n}) \tag{1.22}$$

and the probability that this happens is equal to the probability that exactly k of the numbers $F(X_j)$ fall in the interval $[q, 1]$. Since the $F(X_j)$ are i.i.d. random variables, this probability is equal to the binomial probability:

$$\binom{n}{k} p^k (1-p)^{n-k}.$$

where p denotes the probability that a number $F(X_j)$ belongs to the interval $[q, 1]$. But if we anticipate the results presented in the next section, we learn from Fact 1, that this probability is equal to $(1 - q)$. Consequently, the probability that (1.22) occurs is given by:

$$\binom{n}{k} q^{n-k} (1-q)^k. \tag{1.23}$$

This important result is used in practice to derive confidence intervals for the empirical quantiles of a distribution.

1.5.3 Empirical Q-Q Plots

Figure 1.13 shows two Q-Q plots. It was produced with the commands:

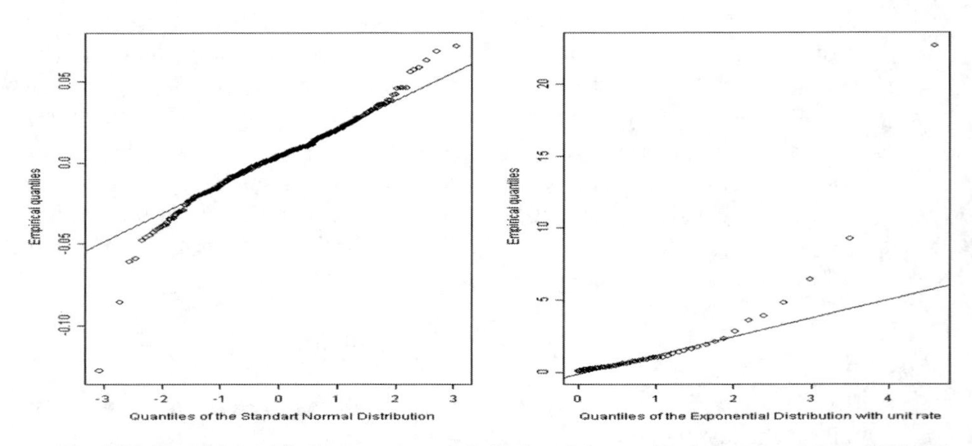

Fig. 1.13. Examples of Q-Q plots: normal Q-Q plot of the weekly log-returns of the S&P 500 (left) and exponential Q-Q plot of the PCS index (right).

```
> par(mfrow=c(1,2))
> myqqnorm(WSPLRet)
> qqexp(PCS.index)
> par(mfrow=c(1,1))
```

The plot on the right shows the empirical percentiles of the PCS index data against the theoretical percentiles of the exponential distribution with unit rate. This plot was obtained with the function `qqexp` which we wrote to this effect. On top of the points whose coordinates are the respective percentiles, it shows a line on which all these points would be found should the distribution be exponential. The slope of this line should be the rate of the distribution (i.e. the inverse of the mean.) The fact that the rightmost points are above the line indicates that the right tail of the distribution is *fatter* than the tail of the exponential distribution. Since this terminology does not appear to be politically correct, we shall use *heavy tail(s)* and *heavier tail(s)* instead. The plot on the left is for the weekly log-returns on the S&P 500 index. It uses the function `myqqnorm` which we wrote to add to the standard `S-Plus` function `qqnorm`, the plot of the line whose slope and intercept determine the parameters of the Gaussian distribution, should all the points sit on this line. This plot shows that both tails are heavier than the tails of the normal distribution.

Q-Q plots are very important when it comes to the comparison of several distributions, especially if tail properties are determining factors.

1.6 Random Generators and Monte Carlo Samples

The commercial distribution of `S-Plus` comes with the commands needed to generate samples from the most common parametric families of distributions. We review some of them later in this section. However, realistic applications require Monte Carlo simulations of quantities with more general (and less generic) statistical properties. We now review the elementary facts which can be used to build random generators, setting the stage for the development of the more sophisticated tools we will build later on. The following simple mathematical facts are fundamental to the discussion of the section.

Fact 1 Given a random variable X with cdf F_X, the random variable $U = F_X(X)$ is uniformly distributed in the interval $[0, 1]$

This is consistent with the fact that the cdf F_X can only take values between 0 and 1. The fact that the values of $F_X(X)$ are uniformly distributed is due precisely to the definition of the cdf. Indeed, if $0 \leq u \leq 1$, we have:

$$\mathbb{P}\{F_X(X) \leq u\} = \mathbb{P}\{X \leq F_X^{-1}(u)\} = F_X(F_X^{-1}(u)) = u. \tag{1.24}$$

This argument is perfectly rigorous when the cdf F_X is strictly increasing (and hence invertible). A little *song and dance* is needed to make the argument mathematically sound in the general case, but we shall not worry about such details here. So we proved that:

$$F_X \text{ cdf of } X \Longrightarrow U = F_X(X) \sim U(0, 1). \tag{1.25}$$

This simple mathematical statement has a far-reaching converse. Indeed, reading (1.24) from right to left we get:

Fact 2 If $U \sim U(0,1)$ and F is a cdf, then if we define the random variable X by $X = F^{-1}(U)$ we necessarily have $F_X = F$.

Indeed:

$$\mathbb{P}\{X \le x\} = \mathbb{P}\{F^{-1}(U) \le x\} = \mathbb{P}\{U \le F(x)\} = F(x). \qquad (1.26)$$

Consequently, if we want to generate a sample of size n from a distribution for which we do not have a random number generator, but for which we can compute the inverse F^{-1} of the cdf F, we only need to generate a sample from the uniform distribution, say

$$u_1, u_2, \dots\dots, u_n$$

(as we already pointed out, any computing environment will provide a random generator capable of producing such samples) and then compute the inverse of the target cdf for each of these outcomes. Because of Fact 2, the sample

$$x_1 = F^{-1}(u_1), x_2 = F^{-1}(u_2), \dots\dots, x_n = F^{-1}(u_n),$$

is a typical sample from the distribution determined by the given cdf F.

This very elementary fact is extremely useful, and we shall use it to simulate extreme events. Its only limitation lies in the actual numerical computation of the inverse cumulative distribution function F^{-1}. For example, this method is not used to generate samples from the normal distribution because the inversion of the cdf Φ is numerically too costly. On the other hand, it is routinely used for the generation of exponential samples. See Problem 1.5 for details. We illustrate the details of this random generation procedure on the example of the Cauchy distribution.

Example. Recall that a random variable X is said to have a Cauchy distribution (with location parameter m and scale parameter λ) if X has density:

$$f_{m,\lambda}(x) = \frac{1}{\pi} \frac{\lambda}{\lambda^2 + (x-m)^2}, \qquad x \in \mathbb{R}$$

which was defined in (1.11). Its cdf

$$F_{m,\lambda}(x) = \frac{1}{\pi}\left[\tan^{-1}\frac{x-m}{\lambda} + \frac{1}{2}\right]$$

was already computed in formula (1.12). As explained earlier, this distribution belongs to the family of generalized Pareto distributions (for which we use the abbreviation GDP) which we will introduce later in our discussion of heavy tail distributions. From the expression of this cdf, we compute the quantile function:

$$F^{-1}_{m,\lambda}(q) = m + \lambda \tan\left(q\pi - \frac{\pi}{2}\right)$$

already defined in (1.17), and consequently, the `S-Plus` command:

Distribution	Random Samples	Density	cdf	Quantiles
uniform distribution	runif	dunif	punif	qunif
univariate normal distribution	rnorm	dnorm	pnorm	qnorm
exponential distribution	rexp	dexp	pexp	qexp
Cauchy distribution	rcauchy	dcauchy	pcauchy	qcauchy

Table 1.2. S-Plus commands for the manipulation of the classical probability distributions.

```
> CAUCHY <- M +LAMBDA*tan(PI*(runif(N)-.5))
```

will produce a vector CAUCHY of length N of Cauchy variates with location M and scale parameter LAMBDA. We shall not use this command in the sequel because S-Plus has a special command rcauchy for the generation of Cauchy random samples. The above exercise was motivated by our desire to *open the box* on how random number generators are designed.

We gather all the S-Plus commands which can be used to produce samples from (i.e. realizations of i.i.d. random variables with the) classical probability distributions singled out in this chapter in Table 1.2. We also include the commands computing the density functions and the cdf's of these probability distributions.

As illustration, we produce and plot samples from the Cauchy and normal distributions. We use the commands:

```
> GWN <- rnorm(1024)
> CWN <- rcauchy(1024)
> par(mfrow=c(2,1))
> tsplot(GWN)
> tsplot(CWN)
> par(mfrow=c(1,1))
```

The corresponding plot is given in Figure 1.14. The S-Plus function tsplot produces a sequential plot of the entries of a numerical vector against the successive integers. As we already mentioned, these two samples look very different. Indeed the Cauchy distribution produces positive and negative numbers with very large absolute values. To really grasp the huge differences between these two samples, one needs to integrate the differences in scales on the vertical axes. Since we claim that the differences are mostly in the sizes of the tails, a Q-Q plot is in order. It is given in Figure 1.15. The discrepancy between the ranges of the two samples is reflected by the fact that the line is essentially horizontal. Moreover, the points on the right and left most parts of the plot depart from this line, showing that the tails of the distribution implied by the Cauchy sample are much thicker than the tails of of the distribution implied by the normal sample.

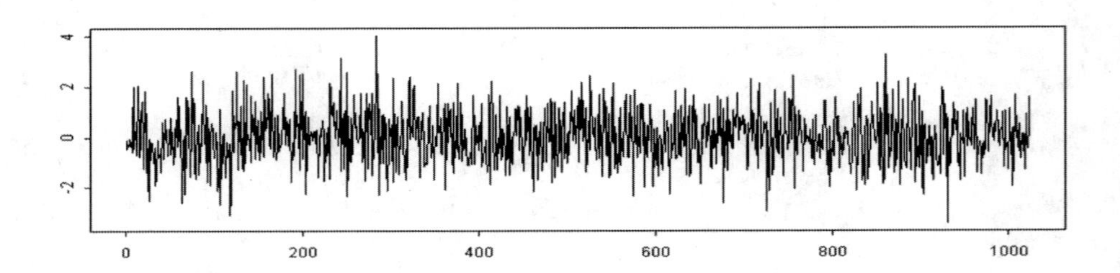

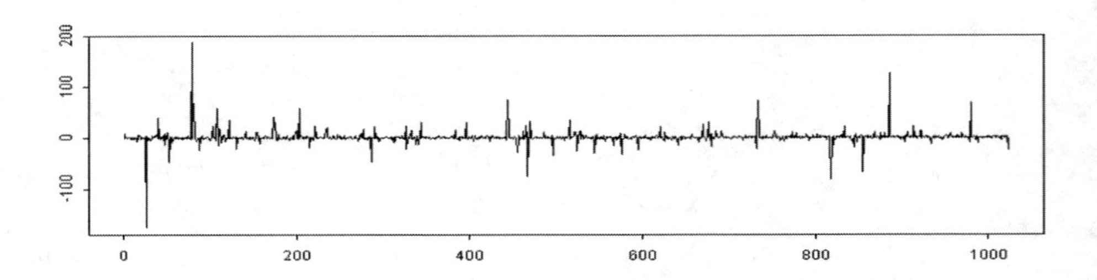

Fig. 1.14. Sequential plot of 1024 i.i.d. samples from the Gaussian distribution $N(0, 1)$ (top) and the Cauchy distribution $C(0, 1)$ (bottom).

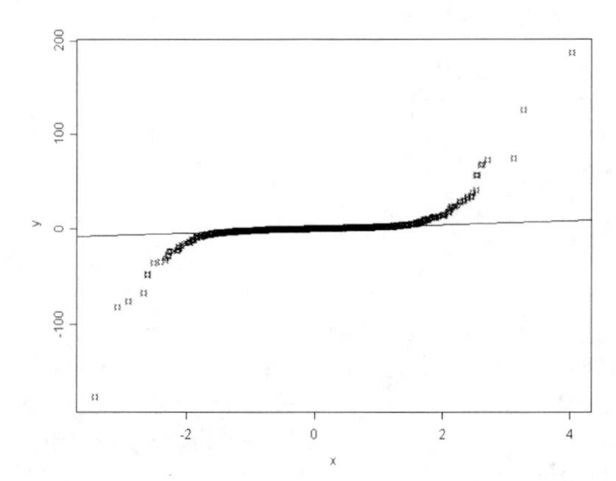

Fig. 1.15. Q-Q plot of 1024 i.i.d. samples from the Gaussian distribution $N(0, 1)$ and the Cauchy distribution $C(0, 1)$ plotted sequentially in Figure 1.14

Sequences of mean zero i.i.d. random variables will be used extensively in the chapters dealing with time series. There, such sequences will be called a white noise sequences.

1.7 EXTREMES AND HEAVY TAIL DISTRIBUTIONS

We know by experience that normal (which we use interchangeably with the adjective Gaussian) variates will very rarely be found further than two standard deviations away from their mean. As we already explained, this is a property of the tail of the distribution. Indeed, the density of the normal distribution is the exponential of the negative square normalized distance to the mean, so it decays very fast as one moves away from the mean. Figure 1.14 emphasizes this fact with a comparison with the Cauchy distribution, the tails of which are much thicker. Even though the density of the exponential distribution involves only the exponential of a negative linear function, it also decays quickly at ∞. These fast decays of the densities in the tails of the distributions guarantee the absence of many large or extreme values. As we noticed in the Q-Q plots of the weekly log-returns of the S&P 500, financial data exhibit a different type of behavior. In fact it is not uncommon to observe data values which are several standard deviations away from the mean.

1.7.1 S&P Daily Returns, Once More

We now revisit the example of the daily closing values of the S&P 500 index, data which we introduced earlier in Subsection 1.3.3. We already argued that the histogram and the kernel density estimators could not give a good account of the tail properties of the distribution, and we insisted that a Q-Q plot was the best way to get a reasonable feeling for these properties. We emphasize one more time the non-normality of the distribution of the daily log-returns by considering their extreme values and their relative distances from the mean of the distribution. Using the functions `mean` and `var`, we compute the mean and the standard deviation of the daily log-returns. We use the function `sqrt` to compute the square root of the variance.

```
> mean(DSPLR)
[1] 0.0002717871
> sqrt(var(DSPLR))
[1] 0.009175474
```

Looking at the sequential plot of the daily log-return (as reproduced in the right pane of Figure 1.10) we notice a few very large negative values. Looking more closely at the largest of these down-moves we see that:

```
> min(DSPLR)
[1]  -0.2289972
> (min(DSPLR)-mean(DSPLR))/sqrt(var(DSPLR))
[1]  -24.98716
```

which shows that this down move was essentially

25 standard deviations away from the mean

daily move ! So much for the normal distribution as a model for the daily moves of this index. The log-return on this single day of October 1987, as well as many others since then (though less dramatic in sizes) cannot be accounted for by a normal model for the daily log-returns. The tails of the normal distribution are too thin to produce such extreme values. However, other families of distributions could be used instead, and stable or Pareto distributions have been proposed with reasonable success. With the possible exception of the Cauchy distribution, the main drawback of these distributions is the lack of a simple formula for the density and/or the cdf. Recall formula (1.11) for the definition of the Cauchy density. Like the normal density, it is bell-shaped, but unlike the normal density, the tails are so *thick* that the moments of the distribution such as the mathematical expectation (or mean) and the variance do not even exist.

The theory of distributions giving rise to unusually large numbers of extremes in each sample is called the theory of extreme-value distributions, or extreme-value theory. It is a well developed mathematical theory which is beyond the scope of this book. We shall nevertheless use some of the estimation tools of this theory to populate the graphics compartment of our exploratory data analysis toolbox.

One way to identify classes of distributions producing wild extremes, is to show that the density of the distribution decays polynomially, and then to estimate the degree of such a polynomial decay. Such a class of distributions will be called the class of Generalized Pareto Distributions (GPD for short.) The main difficulty in identifying membership in this class is that the traditional density estimation procedures (such as histograms or kernel density estimators) cannot estimate the tails of the distribution precisely enough. Indeed, even though they can be good estimators in the center of a distribution where most of the data is to be found, they are rather poor estimators of the tails because there are not enough data points there, by the very nature of what we call the tails of a distribution. So since the definition of a GPD depends only on properties of the tails, the estimation of the sizes of these tails needs to be done in a parametric way. Most of the estimators we use are built on a combination of nonparametric (for the center of the distribution) and parametric (for the tails of the distribution). This semi-parametric estimation procedure can be summarized as follows:

- use standard nonparametric techniques (such as histograms or kernel density estimators) to estimate the center of the distribution:
- use parametric techniques to estimate the polynomial decay of the density in the tails.

Notice that one major issue was *brushed under the rug* in the above discussion. It concerns the determination of when and where the tails of a distribution start. This problem is rather delicate, and an incorrect choice can have disastrous consequences on the performance of the estimator. For that reason, we shall make practical recommendations on how to choose this starting point, but unfortunately, there is no universal way to determine its value.

Membership in the class of generalized Pareto distributions is checked empirically by the estimation of the parameter ξ which controls the polynomial decay of the density. ξ is called the shape parameter, or shape index of the tail. The Cauchy distribution is a particular case of the family of generalized Pareto distributions. It corresponds to the shape index $\xi = 1$. If the density of a random variable behaves as a multiple of $x^{-(1+1/\xi)}$ as $x \to \infty$ (resp. $x \to -\infty$) we say that the shape index of the distribution of the right tail (resp. left tail) of the distribution is ξ. This index ξ can be estimated parametrically by approximate maximum likelihood methods, such as the Hill's estimator. But the resulting estimates are very unstable, and horror stories documenting erroneous results have been reported in the literature, and have attracted the attention of many practitioners. So in order to avoid these problems, we use a nonparametric approach, basing the estimator on the analysis of the frequency with which data points exceed prescribed levels. This method is known as the Peak Over Threshold (POT for short) method. We implemented it in `S-Plus` under the name `pot`.

1.7.2 The Example of the PCS Index

It is possible to propose mathematical models for the time evolution of the PCS index. We describe one of them in the Notes & Complements at the end of the chapter. These models are most often quite sophisticated, and they are difficult to fit and use in practice. Instead of aiming at a theory of the dynamics of the index, a less ambitious program is to consider the value of the index on any given day, and to perform a static analysis of its distribution. This gives us a chance to illustrate how it is possible to fit a generalized Pareto distribution. Notice that, because of the positivity of the data, we shall only have to deal with the right tail of the distribution. The exponential Q-Q plot appearing in the left pane of Figure 1.13 leads us to suspect a heavier than normal tail. We should examine this Q-Q plot carefully to decide which cut-off value should separate the tail from the bulk of the distribution. This choice should be driven by the following two seemingly contradictory requirements. The cut-off point should be large enough so that the behavior of the tail is homogeneous beyond this threshold. But at the same time, it should not be too large, as we need enough data points beyond the threshold to guarantee a reasonable estimation of ξ by the POT method. As per Figure 1.13, we find that .15 could do the trick, so we run the command:

```
> PCS.est <- gpd.tail(PCS.index,one.tail=T,upper=0.15)
```

This command creates an object `PCS.est` which contains all we need to know about the estimation results. As a side effect, it also generates a plot. We reproduce the latter in Figure 1.16. We also give examples of ways to extract information from the objects thus created. We choose to use the command `one.tail=T` because the distribution does not have a lower/left tail (remember that all the values of the index are positive). The fact that the points appearing in the left part of the plot in

Figure 1.16 are essentially in a straight line is an indication that a generalized Pareto distribution may be appropriate.

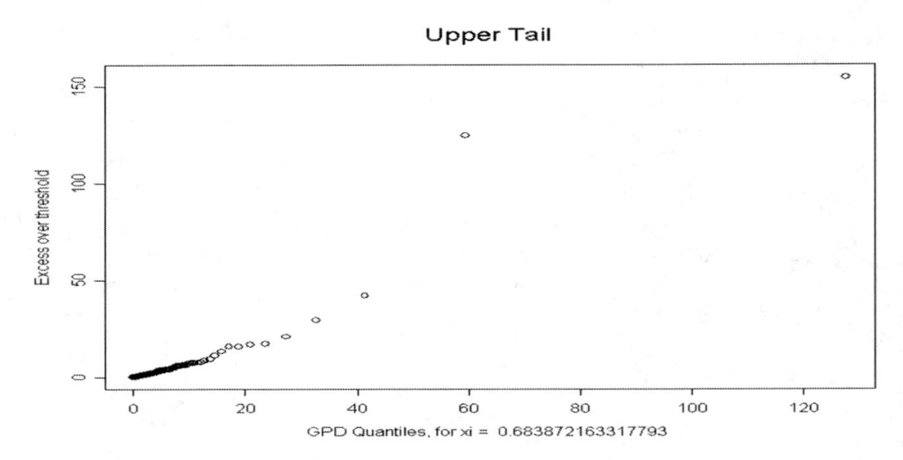

Fig. 1.16. Right tail quantile plot resulting from the fit of a GPD to the PCS index data.

Plotting an object of class gpd with the command `plot(PCS.est)` would produce four plots: a plot of the excesses, a plot of the tail of the underlying distribution, and also a scatterplot and a Q-Q plot of the residuals. Since we are mostly interested in the second of these plots, we use instead the command `tailplot` to visualize the quality of the fit. For the sake of illustration we run the commands:

```
> par(mfrow=c(2,1))
> tailplot(PCS.est,optlog="")
> tailplot(PCS.est)
> par(mfrow=c(1,1))
```

and reproduce the result given in Figure 1.17. Notice that the vertical axis is for the *survival function* $1 - F(x)$, instead of the cdf $F(x)$. The use of the option `optlog` forces `S-Plus` to use the natural scale instead of the logarithmic scale which is used by default. This is done for the first plot reproduced on the top of Figure 1.17. Unfortunately, the curve sticks very early to the horizontal axis and it is extremely difficult to properly quantify the quality of the fit. This plot is not very instructive. It was given for illustration purposes only. Plotting both the values of the index, and the values of the survival function on a logarithmic scale makes it easier to see how well (or possibly how poorly) the fitted distribution gives an account of the data. The second command (using the default value of the parameter `optlog`) gives the plot of the survival function in logarithmic scales. Both plots show that the fit is very good. Our next inquiry concerns the value of the shape parameter ξ. Remember that this number is what controls the *polynomial decay* of the density in the tail of the distribution at ∞. The choice of a threshold indicating the beginning of the tail,

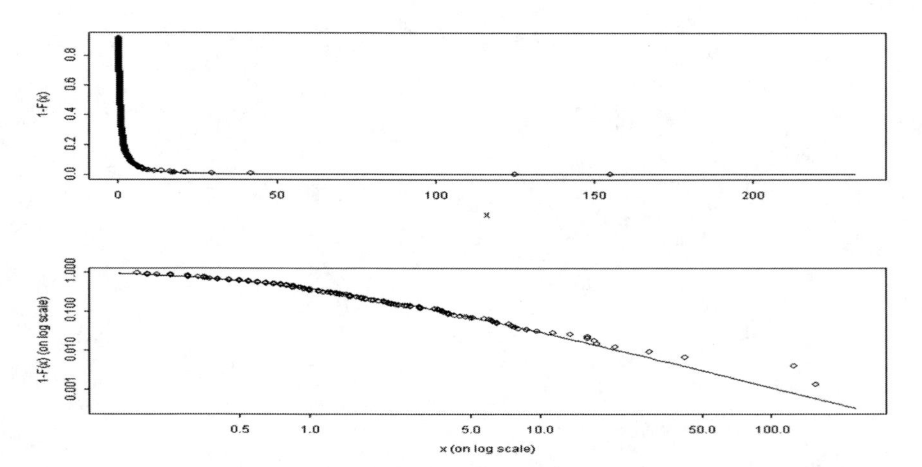

Fig. 1.17. Plot of the tail of the GPD fitted to the PCS data together with the empirical tail given by the actual data points, in the natural scale (top) and in logarithmic scale (bottom).

forces an estimate of ξ. The value of this estimate is printed on the plot produced by the function `gpd.tail` and it can be read off Figure 1.16. This estimated parameter value is part of the object `PCS.est`. Since it is the first of the two parameters defining the upper tail, it can be extracted using the extension `$upper.par.ests[1]` in the following way:

```
> PCS.est$upper.par.ests[1]
        xi
 0.6838722
```

Changing the value of the threshold `upper` in the call of the function `gpd.tail` changes the value of the estimate for ξ, so we should be concerned with the stability of the result: we would not want to rely on a procedure that is too sensitive to small changes in the choice of the threshold. Indeed, since there is no clear procedure to choose this threshold with great precision, the result of the estimation of the shape parameter should remain robust to reasonable errors/variations in the choice of this threshold. The best way to check that this is indeed the case is graphical. It relies on the use of the function `shape.plot` which gives a plot of the estimates of the shape parameter ξ as they change with the values of the threshold used to produce these estimates. Using the command:

```
> shape.plot(PCS.index)
```

we get a plot of all the different estimates of ξ which can be obtained by varying the threshold parameter `upper`. This plot is reproduced in Figure 1.18. The leftmost part of the plot should be ignored because, if the threshold is too small, too much of the bulk of the data (which should be included in the center of the distribution)

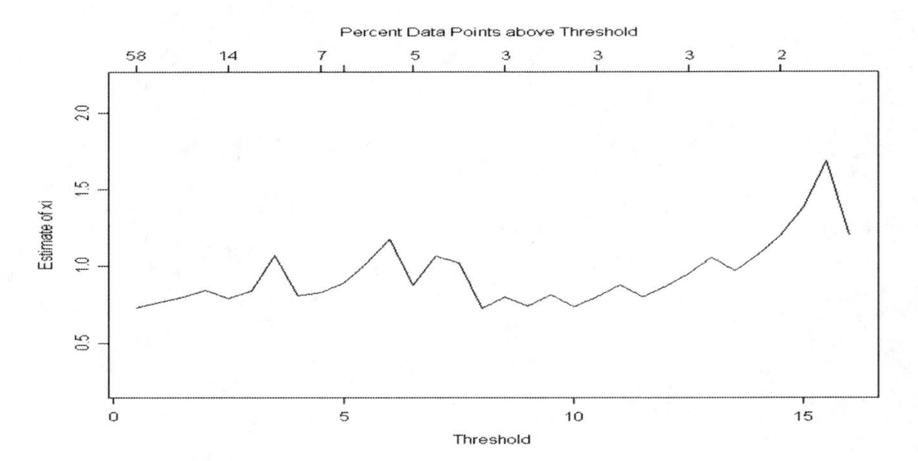

Fig. 1.18. PCS data shape parameter ξ.

contributes to the estimate of the tail, biasing the result. Similarly, the rightmost part of the plot should be ignored because, if the threshold is too large, not enough points contribute to the estimate. A horizontal axis was added to the upper part of the plot to give the proportion of points included in the estimate. This information is extremely useful when it comes to deciding whether or not to take seriously some of the estimates of ξ which appear on the left and right ends of the plot. The central part of the graph should be essentially horizontal (though not always a straight line) when the empirical distribution of the data can be reasonably well explained by a GPD. This is indeed the case in the present situation, and a value of $\xi = .8$ seems to be a reasonable estimate for the intercept of a horizontal line fitting the central part of the graph.

Our last test of the efficiency of our extreme value toolbox is crucial for risk analysis and stress testing of stochastic systems suspected to carry extreme rare events: can we generate random samples from a generalized Pareto distribution fitted to a data set? We now show how this can be done in the case of the PCS index. This is done using the function `gpd.1q`. Indeed, if `X` is a vector of numerical values, and `gpd.object` is a gpd.object, then `gpd.1q(X,gpd.object)` gives the vector of the values computed at the entries of `X`, of the quantile function (i.e. the inverse of the cdf) of the GPD whose characteristics are given by `gpd.object`. If we recall our discussion of the way to generate a Monte Carlo sample from a given distribution, we see that, replacing the numerical vector `X` by a sample from the uniform distribution will give a sample from the desired distribution. Consequently,

```
> PCSsim <- gpd.1q(runif(length(PCS.index)),PCS.est)
```

produces a random sample of the same size as the original PCS data, and hopefully with the same distribution. The plots produced by the following commands are reproduced in Figure 1.19.

```
> par(mfrow=c(1,2))
> plot(PCS[,1],PCS.index)
> plot(PCS[,1],PCSsim)
> par(mfrow=c(1,1))
```

When the S-Plus function plot is called with a couple of numerical vectors with the same numbers of rows say n, as arguments, it produces a plot of n points whose coordinates are the entries found in the rows of the two vectors. Putting next to each other the sequential plots of the original data, and of this simulation, shows that our simulated sample seems to have the same features as the original data. This claim is not the result of a rigorous test, but at this stage, we shall consider ourselves as satisfied ! See nevertheless Problem 1.6 for an attempt at quantifying the goodness of fit.

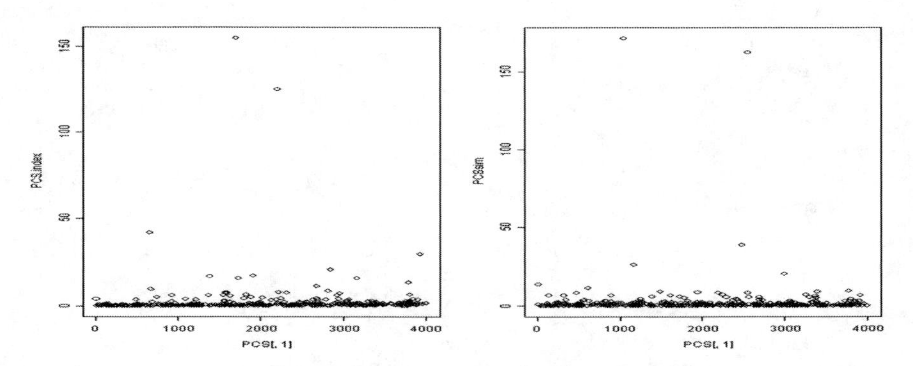

Fig. 1.19. PCS original data (left) and simulated sample (right).

1.7.3 The Example of the Weekly S&P Returns

The following analysis is very similar to the previous one, the main difference being the presence of two tails instead of one. We include it in the text to show the details of all the steps necessary to perform a complete analysis in this case, i.e. when the distribution is unbounded above *and* below. The fundamental object is obtained using the command:

```
> WSPLRet.est <- gpd.tail(WSPLRet,lower=-0.02,upper=0.02)
```

In this command, we chose the values of the thresholds lower and upper in the same way as before: we looked at the Q-Q plot of WSPLRet, and especially at the vertical axis of this plot, to decide that the lower/left tail could start at -0.02, and that the upper/right tail could start at 0.02. The threshold parameters lower and upper do not have to be "opposite" of each other, i.e. they do not need to have the same absolute values. This is likely to be the case for symmetric distributions, but it does not have to be the case in general. Finally, notice that we did not have to set the parameter `one.tail` by including `one.tail=F` in the command because this is done by default. The above command produced the two plots given in Figure 1.20. Both sets of points appear to be essentially in a straight line, so a generalized

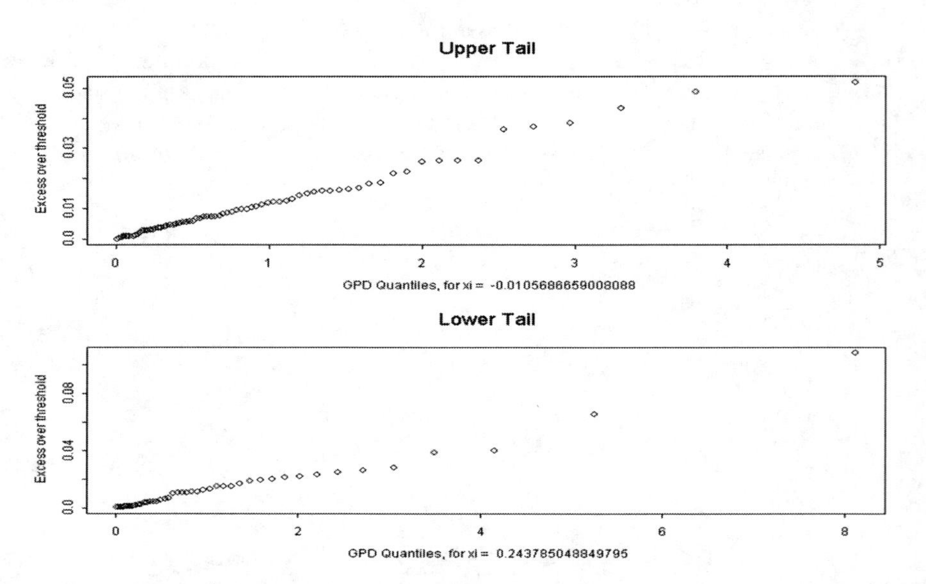

Fig. 1.20. Quantile plot for the right/upper tail (top) and left/lower tail (bottom) resulting from the fit of a GPD distribution to the weekly S&P log-return data.

Pareto distribution is a reasonable guess. Notice that the two estimates of the shape parameter ξ are not the same. The estimates obtained from the particular choices of the threshold parameters `lower` and `upper` are $\xi_{left} = 0.24$ and $\xi_{right} = -0.01$. If the distribution is not symmetric, there is no special reason for the two values of ξ to be the same, in other words, there is no particular reason why in general the polynomial decays of the right and left tails should be identical ! The results of the commands:

```
> par(mfrow=c(2,1))
> shape.plot(WSPLRet)
> shape.plot(-WSPLRet)
> par(mfrow=c(1,1))
```

are reproduced in Figure 1.21. They confirm the differences in the sizes of the left and right tails: the frequency and the size of the negative weekly log-returns are not the same as the positive ones. Also, these plots are consistent with the values we obtained for the estimates of $\hat{\xi}_{left}$ and $\hat{\xi}_{right}$, though they raise the difficult question of knowing if these estimates are *significantly* different from 0. Unfortunately at this stage, we do not have tools to compute confidence intervals for these estimates and settle this question. As before, we can check visually the quality of the fit by superim-

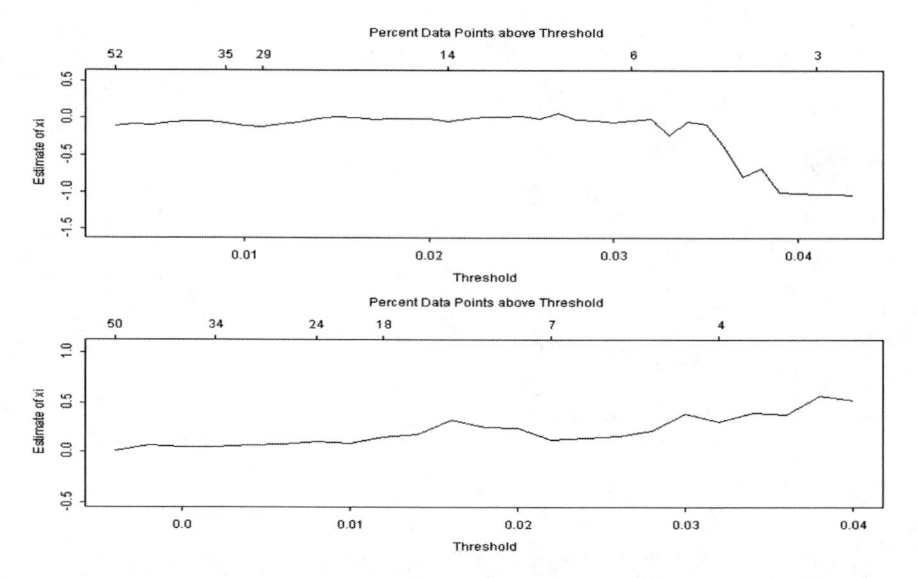

Fig. 1.21. Shape parameter ξ for the right tail (top) and left tail (bottom) of the distribution of the weekly log-returns of the S&P 500 index.

posing the empirical distribution of the points in the tails onto the theoretical graphs of the tails of the fitted distributions. This is done with the command `tailplot`. The commands:

```
> par(mfrow=c(2,1))
> tailplot(WSPLRet.est,tail="upper")
> tailplot(WSPLRet.est,tail="lower")
> par(mfrow=c(1,1))
```

produce the plot given in Figure 1.22, showing the results (in logarithmic scale) for both tails. Using the quantile function `gpd.1q( . ,WSPLRet.est)` as before, we can generate a sample of size N from the fitted distribution with the command:

```
> WSPLRetsim <- gpd.1q(runif(N),WSPLRet.est)
```

PROBLEMS

Ⓔ Ⓢ **Problem 1.1** *1. Create a sample of size* $N = 128$ *from the standard normal distribution and use Q-Q plots to assess the normality of the data.*

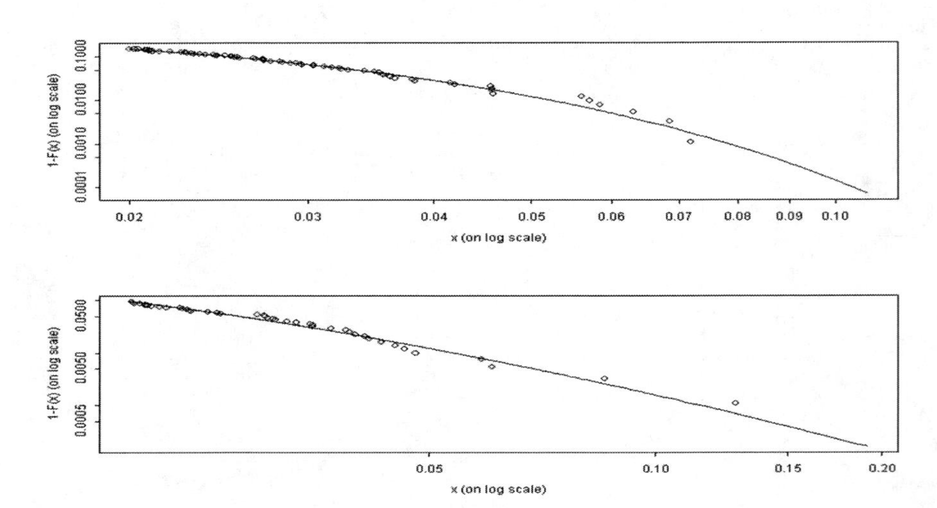

Fig. 1.22. Plot of the tails of the GPD fitted to the WSPLRet data together with the empirical tails given by the actual data points, for the upper tail (top) and the lower tail (bottom).

2. Create a sample of size $N = 128$ *from the exponential distribution with parameter* 1*, and use Q-Q plots to assess the normality of the data. Describe and explain the differences with the results of question 1.*

Ⓔ Ⓢ **Problem 1.2** *1. Generate a sample of size N from the exponential distribution with parameter (rate)* $\lambda = .2$. *Call* `X` *the vector containing the sample values.*
2. Plot on the same graph, the exact (theoretical) density of the distribution of `X`*, and a histogram of* `X`*. It is recommended to try several values for the numbers of bins, and to report only the result found most satisfactory.*
3. Plot on the same graph, the same density as before, together with a kernel density estimate of the distribution of `X`*. Again, it is recommended to try several values of the bandwidth, and to report only the result found most satisfactory.*
4. Compare the two plots and explain the reasons for the differences. Say which estimate of the density you prefer, and explain why.

Ⓔ **Problem 1.3** *1. Open and run the script file* `dsp.asc` *to create the numeric vector* `DSP`*. The entries of this vector represent the daily closing values of the* `S&P 500` *index between the beginning of January 1960 and September 18, 2001. Compute the vector of log-returns and call it* `DSPLR`.
2. Open and run the script file `intrasp.asc` *to create the numeric vector* `MSP`*. The entries of this vector represent minute by minute quotes of the* `S&P 500` *on September 10, 1998. Compute the corresponding log-return vector and call it* `MSPLR`.
3. Produce a Q-Q plot of the empirical distributions of the two log-return vectors, and comment. In particular, say if what you see is consistent with the claim that the properties of the daily series are shared by the minute by minute series. Such an invariance property is called self-similarity. *It is often encountered when dealing with fractal objects.*

4. Fit a GPD to the `DSPLR` *and* `MSPLR` *data, and compare the distributions one more time by comparing the shape parameters.*

Ⓔ **Problem 1.4** *1. Open and run the script file* `powerspot.asc` *to create the numeric vector* `PSPOT`*. The entries of this vector represent the daily Palo Verde (firm on peak) spot prices of electricity between January 4, 1999 and August 19, 2002.*
2. Use exploratory data analysis tools to argue that the tails of the distribution are heavy, fit a GPD to the data, and provide estimates of the shape parameters.

Ⓢ **Problem 1.5** *The goal of this problem is to design and use a home-grown random number generator for the exponential distribution.*
1. Compute the cdf $F_\lambda(x) = \mathbb{P}\{X \leq x\}$ *of a random variable* X *with an exponential distribution with parameter* $\lambda > 0$.
2. Compute its inverse F_λ^{-1} *and write an* `S-Plus` *function* `myrexp` *which takes the parameters* `N`*, and* `LAMBDA`*, and which returns a numeric vector of length* `N` *containing* `N` *samples of random variates from the exponential distribution with parameter* `LAMBDA`.
3. Use your function `myrexp` *to generate a sample of size* $N = 1024$ *from the exponential distribution with mean* 1.5*, use the function* `rexp` *from the* `S-Plus` *distribution to generate a sample of the same size from the same distribution, and produce a Q-Q plot of the two samples. Are you satisfied with the performance of your simulation function* `myrexp`*? Explain.*

Ⓔ **Problem 1.6** *This problem attempts to quantify the goodness of fit resulting from our GPD analysis of samples with heavy tails.*
1. Use the method described in the text to fit a GPD to the PCS index, and simulate a random sample from the fitted distribution with the same size as the original data sample.
2. Produce a Q-Q plot to compare the two samples and comment.
3. Use a two-sample Kolmogorov-Smirnov goodness-of-fit test to quantify the qualitative results of the previous question.

NB*: Such a test is performed in* `S-Plus` *with the command* `ks.gof`*. Check the help files for details on its use.*
4. Same questions as above for the weekly log-returns on the S&P data.

Ⓔ **Problem 1.7** *This problem deals with the analysis of the daily S&P 500 closing values. We assume that these data were read in* `S-Plus` *in a numeric vector, say* `DSP`*.*
1. Create a vector `DSPRET` *containing the daily raw returns. Compute the mean and the variance of this daily raw return vector, produce its* `eda.shape` *plot, and discuss the features of this plot which you find remarkable.*
NB: Recall that the code of the function `eda.shape` is contained in the Appendix devoted to the Introduction to `S-Plus`
2. Fit a Generalized Pareto Distribution (GPD for short) to the daily raw returns, give detailed plots of the fit in the two tails, and discuss your results.
3. Generate a sample of size 10000 *from the GPD fitted above. Call this sample* `SDSPRET`*, produce a Q-Q plot of* `DSPRET` *against* `SDSPRET`*, and comment.*
4. Compute the VaR (expressed in units of the current price) for a horizon of one day, at the level $\alpha = .005$ *in each of the following cases:*

- *assuming that the daily raw return is normally distributed;*
- *using the GPD distribution fitted to the data in the previous question;*
- *compare the results (think of a portfolio containing a VERY LARGE number of contracts).*

Ⓔ **Problem 1.8** *Redo the four questions of Problem 1.7 above after replacing the vector* `DSP` *with a vector* `SDSP` *containing only the first* 6000 *entries of* `DSP`. *Compare the results, and especially the VaR's. Explain the differences.*

NOTES & COMPLEMENTS

The emphasis of this book is on graphical and computational methods for data analysis, with a view toward financial applications. Most introductory statistical textbooks spend a fair amount of time discussing the exploratory data analysis tools introduced in this chapter. An excellent reference in this spirit is the book of Bill Cleveland [26]. For books with applications using `S-Plus`, we refer the interested reader to the book of Venables and Ripley [85], for a thorough discussion of the properties of histograms and their implementations in `S-Plus`. A detailed discussion of the `ASH` variation can be found there. The book by Krause and Olson [55] gives an exhaustive list of tools for the manipulation of histograms with the `Trellis` display graphics of `S-Plus`.

What distinguishes this chapter from the treatments of similar material found in most introductory statistics books, is our special emphasis on heavy tail distributions. Mandelbrot was the first author to stress the importance of the lack of normality of the financial returns. See [58], [59], and also his book [60]. He proposed the Pareto distribution as an alternative to the normal distribution. The theory of extreme value distributions is an integral part of classical probability calculus, and there are many books on the subject. We refer the interested reader to [31] because of its special emphasis on insurance applications. In particular, the following discussion of a possible mathematical model for the PCS index dynamics fits well in the spirit of [31].

Following its introduction in the mid 90's, several mathematical models for the dynamics of the PCS index have been proposed in the literature. Most of these models try to capture the catastrophe arrivals, the initial shocks, and the possible revisions to the damage estimates that arise in the ensuing weeks after a catastrophe occurs. Standard probabilistic arguments suggest to use a model of the form:

$$S(t) = \sum_{i=0}^{N(t)} X_i(t),$$

where $S(t)$ is the PCS (total) index, $X_i(t)$ is the total damage caused by the i-th catastrophe, and $N(t)$ is a Poisson process modeling catastrophe arrival. The catastrophe damage is further modeled as:

$$X_i(t) = \zeta + \theta(t - (T_i - T_\theta)),$$

where ζ is the time of the initial shock (it is usually assumed to be exponentially distributed), $\theta(t)$ is the revision function which is zero for $t < 0$ and equal to κ for $t \geq 0$, T_i is the arrival time of the i-th catastrophe, T_θ is the length of time until the catastrophe damage amount is revised (it is usually assumed to be a random variable with a Poisson distribution), and finally κ is a random variable used for the amount of revision of the index. In order to fit such a model, four parameters need to be estimated:

- the catastrophe arrival time: it is usually modeled as a Poisson process;

- the initial damage of the catastrophe: it is often modeled as an exponentially distributed random variable, but it may also be modeled with a long tail distribution;
- the delay time for the revision: it is usually modeled as an exponential distribution;
- the revision amount.

We shall not discuss any further this ambitious, though realistic, model.

Financial institutions started worrying about risk exposures long before regulators got into the act. The most significant initiative was RiskMetrics' spin off by J.P. Morgan in 1994. Even though the original methodology was mostly concerned with market risk, and limited to Gaussian models, the importance of VaR calculations were clearly presented in a set of technical documents made available on the web at the URL `www.riskmetrics.com`. A more academic discussion of the properties of VaR can be found in the recent book by C. Gourieroux and J. Jasiak [41], and the less technical book by Jorion [50]. A clear exposé of risk measures and their mathematical theory can be found in the recent book [37] by Föllmer and Schied where the reader will find a rigorous discussion of the risk measures encouraging diversification.

The `S-Plus` methods used in this chapter to estimate heavy tail distributions, and to simulate random samples from these distributions, are taken from the `S+FinMetrics` toolbox of `S-Plus`. This toolbox contains two sets of functions for that purpose. The first set is borrowed from a library developed at ETH in Zürich by Mc Neil. The second set of functions is borrowed from a library called `EVANESCE`, developed at Princeton University by Julia Morrisson and the author. It came with a manual which should have been entitled *Extremes and Copulas for Dummies* if we hadn't been afraid of a lawsuit for copyright infringement. Obviously, the analysis of extreme value distributions presented in this chapter is based on the functions of the `EVANESCE` library. Unfortunately, some of the original function names have been changed by the staff in charge of `S+FinMetrics`. For the sake of uniformity, we reluctantly abandoned the original terminology of `EVANESCE` and adopted the function names of `S+FinMetrics`. The following correspondence table gives a way to translate commands/programs written with `EVANESCE` into code which can be used with `S+FinMetrics` and vice-versa.

EVANESCE name	S+FinMetrics **name**
cdf.estimate.1tail	gpd.1p
cdf.estimate.2tails	gpd.2p
quant.estimate.1tail	gpd.1q
quant.estimate.2tails	gpd.2q
jointcdf.estimate.1tail	gpdjoint.1p
cdf.estimate.2tails	gpdjoint.2p
estimate.lmom.gpd	gpd.lmom
estimate.mle.gpd	gpd.ml
pot.2tails.est	gpd.tail

The library EVANESCE, together with its manual are available on the web at the URL

`http:\\www.princeton.edu\~rcarmona\downloads`

2

MULTIVARIATE DATA EXPLORATION

This chapter contains our first excursion away from the simple problems of univariate samples and univariate distribution estimation. We consider samples of simultaneous observations of several numerical variables. We generalize some of the exploratory data analysis tools used in the univariate case. In particular, we discuss histograms and kernel density estimators. Then we review the properties of the most important multivariate distribution of all, the normal or Gaussian distribution. For jointly normal random variables, dependence can be completely captured by the classical Pearson correlation coefficient. In general however, the situation can be quite different. We review the classical measures of dependence, and emphasize how inappropriate some of them can become in cases of significant departure from the Gaussian hypothesis. In such situations, quantifying dependence requires new ideas, and we introduce the concept of copula as a solution to this problem. We show how copulas can be estimated, and how one can use them for Monte Carlo computations and random scenarios generation. We illustrate all these concepts with an example of coffee futures prices. The last section deals with principal component analysis, a classical technique from multivariate data analysis, which is best known for its use in dimension reduction. We demonstrate its usefulness on data from the fixed income markets.

2.1 MULTIVARIATE DATA AND FIRST MEASURE OF DEPENDENCE

We begin the chapter with an excursion into the world of multivariate data, where dependencies between variables are important, and where analyzing variables separately would cause significant features of the data to be missed. We try to illustrate this point by means of several numerical examples, but we shall focus most of our discussion on the specific example of the daily closing prices of futures contracts on Brazilian and Colombian coffee which we describe in full detail in Subsection 2.2.4 below. We reproduce the first seven rows to show how the data look like after computing the daily log-returns.

```
        [,1]    [,2]
[1,] -0.0232 -0.0146
[2,] -0.0118 -0.0074
[3,] -0.0079 -0.0074
[4,]  0.0275  0.0258
[5,] -0.0355 -0.0370
[6,]  0.0000  0.0000
[7,]  0.0000 -0.0038
```

Each row corresponds to a given day, the log return on the Brazilian contract being given in the first column of that row, the log return of the Colombian contract being given in the second one. As we shall see in Subsection 2.2.4, the original data came with time stamps, but as we already explained in the previous chapter, the latter are irrelevant in the static analysis of the marginal distributions. Indeed, for that purpose, the dependence of the measurements upon time does not play any role, and we could shuffle the rows of the data set without affecting the results of this static analysis.

The data set described above is an example of *bivariate* data. We consider examples of multivariate data sets in higher dimensions later in the chapter, but in the present situation, the data can be abstracted in the form of a bivariate sample:

$$(x_1, y_1),\ (x_2, y_2),\ (x_3, y_3), \ldots\ldots, (x_n, y_n),$$

which is to be understood as a set of realizations of n independent couples

$$(X_1, Y_1),\ (X_2, Y_2),\ \ldots\ldots,\ (X_n, Y_n)$$

of random variables with the same joint probability distribution. The goal of this chapter is the analysis of the statistical properties of this joint distribution, and in particular of the dependencies between the components X and Y of these couples. Recall from Chapter 1 that if X and Y are real valued random variables, then their joint distribution is characterized by their joint cdf which is defined by:

$$(x, y) \hookrightarrow F_{(X,Y)}(x, y) = \mathbb{P}\{X \le x,\ Y \le y\}. \tag{2.1}$$

This joint distribution has a density $f_{(X,Y)}(x, y)$ if the joint cdf can be written as an indefinite (double) integral:

$$F_{(X,Y)}(x, y) = \int_{-\infty}^{x} \int_{-\infty}^{y} f_{(X,Y)}(x', y')\, dx' dy',$$

in which case the density is given by the (second partial) derivative:

$$f_{(X,Y)}(x, y) = \frac{\partial^2 F_{(X,Y)}(x, y)}{\partial x \partial y}.$$

Setting $y = +\infty$ in (2.1) leads to a simple expression for the marginal density $f_X(x)$ of X. It reads:

$$f_X(x) = \int_{-\infty}^{+\infty} f_{(X,Y)}(x, y')dy'$$

and similarly

$$f_Y(y) = \int_{-\infty}^{+\infty} f_{(X,Y)}(x', y)dx'.$$

2.1.1 Density Estimation

The notions of histogram and empirical cdf used in the previous chapter can be generalized to the multivariate setting. Let us discuss the bivariate case for the sake of definiteness. Indeed, one can divide the domain of the couples (x_i, y_i) into plaquettes or rectangular bins, and create a surface plot by forming cylinders above these plaquettes, the height of each cylinder being proportional to the number of couples (x_i, y_i) falling into the base. If the lack of smoothness of the one-dimensional histograms was a shortcoming, this lack of smoothness is even worse in higher dimensions. The case of the empirical cdf is even worse: the higher the dimension, the more difficult it becomes to compute it, and use it in a reliable manner. The main drawback of both the histogram and the empirical cdf is the difficulty in adjusting to the larger and larger proportions of the space without data points. However, they can still be used effectively in regions with high concentrations of points. As we shall see later in this chapter, this is indeed the case in several of the `S-Plus` objects used to code multivariate distributions.

The Kernel Estimator

The clumsiness of the multivariate forms of the histogram is one of the main reasons for the extreme popularity of kernel density estimates in high dimension. Given a sample $(x_1, y_1), \ldots, (x_n, y_n)$ from a distribution with (unknown) density $f(x, y)$, the formal kernel density estimator of f is the function $\hat{f}_b$ defined by:

$$\hat{f}_b(x, y) = \frac{1}{nb^2} \sum_{i=1}^{n} K\left(\frac{1}{b}[(x, y) - (x_i, y_i)]\right) \tag{2.2}$$

where the function K is a given non-negative function of the couple (x, y) which integrates to one (i.e. a probability density function) which we call the kernel, and $b > 0$ is a positive number which we call the bandwidth. The interpretation of formula (2.2) is exactly the same as in the univariate case. If (x, y) is in a region with many data points (x_i, y_i), then the sum in the right hand side of (2.2) will contain many terms significantly different from 0 and the resulting density estimate $\hat{f}_b(x, y)$ will be large. On the other hand, if (x, y) is in a region with few or no data points (x_i, y_i), then the sum in the right hand side of (2.2) will contain only very small numbers and the resulting density estimate $\hat{f}_b(x, y)$ will be very small. This intuitive explanation of the behavior of the kernel estimator is exactly what is expected from any density estimator. Notice that the size of the bandwidth b regulates the extent to which this statement is true by changing how much the points (x_i, y_i) will contribute to the sum.

`S-Plus` *Implementation*

There is no function for multivariate histogram or kernel density estimation in the commercial distribution of `S-Plus`, so we added to our library the function `kdest` which takes a bivariate sample as argument, and produces an `S` object (a data frame to be specific) containing a column for the values of the density estimator, and two columns for the values of the coordinates of the points of the grid on which the estimate is computed. To be specific, if `X` and `Y` are numeric vectors with equal lengths, the command:

```
> DENS <- kdest(X,Y)
```

produces a data frame with three columns. Selecting these three columns and using one of the 3-D surface plot commands will produce a surface plot of the values of the kernel density estimate over a regular grid of 256×256 points covering the range of the bivariate vector `(X,Y)`, i.e. at points of the (x, y)-plane for which x grows from `xmin` to `xmax` and y grows from `ymin` to `ymax` in $n = 256$ regular increments. The size of the grid and the default values of the bandwidth parameters can be specified by the user. We illustrate the results of the (bivariate) kernel density estimation with a couple of examples.

• The first example concerns part of a data set which we will study thoroughly in the next chapter. The surface plot of Figure 2.1 is the result of running the command `kdest` on two data vectors `X` and `Y` derived from the values of indexes computed from the share values and the capitalizations of ENRON and DUKE over the period ranging from January 4, 1993 to December 31, 1993. The well-separated bumps show clearly that the observations (x_i, y_i) can be divided into several subsets which can be discriminated from each other on the basis of the values of the two variables. This situation is very much sought after in pattern recognition applications where the goal is to subdivide the population into well-defined, and hopefully well separated, clusters which can be identified by their local means, for example.

• Our second example concerns, once more, the daily closing values of the S&P 500 index. The goal is to estimate the joint probability density of the log-return computed on a period of 5 days starting on a given day, and the log-return computed on a period of 15 days ending the same day. The scatterplot of these two variables is given in the left pane of Figure 2.2. From a central blob of points two sparse clouds extend in the direction of the negative x-axis and the positive y-axis. The most interesting feature of this scatterplot, however, is the following: the large positive values of the 5 days log-returns follow large negative values of the 15 days log-returns. Anticipating the discussion of the correlation coefficient introduced in the next subsection, we suspect there being a negative correlation between the two returns: indeed computing the correlation between these two variables gives a value approximately equal to

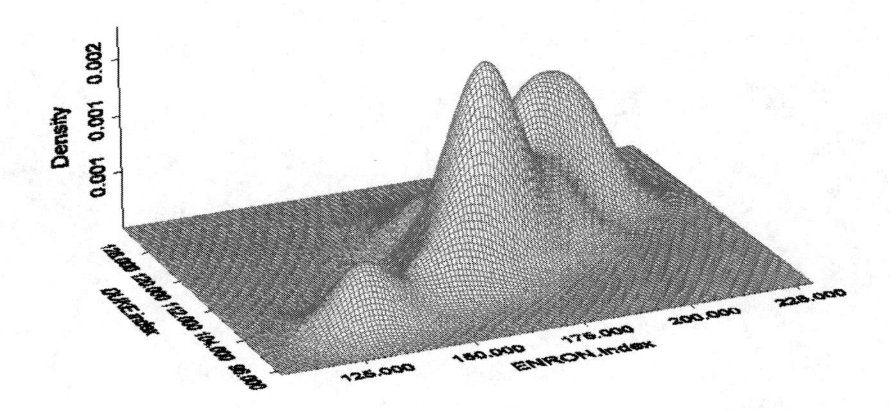

Fig. 2.1. Kernel density estimate for the utility data

$-.5$. The density estimate reproduced in the right pane shows that the central blob of points appearing in the scatterplot is in fact formed by two separate narrow bumps. But this density estimate fails to reproduce the trail of points in the right part of the scatterplot. As we explained earlier, we believe that these points are responsible for the significant negative correlation, and we do not like seeing them ignored by the kernel density estimator. The problem is very delicate. A smaller bandwidth restores the presence of these points, but the surface would be so rough that the density estimate would be less instructive than the scatterplot itself. On the other hand a larger bandwidth gives a smoother surface, but the latter becomes unimodal, wiping out the signs of possible separate bumps near the center of the distribution. We chose the bandwidth to reach a compromise between these extremes, but as we already explained, we lost the trail of days responsible for the negative correlation. Unfortunately, the serious difficulties experienced in the analysis of this example are typical of many of the real-life applications in which one would like to use density estimation.

2.1.2 The Correlation Coefficient

Motivated by the previous discussion of the evidence of a possible linear dependence between variables, we introduce the correlation coefficient between two random variables. This theoretical concept and its empirical counterpart are designed to capture this type of linear dependence. It is the most widely-used measure of dependence between two random variables. It is called the Pearson correlation coefficient. For

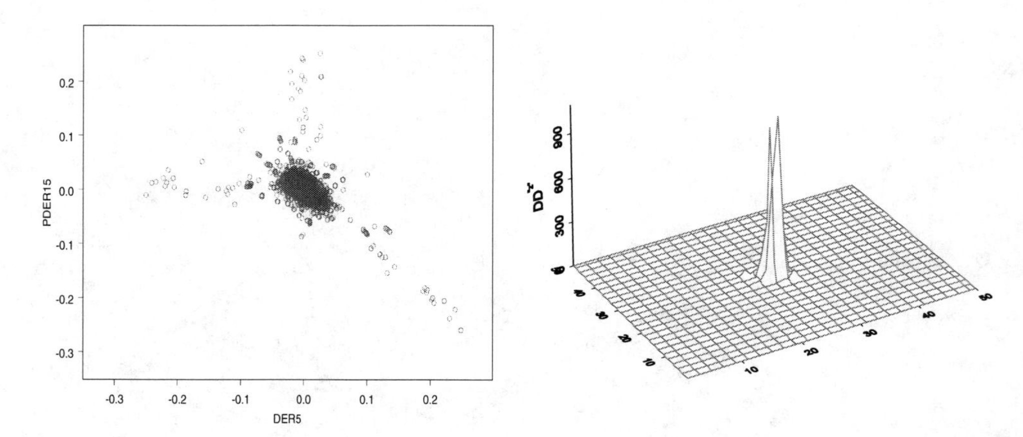

Fig. 2.2. Scatterplot (left) and kernel density estimate (right) for the 5 days & 15 days S&P log-returns

random variables X and Y it is defined as:

$$\rho_P\{X,Y\} = \frac{\text{cov}\{X,Y\}}{\sigma_X \sigma_Y} \tag{2.3}$$

where the covariance $\text{cov}\{X,Y\}$ is defined by:

$$\text{cov}\{X,Y\} = \mathbb{E}\{(X - \mathbb{E}\{X\})(Y - \mathbb{E}\{Y\})\} = \mathbb{E}\{XY\} - \mathbb{E}\{X\}\mathbb{E}\{Y\} \tag{2.4}$$

and where σ_X and σ_Y denote as usual the standard deviations of X and Y, respectively, i.e.

$$\sigma_X = \sqrt{\mathbb{E}\{(X - \mathbb{E}\{X\})^2} = \sqrt{\mathbb{E}\{X^2\} - \mathbb{E}\{X\}^2} \tag{2.5}$$

and similarly for σ_Y. If X and Y have a joint density $f(x,y)$ then the definition of the covariance can be rewritten in terms of a double integral as:

$$\text{cov}\{X,Y\} = \int\int xyf(x,y)\,dxdy - \left(\int xf_X(x)dx\right)\left(\int yf_Y(y)dy\right).$$

Because of its frequent use, the subscript P is often dropped from the notation, and the Pearson correlation coefficient is commonly denoted by ρ. The empirical analog of this measure of dependence is defined for samples $x_1, \ldots, x_n$ and $y_1, \ldots, y_n$. By analogy with formula (2.3) it is defined as:

$$\hat{\rho}\{X,Y\} = \frac{\widehat{\text{cov}\{X,Y\}}}{\hat{\sigma}_X \hat{\sigma}_Y} \tag{2.6}$$

and it is called the empirical correlation between the samples. Here, the empirical covariance $\widehat{\text{cov}\{X,Y\}}$ is defined by:

$$\widehat{\text{cov}\{X,Y\}} = \frac{1}{n}\sum_{i=1}^{n}(x_i - \overline{x})(y_i - \overline{y}) = \frac{1}{n}\sum_{i=1}^{n} x_i y_i - \overline{x}\,\overline{y} \tag{2.7}$$

where we used the notations $\overline{x}$ and $\overline{y}$ for the sample means of x and y defined by:

$$\overline{x} = \frac{1}{n}\sum_{i=1}^{n} x_i \qquad \text{and} \qquad \overline{y} = \frac{1}{n}\sum_{i=1}^{n} y_i, \tag{2.8}$$

and where the sample standard deviations $\hat{\sigma}_X$ and $\hat{\sigma}_Y$ are defined by:

$$\hat{\sigma}_X = \sqrt{\frac{1}{n}\sum_{i=1}^{n}(x_i - \overline{x})^2} = \sqrt{\frac{1}{n}\sum_{i=1}^{n} x_i^2 - \overline{x}^2} \tag{2.9}$$

and similarly for $\hat{\sigma}_Y$. Some of the properties of these correlation coefficients are well known. Others are less so. We review them in order to emphasize the usefulness of the correlation coefficient, and at the same time to stress its limitations.

Properties of the Correlation Coefficient

The most immediate properties of the correlation coefficient are:

- The real numbers ρ and $\hat{\rho}$ are always between -1 and $+1$
- $\rho = 0$ when the random variables X and Y are independent
- $\rho = 1$ when Y is a linear function of X.

These simple properties have lead to the following usage of the sample correlation coefficient $\hat{\rho}$. The samples are regarded as independent when $\hat{\rho}$ is small, while the samples are regarded as strongly dependent when $\hat{\rho}$ is close to 1 or -1. We shall see below that this practice is okay when the samples come from a multivariate normal distribution, but it can be very misleading for other distributions.

The properties listed in the three bullets above are well known. Their intuitive content is the main reason for the enormous popularity of the correlation coefficient as a measure of dependence.

What is often overlooked is the fact that the Pearson correlation coefficient is only a measure of linear dependence between two random variables. In fact, ρ measures the relative reduction of the response variation by a linear regression. Indeed, anticipating our upcoming discussion on least squares linear regression, we can use the following general formula

$$\rho\{X,Y\} = \frac{\sigma^2\{Y\} - \min_{\beta_0,\beta_1}\mathbb{E}\{|Y - \beta_0 - \beta_1 X|^2\}}{\sigma^2\{Y\}}$$

to justify this claim. The numerator of the right hand side is the difference between the variation in the variable Y, and the smallest possible remaining variation after removing a linear function $\beta_0 + \beta_1 X$, of X. This formula gives the slope of the least squares regression line of Y against X in terms of ρ_P.

Finally, we close this section with a very surprising property of the Pearson correlation coefficient. Strangely enough, this property is little known despite its important practical implications, especially in the world of financial models. If the marginal distributions of X and Y are given, but no information is given on the nature of their dependence or lack thereof, the possible values of the correlation coefficient ρ are limited to an interval $[\rho_{min}, \rho_{max}]$. However, contrary to popular belief, this interval is not always the whole interval $[-1, +1]$. There are cases for which this interval is much smaller, even for frequently-used distributions. See for example Problems 2.3 and 2.7 at the end of this chapter, where the case of lognormal random variables is analyzed in detail.

2.2 The Multivariate Normal Distribution

We start our analysis of multivariate statistical distributions with the case of the well-known normal family. All the reasons we gave for the popularity of the univariate normal distribution still hold in the multivariate case. Moreover, the possible competition from other distribution families vanishes. Indeed, the normal family is essentially the only one for which explicit analytic computations are possible. We first give an abstract definition and concentrate on the interpretation of the consequences of such a definition. Even though most of the explicit computations done in the book will be limited to the bivariate case, we start with the general definition of the multivariate normal distribution because of its widespread use in portfolio theory where realistic situations involve very large numbers of instruments. Because of this general setup, the discussion which follows is of rather abstract nature, and a quick look at the contents of Appendix 1 at the end of the chapter may help with some of the mathematics.

One says that k real valued random variables $Z_1, \ldots, Z_k$ are jointly normal, or that their distribution is a multivariate normal distribution, if the joint density of Z_1, ..., Z_k is given by:

$$f_{(Z_1,\ldots,Z_k)}(z_1,\ldots,z_k) = \frac{1}{\sqrt{(2\pi)^k \det(\mathbf{\Sigma})}} \exp\left(-\frac{1}{2}[\mathbf{z}-\mu]^t \mathbf{\Sigma}^{-1}[\mathbf{z}-\mu]\right) \quad (2.10)$$

for some $k \times k$ invertible matrix $\mathbf{\Sigma}$ and a k-dimensional vector μ. In this formula we used the notation $\mathbf{Z}$ for the k-dimensional vector whose components are the Z_i's. The above definition is usually encapsulated in the notation:

$$\mathbf{Z} \sim N_k(\mu, \mathbf{\Sigma})$$

to signify that the random vector has the k-variate normal distribution with mean vector μ and variance/covariance matrix $\mathbf{\Sigma}$. This terminology is consistent with the standard practice of probability calculus with random vectors and matrices, which we recall in Appendix 1 at the end of the chapter. μ is the $k \times 1$ vector of means

$\mu_i = \mathbb{E}\{Z_i\}$ and $\mathbf{\Sigma}$ is the variance/covariance matrix whose entries are $\mathbf{\Sigma}_{i,j} = \text{cov}\{Z_i, Z_j\}$. Using the convention introduced in the appendix, this reads:

$$\mathbb{E}\{\mathbf{Z}\} = \mu \qquad \text{and} \qquad \Sigma_{\mathbf{Z}} = \mathbf{\Sigma}.$$

According to its definition (2.17), the entries of the covariance matrix $\Sigma_{\mathbf{Z}}$ are the covariances $\text{cov}\{Z_i, Z_j\}$, and consequently, the knowledge of all the marginal (bivariate) distributions of the couples (Z_i, Z_j) is enough to determine the entire joint distribution. This particular property is specific to the multivariate normal distribution. It does not hold for general distributions. Moreover, the matrix calculus developed for random vectors in Appendix 1 implies that:

$$\mathbf{Z} \sim N_k(\mu, \mathbf{\Sigma}) \quad \text{when} \quad \mathbf{Z} = \mu + \mathbf{\Sigma}^{1/2}\mathbf{X} \quad \text{and} \quad \mathbf{X} \sim N_k(0, \mathbf{I}_k) \tag{2.11}$$

where $\mathbf{I}_k$ denotes the $k \times k$ identity matrix, and $\mathbf{\Sigma}^{1/2}$ denotes the square root of the symmetric nonnegative-definite matrix $\mathbf{\Sigma}$. See Problem 2.8 at the end of the chapter for details on the definition and the first properties of this square root matrix. In other words, starting from a random vector $\mathbf{X}$ with independent $N(0,1)$ components (see the following remark) we can get to a vector $\mathbf{Z}$ with the most general multivariate normal distribution just by linear operations: multiplying by a matrix and adding a vector. This simple fact is basic for the contents of the following subsection.

Important Remark about Independence

If the random variables Z_i are independent, then obviously all the covariances $\text{cov}\{Z_i, Z_j\}$ are zero when $i \neq j$, and the variance/covariance matrix $\Sigma_{\mathbf{Z}}$ is diagonal. The converse is not true in general, *even when the random variable Z_i's are normal !* See for example Problem 2.5 for a counter-example. But *the converse is true when the Z_i's are jointly normal !* This striking fact highlights what a difference it makes to assume that the marginal distributions are normal, versus assuming that the joint distribution is normal. The proof of this fact goes as follow: if $\Sigma_{\mathbf{Z}}$ is diagonal, then:

$$\frac{1}{2}[\mathbf{z} - \mu]^t \mathbf{\Sigma}^{-1}[\mathbf{z} - \mu] = \frac{(z_1 - \mu_1)^2}{2\sigma_1^2} + \frac{(z_2 - \mu_2)^2}{2\sigma_2^2} + \cdots\cdots + \frac{(z_k - \mu_k)^2}{2\sigma_k^2},$$

if we denote by $\sigma_1^2, \sigma_2^2, \ldots, \sigma_k^2$ the elements which appear on the diagonal of $\Sigma_{\mathbf{Z}}$. So using the definition (2.10) of the multivariate normal distribution, this implies that:

$$f_{(Z_1, Z_2, \ldots, Z_k)}(z_1, z_2, \ldots, z_k) = f_{Z_1}(z_1) f_{Z_2}(z_2) \cdots f_{Z_k}(z_k),$$

which in turn implies that the joint cdf is the product of the marginal cdf's, proving the desired independence property. So the conclusion is that:

For jointly normal random variables, independence is equivalent to the variance/covariance matrix being diagonal !

2.2.1 Simulation of Random Samples

We now show how one can use formula (2.11) to generate random samples from a multivariate normal distribution. To that end, we assume that we are given a $k \times 1$

vector μ of means, and a $k \times k$ variance/covariance matrix $\mathbf{\Sigma}$, and that we want to generate a sample of size N from the distribution $N_k(\mu, \mathbf{\Sigma})$. We proceed as follows:

1. We create a $k \times N$-matrix whose columns are all identical, any column being a copy of the mean vector μ;
2. We generate a sample of size Nk from the standard normal distribution and reshape this $(N \times k) \times 1$ vector into a $k \times N$-matrix;
3. We compute a square root for the variance/covariance matrix, then we multiply each column of the random matrix constructed in Step 2, by the square root of the variance/covaiance matrix;
4. We add the matrix of means constructed in Step 1 to the random matrix constructed in Step 3 above.

Notice that Step 3 (which is the most involved) is not needed when $\mathbf{\Sigma} = \mathbf{I}_k$. The details of this random generation algorithm are given in Problem 2.8 at the end of the chapter. There, we show how to compute the square root of a covariance matrix and we develop the code for a *home grown* function capable of generating samples from a multivariate normal distribution. Writing this code is just for the sake of illustration, since `S-Plus` provides the function `rmvnorm` whose use we illustrate in Subsection 2.2.3 below.

2.2.2 **The Bivariate Case**

In the bivariate case we have:

$$\mu = \begin{bmatrix} \mu_1 \\ \mu_2 \end{bmatrix} \qquad \text{and} \qquad \mathbf{\Sigma} = \begin{bmatrix} \sigma_1^2 & \rho\sigma_1\sigma_2 \\ \rho\sigma_1\sigma_2 & \sigma_2^2 \end{bmatrix}$$

and consequently, the joint density can be written as:

$$f_{(Z_1,Z_2)}(z_1, z_2) = \frac{1}{2\pi\sigma_1\sigma_2\sqrt{1-\rho^2}}$$
$$\exp\left(-\frac{1}{1-\rho^2}\left[\frac{(z_1-\mu_1)^2}{2\sigma_1^2} - \rho\frac{(z_1-\mu_1)(z_2-\mu_2)}{\sigma_1\sigma_2} + \frac{(z_2-\mu_2)^2}{2\sigma_2^2}\right]\right).$$

This formula shows that, if we know the marginal distributions of Z_1 and Z_2, in other words, if we know μ_1, σ_1, μ_2 and σ_2, then the joint distribution is entirely determined by the correlation coefficient ρ. Also, we clearly see from this formula that when $\rho = 0$ we have:

$$\begin{aligned} f_{(Z_1,Z_2)}(z_1, z_2) &= \frac{1}{2\pi\sigma_1\sigma_2}\exp\left(-\frac{(z_1-\mu_1)^2}{2\sigma_1^2} - \frac{(z_2-\mu_2)^2}{2\sigma_2^2}\right) \\ &= \frac{1}{\sqrt{2\pi}\sigma_1}\exp\left(-\frac{(z_1-\mu_1)^2}{2\sigma_1^2}\right)\frac{1}{\sqrt{2\pi}\sigma_1}\exp\left(-\frac{(z_2-\mu_2)^2}{2\sigma_2^2}\right) \\ &= f_{Z_1}(z_1)f_{Z_2}(z_2) \end{aligned}$$

which shows the independence of Z_1 and Z_2. So we recover the fact that if Z_1 and Z_2 are jointly Gaussian, their independence is equivalent to their being uncorrelated. As we already pointed out, this fact is not true in general, not even when Z_1 and Z_2 are (separately) Gaussian. See Problem 2.5 for a counterexample.

2.2.3 A Simulation Example

For the sake of illustration, we consider the case of the distribution of a couple of (slightly correlated) normal random variables X and Y, and we generate one sample of size $n = 128$ from the joint distribution of (X, Y). We use the `S-Plus` command:

```
> TSAMPLE<-rmvnorm(n=128,mean=rep(0,2),sd=rep(1,2),rho=.18)
> TDENS <- kdest(TSAMPLE[,1],TSAMPLE[,2])
```

The function `rmvnorm` is the multivariate analog of `rnom`. It is designed to generate multivariate normal samples. We chose the vector $[0, 0]$ for the mean by using the command `rep(0,2)` which creates a vector by repeating the number zero twice, and we used the command `rep(1,2)` to specify that both components have standard deviations equal to one. Finally, we decided on the correlation coefficient $\rho = .18$ by setting the parameter `rho`. We could have given the entire variance/covariance matrix by specifying the parameter `cov` instead of giving the vector of standard deviations and the correlation coefficient separately. See the help file for details. The second command computes a kernel density estimator (with the default kernel and bandwidth choices) and plots the resulting *surface*. The output

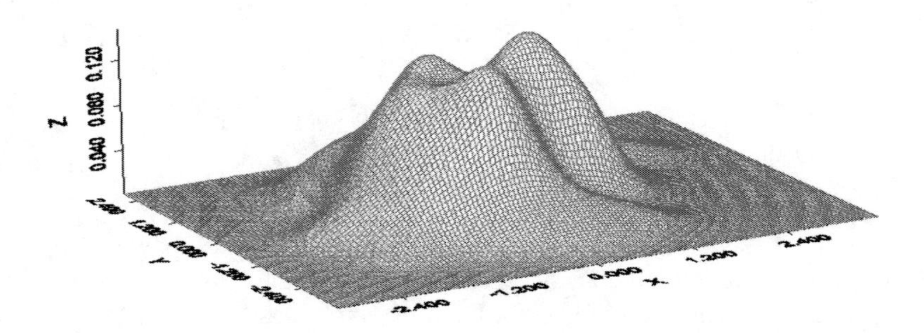

Fig. 2.3. Kernel Density Estimator for a (Bivariate) Normal Sample

is given in Figure 2.3. We see that the unimodality of the density is violated by the estimate which seems to indicate the presence of several bumps. Increasing the bandwidth would resolve this problem, at the cost of a looser fit, by somewhat flattening the central bump. The poor quality of the estimation is to be blamed on the small size of the sample: in general, the higher the dimension, the larger is the sample size needed to get reasonable density estimates.

2.2.4 **Let's Have Some Coffee**

We use Paul Erdös' famous quote:

A mathematician is a machine that turns caffeine into theorems

as a justification for our interest in the price of coffee. As explained in the abstract at the beginning of the chapter, we chose to illustrate the analysis of multivariate distributions with a simple example of two quantities which are obviously correlated. We use samples of log-returns of Brazilian and Colombian coffee spot prices. The original data are plotted in Figure 2.4, which shows the daily spot prices of coffee in Brazil and Colombia between January 9, 1986 and January 1, 1999. We do not

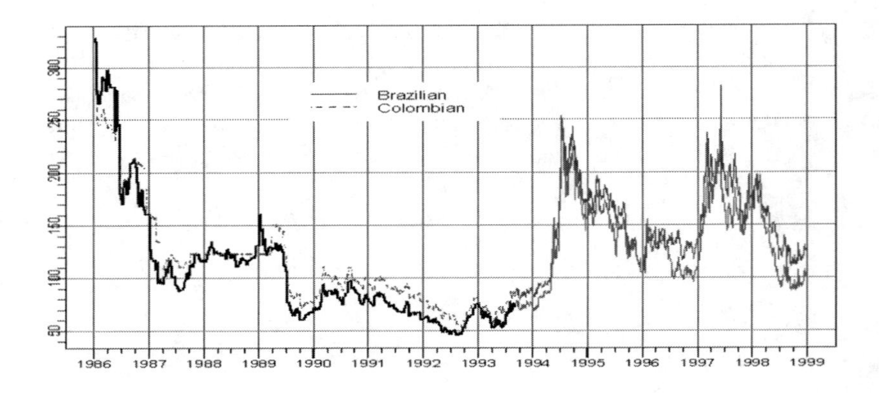

Fig. 2.4. Sequential plot of the daily prices of coffee in Brazil and Colombia from January 9, 1986 to January 1, 1999.

give the `S-Plus` commands used to produce Figure 2.4 as they involve time series objects which we will consider only in Part III of the book. The data of interest to us in this section are contained in the `S` objects `BCofLRet` and `CCofLRet`. They are the two columns of a data matrix given at the beginning of the first section of this chapter. For each day of the period starting January 10, 1986 and ending January 1, 1999, we computed the logarithms of the daily returns from the nearest futures contract active on that day. The scatterplot of these two variables is given in Figure 2.5. A close look at this scatterplot shows that many points are on the vertical axis and on the horizontal axis. This means that quite often, the price does not change from one day to the next, forcing the log returns to vanish on these days. The presence

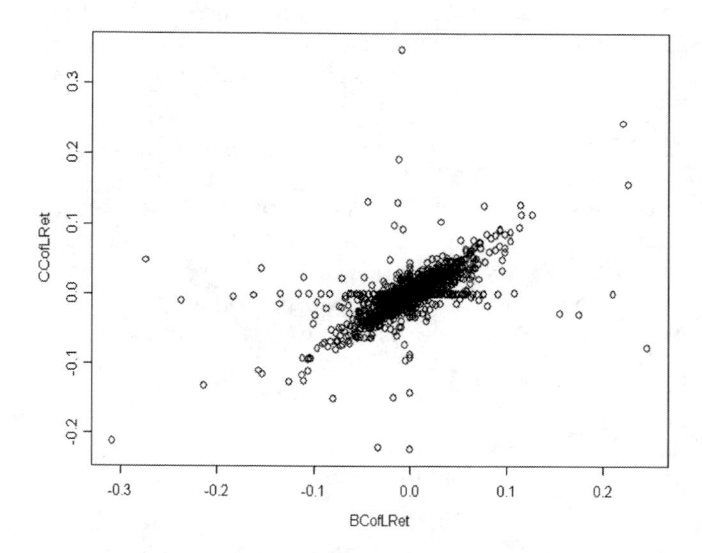

Fig. 2.5. Scatterplot of the daily log-returns of the coffee futures contracts in Brazil and Colombia from January 10, 1986 to January 1, 1999.

of so many of these zeroes indicates that the probability distributions are singular, in the sense that the cumulative distribution functions have jumps at 0. These jumps can be a hindrance to the analysis, so we choose to remove them by removing the zeroes from the data samples.

```
> NZ <- (BCofLRet != 0 & CCofLRet  != 0)
> BLRet <- BCofLRet[NZ]
> CLRet <- CCofLRet[NZ]
```

The vector NZ created by the first command is a boolean vector with the same length as BCofLRet and CCofLRet. It is true (equal to T) when both the daily Brazilian and Colombian log-returns are non-zero. The next two commands show the power of the sub-scripting capabilities of the S language. BLRet and CLRet are the vectors obtained by keeping the entries of BCofLRet and CCofLRet whose indices are those for which the value of NZ is T. The scatterplot of BLRet and CLRet is reproduced in the left pane of Figure 2.6.

We shall work with this new bivariate sample from now on, but we should keep in mind that, if we want to compute statistics of the actual log-returns, we need to put the zeroes back. The command

```
> PNZ <- mean(NZ)
> PNZ
[1] 0.4084465
```

gives the proportion of T's in the vector NZ, and it should be viewed as an estimate of the probability not to have a zero in the data. This probability could be used to add a random number of zeroes should we need to create random samples from the original distribution using samples from the modified distribution.

2.2.5 Is the Joint Distribution Normal?

The first question we address is the normality of the joint distribution of BLRet and CLRet. Our first step is quite mundane: we compare graphically the joint (empirical) distribution of BLRet and CLRet to the distribution of a bivariate normal sample which has the same five parameters (i.e. means, standard deviations and correlation coefficient). In order to do so, we first compute these parameters. We use the commands:

```
> XLIM <- c(-.4,.3)
> YLIM <- c(-.2,.4)
> Mu <- c(mean(BLRet), mean(CLRet))
> Sigma <- var(cbind(BLRet,CLRet))
> N <- length(BLRet)
```

We defined the vectors XLIM and YLIM as the limits of the ranges of BLRet and CLRet, respectively. We use these values each time we want to make sure that the scatterplot of BLRet and CLRet on one hand and the scatterplot of the simulated data on the other are on the same scale. As defined, Mu is the mean vector since it is defined as the vector of the means. Next we use the S-Plus function cbind to bind the columns BLRet and CLRet into one single matrix, then applying the function var to this data matrix produces the variance/covatiance matrix of the columns. So Sigma is the variance/covariance matrix, and N is the sample size. We use the S-Plus function rmvnorm to generate the desired bivariate sample $(x_1, y_1), \ldots, (x_N, y_N)$ of size N from the bivariate Gaussian distribution with mean Mu and variance/covariance matrix Sigma.

```
> CNsim <- rmvnorm(N,mean=Mu,cov=Sigma)
> par(mfrow=c(2,1))
> plot(BLRet,CLRet, xlim=XLIM, ylim=YLIM)
> plot(CNsim,xlim=XLIM, ylim=YLIM)
> par(mfrow=c(1,1))
```

Notice that we could just as well have used as well the home-grown function vnorm developed in Problem 2.8 at the end of the chapter. The results are given in Figure 2.6. Both scatterplots comprise an ellipsoidal cloud of points around the origin. Clearly, this cloud seems to be thinner for the coffee data. However, even if we were to consider that the bulk of the distribution had been reproduced in a reasonable manner, the presence of isolated points in the empirical coffee data is a distinctive feature which has not been reproduced by the simulation. This is a clear indication that the joint distribution of BLRet and CLRet is not normal. There are many reasons why

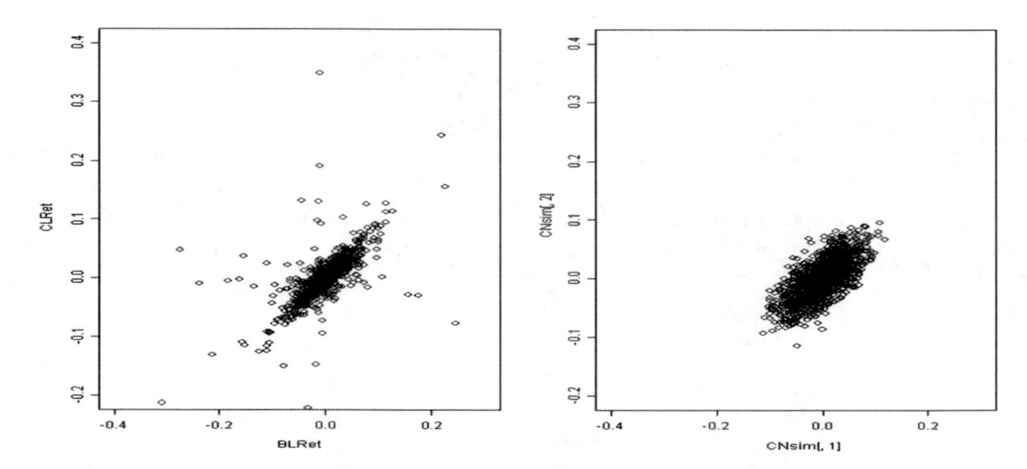

Fig. 2.6. Comparison of the empirical scatterplot of the coffee log-returns after the removal of the zeroes (left) and of the scatterplot of the sample of the same size simulated from a jointly Gaussian distribution with the same mean and covariance structure (right).

this could be the case. In general, it is because at least one of the variables, `BLRet` or `CLRet`, is not normal. But these variables could be normally distributed even when the joint distribution is not normal. We now check that this is not the case by showing that the marginal distributions of `BLRet` and `CLRet` are not normal either.

2.3 Marginals and More Measures of Dependence

Trying to fit a multivariate distribution to a multivariate sample, as we did when trying to fit a bivariate normal distribution to the sample of `BLRet` and `CLRet`, may be trying to tackle all the difficulties at once, and may be overwhelming. So instead, we try the *divide and conquer* approach: we first consider the estimation of the univariate marginal distributions separately, and only after this has been done, do we consider the issue of dependence. In the case of a bivariate sample from a normal distribution, this would amount to first estimating the means and the variances of the two variables separately, and then estimating the correlation coefficient. Since we are interested in more general distributions, possibly with marginals having heavy tails, we may not be able to use the correlation coefficient as a way to quantify dependence. To this end, we review the most commonly used statistics measuring the dependence of two samples, and we prepare for the concept of copula which will be introduced and analyzed in the next section.

As before, we try to sprinkle the presentation of the mathematical concepts with numerical examples, and we still use the example of the coffee data for that.

2.3.1 Estimation of the Coffee Log-Return Distributions

We use the graphical tools introduced in Chapter 1, as encapsulated in the `S-Plus` function `eda.shape` defined in appendix at the end of the book, as part of the introduction to `S-Plus`. The results of the applications of the function `eda.shape` to `BLRet` and `CLRet` are reproduced in Figure 2.7. In both cases one sees clearly that they differ significantly from the results of Figure 7.27 obtained from a normal sample, in the introductory session to `S-Plus` reproduced in appendix to this book. The histograms and kernel density estimates vouch for a unimodular distribution, possibly with extended tails on both sides. The presence of tails which are heavier than normal is confirmed by the boxplots, which show a very large number of observations outside the box. However, when it comes to the tails, the clearest diagnostic is given by the Q-Q plots. The departures from the Q-Q lines are a clear indication that the tails are much heavier than the tails of the normal distributions with the same means and variances, so fitting of a generalized Pareto distribution may be appropriate. Since the analysis of the marginal distribution of the daily log returns of the Colombian coffee is essentially the same, we only report the analysis of the Brazilian coffee.

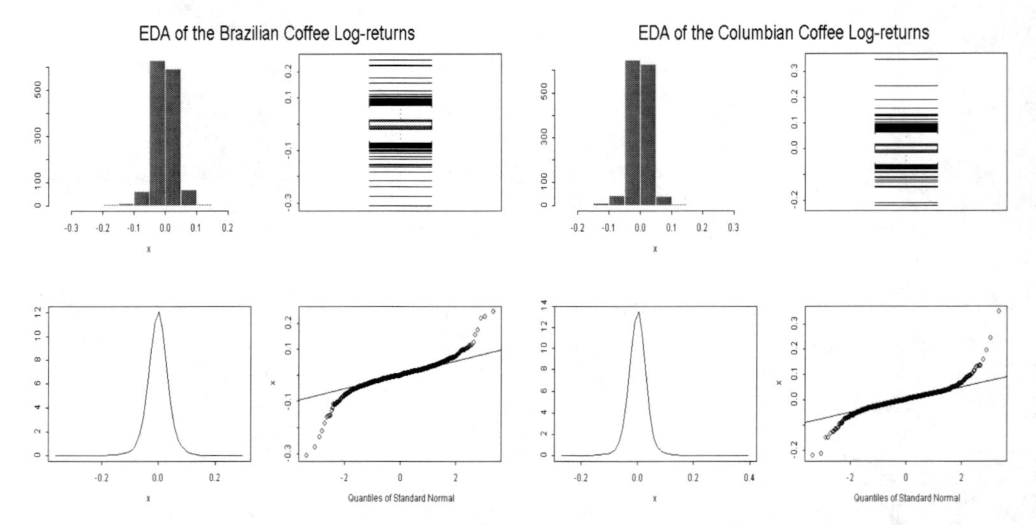

Fig. 2.7. Exploratory Data Analysis of the Brazialian (left) and Colombian (right) coffee daily log-returns for the period from January 9, 1986 to January 1, 1999 after removal of the zeroes.

Remark. Similar results would have been obtained with the original log return samples (prior to the removal of the zeroes) since the zeroes affect only the center of the distribution and not the tails. The main difference would appear in the density plots. Indeed, these density plots would seem *absent* and this requires an explanation.

A closer look at the command used in the code of the function `eda.shape` reminds us that the plot of the density estimate is restricted to the inter-quantile interval. If we compute this interval for the data at hand, we see that the lower and the upper quantile are essentially 0. This explains why no graph would be produced by the density estimator.

As in the case of our analysis of the S&P 500 daily log-returns, we use the function `gpd.tail` to fit a generalized Pareto distribution to both `BLRet` and `CLRet`. This function requires information on the locations of the tails in the form of two parameters telling the program where these tails should start. We explained how to choose these thresholds from a Q-Q plot of the data. We emphasize now the use of the function `shape.plot` as an alternative tool. In practice we recommend a combination of the two approaches to pick values for these thresholds. The commands

```
> par(mfrow=c(1,2))
> shape.plot(BLRet,tail="lower")
> shape.plot(BLRet,tail="upper")
> par(mfrow=c(1,1))
```

produce the results given in Figure 2.8. From these plots we decide that the estima-

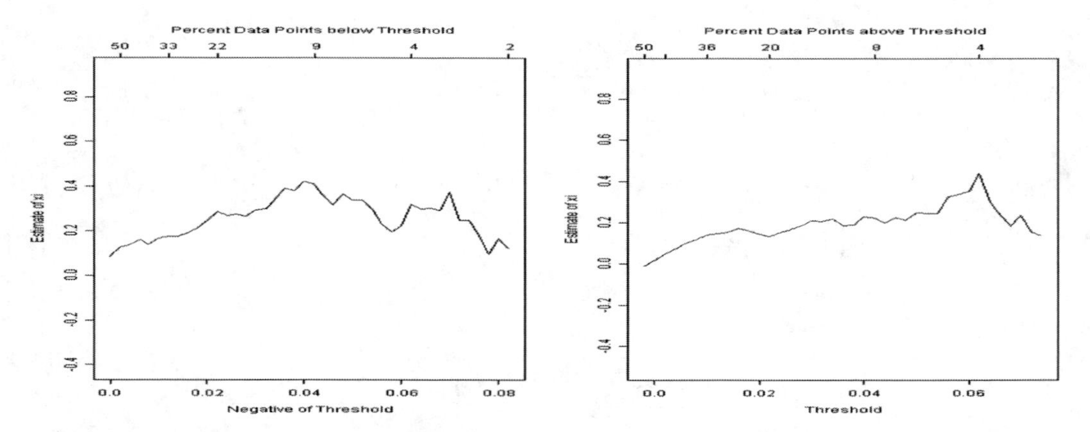

Fig. 2.8. Maximum likelihood estimates of the shape parameters as functions of the thresholds for the left tail (left pane) and right tail (right pane) of the distribution of the daily log returns of the Brazilian coffee.

tion of the upper tail could be done to the right of the value .04, while the estimation of the lower tail could be done to the left of the value $-.045$. Among other things, this will guarantee that each tail contains a bit more than 8% of the data, namely more than 115 points, which is not unreasonable. With this choice we proceed to the actual estimate of the GPD with the command:

```
> B.est <- gpd.tail(BLRet, upper = 0.04, lower = -0.045)
```

The function `gpd.tail` creates Q-Q plots (reproduced in Figure 2.9) of excesses over lower and upper thresholds against the quantiles of a GPD. As we have mentioned several times, if the left parts of these plots are approximately linear, the estimation of the tail is expected to be good. These empirical facts can be justified by mathematical results which are beyond the scope of this book. Also, these mathematical results require the independence of the successive entries in the data set. In other words, our analysis ignores the serial dependence between the successive log returns. This is usually not a great mistake, and also we have rigorous results to back up this claim. We check graphically the quality of the fit with the plots of the tails on

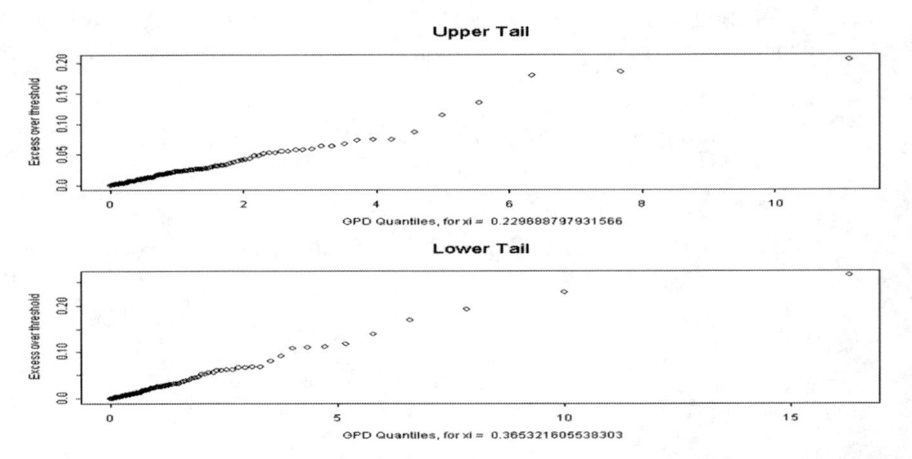

Fig. 2.9. Estimation of the distribution of the daily log returns of the Brazilian coffee by a generalized Pareto distribution.

a logarithmic scale.

```
> par(mfrow=c(1,2))
> tailplot(B.est,tail="lower")
> title("Lower tail Fit")
> tailplot(B.est,tail="upper")
> title("Upper tail Fit")
> par(mfrow=c(1,1))
```

The results are given in Figure 2.10. Given the point patterns, the fit looks very good. We perform the analysis of the heavy tail nature of the distribution of the daily log-returns of the Colombian coffee in exactly the same way.

First Monte Carlo Simulations

Motivated by the desire to perform a simulation analysis of the risk associated with various coffee portfolios containing both Brazilian and Colombian futures contracts,

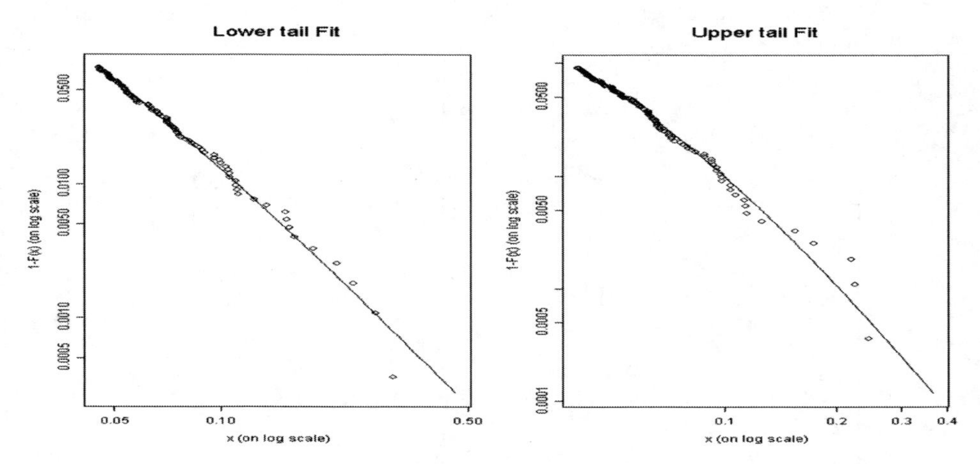

Fig. 2.10. Goodness of the fits for the left tail (left pane) and the right tail (right pane).

we proceed to the Monte Carlo simulation of samples of log-returns using the tools developed in the previous chapter. The commands:

```
> BLRet.sim <- gpd.2q(runif(length(BLRet)),B.est)
> CLRet.sim <- gpd.2q(runif(length(CLRet)),C.est)
```

generate samples `BLRet.sim` and `CLRet.sim` of the same sizes as the original data, and from the GPD's fitted to the data. To make sure that these samples have the right distributions, we check that their Q-Q plots against the empirical data are concentrated along the main diagonal. This is clear from Figure 2.11 which was produced with the `S-Plus` commands:

```
> qqplot(BLRet,BLRet.sim)
> abline(0,1)
> qqplot(CLRet,CLRet.sim)
> abline(0,1)
```

So it is clear that the distributions of the simulated samples are as close as we can hope for from the empirical distributions of the Brazilian and Colombian coffee log-returns. Since they capture the marginal distributions with great precision, these simulated samples can be used for the computations of statistics involving the log-returns separately. However, they cannot be used for the computations of joint statistics since they do not capture the dependencies between the two log-returns. Indeed, the simulated samples are statistically independent. This is clearly illustrated by plotting them together in a scatterplot as in Figure 2.12. We need to work harder to understand better the dependencies between the two log-return variables, and to be able to include their effects in Monte Carlo simulations.

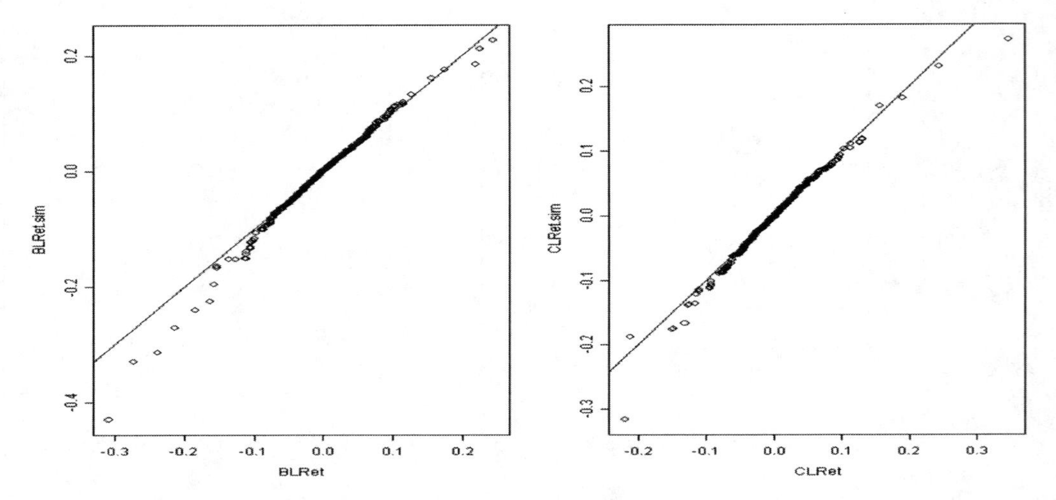

Fig. 2.11. Empirical Q-Q plot of the Monte Carlo sample against the empirical coffee log-return sample in the case of the Brazilian futures (left pane) and the Colombian futures prices (right pane).

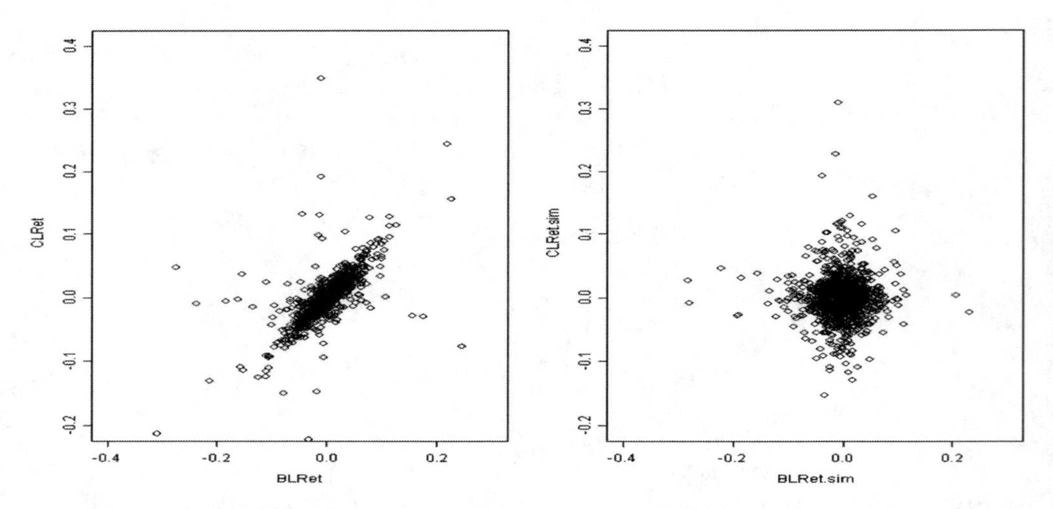

Fig. 2.12. Scatterplot of the Colombian coffee log-returns against the Brazilian ones (left pane), and scatterplot of the Monte Carlo samples (right pane).

2.3.2 **More Measures of Dependence**

Because of the limitations of the correlation coefficient ρ as a measure of the dependence between two random variables, other measures of dependence have been proposed and used throughout the years. They mostly rely on sample order statistics. For the sake of completeness, we shall quote two of the most commonly used: the Kendall's τ and the Spearman's ρ. Given two random variables X and Y, their Kendall's correlation coefficient $\rho_K(X,Y)$ is defined as:

$$\rho_K(X,Y) = \mathbb{P}\{(X_1 - X_2)(Y_1 - Y_2) > 0\} - \mathbb{P}\{(X_1 - X_2)(Y_1 - Y_2) < 0\} \quad (2.12)$$

provided (X_1,Y_1) and (X_2,Y_2) are independent random couples with the same joint distribution as (X,Y). Even though the notation ρ_K should be used for consistency, we shall often use the notation $\tau(X,Y)$ because this correlation coefficient is usually called Kendall's tau.

The dependence captured by Kendall's tau is better understood on sample data. Given samples $x_1,\ldots,x_n$ and $y_1,\ldots,y_n$, the empirical estimate of the Kendall correlation coefficient is given by:

$$\hat{\rho}_K(X,Y) = \frac{1}{\binom{n}{2}} \sum_{1 \le i \le j \le n} \text{sign}\left((x_i - x_j)(y_1 - y_j)\right)$$

which shows clearly that what is measured here is merely the relative frequency with which a change in one of the variables is accompanied by a change in the same direction of the other variable. Indeed, the `sign` appearing in the right hand side is equal to one when $x_i - x_j$ has the same sign as $y_i - y_j$, whether this sign is plus or minus, independently of the actual sizes of these numbers. Computing this coefficient with `S-Plus` can be done in the following way:

```
> cor.test(BLRet,CLRet,method="k")$estimate
     tau
 0.71358
```

The Spearman rho of X and Y is defined by:

$$\rho_S(X,Y) = \rho\{F_X(X), F_Y(Y)\}, \quad (2.13)$$

and its empirical estimate from sample data is defined as:

$$\hat{\rho}_S(X,Y) = \frac{12}{n(n^2-1)} \sum_{i=1}^{n} \left(\text{rank}(x_i) - \frac{n+1}{2}\right)\left(\text{rank}(y_i) - \frac{n+1}{2}\right).$$

The value of this correlation coefficient depends upon the relative rankings of the x_i and the y_j. However, the interpretation of the definition is better understood from the theoretical definition (2.13). Indeed, this definition says that the Spearman's correlation coefficient between X and Y is exactly the Pearson's correlation coefficient between the uniformly distributed random variables $F_X(X)$ and $F_Y(Y)$. This shows that Spearman's coefficient attempts to remove the relative sizes of the values of X among themselves, similarlty for the relative values of Y, and then to capture what is left of the dependence between the transformed variables. We shall come back to this approach to dependence below. Spearman's rho is computed in `S-Plus` with the command:

```
> cor.test(BLRET,CLRET,method="s")$estimate
   rho
 0.8551515
```

2.4 Copulas and Random Simulations

The first part of this section elaborates on the rationale behind the introduction of Spearman's correlation coefficient. As a warm up to the introduction of the abstract concept of copula, we consider first the practical implication of the first of the two fundamental facts of the theory of random generation as presented in Section 1.6. Because of Fact 1, which reads:

$$X \text{ r.v. with cdf } F_X(\,\cdot\,) \quad \Longrightarrow F_X(X) \text{ uniform on } [0,1],$$

we can transform the original coffee log-return data and create a bivariate sample in the unit square in such a way that both marginal point distributions are uniform. Indeed, the above theoretical result says that this can be done by evaluating each marginal cdf exactly at the sample points. In some sense, this wipes out the dependence of the point cloud pattern, seen for example in the left pane of Figure 2.6, upon the marginal distributions, leaving only the intrinsic dependence between the variables. We use our estimates of the distribution functions of the coffee log-returns as proxies for the theoretical distributions.

```
> U <- gpd.2p(BLRet, B.est)
> V <- gpd.2p(CLRet, C.est)
> plot(U,V)
> EMPCOP <- empirical.copula(U,V)
```

The first two commands use the function `gpd.2p` from the `FinMetrics` module, to compute the estimate of the cdf of the GPD, identified by the object of class `gpd`, at the points given in its first argument. Figure 2.13 shows the result of the above plot command. As expected all the data points are in the unit square. Moreover, the first coordinates of the points seem to be uniformly distributed on the unit interval of the horizontal axis, as they should be according to Fact 1, which was recalled above. Similarly for the second coordinates. The fact that the marginal distributions are now uniform is a sign that the influences of the original marginal distributions have been removed from the data. The only remaining feature is the way the numbers u_i and v_i are paired, and we claim that the dependence between the two log-returns is captured by the way these couplings are done. The dense point concentration around the second diagonal of the unit square is a consequence of this pairing, and it should be viewed as a graphical representation of the intrinsic dependence between the two random variates.

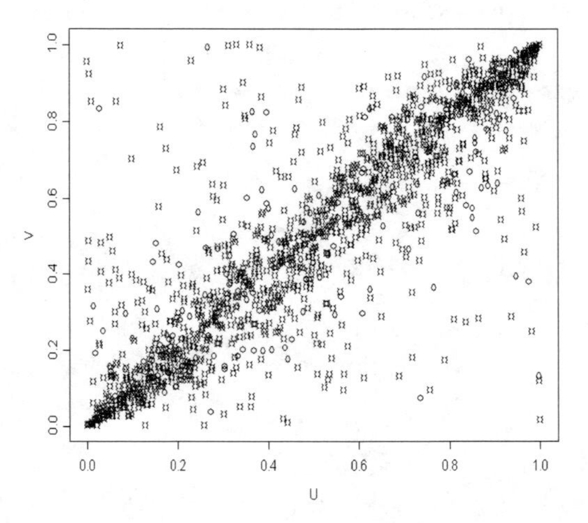

Fig. 2.13. Dependence between the coffee log-returns after removing the effects of the marginal distributions.

2.4.1 **Copulas**

The above discussion motivates the following abstract definition for capturing the dependence between several random variates. The analysis of the coffee data will be continued in Subsection 2.4.4.

Definition 1. *A copula is the joint distribution of uniformly distributed random variables.*

If $U_1, \ldots, U_n$ are $U(0,1)$, then the function C from $[0,1] \times \cdots \times [0,1]$ into $[0,1]$ defined by:

$$C(u_1, \ldots, u_n) = \mathbb{P}\{U_1 \leq u_1, \ldots, U_n \leq u_n\}$$

is a copula. Moreover, if X_1, ..., X_n are r.v.'s with cdf's F_{X_1}, ..., F_{X_n} respectively, then the copula of the uniform random variables

$$U_1 = F_{X_1}(X_1), \ldots, U_n = F_{X_n}(X_n)$$

is called the copula of $(X_1, \ldots, X_n)$. Copulas are typically used in the bivariate case $n = 2$, or for n very large. The case n large is of crucial importance for the analysis of the risk of large portfolios of financial instruments whose distributions are not well accounted for by normal distributions. Unfortunately, except for the normal and the Student copulas no other single copula is available in high dimension for analysis. However, despite this limitation, we present the first properties of copulas in the general case. This choice is justified by the crucial importance of risk-management applications. On the other hand, a complete analysis is possible in the case $n = 2$, that offers a complete description of all the possible forms of dependence between two random variables. We take that up in Subsection 2.4.3 below.

First Properties of Copulas

It is straightforward to check that:

- C does not change if one replaces any of the X_i's by a non-decreasing functions of itself;
- The joint cdf can be recovered from the copula and the marginal cdf's via the formula:
$$F_{(X_1,\dots,X_n)}(x_1,\dots,x_n) = C(F_{X_1}(x_1),\dots,F_{X_n}(x_n));$$
- C is unique if $F_{(X_1,\dots,X_n)}(x_1,\dots,x_n)$ is continuous (no jumps).

Moreover:

- $C(u_1,\dots,u_n)$ is non-decreasing in each variable u_i since it is a cdf;
- $C(1,\dots,1,u_i,1,\dots,1) = u_i$ for all i since the marginal distributions of a copula are uniformly distributed;
- if $u_i \le v_i$ for all i, then:
$$\sum_{1\le i_1\le 2} \cdots \sum_{1\le i_n\le 2} (-1)^{i_1+\cdots+i_n} C(u_{i_1},\dots,u_{i_n}) \ge 0.$$

This last inequality is very technical. It is reproduced here only for the sake of completeness. It essentially formalizes mathematically the fact that a copula is a multivariate cdf, and as such, it has to satisfy some positivity and monotonicity properties. It is instructive to visualize the meaning of this condition in the bivariate case $n = 2$ for which one can easily check that it holds true on a picture!

2.4.2 First Examples of Copula Families

The following are examples of copulas.

◇ ***Independent copula***

$$C_{ind}(u_1,\dots,u_n) = u_1 \cdots u_n.$$

This is the copula of independent random variables.

◇ ***Gaussian copula*** For each $\rho \in [-1,1]$ the function defined by:

$$C_{Gauss,\rho}(u_1,u_2) = \frac{1}{2\pi\sqrt{1-\rho^2}} \int_{-\infty}^{\Phi^{-1}(u_1)} \int_{-\infty}^{\Phi^{-1}(u_2)} e^{-[s^2-2\rho st+t^2]/2(1-\rho^2)} ds\, dt$$

is a copula called the Gaussian copula with parameter ρ. This is the copula of random variables which, whether or not their marginal distributions are Gaussian, depend upon each other as jointly Gaussian random variables do. The family of normal copulas is parameterized by the parameter $\rho \in [-1,+1]$. Figure 2.14 gives the surface plot of this copula when $\rho = .7$, i.e. the plot of the graph of the function $(u, v) \hookrightarrow C_{Gauss,.7}(u, v)$, together with the plot of its density. The fact that the marginals of a copula are uniform is clearly seen on this plot. Indeed, the facts $C(u, 1) = u$ and $C(1, v) = v$ force edges of the surface to be linear (coinciding with the second diagonal) and to meet at height 1 above the point $(1, 1)$.

Varying the parameter ρ is a way of varying the degree of dependence between the two random variables. Notice that two normal random variables X and Y may not be *jointly normal* if their copula is not in the normal family. They are jointly normal when the copula is from the normal family, in which case the parameter ρ has a simple interpretation since it is the correlation coefficient of X and Y. This interpretation is not valid in general. Indeed, if X and Y have Cauchy marginal distributions, their (Pearson) correlation coefficient does not exist since Cauchy random variables do not have means or variances . . ., but nevertheless, it is quite possible for their copula to be in the Gaussian family, i.e. to be equal to $C_{Gauss,\rho}$ for some $\rho \in [-1, +1]$. However, in this case, the parameter ρ cannot have the interpretation of correlation coefficient. Notice that, even though we only gave the formula in the bivariate case,

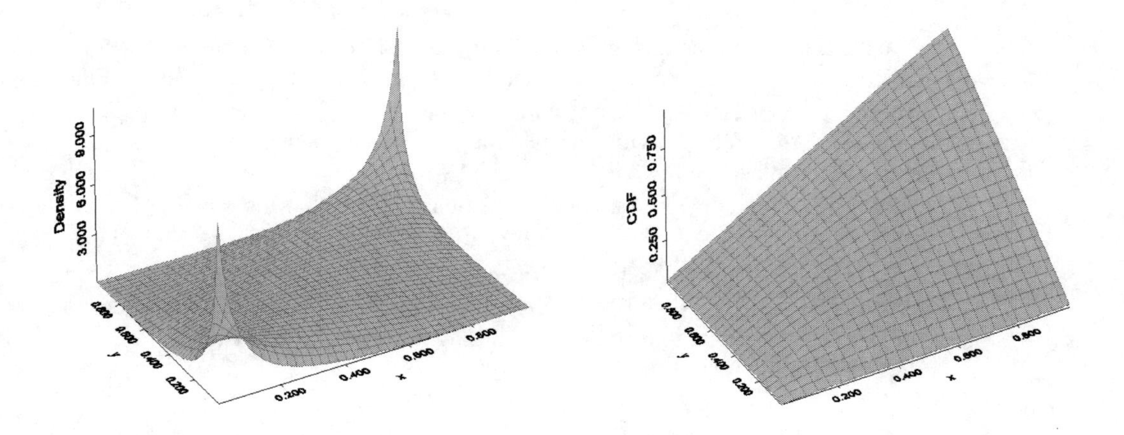

Fig. 2.14. Surface plot of the normal copula with parameter $\rho = .7$ (right), and of its density (left).

it is quite easy to define the Gaussian copula for any number of dimensions, the parameter now being a correlation matrix instead of a mere scalar.

$\diamond$ For each $\beta \in [0, 1]$ the function

$$C_{Gumbel,\beta}(u_1, u_2) = e^{-[(-log u_1)^{1/\beta} + (-log u_2)^{1/\beta}]}$$

is a copula called the ***Gumbel*** (or ***logistic***) copula with parameter β. The Gumbel

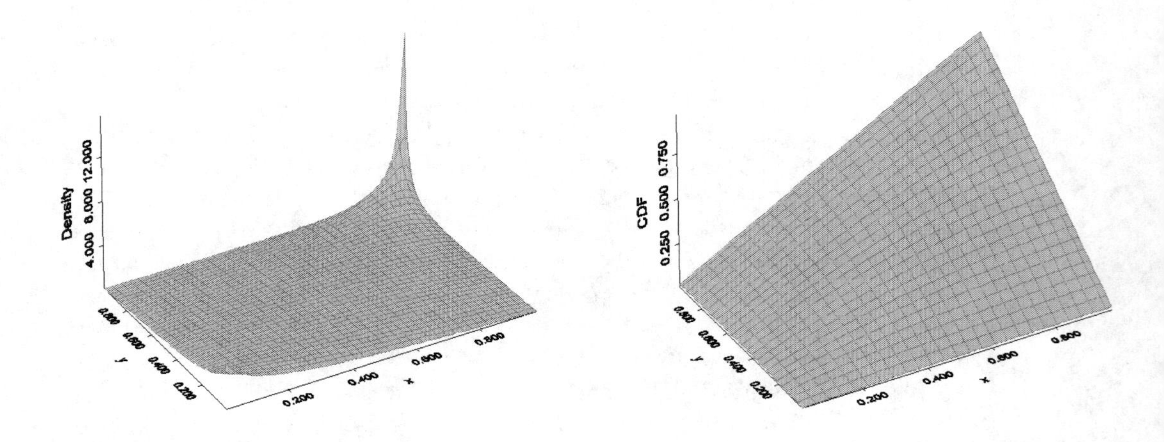

Fig. 2.15. Surface plot of the Gumbel copula with parameter $\beta = 1.5$ (right), and of its density (left).

copula family is parameterized by the parameter β, however the latter does not have as nice an interpretation as the parameter ρ of the Gaussian copula family. Figure 2.15 gives the surface plot of this copula when $\beta = 1.5$ together with the plot of its density. As before, varying the parameter is a way of varying the *strength* of the dependence between two random variables. This family appears naturally in the analysis of extreme events, as it is quite often the case that the copula of random variables with heavy tail distributions is of this family.

A complete list of the parametric copula families supported by the library `EVANESCE` and the module `FinMetrics` is given in Appendix 2, where the reader will also find the defining formulae together with some of the most important properties of these families.

2.4.3 Copulas and General Bivariate Distributions

The goal of this subsection is to show how copulas and univariate cdf's come together to characterize ALL the models of bivariate statistical distributions.

All the copulas which we consider in this book have a density. In other words, the copulas $C(u, v)$ will be differentiable, and the function:

$$c(u, v) = \frac{\partial^2}{\partial u \partial v} C(u, v)$$

will be the density of the copula. Notice that since we are dealing with bivariate distributions, we need a second order derivative in order to get a density from its cdf. Instead of limiting ourselves to distributions with uniform marginals, we can apply this remark to a general bivariate distribution as well. This leads to some interesting formulae.

Let us denote by $F_{(X,Y)}$ the joint cdf of a couple (X, Y) of random variables, and let us denote by $C_{(X,Y)}$ their copula. For the sake of simplicity we momentarily drop the subscript (X, Y) from the notation. According to our definition, we have:

$$F(x, y) = C(F_X(x), F_Y(y)) \tag{2.14}$$

if we denote by F_X and F_Y the cdf's of X and Y respectively. We can compute the joint density $f_{(X,Y)}$ of X and Y by taking partial derivatives on both sides of (2.14). We get:

$$\begin{aligned} f(x, y) &= \frac{\partial^2}{\partial x \partial y} F(x, y) \\ &= \frac{\partial^2}{\partial u \partial v} C(F_X(x), F_Y(y)) \frac{\partial F_X(x)}{\partial x} \frac{\partial F_Y(y)}{\partial y} \end{aligned}$$

which gives the following formula for the joint density of X and Y:

$$f(x, y) = c(F_X(x), F_Y(y)) \, f_X(x) \, f_Y(y) \tag{2.15}$$

in terms of the density of their copula, their marginal cdf's and their marginal densities. Obviously we used the formulae

$$f_X(x) = \frac{d}{dx} F_X(x) \qquad \text{and} \qquad f_Y(y) = \frac{d}{dy} F_Y(y)$$

giving the densities of X and Y in terms of their respective cdf's. Formula (2.15) has the following interesting consequence. Contrary to what can be done with the correlation coefficient (see Problem 2.3 and Problem 2.7 for the striking example of the lognormal distributions), it is *always* possible to specify a bivariate distribution by specifying:

- the marginal distributions
- a copula

without having to worry about the existence of the distribution, Moreover, as formulae (2.14) and (2.15) show, formulae for the components can be used to get formulae for the cdf and the density of the bivariate distribution. Figure 2.16 shows an example where we computed the density of a joint distribution specified by the Gumbel copula with parameter 1.4, and with the normal distribution $N(3, 4)$ and the Student t-distribution $T(3)$ as marginals. A bivariate distribution can be created with the command `bivd`, and the function `persp.dbivd` can be used to produce a 3-D surface plot of the density of a bivariate distribution. The plot of Figure 2.16 was obtained with the `S-Plus` commands:

```
> BIV1 <- bivd(gumbel.copula(1.4), "norm", "t", c(3,4), 3)
> persp.dbivd(BIV1)
```

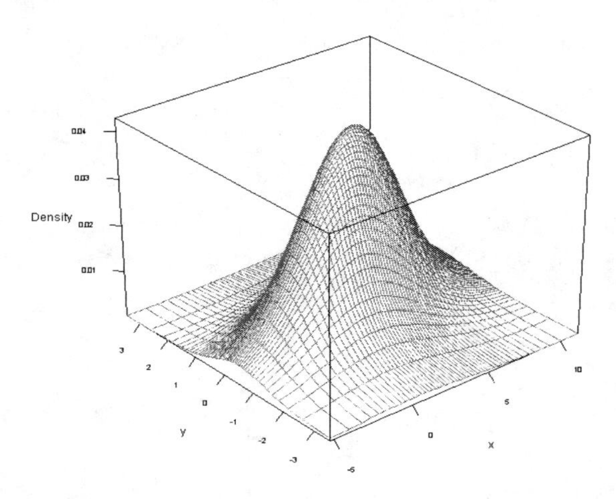

Fig. 2.16. Surface plot of the density of the bivariate distribution with Gumbel copula with parameter 1.4 and marginal distributions $N(3, 4)$ and $T(3)$.

The function `persp.pbivd` produces a surface plot of the cdf of the bivariate distribution, but these plots are not very instructive, especially for copulas, as we can see from Figure 2.14 and Figure 2.15. Indeed, all these copula surface plots show a *tent* tied at the level 0 on the segments going from the origin $(0, 0)$ to the points $(0, 1)$ and $(1, 0)$, and it is also tied at the point $(1, 1)$ where its value is always 1, and it is linear on the two coordinate planes. These properties are mere re-statements of the first properties of copulas given in Subsection 2.4.1. These constraints are common to all the copula surface plots, so it is extremely difficult to differentiate between them from the plots of their cdf's. For this reason, one very often uses contour plots to get a sense of the shape of the distribution and possibly to compare several copulas, or more general bivariate distributions. The commands:

```
> par(mfrow=c(1,2))
> contour.dbivd(BIV1)
> title("Density of the Bivariate Distribution BIV1")
> contour.pbivd(BIV1)
> title("CDF of the Bivariate Distribution BIV1")
> par(mfrow=c(1,1))
```

were used to produce the plots of Figure 2.17.

2.4.4 **Fitting Copulas**

Because of the serious difficulties resulting from the lack of data in the tails of the marginal distributions, copulas are best estimated by parametric methods. In order

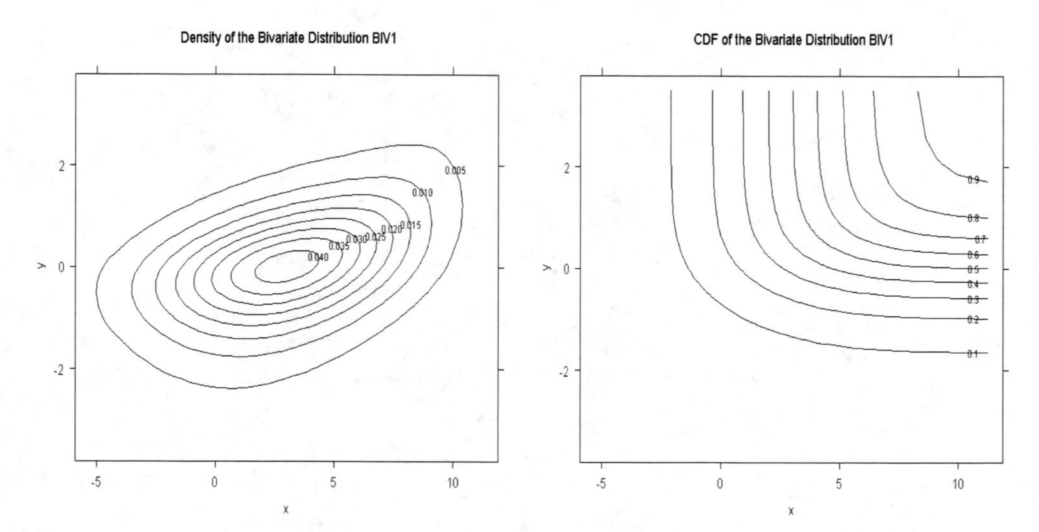

Fig. 2.17. Contour plot of the density (left) and the cdf (right) of the bivariate distribution with Gumbel copula with parameter 1.4 and marginal distributions $N(3,4)$ and $T(3)$.

to do so, we choose a family of copulas (see Appendix 2 for a list of the parametric families supported by the library EVANESCE) and the module FinMetrics, and we estimate the parameter(s) of the family in question by a maximum likelihood standard estimation procedure. The function fit.copula which we use below returns an object of class copula.family and creates a contour plot of level sets of the empirical copula and of the fitted copula. Since it is so difficult to compare copulas using only their graphs, in order to visualize the goodness of the fit, we chose to put a selected ensemble of level sets for the two surfaces on the same plot. Differences in these level curves can easily be interpreted as differences between the two surfaces. The fitting procedure can be implemented in the case of the coffee log-return data by the following commands.

```
> FAM <- "gumbel"
> ESTC <- fit.copula(EMPCOP,FAM)
```

Recall that EMPCOP was the S-Plus object constructed as the empirical copula of the Brazilian and Colombian coffee daily log-returns. The results are shown in Figure 2.18. The level sets of the Gumbel copula fitted to the data are very close to the level sets of the empirical copula. This graphical check shows that the fit is very good.

2.4.5 Monte Carlo Simulations with Copulas

We learned in Chapter 1 how to generate random samples from a univariate distribution, and we introduced and tested a set of tools to do just that, even when the

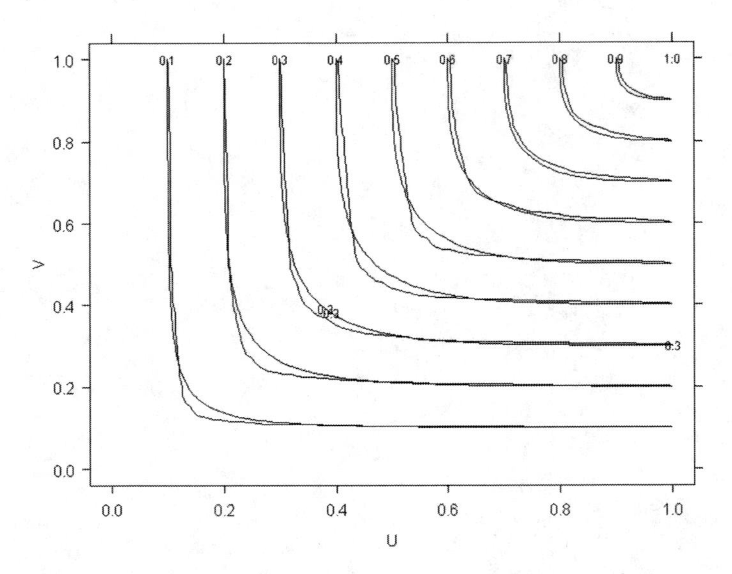

Fig. 2.18. Contour plot of the empirical copula with the contours of the fitted Gumbel copula superimposed.

distribution in question had to be estimated from data, and even when the distribution was suspected to have heavy tails. But as we saw earlier (recall Figure 2.12) having separate univariate random samples is not enough if we want to have a realistic rendering of how the variables in a bivariate sample relate to each other.

In this subsection, we consider the problem of the generation of random samples from a bivariate distribution, which we assume to be given by its marginal distributions and a copula. Let us imagine that we have a tool capable of producing bivariate samples from a copula. We shall not enter into the details of the construction of such a tool, we shall merely indicate that it can be built by aggregation of one dimensional random generators for the various conditional distributions. The gory details of such a construction are too technical for this book, so we shall leave them aside. Armed with such a weapon, it is very simple to generate samples from all the distributions having this specific copula as their own copula. Indeed, the first components of a random sample from a copula form a univariate sample uniformly distributed on $[0, 1]$. So transforming this sample by computing the quantile function of the first marginal distribution will turn this uniform sample into a sample from the first marginal distribution. This is an instance of our favorite method for generating random samples. Similarly, transforming the second components (which also form a uniform sample, by definition of a copula) by computing the quantile function of the second marginal distribution will give a random sample from the second marginal distribution. Now, by the very definition of the copula, these two univariate samples have not only the right marginals, but they also have the right copula! So put together, they form a bivariate sample from the bivariate distribution we started with.

We implement this strategy on our example of the coffee log returns. The following set of commands produces a bivariate sample of same size as the data, from our estimation of the joint distribution of the coffee log-returns. Remember that this estimate is comprised of the estimates of the marginal distributions of the two random quantities together with the parametric estimate of their copula.

```
> N <- length(BLRet)
> SD <- rcopula(ESTC,N)
> Xsim <- gpd.2q(SD$x, B.est)
> Ysim <- gpd.2q(SD$y, C.est)
```

We review the main steps of the simulation before commenting on the plots. The function `rcopula` produces bivariate samples from the copula whose information is encapsulated in the argument `ESTC`, which needs to be an object of class `copula.object`. As usual, we extract the two columns of the $N \times 2$ matrix `SD` by means of the dollar signs $ followed by the lower case `x` for the first column, and by the lower case `y` for the second column. By definition of a copula, `SD$x` and `SD$y` are random samples uniformly distributed over the unit interval. Consequently, the third and fourth commands result in samples `Xsim` and `Ysim` from the GPD's given by the `gpd` objects `BEST` and `CEST`. This is because we compute quantile functions on uniform samples. We already used this trick several times to generate random samples from a given distribution. But the situation is quite different from those earlier in that the uniform samples are paired by the copula used to generate them. So the copula of the resulting bivariate sample is the copula we started from. The loop is closed, and we have produced a bivariate sample with the desired distribution. In the same way we produced Figure 2.12, we can place the scatterplot of the simulated samples `Xsim` and `Ysim` to the right of the scatterplot of the original samples `BLRet` and `CLRet`. The result is reproduced in Figure 2.19. The differences with Figure 2.12 are striking. This plot shows that our model and the ensuing simulations capture rather well the characteristics of the point distribution in the plane. As further evidence, the numerical measures of dependence which we introduced earlier confirm that the results are very satisfactory. This is clear from the comparison of the values of the Kendall's tau and Spearman's rho statistics computed for the empirical copula (directly from the data) and from the fitted copula. We reproduce the commands and the results:

```
> print(ESTC)
Gumbel copula family; Extreme value copula.
Parameters :
   delta  =  2.98875657681924
> Kendalls.tau(EMPCOP)
[1] 0.6881483
> Kendalls.tau(ESTC, tol = 1e-2)
[1] 0.6654127
```

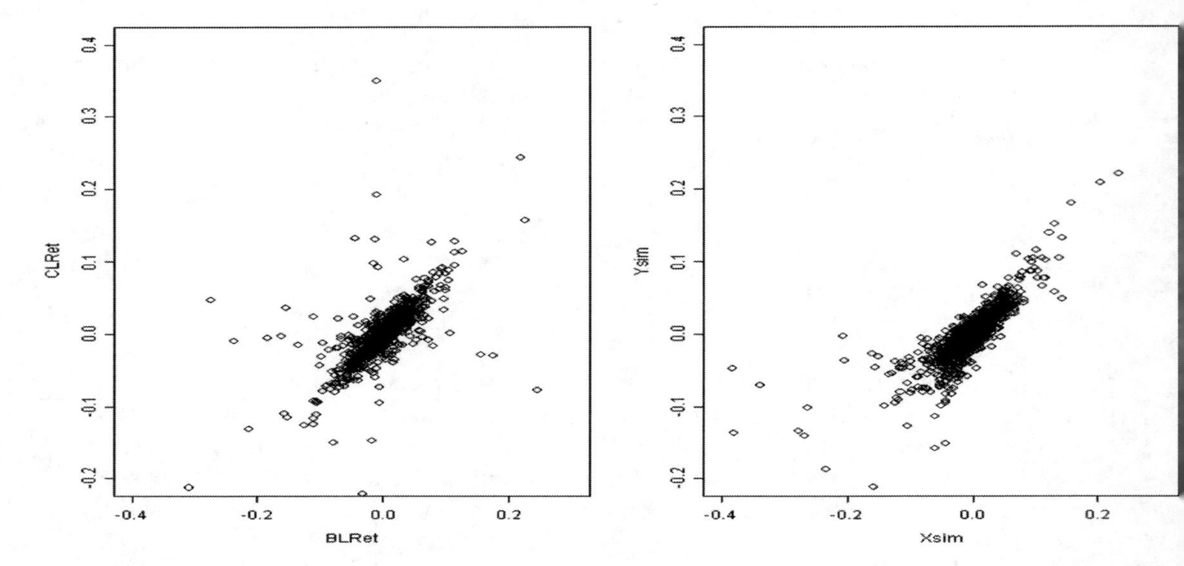

Fig. 2.19. Scatterplot of the Colombian coffee log-returns against the Brazilian ones (left pane), and scatterplot of the Monte Carlo samples produced with the dependence as captured by the fitted copula (right pane).

```
> Spearmans.rho(EMPCOP)
[1] 0.8356747
> Spearmans.rho(ESTC)
[1] 0.8477659
```

2.4.6 A Risk Management Example

Even though the practical example treated in this subsection is limited in scope, it should not be underestimated. Indeed, the risk measures which we compute provide values which are orders of magnitude different from the values obtained from the theory of the Gaussian distribution. Relying on the normal theory leads to overly optimistic figures ... and possibly to very bad surprises. Fitting heavy tail distributions and using copulas give more conservative (and presumably more realistic) quantifications of the risk carried by most portfolios.

We consider a simple example of risk management using the tools developed in this chapter. The goal is to quantify the risk exposure of a portfolio. For the sake of simplicity, we consider only a single period analysis, and we assume that the portfolio is comprised of only two instruments. The initial value of the portfolio is:

$$V_0 = n_1 S_1 + n_2 S_2$$

where n_1 and n_2 are the numbers of units of the two instruments, which are valued at S_1 and S_2 at the beginning of the period. Let us denote by S_1' and S_2' their values at the end of the period. Then the new value of the portfolio is:

$$\begin{aligned} V &= n_1 S_1' + n_2 S_2' \\ &= n_1 S_1 \, e^X + n_2 S_2 \, e^Y, \end{aligned}$$

if we denote by $X = \log(S_1'/S_1)$ and $Y = \log(S_2'/S_2)$ the log returns on the individual instruments. Notice that (just to keep in line with the example discussed in this chapter) we could consider the portfolio to be composed of futures contracts of Brazilian and Colombian coffee, in which case the random log-returns X and Y are just the quantities studied so far. In any case, the log-return of the portfolio is

$$R = \log\left(\frac{V}{V_0}\right) = \log\left(\frac{n_1 S_1}{n_1 S_1 + n_2 S_2} \, e^X + \frac{n_2 S_2}{n_1 S_1 + n_2 S_2} \, e^Y\right) = \log\left(\lambda_1 \, e^X + \lambda_2 \, e^Y\right).$$

if we use the notation λ_1 and λ_2 for the fractions of the portfolio invested in the two instruments.

Value-at-Risk VaR_q

We now compute the value at risk of this portfolio. According to the discussion of Chapter 1, for a given level q, VaR_q was defined in relation to the capital needed to cover losses occurring with frequency q, and more precisely, VaR_q was defined as the $100q$-th percentile of the loss distribution. i.e the solution r of the equation:

$$q = \mathbb{P}\{-R \geq r\} = \mathbb{P}\{R \leq -r\} = F_R(-r).$$

In order to solve for r in the above equation, we need to be able to compute the cdf of the log-return R. The latter can be expressed analytically as:

$$\begin{aligned} \mathbb{P}\{-R \geq r\} &= \mathbb{P}\{\log\left(\lambda_1 \, e^X + \lambda_2 \, e^Y\right) \leq -r\} \\ &= \int\int_{\{(x,y);\lambda_1 e^x + \lambda_2 e^y \leq e^{-r}\}} f_{(X,Y)}(x,y) \, dx dy \\ &= \int_{-\infty}^{-r-\log\lambda_1} dx \int_{-\infty}^{\log(e^{-r}/\lambda_2 - \lambda_1/\lambda_2 \, e^x)} c(F_X(x), F_Y(y)) f_X(x) f_Y(y) dy \\ &= \int_0^{F_X(-r-\log\lambda_1)} du \int_0^{F_Y(\log(e^{-r}/\lambda_2 - \lambda_1/\lambda_2 \, e^{F_X^{-1}(u)}))} dv \; c(u,v) \\ &= \int_0^{F_X(-r-\log\lambda_1)} du \; \frac{\partial}{\partial u} C(u,v)\Big|_{v=F_Y(\log(e^{-r}/\lambda_2 - \lambda_1/\lambda_2 \, e^{F_X^{-1}(u)}))} \end{aligned}$$

where we used several substitutions to change variables in simple and double integrals. Despite all these efforts, and despite the fact that we managed to reduce the computation to the evaluation of a single integral, this computation cannot be pushed further in this generality. Even when we know more about the copula and the marginal distributions, this integral can very rarely be computed explicitly. We need to use numerical routines to compute this integral. In fact, we need to run these routines quite a lot of times to solve the equation giving the desired value of r.

The function VaR.exp.portf was written to compute the value at risk following this strategy. But the user should be aware that this numerical procedure does not converge all the time, and the results can be disappointing. A typical call to this function looks like:

```
> VaR.exp.portf(0.01, range=c(0.016,0.08), copula=ESTC,
      x.est=B.est, y.est=C.est,lambda1=0.5, lambda2=0.5)
```

We give it only for the sake of completeness.

Expected Shortfall $\mathbb{E}\{\Theta_q\}$

We now compute the other measure of risk which we introduced in Chapter 1. The analytic technicalities of the computation of the expected shortfall $\mathbb{E}\{\Theta_q\}$ (recall the definition given in Chapter 1) are even more daunting than for the computation of the value at risk VaR_q. Just to give a flavor of these technical difficulties, we initialize the process by:

$$
\begin{aligned}
\mathbb{E}\{\Theta_q\} &= \mathbb{E}\{-R| - R > VaR_q\} \\
&= \frac{1}{q}\int_{-\infty}^{-VaR_q} -rF_R(dr) \\
&= \int_{-\infty}^{-r-\log\lambda_1} dx\, f_X(x) \int_{-\infty}^{\log(e^{-r}/\lambda_2-\lambda_1/\lambda_2\, e^{x})} \log\left(\lambda_1\, e^{X} + \lambda_2\, e^{Y}\right) \\
&\qquad c(F_X(x), F_Y(y))\, f_Y(y) dy
\end{aligned}
$$

where we have used the same notation as before. Unfortunately, it seems much more difficult to reduce this double integral to a single one, and it seems hopeless to try to derive reasonable approximations of the analytic expression of the expected shortfall which can be evaluated by tractable computations. Following this road, we ended up in a cul-de-sac.

Fortunately, random simulation of large samples from the joint distribution of (X, Y) and Monte Carlo computations can come to the rescue and save the day.

Use of Monte Carlo Computations

We illustrate the use of Monte Carlo techniques by computing the VaR_q and the expected shortfall $\mathbb{E}\{\Theta_q\}$ of a portfolio of Brazilian and Colombian coffee futures contracts. We solve the problem by simulation using historical data on the daily log-returns of the two assets. The strategy consists of generating a large sample from the joint distribution of X and Y as estimated from the historical data, and computing for each couple (x_i, y_i) in the sample, the value of R. Our estimate of the value at risk is simply given by the empirical quantile of the set of values of R thus obtained. We can now restrict ourselves to the values of R smaller than the negative of the VaR_q just computed, and the negative of the average of these R's gives the expected shortfall. This is implemented in the function `VaR.exp.sim` whose use we illustrate in the commands below.

```
> VaR.exp.sim(n=10000, Q=0.01, copula=ESTC, x.est=B.est,
        y.est=C.est, lambda1=0.7, lambda2=0.3)[1]
Simulation size
        10000

> VaR.exp.sim(n=10000, Q=0.01, copula=ESTC, x.est=B.est,
        y.est=C.est, lambda1=0.7, lambda2=0.3)[2]
VaR Q=0.01
0.09290721

> VaR.exp.sim(n=10000, Q=0.01, copula=ESTC, x.est=B.est,
        y.est=C.est, lambda1=0.7, lambda2=0.3)[3]
ES Q=0.01
0.1460994
```

which produce the value at risk and the expected shortfall over a one-period horizon of a unit portfolio with 70% of Brazilian coffee and 30% of Colombian coffee. Notice that the function `VaR.exp.sim` returns a vector with three components. The first one is the number of Monte Carlo samples used in the computation, the second is the estimated value of the VaR while the third one is the estimated value of the expected shortfall.

Comparison with the Results of the Normal Model

For the sake of comparison, we compute the same value at risk under the assumption that the joint distribution of the coffee log-returns is normal. This assumption is implicit in most of the VaR computations done in everyday practice. Our goal here is to show how different the results are.

```
> Port <- c(.7,.3)
> MuP <- sum(Port*Mu)
> MuP
[1] -0.0007028017
> SigP <- sqrt(t(Port) %*% Sigma %*% Port)
> SigP
           [,1]
[1,] 0.03450331
> qnorm(p=.01,mean=MuP,sd=SigP)
[1] -0.08096951
```

For the given portfolio `Port`, we compute the mean `MuP` and the standard deviation `SigP` of the portfolio return, and we compute the one percentile of the corresponding normal distribution. We learn that only one percent of the time will the return be less than 8% while the above computation was telling us that it should be expected to be less than 9.2% with the same frequency. One cent on the dollar is not much, but for a large portfolio, things add up!

2.5 Principal Component Analysis

Dimension reduction without significant loss of information is one of the main challenges of data analysis. The internet age has seen an exponential growth in the research efforts devoted to the design of efficient codes and compression algorithms. Whether the data are comprised of video streams, images, and/or speech signals, or financial data, finding a basis in which these data can be expressed with a small (or at least a smaller) number of coefficients is of crucial importance. Other important domains of applications are cursed by the high dimensionality of the data. Artificial intelligence applications, especially those involving machine learning and data mining, have the same dimension reduction problems. Pattern recognition problems are closer to the hearts of traditional statisticians. Indeed, regression and statistical classification problems have forced statisticians to face the curse of dimensionality, and to design systematic procedures to encapsulate the important features of high dimensional observations in a small number of variables. Principal component analysis as presented in this chapter, offers an optimal solution to these dimension reduction issues.

2.5.1 Identification of the Principal Components of a Data Set

Principal component analysis (PCA, for short) is a data analysis technique designed for numerical data (as opposed to categorical data). The typical situation that we consider is where the data come in the form of a matrix $[x_{i,j}]_{i=1,\dots,N,j=1,\dots,M}$ of real numbers, the entry $x_{i,j}$ representing the value of the i-th observation of the j-th variable. As usual, we follow the convention of labeling the columns of the data matrix by the variables measured, and the rows by the individuals of the population under investigation. Examples are plentiful in most data analysis applications. We give below detailed analyses of several examples from the financial arena.

As we mentioned above, the N members of the population can be identified with the N rows of the data matrix, each one corresponding to an M-dimensional (row) vector of numbers giving the values of the variables measured on this individual. It is often desirable (especially when M is large) to reduce the complexity of the descriptions of the individuals and to replace the M descriptive variables by a smaller number of variables, while at the same time, losing as little information as possible. Let us consider a simple (and presumably naive) illustration of this idea. Imagine momentarily that all the variables measured are scores of the same nature (for examples they are all lengths expressed in the same unit, or they are all prices expressed in the same currency, ...) so that it would be conceivable to try to characterize each individual by the mean, and a few other numerical statistics computed on all the individual scores. The mean:

$$\overline{x_{i\cdot}} = \frac{x_{i1} + x_{i2} + \cdots + x_{iM}}{M}$$

can be viewed as a linear combination of the individual scores with coefficients $1/M$, $1/M$, ... , $1/M$. Principal component analysis, is an attempt to describe the individual features in the population in terms of M linear combinations of the original features, as captured by the variables originally measured on the N individuals. The coefficients used in the example of the mean are all non-negative and sum up to one. Even though this convention is very attractive because of the probabilistic interpretation which can be given to the coefficients, we shall use another convention for the linear combinations. We shall allow the coefficients to be of any sign (positive as well as negative) and we normalize them so that the sum of their squares is equal to 1. So if we were to use the mean, we would use the normalized linear combination (NLC, for short) given by:

$$\widetilde{x_{i\cdot}} = \frac{1}{\sqrt{M}} x_{i1} + \frac{1}{\sqrt{M}} x_{i2} + \cdots + \frac{1}{\sqrt{M}} x_{iM}.$$

The goal of principal component analysis is to search for the main sources of variation in the M-dimensional row vectors by identifying M linearly independent and orthogonal NLC's in such a way that a small number of them capture most of the variation in the data. This is accomplished by identifying the eigenvectors and eigenvalues of the covariance matrix C_x of the M column variables. This covariance matrix is defined by:

$$C_x[j, j'] = \frac{1}{N} \sum_{i=1}^{N} (x_{ij} - \overline{x_{\cdot j}})(x_{ij'} - \overline{x_{\cdot j'}}), \qquad j, j' = 1, \ldots, M,$$

where we used the standard notation:

$$\overline{x_{\cdot j}} = \frac{x_{1j} + x_{2j} + \cdots + x_{Nj}}{N}$$

for the mean of the j-th variable over the population of N individuals. It is easy to check that the matrix C_x is symmetric (hence diagonalizable) and non-negative definite (which implies that all the eigenvalues are non-negative). One usually orders the eigenvalues in decreasing order, say:

$$\lambda_1 \geq \lambda_2 \geq \cdots \geq \lambda_M \geq 0.$$

The corresponding eigenvectors are called the loadings. In practice we choose c_1 to be a normalized eigenvector corresponding to the eigenvalue λ_1, c_2 to be a normalized eigenvector corresponding to the eigenvalue λ_2, ... , and finally c_M to be a normalized eigenvector corresponding to the eigenvalue λ_M,and we make sure that all the vectors c_j are orthogonal to each other. This is automatically true when the eigenvalues λ_j are simple. See the discussion below for the general case. Recall that we say a vector is normalized if the sum of the squares of its components is equal to 1. If we denote by C the $M \times M$ matrix formed by the M column vectors containing the components of the vectors c_1, c_2, . . ., c_M in this order, this matrix is orthogonal (since it it a matrix transforming one orthonormal basis into another) and it satisfies:

$$C_x = C^t DC$$

where we use the notation ^{t} to denote the transpose of a matrix or a vector, and where D is the $M \times M$ diagonal matrix with λ_1, λ_2, . . ., λ_M on the diagonal. Notice the obvious lack of uniqueness of the above decomposition. In particular, if c_j is a normalized eigenvector associated to the eigenvalue λ_j, so is $-c_j$! This is something one should keep in mind when plotting the eigenvectors c_j, and when trying to find an intuitive interpretation for the features of the plots. However, this sign flip is easy to handle, and fortunately, it is the only form of non uniqueness when the eigenvalues are simple (i.e. nondegenerate). The ratio:

$$\frac{\lambda_j}{\sum_{j'=1}^{M} \lambda_{j'}}$$

of a given eigenvalue to the trace of C_x (i.e. the sum of its eigenvalues) has the interpretation of the proportion of the variation explained by the corresponding eigenvector, i.e. the loading c_j. In order to justify this statement, we appeal to the Raleigh-Ritz variational principle from linear algebra. Indeed, according to this principle, the eigenvalues and their corresponding eigenvectors can be characterized recursively in the following way. The largest eigenvalue λ_1 appears as the maximum:

$$\lambda_1 = \max_{x \in \mathbb{R}^M, \|x\|=1} x^t C_x x$$

while the corresponding eigenvector c_1 appears as the argument of this maximization problem:

$$c_1 = \arg \max_{x \in \mathbb{R}^M, \|x\|=1} x^t C_x x.$$

If we recall the fact that $x^t C_x x$ represents the quadratic variation (empirical variance) of the NLC's $x^t x_{1\cdot}$, $x^t x_{2\cdot}$,. . ., $x^t x_{N\cdot}$, λ_1 can be interpreted as the maximal quadratic variation when we consider all the possible (normalized) linear combinations of the M original measured variables. Similarly, the corresponding (normalized) eigenvector has precisely the interpretation of this NLC which realizes the maximum variation.

As we have already pointed out, the first loading is uniquely determined up to a sign change if the eigenvalue λ_1 is simple. If this is not the case, and if we denote by m_1 the multiplicity of the eigenvalue λ_1, we can choose any orthonormal set $\{c_1, \cdot, c_{m_1}\}$ in the eigenspace of λ_1 and repeat the eigenvalue λ_1, m_1 times in the list of eigenvalues (and on the diagonal of D as well). This lack of uniqueness is not a mathematical difficulty, it is merely annoying. Fortunately, it seldom happens in practice ! We shall assume that all the eigenvalues are simple (i.e. non-degenerate) for the remainder of our discussion. If they were not, we would have to repeat them according to their multiplicities.

Next, still according to the Raleigh-Ritz variation principle, the second eigenvalue λ_2 appears as the maximum:

$$\lambda_2 = \max_{x \in \mathbb{R}^M, \|x\|=1, x \perp c_1} x^t C_x x$$

while the corresponding eigenvector c_2 appears as the argument of this maximization problem:

$$c_2 = \arg \max_{x \in \mathbb{R}^M, \|x\|=1, x \perp c_1} x^t C_x x.$$

The interpretation of this statement is the following: if we avoid any overlap with the loading already identified (i.e. if we restrict ourselves to NLC's x which are orthogonal to c_1), then the maximum quadratic variation will be λ_2 and any NLC realizing this maximum variation can be used for c_2. We can go on and identify in this way all the eigenvalues λ_j (having to possibly repeat them according to their multiplicities) and the loadings c_j's.

Armed with a new basis of $\mathbb{R}^M$, the next step is to rewrite the data observations (i.e. the N rows of the data matrix) in this new basis. This is done by multiplying the data matrix by the *change of basis* matrix (i.e. the matrix whose columns are the eigenvectors identified earlier). The result is a new $N \times M$ matrix whose columns are called *principal components*. Their relative importance is given by the proportion of the variance explained by the loadings, and for that reason, one typically considers only the first few principal components, the remaining ones being ignored and/or treated as noise.

2.5.2 PCA with `S-Plus`

The principal component analysis of a data set is performed in `S-Plus` with the function `princomp`, which returns an object of class `princomp` that can be printed and plotted with generic methods. Illustrations of the calls to this function and of the interpretation of the results are given in the next subsections in which we discuss several financial applications of the PCA.

2.5.3 Effective Dimension of the Space of Yield Curves

Our first application concerns the markets of fixed income securities which we will introduce in Section 3.8. The term structure of interest rates is conveniently captured by the daily changes in the yield curve. The dimension of the space of all possible yield curves is presumably very large, potentially infinite if we work in the idealized world of continuous-time finance. However, it is quite sensible to try to approximate these curves by functions from a class chosen in a parsimonious way. Without any a priori choice of the type of functions to be used to approximate the yield curve, PCA can be used to extract, one by one, the components responsible for the variations in the data.

PCA of the Yield Curve

For the purposes of illustration, we use data on the US yield curve as provided by the Bank of International Settlements (BIS, for short). These data are the result of a nonparametric processing (smoothing spline regression, to be specific) of the raw data. The details will be given in Section 4.4 of Chapter 4, but for the time being, we shall ignore the possible effects of this pre-processing of the raw data. The data are imported into an S-object named us.bis.yield which gives, for each of the 1352 successive trading days following January 3rd 1995, the yields on the US Treasury bonds for times to maturity

$$x = 0, 1, 2, 3, 4, 5, 5.5, 6.5, 7.5, 8.5, 9.5 \quad \text{months.}$$

We run a PCA on these data with the following S-Plus commands:

```
> dim(us.bis.yield)
[1] 1352   11
> us.bis.yield.pca <- princomp(us.bis.yield)
> plot(us.bis.yield.pca)
[1]   0.700000   1.900000   3.100000   4.300000   5.500000
[6] 6.700000   7.900000   9.099999  10.299999 11.499999
> title("Proportions of the Variance Explained
                                  by the Components")
```

The results are reproduced in Figure 2.20 which gives the proportions of the variation explained by the various components. The first three eigenvectors of the covariance matrix (the so-called loadings) explain 99.9% of the total variation in the data. This suggests that the effective dimension of the space of yield curves could be three. In other words, any of the yield curves from this period can be approximated by a linear combination of the first three loadings, the relative error being very small. In order to better understand the far reaching implications of this statement we plot the first four loadings.

```
> X <- c(0,1,2,3,4,5,5.5,6.5,7.5,8.5,9.5)
> par(mfrow=c(2,2))
> plot(X,us.bis.yield.pca$loadings[,1],ylim=c(-.7,.7))
> lines(X,us.bis.yield.pca$loadings[,1])
> plot(X,us.bis.yield.pca$loadings[,2],ylim=c(-.7,.7))
> lines(X,us.bis.yield.pca$loadings[,2])
> plot(X,us.bis.yield.pca$loadings[,3],ylim=c(-.7,.7))
> lines(X,us.bis.yield.pca$loadings[,3])
> plot(X,us.bis.yield.pca$loadings[,4],ylim=c(-.7,.7))
```

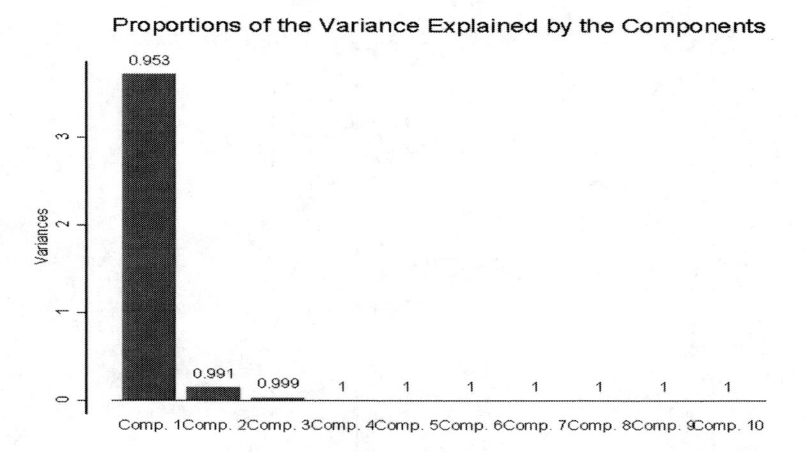

Fig. 2.20. Proportions of the variance explained by the components of the PCA of the daily changes in the US yield curve.

```
> lines(X,us.bis.yield.pca$loadings[,4])
> par(mfrow=c(1,1))
> title("First Four Loadings of the US Yield Curves")
```

The results are reproduced in Figure 2.21. The first loading is essentially flat, so a component on this loading will essentially represent the average yield over the maturities, and the effect of this most-important component on the actual yield curve is a parallel shift. Because of the monotone and increasing nature of the second loading, the second component measures the upward trend (if the component is positive, and the downward trend otherwise) in the yield. This second factor is interpreted as the tilt of the yield curve. The shape of the third loading suggests that the third component captures the curvature of the yield curve. Finally, the shape of the fourth loading does not seem to have an obvious interpretation. It is mostly noise (remember that most of the variations in the yield curve are explained by the first three components). These features are very typical, and they should be expected in most PCA's of the term structure of interest rates.

The fact that the first three components capture so much of the yield curve may seem strange when compared to the fact that some estimation methods, which we discuss later in the book, use parametric families with more than three parameters! There is no contradiction there. Indeed, for the sake of illustration, we limited the analysis of this section to the first part of the yield curve. Restricting ourselves to short maturities makes it easier to capture all the features of the yield curve in a small number of functions with a clear interpretation.

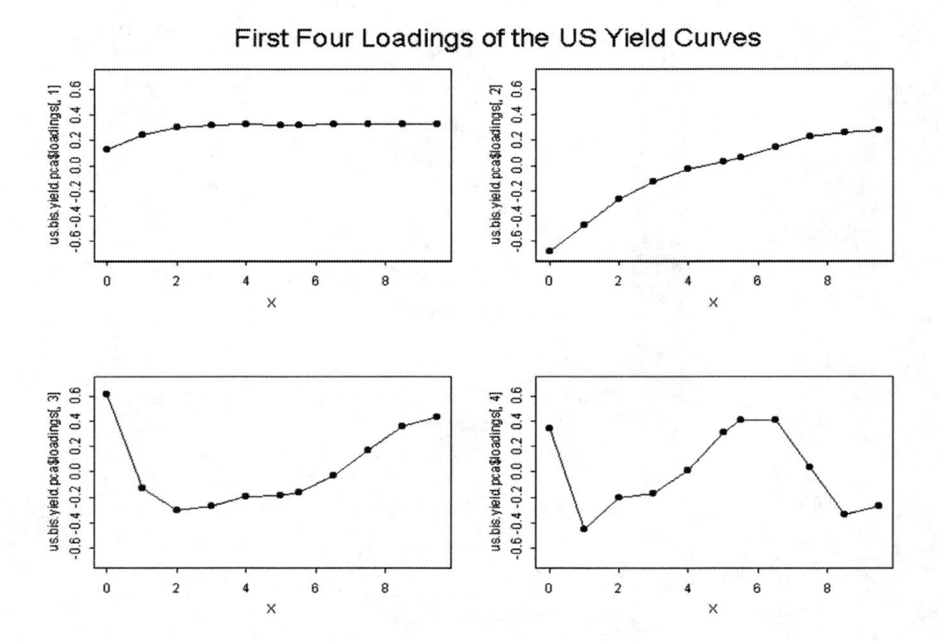

Fig. 2.21. From left to right and top to bottom, sequential plots of the first four US yield loadings.

2.5.4 **Swap Rate Curves**

Swap contracts have been traded publicly since 1981. As of today, they are the most popular fixed income derivatives. Because of this popularity, the swap markets are extremely liquid, and as a consequence, they can be used to hedge interest-rate risk of fixed income portfolios at a low cost. The estimation of the term-structure of swap rates is important in this respect and the PCA which we present below is the first step toward a better understanding of this term structure.

Swap Contracts and Swap Rates

As implied by its name, a swap contract obligates two parties to exchange (or swap) some specified cash flows at agreed upon times. The most common swap contracts are interest rate swaps. In such a contract, one party, say counter-party A, agrees to make interest payments determined by an instrument P_A (say, a 10 year US Treasury bond rate), while the other party, say counter-party B, agrees to make interest payments determined by another instrument P_B (say, the London Interbank Offer Rate – LIBOR for short) Even though there are many variants of swap contracts, in a typical contract, the principal on which counter-party A makes interest payments is equal to the principal on which counterparty B makes interest payments. Also, the payment schedules are identical and periodic, the payment frequency being quarterly, semi-annually, ...

It is not difficult to infer from the above discussion that a swap contract is equivalent to a portfolio of forward contracts, but we shall not use this feature here. In this section, we shall restrict ourselves to the so-called plain vanilla contracts involving a fixed interest rate and the 3 or 6 months LIBOR rate.

We will not attempt to derive here a formula for the price of a swap contract, neither will we try to define rigorously the notion of swap rate. These derivations are beyond the scope of this book. See the Notes & Complements at the end of the chapter for references to appropriate sources. We shall use only the intuitive idea of the swap rate being a rate at which both parties will agree to enter into the swap contract.

PCA of the Swap Rates

Our second application of principal component analysis concerns the term structure of swap rates as given by the swap rate curves. As before, we denote by M the dimension of the vectors. We use data downloaded from `Data Stream`. It is quite likely that the raw data have been processed, but we are not quite sure what kind of manipulation is performed by `Data Stream`, so for the purposes of this illustration, we shall ignore the possible effects of the pre-processing of the data. In this example, the day t labels the rows of the data matrix. The latter has $M = 15$ columns, containing the swap rates with maturities T conveniently labeled by the times to maturity $x = T - t$, which have the values $1, 2, \dots, 10, 12, 15, 20, 25, 30$ years in the present situation. We collected these data for each day t of the period from May 1998 to March 2000, and we rearranged the numerical values in a matrix $R = [r_{i,j}]_{i=1,\dots,N,\ j=1,\dots,M}$. Here, the index j stands for the time to maturity, while the index i codes the day the curve is observed.

The data is contained in the `S` object `swap`. The PCA is performed in `S-Plus` with the command:

```
> dim(swap)
[1] 496  15
> swap.pca <- princomp(swap)
> plot(swap.pca)
[1]  0.700000  1.900000  3.100000  4.300000  5.500000
[6] 6.700000  7.900000  9.099999  10.299999 11.499999
> title("Proportions of the Variance Explained by
                                    the Components")
> YEARS <- c(1,2,3,4,5,6,7,8,9,10,12,15,20,25,30)
> par(mfrow=c(2,2))
> plot(YEARS,swap.pca$loadings[,1],ylim=c(-.6,.6))
> lines(YEARS,swap.pca$loadings[,1])
> plot(YEARS,swap.pca$loadings[,2],ylim=c(-.6,.6))
> lines(YEARS,swap.pca$loadings[,2])
> plot(YEARS,swap.pca$loadings[,3],ylim=c(-.6,.6))
```

```
> lines(YEARS,swap.pca$loadings[,3])
> plot(YEARS,swap.pca$loadings[,4],ylim=c(-.6,.6))
> lines(YEARS,swap.pca$loadings[,4])
> par(mfrow=c(1,1))
> title("First Four Loadings of the Swap Rates")
```

Figure 2.22 gives the proportions of the variation explained by the various components, while Figure 2.23 gives the plots of the first four eigenvectors.

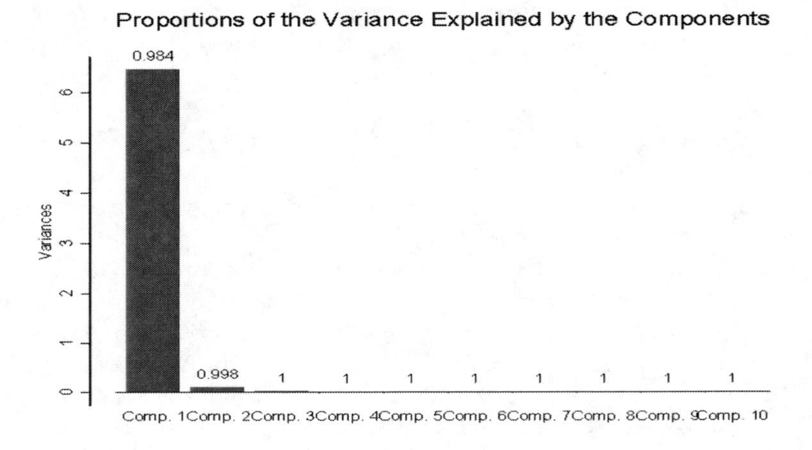

Fig. 2.22. Proportions of the variance explained by the components of the PCA of the daily changes in the swap rates for the period from May 1998 to March 2000.

Looking at Figure 2.23 one sees that the remarks made above, for the interpretation of the results in terms of a parallel shift, a tilt and a curvature component, do apply to the present situation as well.

Since such an overwhelming proportion of the variation is explained by one single component, it is often recommended to remove the effect of this component from the data, (here, that would amount to subtracting the overall mean rate level) and to perform the PCA on the transformed data (here, the fluctuations around the mean rate level).

APPENDIX 1: CALCULUS WITH RANDOM VECTORS AND MATRICES

The nature and technical constructs of this chapter justify our spending some time discussing the properties of random vectors (as opposed to random variables) and reviewing the fundamental results of the calculus of probability with random vectors. Their definition is very natural: a random vector is a vector whose entries are random

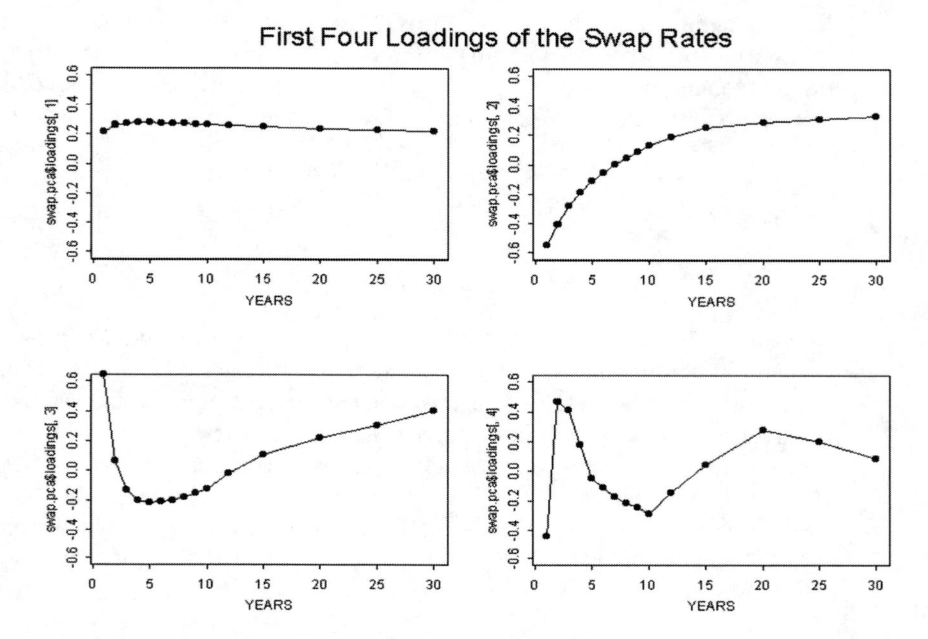

Fig. 2.23. From left to right and top to bottom, sequential plots of the eigenvectors (loadings) corresponding to the 4 largest eigenvalues. Notice that we changed the scale of the horizontal axis to reflect the actual times to maturity.

variables. With this definition in hand, it is easy to define the notion of expectation. The expectation of a random vector is the (deterministic) vector whose entries are the expectations of the entries of the original random vector. In other words,

$$\text{if} \quad \mathbf{X} = \begin{bmatrix} X_1 \\ \vdots \\ X_n \end{bmatrix}, \qquad \text{then} \quad \mathbb{E}\{\mathbf{X}\} = \begin{bmatrix} \mathbb{E}\{X_1\} \\ \vdots \\ \mathbb{E}\{X_n\} \end{bmatrix}.$$

Notice that, if $\mathbf{B}$ is an n-dimensional (deterministic) vector then:

$$\mathbb{E}\{\mathbf{X}+\mathbf{B}\} = \mathbb{E}\{\mathbf{X}\} + \mathbf{B}. \tag{2.16}$$

Indeed:

$$\mathbb{E}\{\mathbf{X}+\mathbf{B}\} = \begin{bmatrix} \mathbb{E}\{X_1+b_1\} \\ \vdots \\ \mathbb{E}\{X_n+b_n\} \end{bmatrix} = \begin{bmatrix} \mathbb{E}\{X_1\}+b_1 \\ \vdots \\ \mathbb{E}\{X_n\}+b_n \end{bmatrix} = \mathbb{E}\{\mathbf{X}\} + \mathbf{B}$$

where we used the notation b_i for the components of the vector $\mathbf{B}$. The notion of variance (or more generally of second moment) appears somehow less natural at first. We define the variance/covariance matrix of a random vector to be the (deterministic) matrix whose entries are the variances and covariances of the entries of the original random vector. More precisely, if $\mathbf{X}$ is a random vector as above, then its variance/covariance matrix is the matrix $\Sigma_{\mathbf{X}}$ defined by:

$$\Sigma_{\mathbf{X}} = \begin{bmatrix} \sigma_1^2 & \text{cov}(X_1, X_2) & \cdots & \text{cov}(X_1, X_n) \\ \text{cov}(X_2, X_1) & \sigma_2^2 & \cdots & \text{cov}(X_2, X_n) \\ \vdots & \vdots & \vdots & \vdots \\ \text{cov}(X_n, X_1) & \text{cov}(X_n, X_2) & \cdots & \sigma_n^2 \end{bmatrix}, \tag{2.17}$$

In other words, on the i-th row and the j-th column of $\Sigma_{\mathbf{X}}$ we find the covariance of X_i and X_j. For the purposes of this appendix, we limit ourselves to random variables of order 2 (i.e. for which the first two moments exist) so that all the variances and covariances make perfectly good mathematical sense. Note that this is not the case for many generalized Pareto distributions, and especially for our good old friend the Cauchy distribution.

The best way to look at the variance/covariance matrix of a random vector is the following. Using the notation $\mathbf{Z}^t$ for the transpose of the vector or matrix $\mathbf{Z}$, we notice that:

$$[\mathbf{X} - \mathbb{E}\{\mathbf{X}\}][\mathbf{X} - \mathbb{E}\{\mathbf{X}\}]^t = \begin{bmatrix} X_1 - \mu_1 \\ \vdots \\ X_n - \mu_n \end{bmatrix} [X_1 - \mu_1, \ldots, X_n - \mu_n]$$

$$= \begin{bmatrix} (X_1 - \mu_1)^2 & (X_1 - \mu_1)(X_2 - \mu_2) & \cdots & (X_1 - \mu_1)(X_n - \mu_n) \\ (X_2 - \mu_2)(X_1 - \mu_1) & (X_2 - \mu_2)^2 & \cdots & (X_2 - \mu_2)(X_n - \mu_n) \\ \vdots & \vdots & \vdots & \vdots \\ (X_n - \mu_n)(X_1 - \mu_1) & (X_n - \mu_n)(X_2 - \mu_2) & \cdots & (X_n - \mu_n)^2 \end{bmatrix}$$

if we use the notation $\mu_j = \mathbb{E}\{X_j\}$ to shorten the typesetting of the formula. The variance/covariance matrix $\Sigma_{\mathbf{X}}$ is nothing more than the expectation of this random matrix, since the expectation of a random matrix is defined as the (deterministic) matrix whose entries are the expectations of the entries of the original random matrix. Consequently we have proven that, for any random vector $\mathbf{X}$ of order 2, the variance/covariance matrix $\Sigma_{\mathbf{X}}$ is given by the formula:

$$\Sigma_{\mathbf{X}} = \mathbb{E}\{[\mathbf{X} - \mathbb{E}\{\mathbf{X}\}][\mathbf{X} - \mathbb{E}\{\mathbf{X}\}]^t\}. \tag{2.18}$$

Notice that, if the components of a random vector are independent, then the variance/covariance matrix of this random vector is diagonal since all the entries off the diagonal must vanish due to the independence assumption.

Some Useful Formulae

If $\mathbf{X}$ is an n-dimensional random vector, if $\mathbf{A}$ is an $m \times n$ deterministic matrix and $\mathbf{B}$ is an m-dimensional deterministic vector, then:

$$\mathbb{E}\{\mathbf{AX} + \mathbf{B}\} = \mathbf{A}\mathbb{E}\{\mathbf{X}\} + \mathbf{B} \tag{2.19}$$

as can be checked by computing the various components of the m-dimensional vectors on both sides of the equality sign. Notice that formula (2.16) is merely a particular case of (2.19) when $m = n$ and $\mathbf{A}$ is the identity matrix. In fact, formula (2.19) remains true when $\mathbf{X}$ is an $n \times p$ random matrix and $\mathbf{B}$ is an $m \times p$ deterministic matrix. By transposition one gets that

$$\mathbb{E}\{\mathbf{XA} + \mathbf{B}\} = \mathbb{E}\{\mathbf{X}\}\mathbf{A} + \mathbf{B} \tag{2.20}$$

holds whenever $\mathbf{X}$ is an $n \times p$ random matrix and $\mathbf{A}$ and $\mathbf{B}$ are deterministic matrices with dimensions $p \times m$ and $n \times m$ respectively. Using (2.19) and (2.20) we get:

$$\Sigma_{\mathbf{AX}+\mathbf{B}} = \mathbf{A}\Sigma_{\mathbf{X}}\mathbf{A}^t \tag{2.21}$$

A proof of this formula goes as follows:

$$\begin{aligned}
\Sigma_{\mathbf{AX}+\mathbf{B}} &= \mathbb{E}\{[\mathbf{AX} + \mathbf{B} - \mathbb{E}\{\mathbf{AX} + \mathbf{B}\}][\mathbf{AX} + \mathbf{B} - \mathbb{E}\{\mathbf{AX} + \mathbf{B}\}]^t\} \\
&= \mathbb{E}\{[\mathbf{A}(\mathbf{X} - \mathbb{E}\{\mathbf{X}\})][\mathbf{A}(\mathbf{X} - \mathbb{E}\{\mathbf{X}\})]^t\} \\
&= \mathbb{E}\{\mathbf{A}[\mathbf{X} - \mathbb{E}\{\mathbf{X}\}][\mathbf{X} - \mathbb{E}\{\mathbf{X}\}]^t\mathbf{A}^t\} \\
&= \mathbf{A}\mathbb{E}\{[\mathbf{X} - \mathbb{E}\{\mathbf{X}\}][\mathbf{X} - \mathbb{E}\{\mathbf{X}\}]^t\}\mathbf{A}^t \\
&= \mathbf{A}\Sigma_{\mathbf{X}}\mathbf{A}^t
\end{aligned}$$

Similar formulae can be proven for the variance/covariance matrix of expressions of the form $\mathbf{AXB} + \mathbf{C}$ when $\mathbf{X}$ is a random vector or a random matrix and when $\mathbf{A}$, $\mathbf{B}$ and $\mathbf{C}$ are deterministic matrices or vectors whose dimensions make the product meaningful.

Warning: Remember to be very cautious with the order in a product of matrices. Just because one can change the order in the product of numbers, does not mean that it is a good idea to do the same thing with a product of matrices, as the results are in general (very) different !!!

APPENDIX 2: FAMILIES OF COPULAS

There are many parametric families of copulas, and new ones are created every month. For the sake of completeness, we review those implemented in the `S-Plus` library `EVANESCE`. They can be organized in two main classes.

◇ **Extreme value copulas** are copulas of the form

$$C(x, y) = \exp\left[\log(xy)\, A\left(\frac{\log(x)}{\log(xy)}\right)\right],$$

where $A : [0, 1] \to [0.5, 1]$, is a convex function, and $\max(t, 1 - t) \le A(t) \le 1$ for all $t \in [0, 1]$.

◇ **Archimedean copulas** are copulas of the form

$$C(x,y) = \phi^{-1}\left[(\phi(x)+\phi(y))\,A\left(\frac{\phi(x)}{\phi(x)+\phi(y)}\right)\right],$$

where $\phi(t)$ is a valid Archimedean generator (convex and decreasing on $(0,1)$), and A is as above.

We now list the most commonly used parametric families of copulas:

• **Bivarate Normal**, `"normal"`
```
> normal.copula(delta)
```

$$C(x,y) = \Phi_\delta\left(\Phi^{-1}(x), \Phi^{-1}(y)\right),$$

$0 \leq \delta \leq 1$, where Φ^{-1} is the quantile function of the standard normal distribution, and Φ_δ is the cdf of the standard bivariate normal with correlation δ
• **Frank Copula**, `"frank"`
```
> frank.copula(delta)
```

$$C(x,y) = -\delta^{-1}\log\left(\frac{\eta-(1-e^{-\delta x})(1-e^{-\delta x})}{\eta}\right)$$

$0 \leq \delta < \infty$, and $\eta = 1-e^{-\delta}$.
• **Kimeldorf and Sampson copula**, `"kimeldorf.sampson"`
```
> kimeldorf.sampson.copula(delta)
```

$$C(x,y) = (x^{-\delta}+y^{-\delta}-1)^{-1/\delta},$$

$0 \leq \delta < \infty$.
• **Gumbel copula**, `"gumbel"`
```
> gumbel.copula(delta)
```

$$C(x,y) = \exp\left(-\left[(-\log x)^\delta + (-\log y)^\delta\right]^{1/\delta}\right),$$

$1 \leq \delta < \infty$. This is an extreme value copula with the dependence function

$$A(t) = (t^\delta + (1-t)^\delta)^{1/\delta}.$$

• **Galambos** , `"galambos"`
```
> galambos.copula(delta)
```

$$C(x,y) = xy\exp\left(\left[(-\log x)^{-\delta} + (-\log y)^{-\delta}\right]^{-1/\delta}\right),$$

$0 \leq \delta < \infty$. This is an extreme value copula with the dependence function

$$A(t) = 1-(t^{-\delta}+(1-t)^{-\delta})^{-1/\delta}.$$

• **Hüsler and Reiss**, `"husler.reiss"`

```
> husler.reiss.copula(delta)
```

$$C(x, y) = \exp\left(-\tilde{x}\,\Phi\left[\frac{1}{\delta} + \frac{1}{2}\delta\log\left(\frac{\tilde{x}}{\tilde{y}}\right)\right] - \tilde{y}\,\Phi\left[\frac{1}{\delta} + \frac{1}{2}\delta\log\left(\frac{\tilde{y}}{\tilde{x}}\right)\right]\right)$$

$0 \le \delta < \infty$, $\tilde{x} = -\log x$, $\tilde{y} = -\log y$, and Φ is the cdf of the standard normal distribution. This is an extreme value copula with the dependence function

$$A(t) = t\,\Phi\left[\delta^{-1} + \frac{1}{2}\delta\log\left(\frac{t}{1-t}\right)\right] + (1-t)\,\Phi\left[\delta^{-1} - \frac{1}{2}\delta\log\left(\frac{t}{1-t}\right)\right]$$

• **Twan**, `"twan"`

```
> twan.copula(alpha, beta, r)
```

This is an extreme value copula with the dependence function

$$A(t) = 1 - \beta + (\beta - \alpha) + \{\alpha^r t^r + \beta^r (1-t)^r\}^{1/r},$$

$0 \le \alpha,\ \beta \le 1,\ 1 \le r < \infty$.

• **BB1**, `"bb1"`

```
> bb1.copula(theta, delta)
```

$$C(x, y) = \left(1 + \left[(x^{-\theta} - 1)^{\delta} + (y^{-\theta} - 1)^{\delta}\right]^{1/\delta}\right)^{-1/\theta}$$

$\theta > 0, \delta \ge 1$. This is an Archimedean copula with the Archimedean generator $\phi(t) = \left(t^{-\theta} - 1\right)^{\delta}$.

• **BB2**, `"bb2"`

```
> bb2.copula(theta, delta)
```

$$C(x, y) = \left[1 + \delta^{-1}\log\left(e^{\delta(x^{-\theta})} + e^{\delta(y^{-\theta})} - 1\right)\right]^{1/\theta},$$

$\theta > 0, \delta > 0$. This is an Archimedean copula with the Archimedean generator $\phi(t) = e^{\delta(t^{-\theta} - 1)} - 1$.

• **BB3**, `"bb3"`

```
> bb3.copula(theta, delta)
```

$$C(x, y) = \exp\left(-\left[\delta^{-1}\log\left(e^{\delta\tilde{x}^{\theta}} + e^{\delta\tilde{y}^{\theta}} - 1\right)\right]^{1/\theta}\right),$$

$\theta \ge 1, \delta > 0$, $\tilde{x} = -\log x$, $\tilde{y} = -\log y$. This is an Archimedean copula with the Archimedean generator $\phi(t) = \exp\left\{\delta\left(-\log t\right)^{\theta}\right\} - 1$.

• **BB4**, `"bb4"`

```
> bb4.copula(theta, delta)
```

$$C(x, y) = \left(x^{-\theta} + y^{-\theta} - 1 - \left[\left(x^{-\theta} - 1\right)^{-\delta} + \left(y^{-\theta} - 1\right)^{-\delta}\right]^{-\frac{1}{\delta}}\right)^{-\frac{1}{\theta}}$$

$\theta \geq 0, \delta > 0$. This is an Archimedean copula with the Archimedean generator $\phi(t) = t^{-\theta} - 1$ and the dependence function $A(t) = 1 - (t^{-\delta} + (1-t)^{-\delta})^{-1/\delta}$ (same as for B7 family).

• **BB5**, "bb5"

```
> bb5.copula(theta, delta)
```

$$C(x,y) = \exp\left(-\left[\tilde{x}^{\theta} + \tilde{y}^{\theta} - \left(\tilde{x}^{-\theta\delta} + \tilde{y}^{-\theta\delta}\right)^{-1/\delta}\right]^{1/\theta}\right),$$

$\delta > 0$, $\theta \geq 1$, $\tilde{x} = -\log x$, $\tilde{y} = -\log y$. This is an extreme value copula with the dependence function

$$A(t) = \left[t^{\theta} + (1-t)^{\theta} - \left(t^{-\delta\theta} + (1-t)^{-\delta\theta}\right)^{-1/\delta}\right]^{1/\theta}.$$

• **BB6**, "bb6"

`> bb6.copula(theta, delta)` This is an Archimedean copula with the generator $\phi(t) = \left[-\log\left(1-(1-t)^{\theta}\right)\right]^{\delta}$ $\qquad \theta \geq 1, \delta \geq 1$.

• **BB7**, "bb7"

```
> bb7.copula(theta, delta)
```

This is an Archimedean copula with the generator
$\phi(t) = \left(1-(1-t)^{\theta}\right)^{-\delta} - 1, \qquad \theta \geq 1, \delta > 0.$

• **B1Mix**, "normal.mix"

```
> normal.mix.copula(p, delta1, delta2)
```

$$C(x,y) = p\, C^{(\mathrm{B1})}_{\delta_1}(x,y) + (1-p)\, C^{(\mathrm{B1})}_{\delta_2}(x,y),$$

$0 \leq p, \delta_1, \delta_2 \leq 1$, where $C^{(\mathrm{B1})}_{\delta}(x,y)$ is a bivariate normal copula (family "normal").

PROBLEMS

Ⓔ **Problem 2.1** *This problem is based on the data contained in the script file* `utilities.asc`. *Opening it in* `S-Plus`, *and running it as a script creates a matrix with two columns. Each row corresponds to a given day. The first column gives the log of the weekly return on an index based on Southern Electric stock value and capitalization, (we'll call that variable X), and the second column gives, on the same day, the same quantity for Duke Energy (we'll call that variable Y), another large utility company.*

1. Compute the means and the standard deviations of X and Y, and compute their correlation coefficients.

2. We first assume that X and Y are samples from a jointly Gaussian distribution with parameters computed in question 1. Compute the q-percentile with $q = 2\%$ of a the variables $X + Y$ and $X - Y$.

3. Fit a generalized Pareto distribution (GPD) to X and Y separately, and fit a copula of the Gumbel family to the empirical copula of the data.

4. Generate a sample of size N (where N is the number of rows of the data matrix) from the joint distribution estimated in question 3.

4.1. Use this sample to compute the same statistics as in question 1 (i.e. means and standard deviations of the columns, as well as their correlation coefficients), and compare the results to the numerical values obtained in question 1.

4.2. Compute, still for this simulated sample, the two percentiles considered in question 2, compare the results, and comment.

Ⓔ **Problem 2.2** *This problem is based on the data contained in the script file* `SPfutures.asc` *which creates a matrix* `SPFUT` *with two columns, each row corresponding to a given day. The first column gives, for each day, the log return of a futures contract which matures three weeks later, (we'll call that variable X), and the second column gives, on the same day, the log return of a futures contract which matures one week later (we'll call that variable Y).*

Question 2 is not required for the rest of the problem. In other words, you can answer questions 3 and 4 even if you did not get question 2.

1. Compute the means and the standard deviations of X and Y, and compute their correlation coefficients.

2. We first assume that X and Y are samples from a jointly Gaussian distribution with parameters computed in part 1. For each value $\alpha = 25\%$, $\alpha = 50\%$ and $\alpha = 75\%$ of the parameter α, compute the q-percentile with $q = 2\%$ of the variable $\alpha X + (1-\alpha)Y$.

3. Fit a generalized Pareto distribution (GPD) to X and Y separately, and fit a copula of the Gumbel family to the empirical copula of the data.

4. Generate a sample of size N (where N is the number of rows of the data matrix) from the joint distribution estimated in question 3.

4.1. Use this sample to compute the same statistics as in question 1 (i.e. means and standard deviations of the columns, as well as their correlation coefficients) and compare to the numerical values obtained in question 1.

4.2. Compute, still for this simulated sample, the three percentiles considered in question 2, and compare the results.

Ⓢ **Problem 2.3** *1. Construct a vector of 100 increasing and regularly spaced numbers starting from .1 and ending at 20. Call it* `SIG2`. *Construct a vector of 21 increasing and regularly spaced numbers starting from -1.0 and ending at 1.0. Call it* `RHO`.

2. For each entry σ^2 of `SIG2` *and for each entry ρ of* `RHO`*:*

- *Generate a sample of size $N = 500$ from the distribution of a bivariate normal vector $Z = (X, Y)$, where $X \sim N(0,1)$, and $Y \sim N(0, \sigma^2)$, and the correlation coefficient of X and Y is ρ (the* `S` *object you create to hold the values of the sample of Z's should be a 500×2 matrix);*
- *Create a 500×2 matrix, call it* `EXPZ`, *with the exponentials of the entries of Z (the distributions of these columns are lognormal as defined in Problem 2.7);*
- *Compute the correlation coefficient, call it $\tilde{\rho}$, of the two columns of* `EXPZ`

3. Produce a scatterplot of all the points $(\sigma^2, \tilde{\rho})$ so obtained. Comment.

Ⓣ **Problem 2.4** *This elementary exercise is intended to give an example showing that lack of correlation does not necessarily mean independence!*

Let us assume that $X \sim N(0,1)$ and let us define the random variable Y by:

$$Y = \frac{1}{\sqrt{1-2/\pi}}(|X| - \sqrt{2/\pi})$$

1. Compute $\mathbb{E}\{|X|\}$
2. Show that Y *has mean zero, variance* 1*, and that it is uncorrelated with* X*.*

Ⓣ **Problem 2.5** *The purpose of this problem is to show that lack of correlation does not imply independence, even when the two random variables are Gaussian !!!*
We assume that X*,* ϵ_1 *and* ϵ_2 *are independent random variables, that* $X \sim N(0,1)$*, and that* $\mathbb{P}\{\epsilon_i = -1\} = \mathbb{P}\{\epsilon_i = +1\} = 1/2$ *for* $i = 1, 2$*. We define the random variable* X_1 *and* X_2 *by:*

$$X_1 = \epsilon_1 X, \qquad \text{and} \qquad X_2 = \epsilon_2 X.$$

1. Prove that $X_1 \sim N(0,1)$*,* $X_2 \sim N(0,1)$ *and that* $\rho\{X_1, X_2\} = 0$*.*
2. Show that X_1 *and* X_2 *are not independent.*

Ⓣ **Problem 2.6** *The goal of this problem is to prove rigorously a couple of useful formulae for normal random variables.*
1. Show that, if $Z \sim N(0,1)$*, if* $\sigma > 0$*, and if* f *is ANY function, then we have:*

$$\mathbb{E}\{f(Z)e^{\sigma Z}\} = e^{\sigma^2/2}\mathbb{E}\{f(Z+\sigma)\},$$

and use this formula to recover the well known fact

$$\mathbb{E}\{e^X\} = e^{\mu+\sigma^2/2}$$

whenever $X \sim N(\mu, \sigma^2)$*.*
2. We now assume that X *and* Y *are jointly-normal mean-zero random variables and that* h *is ANY function. Prove that:*

$$\mathbb{E}\{e^X h(Y)\} = \mathbb{E}\{e^X\}\mathbb{E}\{h(Y + cov\{X, Y\})\}.$$

Ⓣ **Problem 2.7** *The goal of this problem is to prove rigorously the theoretical result illustrated by the simulations of Problem 2.3.*
1. Compute the density of a random variable X *whose logarithm* $\log X$ *is* $N(\mu, \sigma^2)$*. Such a random variable is usually called a lognormal random variable with mean* μ *and variance* σ^2*.*

Throughout the rest of the problem we assume that X *is a lognormal random variable with parameters* 0 *and* 1 *(i.e.* X *is the exponential of an* $N(0,1)$ *random variable) and that* Y *is a lognormal random variable with parameters* 0 *and* σ^2 *(i.e.* Y *is the exponential of an* $N(0,\sigma^2)$ *random variable). Moreover, we use the notation* ρ_{min} *and* ρ_{max} *introduced in the last paragraph of Subsection 2.1.2.*
2. Show that $\rho_{min} = (e^{-\sigma} - 1)/\sqrt{(e-1)(e^{\sigma^2} - 1)}$*.*
3. Show that $\rho_{max} = (e^{\sigma} - 1)/\sqrt{(e-1)(e^{\sigma^2} - 1)}$*.*
4. Check that $\lim_{\sigma\to\infty} \rho_{min} = \lim_{\sigma\to\infty} \rho_{max} = 0$*.*

ⓈⓉ **Problem 2.8** *The first question concerns the computation in* `S-Plus` *of the square root of a symmetric nonnegative-definite square matrix.*
1. Write an `S`*-function, call it* `msqrt`*, with argument A which:*

- *checks that A is a square matrix and exits if not;*
- *checks that A is symmetric and exits if not;*
- *diagonalizes the matrix by computing the eigenvalues and the matrix of eigenvectors (hint: check out the help file of the function* `eigen` *if you are not sure how to proceed);*
- *checks that all the eigenvalues are nonnegative and exits, if not;*

• returns a symmetric matrix of the same size as A, with the same eigenvectors, the eigenvalue corresponding to a given eigenvector being the square root of the corresponding eigenvalue of A.
The matrix returned by such a function `msqrt` *is called the square root of the matrix A and it will be denoted by $A^{1/2}$.*

The second question concerns the generation of normal random vectors in `S-Plus`. *In other words, we write an* `S-Plus` *function to play the role of the function* `rnorm` *in the case of multidimensional random vectors. Such a function does exist in the* `S-Plus` *distribution. It is called* `mvrnorm`. *The goal of this second question is to understand how such a generation method works.*

2. Write an S-function, call it `vnorm`, *with arguments* `Mu`, `Sigma` *and* `N` *which:*

• checks that `Mu` *is a vector, exits if not, and otherwise reads its dimension, say* `L`;

• checks that `Sigma` *is an* `LxL` *symmetric matrix with nonnegative eigenvalues and exits, if not;*

• creates a numeric array with `N` *rows and* `L` *columns and fills it with independent random numbers with the standard normal distribution $N(0,1)$;*

• treats each row of this array as a vector, and multiplies it by the square root of the matrix `Sigma` *(as computed in question* **I.1** *above) and adds the vector* `Mu` *to the result;*

• returns the random array modified in this way.

The array produced by the function `vnorm` *is a sample of size* `N` *of* `L`*-dimensional random vectors (arranged as rows of the matrix outputted by* `vnorm`*) with the normal distribution with mean* `Mu` *and variance/covariance matrix* `Sigma`. *Indeed, this function implements the following simple fact reviewed during the lectures:*

If X is an L-dimensional normal vector with mean 0 and covariance matrix given by the $L \times L$ identity matrix (i.e. if all the L entries of X are independent $N(0,1)$ random variables), then:

$$Y = \mu + \Sigma^{1/2} X$$

is an L-dimensional normal vector with mean μ and variance/covariance matrix Σ.

NOTES & COMPLEMENTS

This chapter concentrated on multivariate distributions and on dependence between random variables. The discussion of the various correlation coefficients is modeled after the standard treatments which can be found in most multivariate statistics books. The originality of this chapter lies in the statistical analysis of the notion of dependence by way of copulas. The latter are especially important when the marginal distributions have heavy tails, which is the case in most financial applications as we saw in the first chapter. The recent renewal of interest in the notion of copula prompted a rash of books on the subject. We shall mention for example the monograph [65] of R.B. Nelsen, or the book of D. Drouet Mari and S. Kotz [29]. We refer the interested reader to their bibliographies for further references on the various notions of dependence and copulas.

The `S-Plus` methods used in this chapter to estimate copulas and generate random samples from multivariate distributions identified by their copulas were originally developed for the library EVANESCE [17] developed by J. Morrisson and the author. As explained in the

Notes & Complements of Chapter 1 this library has been included in the `S+FinMetrics` module of the commercial `S-Plus` distribution.

To the best of my knowledge, the first attempt to apply principal component analysis to the yield curve is due to Litterman and Scheinkmann [57]. Rebonato's book [70], especially the short second chapter, and the book [3] by Anderson, Breedon, Deacon, Derry, and Murphy, are good sources of information on the statistical properties of the yield curve. Discussions of interest rate swap contracts and their derivatives can also be found in these books. The reader interested in a statistical discussion of the fixed income markets with developments in stochastic analysis including pricing and hedging of fixed income derivatives, can also consult the monograph [21]. An application of PCA to variable selection in a regression model is given in Problem 4.17 of Chapter 4.

The decomposition of a data set into its principal components is known in signal analysis as the Karhunen-Loève decomposition, and the orthonormal basis of principal components is called the Karhunen-Loève basis. This basis was identified as optimal for compression purposes. Indeed, once a signal is decomposed on this basis, most of the coefficients are zero or small enough to be discarded without significantly affecting the information contained in the signal. Not surprisingly, the optimality criterion is based on a form of the entropy of the set of coefficients. PCA is most useful for checking that data do contain features which are suspected to be present. For this reason, some authors suggest to remove by regression the gross features identified by a first PCA run, and to then run PCA on the residuals. PCA has been successfully used in many applications, especially in signal and image analysis. For example, the Notes & Complements to Chapter 5 contain references to studies using PCA in brain imaging.

Part II

REGRESSION

3

PARAMETRIC REGRESSION

This chapter gives an introduction to several types of regression: simple and multiple linear, as well as simple polynomial and nonlinear regression. In all cases we identify the regression function in a parametric family, thus the title of the chapter. We also address issues of robustness, and we illustrate these concepts with a parallel comparison of the least squares and the least absolute deviations regression methods. Even though we introduce regression from a data smoothing point of view, we interpret the results in terms of statistical models, and we derive the statistical inference and diagnostic tools provided by the theory behind these statistical models. The chapter ends with a thorough discussion of the parametric estimation of the term structure of interest rates based on the Nelson-Siegel and Swensson families. As before, we try to work from examples, introducing theoretical results as they become relevant to the discussions of the analyzes which are used as illustrations.

3.1 SIMPLE LINEAR REGRESSION

Although multiple linear regression is ubiquitous in economic and in econometric applications, simple linear regression does not play a very central role in the financial arena. Nevertheless, its mathematical underpinnings and inferential results are so important that we must present them, even if the conclusions drawn from its use in practical financial applications are not always earth-shattering. As in earlier chapters, we choose particular data sets to illustrate the statistical concepts and techniques which we introduce. In the first part of this chapter we choose to work with the values of an energy index, as they relate to the values of several utilities, but as a disclaimer, it is important to emphasize that this choice was made for illustration purposes only. Most financial data come naturally in the form of time series, and the serial dependencies contained in the data may not be appropriate for some forms of regression analysis. With this in mind, the reader should be prepared to have a critical attitude toward the results produced by the algorithms introduced in this chapter.

Case in point, we shall abandon the original form of our first regression example after a couple of sections. We will switch to a form of the data more amenable to regression as soon as we understand the theoretical underpinnings of a regression model. Our choice to work by trial and error was made for didactic reasons.

The original index data were computed following the methodology and data used by Dow Jones to produce its Total Market Index. Our goal is to investigate to which extent two large companies can influence the index of an entire economic sector. We chose the energy/utility sector because of its tremendous growth in the second half of the nineties. However, we will stay away from what happened during the California energy crisis and after ENRON's bankruptcy.

3.1.1 Getting the Data

A very convenient way to store data in an `S` object is to use the structure of `data frame`. Whether the data set is already part of the `S-Plus` distribution or it has been read from a disk file, or downloaded from the internet, (see the `S-Plus` tutorial at the end of the book for details), the way to make such a data set available to the current `S-Plus` session is to `attach` the `data frame` in question. This is accomplished with the command:

```
> attach(UTIL.index)
```

Once this has been done, the values which are stored in the columns of the data frame `UTIL.index` can be manipulated by `S-Plus` commands using their names. Typing the name of the data frame displays the data in the command window, so the command:

```
> UTIL.index
```

produces the output:

```
           ENRON.index DUKE.index UTILITY.index
01/04/1993    135.0000   104.2857      99.94170
01/05/1993    135.3714   103.5714      99.49463
01/06/1993    132.8571   104.2857      99.86034
01/07/1993    130.7143   103.5714      98.70023
    ......      ......     ......        ......
12/28/1993    166.4571   125.0000     107.15202
12/29/1993    170.7429   123.9429     106.75023
12/30/1993    169.3143   124.2857     106.12351
12/31/1993    165.7143   121.0857     104.95227
```

As always in a `data.frame` object, the first row contains the names of the column variables, which in the case at hand, are the indexes computed from the share values and the capitalizations of ENRON and DUKE Energy, and a utility index computed from all the publicly traded large utilities. The left most column contains the names of the row variables, which in the present situation, are the dates of the quotes. As we can see, they cover the year 1993. I chose this year in order to stay away from the speculative period of the late nineties during which the energy sector heated up to unhealthy levels. The actual data consist of the entries of the numerical matrix whose rows and columns we just described. The dimensions (i.e. the number of rows and the number of columns) of this numerical matrix can be obtained with the `S-Plus` command `dim`. For example, the result of the command:

```
> dim(UTIL.index)
[1] 260   3
```

tells us that the data frame `UTIL.index` has 260 rows and 3 columns.

3.1.2 First Plots

It is always a good idea to look at graphical representations of the data whenever possible. In the present situation one can split the graphic window in two columns and place a scatterplot of the `UTILITY.index` variable against the `ENRON.index` variable on the left entry of this 1 by 2 matrix of plots, and the scatterplot of the `DUKE.index` and the `UTILITY.index` variables on the right. This is done with the commands:

```
> par(mfrow=c(1,2))
> plot(ENRON.index,UTILITY.index)
> plot(DUKE.index,UTILITY.index)
> par(mfrow=c(1,1))
```

The last command resets the graphics window to a 1 by 1 matrix of plots for future use. The results are shown in Figure 3.1.

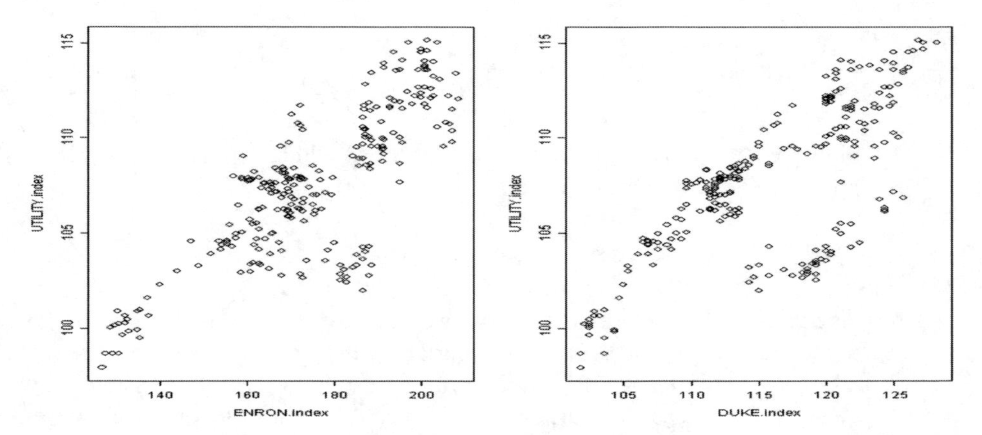

Fig. 3.1. Scatterplot of `UTILITY.index` against `ENRON.index` (left) and of `UTILITY.index` against `DUKE.index` (right) of the `UTIL.index` data set.

Plots of this type (one variable versus another) are called *scatterplots.* They are very convenient for getting a feeling of the dependence/independence of two variables, as we saw in Chapter 2. When the data frame contains more than two variables, it is possible to get all these 2 by 2 plots simultaneously with the command `pairs`

```
> pairs(UTIL.index)
```

The result is shown in Figure 3.2. The plots on the diagonal entries of this matrix of

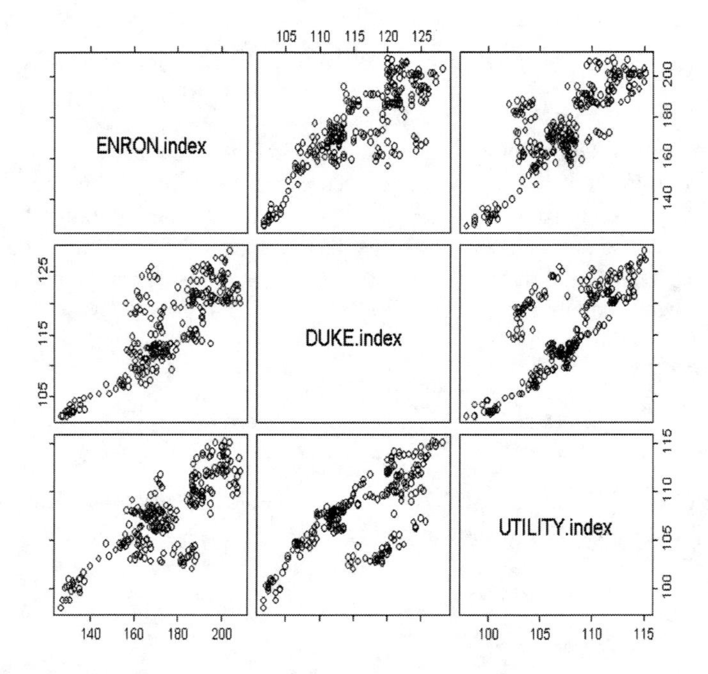

Fig. 3.2. Matrix plot of the pair scatterplots of the variables in the `UTIL.index` data frame.

plots are not given for an obvious reason: they would be un-informative since all the points would have to be on the main diagonal.

3.1.3 **Regression Set-up**

As expected, the scatterplots give strong evidence of relationships between the values of the utility index and the two other variables. Indeed, the points would be scattered all over the plotting area if it weren't for these relationships. Regression is a way to quantify these dependencies, and we proceed to fix the notation which we will use when setting up a regression problem.

The general form of the (simple) regression setup is given by observations

$$y_1, y_2, \ldots\ldots, y_n$$

of one variable, which we call the *response* variable (think for example of the $n = 260$ values of the utility index in the case discussed in this section), and of observations:

$$x_1, x_2, \ldots\ldots, x_n$$

of an explanatory variable, which we call the *regressor*. For the sake of definiteness, we first consider the case of the $n = 260$ values of the index computed for ENRON. Regression is the search for a functional relationship of the form:

$$y \approx \varphi(x)$$

so that the actual observations satisfy:

$$y_i = \varphi(x_i) + \text{error term}, \qquad i = 1, \ldots, n$$

where the error term is hopefully small, but most importantly unpredictable in a sense which we will try to make clear later on.

Warning. The regressor variable is sometimes called the *independent variable* to be consistent with the terminology *dependent variable* often used for the response. We believe that this terminology is very misleading and we shall refrain from using it.

The regression terminology is well established and quite extensive, and we shall conform to the common usage. For example, we talk of

- *simple linear regression* when we have only one regressor and when the dependence is given by an affine function of the form $\varphi(x) = \beta_0 + \beta_1 x$. The regression problem is then to estimate the parameters β_1 giving the slope of the regression line, and β_0 giving the intercept, and possibly some statistical characteristics of the noise such as its variance for example;
- *simple nonlinear regression* when we have only one regressor and when the dependence is given by a general (typically nonlinear) function $\varphi()$;
- *simple spline regression* when we have only one regressor and when the dependence is given by a function φ constructed from spline functions;
- *multiple regression* (whether it is linear or nonlinear) when we have several regressors which we usually bind in a vector of *explanatory variables.*

Coming back to our utility index example, and trying to explain the values of the utility index (representing the entire industrial sector) from the values of the ENRON and DUKE indexes, we can choose the entries of column variable `UTILITY.index` for the values of y_i and

- looking at the scatterplot in the left pane of Figure 3.1, we can decide to choose the entries of `ENRON.index` for the values of x_i and perform a simple linear regression of `UTILITY.index` against `ENRON.index`, searching for real numbers β_0 and β_1 and a regression function φ of the form $\varphi(x) = \beta_0 + \beta_1 x$;

- alternatively, looking at the scatterplot in the right pane of Figure 3.1, we can decide that the variable `DUKE.index` better explains the changes in the response `UTILITY.index`, and perform a simple linear regression of `UTILITY.index` against the regressor `DUKE.index`;
- Finally, suspecting that a cleverly chosen combination of the values of the variables `ENRON.index` and `DUKE.index` could provide a better predictor of the values of `UTILITY.index`, we could decide to opt for a multiple linear regression of `UTILITY.index` against both variables `ENRON.index` and `DUKE.index`. In this case, we would choose to view the entries of the variable `ENRON.index` as observations of a first explanatory variable $x^{(1)}$, the entries of `DUKE.index` as observations of a second explanatory variable $x^{(2)}$, bundle these two explanatory variables into a bivariate vector $\mathbf{x} = (x^{(1)}, x^{(2)})$ and search for real numbers β_0, β_1 and β_2 and a function φ of the form:

$$y = \varphi(x^{(1)}, x^{(2)}) = \beta_0 + \beta_1 x^{(1)} + \beta_2 x^{(2)}.$$

We will do just that in this chapter. Notice that we purposely restricted ourselves to linear regressions. Indeed, the shapes of the clouds of points appearing in the scatterplots of Figure 3.1 and Figure 3.2 are screaming for linear regression, and it does not seem reasonable to embark on a search for nonlinear regression functions for the data at hand.

Going through the program outlined by the first set of bullets will keep us busy for the next two chapters. Despite the classification introduced by this terminology, regressions are most often differentiated by the specifics of the algorithms involved, and regression methods are frequently organized according to the dichotomy *parametric regression methods* versus *nonparametric regression methods* which we shall define later on.

Coming back to the general setup introduced in this subsection, we outline search strategies for the regression function $\varphi()$. A standard approach is to associate a *cost* to each admissible candidate φ, and to choose the candidate (hopefully it will exist and be defined unambiguously) which minimizes this cost. Even though there are many possible choices we shall concentrate on the two most common ones:

$$\mathcal{L}_2(\varphi) = \sum_{j=1}^{n} [y_j - \varphi(x_j)]^2 \tag{3.1}$$

and:

$$\mathcal{L}_1(\varphi) = \sum_{j=1}^{n} |y_j - \varphi(x_j)|. \tag{3.2}$$

Since both cost functions are based on the sizes of the differences $y_i - \varphi(x_i)$, the resulting function φ provides the best possible fit *at the observations* x_i. Given a family $\mathbf{\Phi}$ of functions φ, and given the data, the function $\varphi \in \mathbf{\Phi}$ which minimizes $\mathcal{L}_2(\varphi)$ over $\mathbf{\Phi}$ is called the least squares regression on $\mathbf{\Phi}$, and the function which minimizes $\mathcal{L}_1(\varphi)$ over $\mathbf{\Phi}$ is called the least absolute deviations regression over $\mathbf{\Phi}$.

We shall often use the abbreviations L2 and L1 regression respectively, but the reader should be aware of the fact that the alternative abbreviations LS and LAD are also used frequently in the literature, and we may be using them from time to time. We first consider cases for which $\mathbf{\Phi}$ is a set of linear functions.

Remark. It is important to emphasize that this L2/L1 duality is not new. We already learned it in introductory statistics for the solution of the simpler problem of the statistical estimation of the location of a data sample. Indeed, given a sample x_1, x_2, ..., x_n of real numbers, the classical estimates of the location given by the sample mean $\overline{x}$ and the sample median $\hat{\pi}_{.5}$ are known to be the solutions of the minimization problems:

$$\overline{x} = \arg\min_m \sum_{i=1}^n |x_i - m|^2$$

and

$$\hat{\pi}_{.5} = \arg\min_m \sum_{i=1}^n |x_i - m|$$

respectively. As we are about to see, least squares simple linear regression and least absolute deviations simple linear regression are mere generalizations of the two location estimation problems just mentioned. We shall revisit this example in Subsection 3.2.2 below.

3.1.4 **Simple Linear Regression**

As explained in the introduction, the setup is given by input data of the form

$$(x_1, y_1), \dots, (x_n, y_n)$$

where:

- n denotes the sample size,
- the x_i's denote the observations of the explanatory variable
- the y_i's denote the observations of the response variable,

and the problem is to find a straight line (whose equation will be written as $y = \beta_0 + \beta_1 x$) summarizing the data as faithfully as possible. For the purposes of the present discussion we shall limit ourselves to the two cost functions introduced earlier.

Least Squares (Simple) Regression

We first consider the case of the least squares regression and we illustrate its use in the case of the utility indexes data. We shall see later that `S-Plus` has powerful methods for performing least squares linear regression, but for the time being, we shall restrict ourselves to the function `lsfit` in order to emphasize the parallel with the least absolute deviations regression method. The plots of Figure 3.3 were obtained with the commands:

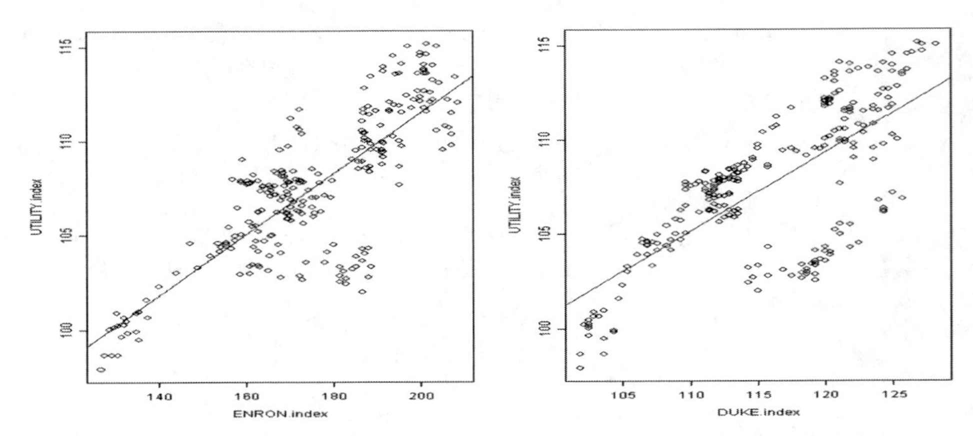

Fig. 3.3. Least Squares Regression of the utility index on the ENRON index (left) and on the DUKE index (right).

```
> UEl2 <- lsfit(UTILITY.index,ENRON.index)
> UDl2 <- lsfit(UTILITY.index,DUKE.index)
> par(mfrow=c(1,2))
> plot(ENRON.index,UTILITY.index)
> abline(UEl2)
> plot(DUKE.index,UTILITY.index)
> abline(UDl2)
> par(mfrow=c(1,1))
```

The function `lsfit` produces objects of class `lsfit` containing all the information necessary to process the results of the regression. The function `abline` adds a line to an existing plot. It is a very convenient way to add a regression line to a scatterplot. In particular, the coefficients β_0 and β_1 of the regression line can be extracted from the object produced by `lsfit` by using the extension `$coef`.

```
> UEl2$coef
 Intercept         X
  79.22646 0.1614243
 > UDl2$coef
 Intercept         X
 59.00217 0.4194389
```

As already explained, we chose to use the `S-Plus` function `lsfit` to emphasize the parallel with the least absolute deviations regression and the function `l1fit` provided by `S-Plus` for that purpose. The use of `lsfit` will eventually be abandoned when we come across the more powerful method `lm` designed to fit more general linear models.

Least Absolute Deviations (Simple) Regression

For the sake of comparison, we produce the results of the least absolute deviations regression in the same format. We shall often use the abbreviation LAD or L1 for *least absolute deviations* while *least squares* is usually abbreviated as LS or L2. The

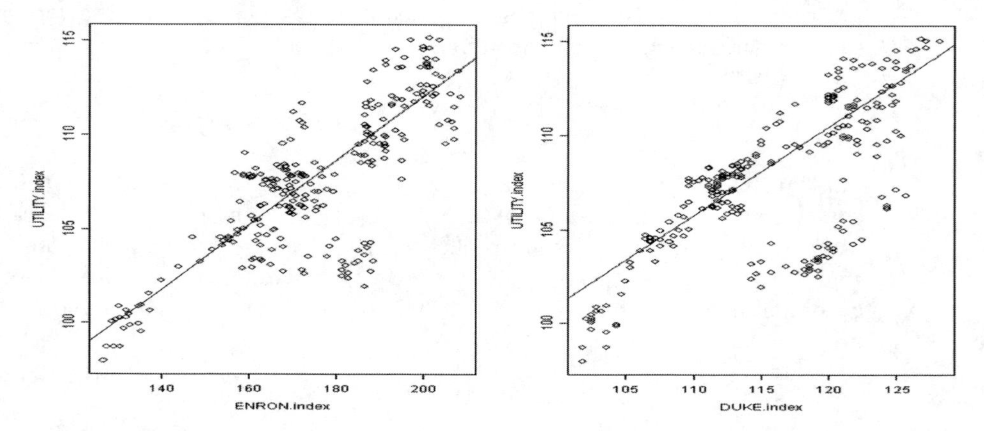

Fig. 3.4. Simple Least Absolute Deviations regressions of the utility index on the ENRON index (left) and on the DUKE index (right).

plots of Figure 3.4 were obtained with the commands:

```
> UEl1 <- l1fit(ENRON.index,UTILITY.index)
> UDl1 <- l1fit(DUKE.index,UTILITY.index)
> par(mfrow=c(1,2))
> plot(ENRON.index,UTILITY.index)
> abline(UEl1)
> plot(DUKE.index,UTILITY.index)
> abline(UDl1)
> par(mfrow=c(1,1))
```

The graphical results seem to be very similar, and we will need to investigate more to understand the differences between these two regression methods. As before we can print the coefficients of the regression line:

```
> UEl1$coef
Intercept         X
  77.92709 0.1706265
 > UDl1$coef
 Intercept         X
  53.98615 0.4706897
```

We notice that the coefficients produced by the least absolute deviations regression are different from those produced by the least squares regression. Nevertheless, it is difficult at this stage to assess how serious these differences are, and how statistically significant they may be. More on that later.

3.1.5 **Cost Minimizations**

In a simple linear regression problem, the least squares regression line is given by its slope $\hat{\beta}_1$ and intercept $\hat{\beta}_0$ which minimize the function:

$$(\beta_0, \beta_1) \hookrightarrow \mathcal{L}_2(\beta_0, \beta_1) = \sum_{i=1}^{n} (y_i - \beta_0 - \beta_1 x_i)^2. \tag{3.3}$$

This function is quadratic in β_0 and β_1, so it is easily minimized. Indeed, one can compute explicitly (in terms of the values of $x_1, \ldots, x_n$ and $y_1, \ldots, y_n$) the partial derivatives of $\mathcal{L}_2(\beta_0, \beta_1)$ with respect to β_0 and β_1, and setting these derivatives to zero gives a system of two equations with two unknowns (often called the first order conditions in the jargon of optimization theory) which can be solved. The solution is given by the formulae:

$$\hat{\beta}_1 = \frac{\mathrm{cov}(x, y)}{\sigma_x^2} \qquad \text{and} \qquad \hat{\beta}_0 = \overline{y} - \frac{\mathrm{cov}(x, y)}{\sigma_x^2}\overline{x}. \tag{3.4}$$

Recall that the empirical means $\overline{x}$ and $\overline{y}$ were already defined in (2.8) of Chapter 2, as well as $\mathrm{cov}(x, y)$ and σ_x which were defined in (2.9). This is in contrast with the LAD regression for which we cannot find formulae for the optimal slope and intercept parameters. The least absolute deviations regression line is determined by the values of the intercept $\hat{\beta}_0$ and the slope $\hat{\beta}_1$ which minimize the function:

$$(\beta_0, \beta_1) \hookrightarrow \mathcal{L}_1(\beta_0, \beta_1) = \sum_{i=1}^{n} |y_i - \beta_0 - \beta_1 x_i|. \tag{3.5}$$

Unfortunately, one cannot expand the absolute value as one does for the squares, and solving for vanishing derivatives cannot be done by closed-form formulae in this case. Nevertheless, it is relatively easy to show that the function $(\beta_0, \beta_1) \hookrightarrow \mathcal{L}_1(\beta_0, \beta_1)$ is convex and hence, that it has at least one minimum. But because it is not strictly convex, uniqueness of this minimum is not guaranteed, and as in the classical case of the median, we may have entire intervals of minima. Efficient algorithms exist to compute the minima of $\mathcal{L}_1(\beta_0, \beta_1)$. They are based on a reduction of the problem to a classical linear programming problem. But the lack of uniqueness and the lack of explicit formulae is still a major impediment to widespread use of the L1 method.

3.1.6 **Regression as a Minimization Problem**

This last subsection does not contain any new technical material, and as a consequence, it can be skipped on a first reading. Its prose is intended to shed some light on the importance of optimization in statistical estimation, and to use this opportunity to highlight some of the nagging problems coming with the territory.

The approach to regression which we advocate in this section is based on the idea of *smoothing* of a cloud of points into a curve that captures the main features of the data. In this spirit, a simple regression problem can be formulated in the following way. We start from observations:

$$(x_1, y_1), \ldots\ldots, (x_n, y_n),$$

and we try to find a function $x \hookrightarrow \varphi(x)$ which minimizes a loss or penalty, say $\mathcal{L}(\varphi)$, associated to each specific choice of the candidate φ. This candidate can be picked in a specific class of functions, say $\mathbf{\Phi}$, e.g. the set of affine functions of x when we consider linear regression, the set of polynomials when we deal with a polynomial regression problem later in this chapter, ... What distinguishes parametric regression from nonparametric regression is the fact that the class $\mathbf{\Phi}$ can be described in terms of a small number of parameters. For example, in the case of simple linear regression, the parameters are usually chosen to be the slope and the intercept, while in the case of polynomial regression the parameters are most often chosen to be the coefficients of the polynomial. See nevertheless Subsection 3.6.3 later in this chapter. We shall use the notation θ for the parameter used to label the candidate (i.e. the set $\Theta = \{\theta\}$ replaces the set $\mathbf{\Phi} = \{\varphi\}$ via a correspondence of the type $\theta \leftrightarrow \varphi_\theta$. In the case of simple linear regression this amounts to setting:

$$\theta = (\beta_0, \beta_1) \qquad \text{and} \qquad \varphi_\theta(x) = \beta_0 + \beta_1 x.$$

Notice that the present discussion is not limited to the case of a single scalar regressor variable. Indeed, the regression variables x_i can be multivariate as when each observed value of x_i is of the form $\mathbf{x}_i = (x_{i,1}, \ldots, x_{i,p})$. As explained earlier, the parameter θ is likely to be multivariate. In the discussion above the regression problem always reduces to minimization of a function of the form:

$$\theta \hookrightarrow \mathcal{L}(\theta). \tag{3.6}$$

This problem does not have a clear-cut answer in general. We say that the problem is *well posed* if there exists at least one value, say $\theta_0 \in \Theta$, such that:

$$\mathcal{L}(\theta_0) = \inf_{\theta \in \Theta} \mathcal{L}(\theta) \tag{3.7}$$

When (as is often the case) Θ is a subset of a vector space, then the problem is often well posed when the loss function (3.6) is convex. Moreover, there is a unique value of θ realizing the minimum (3.7) whenever the loss function (3.6) is actually strictly convex.

The Search for a Minimum

As we saw in the case of least squares regression, the search for an optimal value of the parameter θ is usually replaced by the search for values of θ at which all the partial derivatives of the function $\mathcal{L}$ vanish. In other words one looks for solutions of the equation:

$$\nabla\mathcal{L}(\theta) = 0 \tag{3.8}$$

where the notation ∇ is used for the gradient, i.e. the vector of partial derivatives. Notice that this equation is in fact a system of k-equations when θ is k-dimensional, since in this case the gradient is made of k partial derivatives and equation (3.8) says that all of them should be set to 0. A solution of (3.8) is called a *critical point.* Such a strategy for the search of a minimum is reasonable because, if

$$\theta_0 = \arg\inf_{\theta\in\Theta} \mathcal{L}(\theta)$$

is a point at which the minimum is attained, and if the function $\theta \hookrightarrow \mathcal{L}(\theta)$ is differentiable, then all the partial derivatives of $\mathcal{L}$ vanish for $\theta = \theta_0$. Unfortunately, the converse is not true in the sense that there might be critical points which are not solutions of the minimization problem. Indeed, the gradient of $\mathcal{L}$ vanishes at any local minimum, (or even at any local maximum for that matter) even if it is not a global minimum, and this is a source of serious problems, for there is no good way to find out if a critical point is in fact a global minimum, or even to find out how good or bad a proxy it can be for such a global minimum.

To make matters worse, one is very often incapable of solving equation (3.8). Indeed except for the very special case of least squares linear regression (see the previous subsection for a re-cap of all the formulae derived in this case), it is not possible to find closed form formulae for a solution, and one is led to compute numerical approximations by iterative procedures. As is often the case, stability and convergence problems plague such a strategy.

3.2 Regression for Prediction & Sensitivities

Regression is often performed to explain specific features of the existing (i.e. already collected) data. This involves looking at the values of the response variables given by the evaluation of the regression function $\varphi(x)$ for values of the regressor(s) contained in the data set. However, regression is also often used for prediction purposes. The functional relationship between the response and the regressor variables identified by the regression is taken advantage of to predict what the response would be for values of the regressor(s) for which the responses have not yet been observed. Filling in missing data appears as an intermediate situation between these two uses of regression.

3.2.1 **Prediction**

We give a first informal presentation of the notion of prediction operator. We shall make this concept more precise later in the book. The purpose of a prediction operator is to produce the best *reasonable* guess of a random quantity. It is a good idea to have in mind the least squares error criterion as a measure of how *reasonable* a guess can be. So such a prediction operator, say P, will take a random variable, say Y, into a number $P(Y)$ which serves as this best guess. In the absence of any other information, and if one uses the least squares criterion, the operator P is given by the expectation operator in the sense that $P(Y) = \mathbb{E}\{Y\}$. That is, the number $m = \mathbb{E}\{Y\}$ is the number for which the criterion $\mathbb{E}\{|Y - m|^2\}$ is minimum. Other criteria lead to other prediction operators. It is a very intuitive fact (with which we have been familiar since our first introductory statistics class) that, in the absence of any extra information, the best (in the least squares sense) predictor of a random quantity is its expected value. We shall revisit (and prove) this statement in the next subsection.

We are interested in prediction when partial information is available. Apart from the fact that we will thus work with conditional expectations instead of the usual expectations, nothing should be different, If the information x is available, we shall denote a prediction operator by P_x. As explained above, we should think of $P_x(Y)$ as the best predictor of the random variable Y in the presence of the information x. The following properties are imposed on such an operator:

- $P_x(Y)$ should be a *linear* function of Y
- If Y is not really random, and it is known that $Y = y$ for a given number y, then one should have $P_x(Y) = y$.

Obviously, the expected valued (conditioned on the information x) is a prediction operator which satisfies these properties. But many other operators are possible. Very often, we will also demand that $P_x(Y) = 0$ whenever Y represents a noise term, and this is often modeled as Y having mean zero. For most of the applications discussed in this book, the prediction operators will be given by expectations and conditional expectations.

We will still use the data on the utility indexes for illustration purposes, but we will concentrate for the next little while, on the analysis of the dependence of `UTILITY.index` upon `DUKE.index`. So for the time being, we might as well forget that we also have the variable `ENRON.index` to explain the values of the overall utility index.

For an example of prediction based on regression results, let us imagine that back in January 1994, we looked into our crystal ball, and we discovered that the stock of DUKE energy was going to appreciate in a dramatic fashion over the next two years. The DUKE index would increase accordingly and this would presumably imply a significant increase in the utility index as well. But how could we quantify these qualitative statements. To be more specific, could we predict the value of the utility index if we knew that the value of the DUKE index was 150?

Using the results of the regressions performed earlier, one needs only to compute the value of $\beta_0 + \beta_1 x$ for $x = 150$ and β_0 and β_1 determined by each regression. We compute predictions both for the least squares and the least absolute deviations regressions.

```
> NEWDUKE <- 150
> PRED2 <- UDl2$coef[1]+UDl2$coef[2]*NEWDUKE
> PRED2
  121.918
> PRED1 <- UDl1$coef[1]+UDl1$coef[2]*NEWDUKE
> PRED1
  124.5896
```

One can see that the two predictions are different and deciding which one to trust is a touchy business. We cannot resolve this issue at this stage, especially since we cannot talk about confidence intervals yet. Nevertheless, even though we do not have the tools to justify the claims we are about to make, and in order to shed some light on the reasons why the results are different, we do venture the following explanations: The value of the explanatory variable for which we seek a value of the response variable, i.e. 150, is far from the bulk of the data from which the regression lines were constructed. Indeed the values of the explanatory variable ranged between 101 and 128 during the year 1993. Predictions that far from the available data can be very unstable.

3.2.2 **Introductory Discussion of Sensitivity and Robustness**

Before relying on the results of a regression for decision making, it is a good idea to understand how stable and/or reliable the coefficients of a regression are. In each case (i.e. for the least squares and least absolute deviations regressions) we investigate the sensitivity of the values of the slope and the intercept when variations in the observations are present (or introduced). We shall also illustrate the sensitivity of our predictions.

A (simple) linear regression is said to be significant when the results of the regression confirm the influence of the explanatory variable on the outcome of the response, in other words when the slope is determined to be nonzero. When this is not the case, the regression line is horizontal and the value of the explanatory variable has no effect on the response, the latter being merely explained by the measure of location given by the intercept. We shall introduce the notion of robustness by first discussing this case.

For a set of n real numbers $y_1, \ldots, y_n$, the most common measure of location is the sample mean:

$$\overline{y} = \frac{1}{n}\sum_{i=1}^{n} y_i.$$

As we already pointed out in Subsection 3.1.5, the sample mean $\overline{y}$ is the solution of a least squares minimization problem since:

$$\overline{y} = \arg\min_{m} \sum_{i=1}^{n} |y_i - m|^2.$$

In other words, the sample mean is to the location problem what the slope and the intercept of the least squares regression line are to the (simple) regression problem.

Next to the sample mean, the median also enjoys great popularity. Bearing some annoying non-uniqueness problems when n is even (problems which are solved by agreeing on a specific convention to handle this case), the median $\hat{\pi}_{.5}$ is an element of the data set which splits the data into two subsets of approximately the same size. Like the mean, it also appears as the solution of a minimization problem since:

$$\hat{\pi}_{.5} = \arg\min_{m} \sum_{i=1}^{n} |y_i - m|.$$

In other words, the sample median is to the location problem what the slope and the intercept of the least absolute deviations regression line are to the (simple) regression problem. We claim that the median is much less sensitive than the mean to perturbations or errors in the data, and we illustrate this fact on the following simple example. Let us consider the data:

$$y_1 = 1.5 \quad y_2 = 0.7 \quad y_3 = 5.1, \quad y_4 = 2.3, \quad y_5 = 3.4.$$

The mean of this sample is:

$$\overline{y} = (y_1 + y_2 + y_3 + y_4 + y_5)/5 = (1.5 + 0.7 + 5.1 + 2.3 + 3.4)/5 = 13/5 = 2.6$$

while the median is obtained by first ordering the data:

$$y_{(1)} = .7 < y_{(2)} = 1.5 < y_{(3)} = 2.3 < y_{(4)} = 3.4 < y_{(5)} = 5.1$$

and then picking the *mid-value* $\hat{\pi}_{.5} = 2.3$. Let us now assume the value $y_5 = 3.4$ is erroneously recorded as $y_5 = 13.4$. In this case the new mean is increased by 2 since:

$$\overline{y} = 1.5 + 0.7 + 5.1 + 2.3 + 13.4)/5 = 23/5 = 4.6$$

while the *value of the median does not change!* In fact the median will not change as long as the changes do not affect the number of values greater than the median. This feature is extremely useful in the case of undesirable erroneous measurements, and/or uncontrollable perturbations of the data. Indeed, it is plain to see that the mean can be made arbitrarily large or small by appropriately changing a single measurement! With this introductory example in mind, we compare the robustness of the least squares regression to the least absolute deviations regression.

3.2.3 **Comparing L2 and L1 Regressions**

As in the case of the discussion of the robustness of the measures of location (mean and median) we try to quantify (or at least visualize) the effects that perturbations of the data have on the regression lines.

In order to do so, we create a new data set, say `NEWUTIL.index`, by modifying the first twenty entries of `UTILITY.index`, we then perform simple least squares and least absolute deviations regressions of this new index against the DUKE index, and we compare the resulting regression lines to the lines obtained with the original data. The results are given in Figure 3.5. They were produced using the following commands:

```
> GOOD <- 21:260
> NEWUTIL.index <- c(UTILITY.index[-GOOD]-10,
                          UTILITY.index[GOOD])
> NUDl2 <- lsfit(DUKE.index,NEWUTIL.index)
> NUDl1 <- l1fit(DUKE.index,NEWUTIL.index)
> par(mfrow=c(1,2))
> plot(DUKE.index,NEWUTIL.index)
> abline(UDl2,lty=4)
> abline(NUDl2)
> plot(DUKE.index,NEWUTIL.index)
> abline(UDl1,lty=4)
> abline(NUDl1)
> par(mfrow=c(1,1))
```

The first command defines the row numbers for which we are not going to change the value of `UTILITY.index`. The second command actually creates the new utility index by concatenating two vectors with the `S-Plus` function `c`. The first vector is formed by the first twenty entries of `UTILITY.index` from which we subtract 10, while the second vector is formed by the remaining entries of `UTILITY.index`. Notice how we used the subscript `-GOOD` to extract the entries whose row numbers are not in `GOOD`. The next two commands perform the least squares and the least absolute deviations regressions of the new utility index against the old DUKE index, and the remaining commands produce the plots of Figure 3.5. We used the parameter `lty` to draw the original L2 and L1 regressions as dash lines. The differences between the two regression methods are clearly illustrated on these plots. The L2 line changed dramatically while the L1 line remained the same. As in the case of the median, changing a small number of values did not affect the result of the L1 fitting/estimation procedure. This robustness of the least absolute deviations regression can be extremely useful. Indeed, there are times when one wants the estimations and predictions to change with changing data, however, with noisy data, it is generally not a good idea to use estimation and prediction procedures which are too sensitive to small changes, mostly because the latter are very likely due to the noise, and for this reason, they should be ignored.

We can also compare the effects of data perturbations on the performance of the predictions produced by the regressions. If we revisit the prediction of the value of the utility index when the DUKE index reaches the value 150, we now find:

```
> NEWDUKE <- 150
> NPRED2 <- NUDl2$coef[1]+NUDl2$coef[2]*NEWDUKE
> NPRED2
    128.8155
```

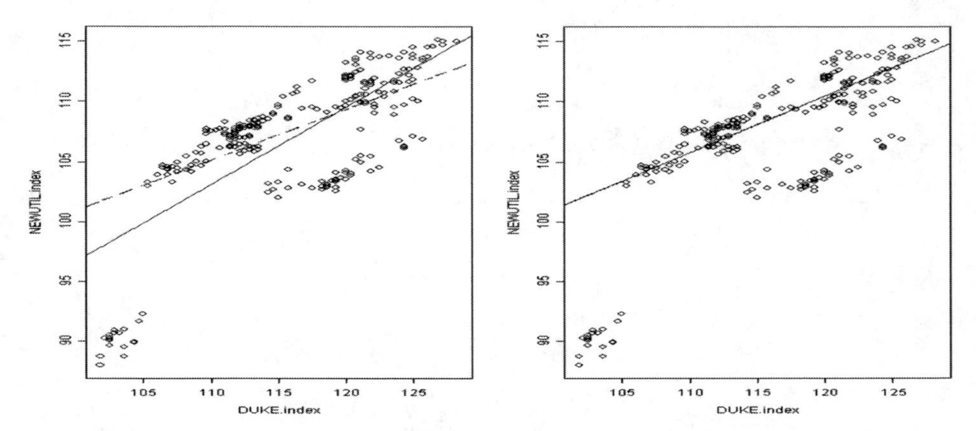

Fig. 3.5. Left: Simple Least Squares Regression of the modified utility index against the DUKE index, with the original least squares regression line superimposed as a dashed line. The two lines are very different. Right: Same thing for the least absolute deviations regressions. The two lines are identical !

```
> NPRED1 <- NUDl1$coef[1]+NUDl1$coef[2]*NEWDUKE
> NPRED1
   124.5896
```

from which we see that the L1 prediction does not change while the L2 prediction goes from 121.9 to 128.8. As we already explained, it is difficult to assess how bad such a fluctuation is, but at a very intuitive level, it may be comforting to see that some prediction systems do not jump all over the place when a few data points change! More on that later when we discuss statistical inference issues.

We now proceed to illustrate the differences between L1 and L2 regression methods with the coffee data. See also the Notes & Complements at the end of the chapter for references to textbooks devoted to the important aspects of robustness in statistical inference.

3.2.4 **Taking Another Look at the Coffee Data**

We revisit the case of the coffee daily price data already considered in the previous chapters. A scatterplot of the daily log-returns of the two commodities shows that there is a strong correlation between the two, and that a linear relation may hold. We compute the least squares and the least absolute deviations regressions of the Brazilian coffee daily log-returns against the Colombian ones and plot the regression lines so-obtained.

```
> plot(CLRet,BLRet)
```

```
> BCL2 <- lsfit(CLRet,BLRet)
> BCL1 <- l1fit(CLRet,BLRet)
> abline(BCL2)
> abline(BCL1,lty=3)
```

We give the results in Figure 3.6. The least absolute deviations regression captures

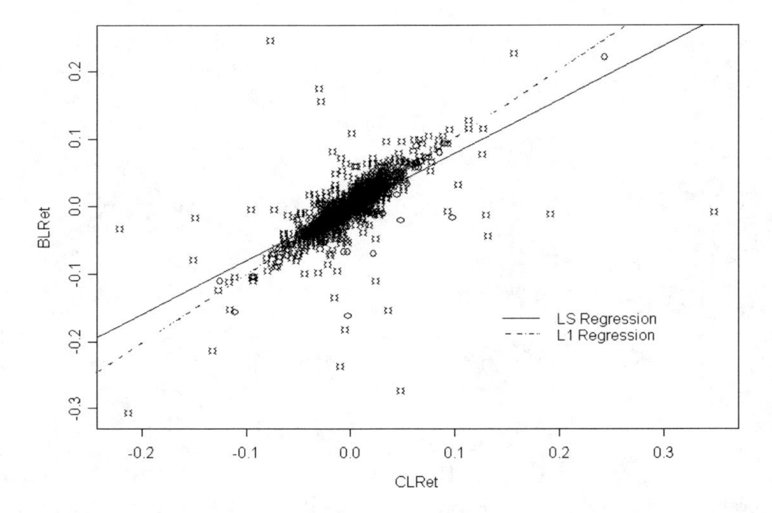

Fig. 3.6. Least squares and least absolute deviations regressions of the Brazilian daily log-returns against the Colombian daily log-returns.

the upward trend of the cloud of points more faithfully than the least squares regression. The few low values of the Colombian coffee seen on days the Brazilian coffee had a large return (i.e. the few points on the middle right part of the scatterplot) are influencing the results of the least squares regression, pulling the regression line down. On the other hand, these points do not seem to have much influence on the least absolute deviations regression line. We shall explain this extra sensitivity of the least squares regression and the robustness of the least absolute deviations regression later when we revisit this issue in the context of the statistical distribution theories associated with these two different types of regression, and when we bring into the picture the tails of the distributions of the variables involved in the regression.

Very Important Remark. It is crucial to understand the differences between the regressions performed on the utility indexes and the regressions of this subsection. The latter are computed from log-returns, and the features of these transformed data are very different from the statistical features of the raw price data. Indeed, these raw prices show strong dependencies among themselves, say from day to day, and regression with this type of data is a very touch business. We warn the reader against the pitfalls of this trade. These regressions will be called time series regressions later. This important remark will be revisited in Subsection 3.3.2 below.

3.3 Smoothing versus Distribution Theory

So far, our approach to regression was based on a *smoothing* philosophy, and our discussion did not rely upon any statistical assumption or principle. For this reason, it belongs more in a course on computer graphics, or a course on function approximation, than in a statistics course.

To remedy this, we assume the existence of a statistical model for the roughness and/or uncertainty in the data. More precisely we assume that the values $x_1, \dots, x_n$ and $y_1, \dots, y_n$ are observed realizations of n random variables $X_1, \dots, X_n$ and Y_1, $\dots, Y_n$. The main assumption of a regression model is not about the overall distributions of the response variables $Y_1, \dots, Y_n$, but rather, their conditional distributions with respect to the explanatory variables $X_1, \dots, X_n$, hence their names.

3.3.1 Regression and Conditional Expectation

Since it happens very often in practice that the observations $x_1, \dots, x_n$ are in fact realizations of random variables $X_1, \dots, X_n$, it is more convenient to consider the observations (x_i, y_i) as realizations of couples (X_i, Y_i) of random variables having the same joint distribution. The conditional expectation of any of the random variables Y_i given the value of the corresponding X_i becomes a deterministic function of X_i. In other words, if one knows that the random variable X_i has the value x, then the conditional expectation of Y_i given $X_i = x$ is a function of x which we could denote by $\varphi(x)$. Notice that this function of x does not depend upon the index i, since we assume that the joint distributions of all the couples (X_i, Y_i) are the same. This function is called the *regression function* of Y against X, and the values y_1, $\dots, y_n$ can be viewed as (noisy) observations of the (expected) values $\varphi(x_1), \dots,$ $\varphi(x_n)$.

It is reasonable to expect the function φ to be nonlinear in general. Nevertheless, when the joint distribution of the (X_i, Y_i) is Gaussian, the function φ is linear (affine, to be precise) and linear regression is fully justified in this case.

As we have already seen, the goal of regression analysis is to quantify the way the values of the response variable are influenced by the values of the explanatory variables (if there is any influence at all). In other words, given the values $x_1, \dots, x_n$ of the explanatory (random) variables, we assume that the means of the respective response variables are of the form $\varphi(x_1), \dots, \varphi(x_n)$. More precisely, we assume:

$$\mathbb{E}\{Y_i | X_i = x_i\} = \varphi(x_i), \qquad i = 1, \dots, n.$$

Still another form of the same assumption is to postulate that for $i = 1, \dots, n$, given that $X_i = x_i$, we have:

$$Y_i = \varphi(x_i) + \epsilon_i, \tag{3.9}$$

for some mean-zero random variable ϵ_i. We shall further assume that all the random variables ϵ_i have the same variance, say σ^2, and that they are uncorrelated. As we have seen in the `S-Plus` tutorial, such a sequence $\{\epsilon_i\}_{i=1,\dots,n}$ is called a white noise. In fact, more than merely assuming that the ϵ_i's are uncorrelated, we shall assume most of the time that they are *independent*. Recall that

- choosing the number of explanatory variables determines if the regression is called a *simple regression* or a *multiple regression*;
- choosing if the function φ is to be restricted to a limited class of functions which are determined by the choice of a small number of parameters as opposed to be allowed to be of a general type, determines if the regression is called a *parametric regression* or a *nonparametric regression*;
- in the case of parametric regression, deciding whether or not the function φ should be linear determines if the regression is called a *linear regression* .

Once these choices have been made, choosing the distribution of the *noise terms* ϵ_i completely determines the statistical model and makes statistical inference possible.

As before, we shall use the comparison of the least squares (simple linear) regression to the least absolute deviations (simple linear) regression as an illustrative example, but one should keep in mind that the conclusions of this discussion are not restricted to these two particular regressions.

3.3.2 Maximum Likelihood Approach

Specifying the common distribution of the noise terms ϵ_i makes it possible to write down the joint distribution of the response variables $Y_1, \dots, Y_n$. Recall that we assume that the ϵ_i are independent and that consequently, the response variables $Y_1, \dots, Y_n$ are, conditionally on the knowledge of the values $X_i = x_i$ of the explanatory variables, independent, with means $\varphi(x_i)$, and with the same variance σ^2 as the ϵ_i's. Given the fact that, in the statistical jargon, the likelihood function of the model, say $\mathcal{L}(y_1, \dots, y_n)$, is nothing but the joint density $f_{(Y_1,\dots,Y_n)}(y_1, \dots, y_n)$ of the observed responses $Y_1, \dots, Y_n$, specifying the distribution of the noise terms ϵ_i determines the likelihood function of the model. Notice that the likelihood function also depends upon the values of the means $\varphi(x_i)$ and the variance σ^2. These dependencies will be emphasized or de-emphasized depending on the goal of the computation.

Next, we re-derive the least squares and the least absolute deviations regression procedures as instances of the general maximum likelihood approach.

Simple Least Squares Regression Revisited

In this subsection we assume that the noise terms ϵ_i are independent and normally distributed:

$$\epsilon_i \sim N(0, \sigma^2)$$

where the common variance σ^2 is assumed to be unknown. In this case we have:

$$\begin{aligned}
\mathcal{L}_2(\beta_0,\beta_1,\sigma^2) &= \mathcal{L}(\beta_0,\beta_1,\sigma^2,y_1,\dots,y_n)\\
&= f_{(Y_1,\dots,Y_n)}(y_1,\dots,y_n)\\
&= f_{Y_1}(y_1)\cdots f_{Y_n}(y_n)\\
&= \frac{1}{\sqrt{2\pi\sigma^2}}e^{-[y_1-(\beta_0+\beta_1x_1)]^2/2\sigma^2}\cdots\frac{1}{\sqrt{2\pi\sigma^2}}e^{-[y_n-(\beta_0+\beta_1x_n)]^2/2\sigma^2}\\
&= \frac{1}{\sqrt{2\pi}^n}\sigma^{-n}e^{-\frac{1}{2\sigma^2}\sum_{i=1}^n[y_i-(\beta_0+\beta_1x_i)]^2}
\end{aligned}$$

where we used the independence of the observation variables Y_i to claim that the joint density of the Y_i's was the product of the individual densities, and we chose the form of the density to be that of the normal density. From this expression it is plain to see (even without computing the partial derivatives of the objective function) that maximizing the likelihood of the observations $y_1, \dots, y_n$ given the values $x_1, \dots, x_n$ of the explanatory variables, is equivalent to minimizing the sum of square errors. This implies that the slope $\hat\beta_1$ and the intercept $\hat\beta_0$ of the least squares regression line appear as the maximum likelihood estimators of the parameters β_1 and β_0 of the model. As such, they can be applied to the inferential results from the theory of normal families. In particular one obtains:

- confidence intervals for the true slope (from which we can test if the slope is zero or not, and in so doing, assess the significance of the regression);
- confidence intervals for the true intercept;
- joint confidence regions (typically ellipsoids) for the couple (β_0, β_1) of parameters;
- maximum likelihood and unbiased estimates of the noise variance σ^2 and corresponding confidence intervals;
- coefficient of determination R^2 giving the proportion of the variation explained by the regression,

and much more. All the statistical properties of normal families can be used for inferential purposes. This abundance of tools is due to the fact that we assumed that the noise terms ϵ_i were mean zero, i.i.d. and normally distributed. Some of the statistical diagnostics mentioned above can be obtained from the output of the function `ls.diag` which takes as argument any object created with the function `lsfit`. We shall see more of the statistical inference tools provided by `S-Plus` when we discuss linear models and the function `lm`.

Simple Least Absolute Deviations Regression Revisited

We now assume that the noise terms ϵ_i have a double exponential distribution (also called Laplace distribution) given by the density:

$$f_{\epsilon_i}(x) = \frac{1}{2\lambda}e^{-\lambda|x|}$$

for some variance-like scale parameter $\lambda > 0$ which is assumed to be unknown. A computation similar to the computation done earlier in the Gaussian case gives:

$$\begin{aligned}\mathcal{L}_1(\beta_0, \beta_1, \lambda) &= \mathcal{L}(\beta_0, \beta_1, \lambda, y_1, \dots, y_n) \\ &= f_{(Y_1,\dots,Y_n)}(y_1, \dots, y_n) \\ &= f_{Y_1}(y_1) \cdots f_{Y_n}(y_n) \\ &= \frac{1}{2\lambda} e^{-\lambda|y_1-(\beta_0+\beta_1 x_1)|} \cdots \frac{1}{2\lambda} e^{-\lambda|y_n-(\beta_0+\beta_1 x_n)|} \\ &= \frac{1}{2^n} \lambda^{-n} e^{-\lambda \sum_{i=1}^n |y_i-(\beta_0+\beta_1 x_i)|}\end{aligned}$$

where as before, we used the independence of the variables Y_i to claim that the joint density of the Y_i's was the product of the individual densities, and we chose the specific form of the density to be that of the double exponential density. From this expression we see that, maximizing the likelihood of the observations $y_1, \dots, y_n$ given the values $x_1, \dots, x_n$ of the explanatory variables, is equivalent to minimizing the sum of absolute deviations. This implies that the slope $\hat{\beta}_1$ and the intercept $\hat{\beta}_0$ of the least absolute deviations regression line appear as the maximum likelihood estimators of the parameters β_1 and β_0 of the model in which the noise terms have a double exponential distribution.

Unfortunately, the usefulness of this result is very limited. Indeed, there are no closed formulae for the estimators, and double exponential families do not have a distribution theory as developed as that of the normal families. As a consequence, we want to use approximations each time we need a confidence interval, a test, ... These approximations are usually derived theoretically from asymptotic results or from Monte Carlo simulations. The latter can be computer intensive and in any case, it is difficult if not impossible to control the extend of the errors produced by the approximations.

Nevertheless, this result sheds light on the robustness of the least absolute deviations regression as compared to the least squares regression. Indeed, the Laplace distribution of the noise has thicker tails (since for large values of $|x|$, the exponential of a negative multiple of $|x|$ is significantly larger than the exponential of a negative multiple of x^2) and consequently, the model allows for larger values of the error terms ϵ_i. In other words, the LAD regression will be more tolerant of points far away from the regression curve $\varphi(x)$, while the LS regression will try harder to get the regression curve $\varphi(x)$ to be closer to these points.

When Can We Perform a Regression ?

In this subsection, we elaborate on the Very Important Remark at the end of Section 3.2. In order to allow for statistical inference, the (simple) regression setup requires that the data $(x_1, y_1), \dots, (x_n, y_n)$ form a sample of observations from identically distributed random couples $(X_1, Y_1), \dots, (X_n, Y_n)$ having the same joint distribution. To be more specific, the statistical models made explicit in this section in order to derive expressions for the likelihood function are based on the following premises:

the residuals $r_i = y_i - \varphi(x_i)$ given by the differences between the observations y_i of the responses and the values $\varphi(x_i)$ of the regression function at the observed explanatory variables, are realizations of independent identically distributed random variables.

The top pane of Figure 3.7 gives the sequential plots of the residuals of the least squares regression of `ENRON.index` against `DUKE.index` while the bottom pane gives the sequential plots of the residuals of the least squares regression of `BLRet` against `CLRet`. These sequential plots were produced with the commands:

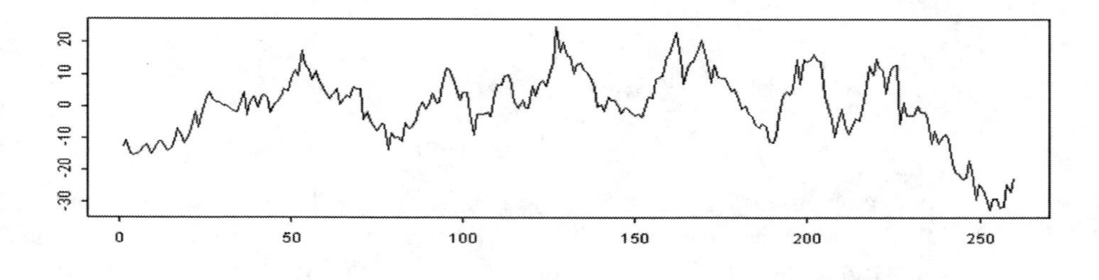

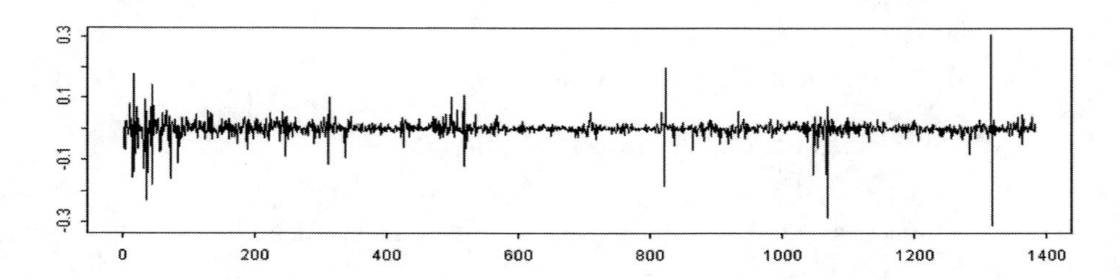

Fig. 3.7. Sequential plot of the residuals of the simple least squares regression of ENRON's index against DUKE's index (top) and of the Brazilian coffee daily log-returns against the Colombian coffee daily log-returns (bottom).

```
> tsplot(EDL2$residuals)
> tsplot(BCL2$residuals)
```

Both residual sequences have mean zero by construction. However, our experience with i.i.d. sequences of mean zero random variables (recall the Introductory Session to `S-Plus` reproduced in Appendix, and Chapter 1) tells us that if the bottom plot can be the plot of a white noise (this is the terminology we use for i.i.d. sequences of mean zero random variables), this is certainly not the case for the plot on the top!

In fact, if we were to compute the auto-correlation function (acf for short) of the utility index residuals, we would see strong correlations which are inconsistent with the assumptions of the regression setup. The notion of acf will be introduced and explained in Chapter 5, while regression diagnostics will be presented later in this chapter.

The serial correlations present in the utility data contradict the regression premises restated above. Not only does the distribution of the couple of random variables DUKE.index and ENRON.index) change over time, but the couples actually observed day after day are not independent. These two facts formed the justification for our analysis of the log-returns of the coffee data instead of the original raw data of the index values. So as announced earlier, we stop using the utility data in its original form. From now on, for the purposes of regression analysis, we replace the actual indexes by their log-returns.

```
> UtilLR <- diff(log(UTILITY.index))
> EnronLR <- diff(log(ENRON.index))
> DukeLR <- diff(log(DUKE.index))
```

As explained earlier, the difference can be undone by a cumulative sum, and the logarithm can be inverted with an exponential function. In this way, conclusions, estimations, predictions, etc. , concerning the log-returns can be re-interpreted as conclusions, estimations, predictions, etc. for the original index values.

When Should We Use Least Absolute Deviations ?

Advantages of the least squares regression:

- existence and uniqueness of the estimators $\hat{\beta}_0$ and $\hat{\beta}_1$;
- existence of explicit formulae for the estimators $\hat{\beta}_0$ and $\hat{\beta}_1$;
- fast and reliable computations;
- existence of a distribution theory leading to convenient statistical inferences, e.g. exact confidence intervals, tests.

Drawbacks of the least squares regression:

- sensitivity to outliers and extreme observations.

Advantages of the least absolute deviations regression:

- existence of the estimators $\hat{\beta}_0$ and $\hat{\beta}_1$ and reasonable computational algorithms;
- extreme robustness to one type of outlier.

Drawbacks of the least squares regression:

- lack of uniqueness of the estimators $\hat{\beta}_0$ and $\hat{\beta}_1$;
- lack of a convenient distribution theory: the estimators, the tests, the confidence intervals, etc., have too often to be computed by lengthy Monte Carlo methods.

So to summarize:

- if the complexity of the computations and/or the computing time is an issue one may want to use least squares regression;
- on the other hand if robustness is important then it is likely that the least absolute deviations regression will give more satisfactory results.

3.4 Multiple Regression

Back to the analysis of our utility data. If our goal is to explain the overall utility index `UTILITY.index` using all the information at our disposal, we may want to use both of the individual indexes, namely `ENRON.index` and `DUKE.index`, together in the same formula. As explained earlier, from now on we work with the log-returns, and restricting ourselves to linear (or affine) functions, we may seek a relationship of the form

$$\texttt{UtilLR} = \beta_0 + \beta_1 * \texttt{EnronLR} + \beta_2 * \texttt{DukeLR} + \texttt{noise}.$$

This is the epitome of a multiple linear regression model.

3.4.1 Notation

As before, we denote by n the sample size and we assume that the observations come as n pairs

$$(\mathbf{x}_1, y_1), \ldots\ldots, (\mathbf{x}_n, y_n),$$

where the last component y_i is the response variable whose value we try to explain from the values of the explanatory variables, i.e. the components of $\mathbf{x}_i$. The main difference is that we now allow the explanatory variable $\mathbf{x}_i$ to be a vector of p different scalar explanatory variables $x_{i,1}, \ldots, x_{i,p}$. In the case of the utility data, the response variable should be $y = \texttt{UtilLR}$, and the explanatory bivariate variable (i.e. $p = 2$) is now $\mathbf{x} = (\texttt{EnronLR}, \texttt{DukeLR})$.

As in the case of simple linear regression, the `S-Plus` functions `lsfit` and `l1fit` can still be used to perform the regression in the sense of least squares and in the sense of least absolute deviations respectively. One simply needs to bind the p explanatory column variables into an $n \times p$ matrix with the command `cbind`. For example, the following commands:

```
> UtilLRS <- cbind(EnronLR,DukeLR)
> UEDls <- lsfit(UtilLRS,UtilLR)
> UEDl1 <- l1fit(UtilLRS,UtilLR)
```

perform the least squares and the least absolute deviations linear regression of the utility index log-return against the ENRON and DUKE log-returns. As before, one can argue that these two regressions are the results of maximum likelihood estimations of the coefficients of the regression "planes" when the noise terms are assumed to be normally and double-exponentially distributed, respectively. Again, inferential tools are not as developed in the case of the least absolute deviations regression, and for this reason, we shall mostly concentrate on the least squares method. See nevertheless the Notes and Complements at the end of this chapter for further information. Although diagnostic tools have been added to the function `lsfit`, statistical inference is best done with the powerful function `lm` which we now introduce.

3.4.2 **The** `S-Plus` **Function** `lm`

As explained in our comparison of the least squares and least absolute deviations regressions, linear models come with a distribution theory providing exact tests of significance, confidence intervals, ..., when the error terms are assumed to be independent and identically distributed (i.i.d. for short) and normally distributed. Because of its weak distribution theory, we shall momentarily refrain from using the least absolute deviations regression, and even though statistical inference can be performed with the function `lsfit`, we shall start using the more powerful function `lm` provided by `S-Plus`. The regression objects produced by `lm` are the result of least squares regression, and the statistical estimates, confidence intervals, p-values, ... are based on the assumption that the error terms ϵ_i are i.i.d. $N(0, \sigma^2)$, for some unknown $\sigma > 0$. The command `lm` gives an implementation of the most general linear models defined below in Section 3.5, but it is still worth using it even in the restrictive framework of plain linear regression.

1. We shall give many examples of uses of the method `lm` and the student is invited to carefully read the help file. Resuming the analysis of the utility indexes data, we first perform the least squares regression of the utility index daily log-return against the ENRON daily log-return. This could have been done with the command:

```
> UeL2 <- lsfit(EnronLR,UtilLR)
```

but we do it now with the commands:

```
> Ue <- lm(UtilLR ~ EnronLR)
> summary(Ue)
```

Notice that the argument of the function `lm` is a formula stating that the variable `UtilLR` which appears on the left of the tilde, has to be expressed as a function of the variable `EnronLR` which appears on its right. The function `lm` has an optional argument which can be set by the parameter `data`. It is very useful when the variables are columns in a data frame. It gives the name of the data frame containing the variables. The object produced by such a command is of class `lm`. Contained in the summary, we find the estimated slope .0784, and the estimated intercept .0001, values which could have been obtained with the function `lsfit`. The numerical results extracted from the `lm` object `Ue` by the command `summary` are very detailed. They contain the coefficients of the simple linear model:

$$\texttt{UtilLR} = .0001 + .0784 \times \texttt{EnronLR} + \epsilon,$$

but also extra information on the model. In particular, they include estimates of the variances of the slope and the intercept (which can be used to compute confidence intervals for these parameters), and a p-value for tests that the parameters are significantly different from 0 (i.e. tests of significance of the regression). We discuss the well known R^2 coefficient in the next subsection.

2. Similarly, in the same way one could have used the command:

```
> UdL2 <- lsfit(DukeLR,UtilLR)
```

one can use the `S-Plus` commands:

```
> Ud <- lm(UtilLR ~ DukeLR)
```

in order to perform the simple least squares linear regression of `UtilLR` against `DukeLR` with the command `lm`. In this case the linear model is:

$$\texttt{UilLR} = -0.0001 + 0.5090 \times \texttt{DukeLR} + \epsilon$$

as we can see by reading the values of estimated slope and estimated intercept from the summary of the `lm` object `Ud`.

3. To perform the multiple least squares linear regression of `UtilLR` against the variables `EnronLR` and `DukeLR` together, we use the function `lm` in the following way:

```
> Ued <- lm(UtilLR~EnronLR+DukeLR)
```

The plus sign "+" in the above formula should not be understood as a sum, but rather as a way to get the regression to use both variables `EnronLR` and `DukeLR`. The estimated coefficients can be read off the summary of the `lm` object `Ued`. The fitted model is now:

$$\texttt{UtilLR} = -0.0001 + 0.0305 \times \texttt{EnronLR} + 0.4964 \times \texttt{DukeLR} + \epsilon.$$

3.4.3 R^2 as a Regression Diagnostic

We will now discuss the well known R^2 coefficient which is used to quantify the quality of a regression. This number gives the proportion of the variance explained by the regression. In the case of a simple least squares regression, it is defined by the formula:

$$R^2 = 1 - \frac{SSE}{TSS} \tag{3.10}$$

where

$$SSE = \sum_{i=1}^{n}(y_i - \hat{y}_i)^2 = \sum_{i=1}^{n}(y_i - \hat{\beta}_0 - \hat{\beta}_1 x_i)^2 \tag{3.11}$$

is the sum of squared residuals from the regression, called the sum of squared errors, and where:

$$TSS = \sum_{i=1}^{n} (y_i - \overline{y})^2 \qquad (3.12)$$

represents the total sum of squares. So the closer R^2 to 1, the better the regression. R^2 is often called the coefficient of determination of the regression. It is not difficult to imagine possible generalizations of this coefficient to the case of least absolute deviations regression. We shall use some of these generalizations in the problems at the end of the chapter.

In the three least squares regressions performed earlier, the values of the R^2 coefficient were

- 0.0606 in the case of `UtilLR` against `EnronLR`;
- 0.5964 in the case of `UtilLR` against `DukeLR`;
- 0.6052 in the case of `UtilLR` against both variables `EnronLR` and `DukeLR`.

By comparing the R^2 values in the cases of the first and second bullets, one could conclude that the variable `EnronLR` is a far worse predictor than `DukeLR` when it comes to explaining the overall sector log-returns `UtilLR`. Indeed, the former has a smaller value of R^2 (0.0606) than the latter (0.5964). In our zealous attempt to show that the raw index data were inappropriate for direct regression analysis, we departed from our tradition, and we performed the regression analysis of the log-returns without plotting the data first. It is time to change that. The scatterplots given in Figure 3.8 will help us understand the values of the R^2. The striking differences

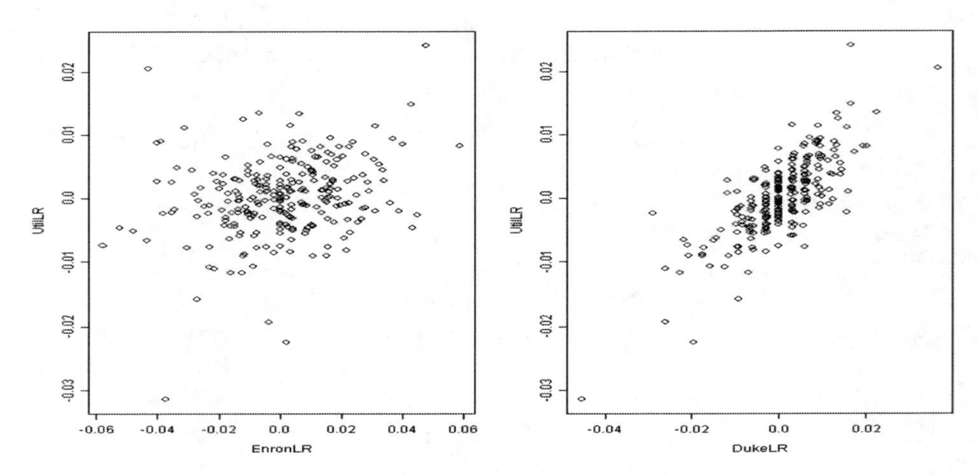

Fig. 3.8. Scatterplot of the utility index daily log-returns against ENRON's daily log-returns (left) and DUKE's daily log-returns (right).

between the scatterplots explain the difference between the two R^2 scores. Indeed, a quick look at the scatterplot in the left pane of Figure 3.8 shows that there does not seem to be any functional relation between the values of the utility index log-returns and ENRON's log-returns. On the other hand, the scatterplot contained in the right pane of the figure where the dependence of `UtilLR` upon `DukeLR` is visualized, shows that the data is reasonably well suited for a linear regression.

However, the most interesting remark prompted by the numerical values of R^2 is the following. Despite the fact that, as expected, the R^2 value of 0.6599 obtained with multiple regression is larger than the values of R^2 obtained with simple regressions, the improvement does not seem to be significant. It seems that adding an extra explanatory variable does not help much with explaining the response. This is presumably due to the fact that the information carried by `EnronLR` is already contained in `DukeLR` for the most part. We shall discuss this issue in more details later.

3.5 MATRIX FORMULATION AND LINEAR MODELS

We will now expand on the notion of linear model alluded to in our discussion of the `S-Plus` function `lm`. The statistical formulation of the simple linear regression problem is based on the model:

$$y_i = \beta_0 + \beta_1 x_i + \epsilon_i, \qquad i = 1, \dots, n,$$

which can be rewritten in matrix notation as:

$$\mathbf{Y} = \mathbf{X}\beta + \epsilon \tag{3.13}$$

provided we set:

$$\mathbf{Y} = \begin{bmatrix} y_1 \\ \vdots \\ y_n \end{bmatrix}, \quad \mathbf{X} = \begin{bmatrix} 1 & x_1 \\ \vdots & \vdots \\ 1 & x_n \end{bmatrix}, \quad \epsilon = \begin{bmatrix} \epsilon_1 \\ \vdots \\ \epsilon_n \end{bmatrix}, \quad \text{and} \quad \beta = \begin{bmatrix} \beta_0 \\ \beta_1 \end{bmatrix}.$$

Similarly, the multiple linear regression model:

$$y_i = \beta_0 + x_{i,1}\beta_1 + \cdots + x_{i,p}\beta_p + \epsilon_i, \qquad i = 1, \dots, n$$

can be recast in the same matrix formulation (3.13) of a linear model provided we set:

$$\mathbf{Y} = \begin{bmatrix} y_1 \\ \vdots \\ y_n \end{bmatrix}, \quad \mathbf{X} = \begin{bmatrix} 1 & x_{1,1} & \cdots & x_{1,p} \\ \vdots & \vdots & & \vdots \\ 1 & x_{n,1} & \cdots & x_{n,p} \end{bmatrix}, \quad \epsilon = \begin{bmatrix} \epsilon_1 \\ \vdots \\ \epsilon_n \end{bmatrix}, \quad \text{and} \quad \beta = \begin{bmatrix} \beta_0 \\ \beta_1 \\ \vdots \\ \beta_p \end{bmatrix}.$$

3.5.1 Linear Models

We promote the notation introduced above to the rank of definition by stating that we have a *linear model* when:

$$\mathbf{Y} = \mathbf{X}\beta + \epsilon \tag{3.14}$$

where:

- $\mathbf{Y}$ is a (random) vector of dimension n;
- $\mathbf{X}$ is an $n \times (p+1)$ matrix (called the *design matrix*);
- β is an $(p+1)$ - vector of (unknown) parameters;
- ϵ is a mean zero random vector of uncorrelated errors with common (unknown) variance σ^2.

From now on, we will assume that the design matrix $\mathbf{X}$ has full rank (i.e. $p+1$).

Normal Linear Models: When the individual error terms ϵ_i are assumed to be jointly normal we have:

$$\epsilon \sim N_n(0, \sigma^2 \mathbf{I}_n)$$

and consequently that:

$$\mathbf{Y} \sim N_n(\mathbf{X}\beta, \sigma^2 \mathbf{I}_n).$$

Since the noise term of a linear model is of mean 0, it is clear that, in the case of a linear model one has:

$$\mathbb{E}\{\mathbf{Y}\} = \mathbf{X}\beta. \tag{3.15}$$

since $\mathbb{E}\{\epsilon\} = 0$. Notice also that, in the case of the noise term ϵ of a linear model, since all the components ϵ_i are assumed to have the same variance σ^2, the variance/covariance matrix of ϵ is in fact σ^2 times the $n \times n$ identity matrix $\mathbf{I}_n$. Finally, since changing the expectation of a random variable does not change its variance, we conclude that the variance/covariance matrix of the observation vector $\mathbf{Y}$ is also $\sigma^2 \mathbf{I}_n$. Hence we have proved that:

$$\Sigma_{\mathbf{Y}} = \Sigma_\epsilon = \sigma^2 \mathbf{I}_n. \tag{3.16}$$

3.5.2 Least Squares (Linear) Regression Revisited

The purpose of least squares linear regression is to find the value of the $(p+1)$-dimensional vector β of parameters minimizing the loss function:

$$\begin{aligned}
\mathcal{L}_2(\beta) &= \sum_{i=1}^{n} [y_i - \beta_0 - \beta_1 x_{i,1} - \cdots - \beta_p x_{i,p}]^2 \\
&= \sum_{i=1}^{n} [y_i - {}^t\mathbf{X}_i \beta]^2 \\
&= \|\mathbf{Y} - \mathbf{X}\beta\|^2
\end{aligned} \tag{3.17}$$

where we use the notation $\|\cdot\|$ for the Euclidean norm of an n-dimensional vector, i.e. $\|\mathbf{y}\| = \left(\sum_{i=1}^n y_i^2\right)^{1/2}$. Notice that we are now using the notation β for the parameter which we had called θ in our general discussion of Subsection 3.1.6. As announced earlier, there is a unique solution to the above problem (although we need to use the full rank assumption to rigorously justify this statement) and this solution is given by an explicit formula which we give here without proof:

$$\hat{\beta} = [\mathbf{X}^t\mathbf{X}]^{-1}\mathbf{X}^t\mathbf{Y}. \tag{3.18}$$

A two-line proof of this result can be given if one is familiar with vector differential calculus. Indeed, the minimum $\hat{\beta}$ can be obtained by solving for the zeroes of the gradient vector of the loss function $\mathcal{L}_2(\beta)$ as given by (3.17), and this can be done in a straightforward manner. The interested reader is invited to consult the references given in the Notes & Complements at the end of this chapter.

Properties of $\hat{\beta}$

The following list summarizes some of the most important properties of the least squares estimator $\hat{\beta}$. The first property is a simple remark on formula (3.18), while the other ones rely on the distribution properties of the linear model (3.14).

- $\hat{\beta}$ is *linear* in the sense that it is a linear function of the observation vector $\mathbf{Y}$;
- $\hat{\beta}$ is *unbiased* in the sense that it is equal on the average to the true (and unknown) value of β, whatever this value is. Mathematically this is expressed by the formula $\mathbb{E}\{\hat{\beta}\} = \beta$;
- The variance/covariance matrix of this estimator can be computed explicitly. It is given by the formula $\Sigma_{\hat{\beta}} = \sigma^2[\mathbf{X}^t\mathbf{X}]^{-1}$;
- $\hat{\beta}$ is optimal in the sense that it has *minimum variance* among all the linear unbiased estimators of β.

Properties of the Fitted Values $\hat{Y}$

The formula giving $\hat{\beta}$ implies that the fitted values are given by:

$$\hat{\mathbf{Y}} = \mathbf{X}\hat{\beta} = \mathbf{X}[\mathbf{X}^t\mathbf{X}]^{-1}\mathbf{X}^t\mathbf{Y} = \mathbf{H}\mathbf{Y}$$

if we introduce the notation

$$\mathbf{H} = \mathbf{X}[\mathbf{X}^t\mathbf{X}]^{-1}\mathbf{X}^t.$$

As for the parameter estimate(s), the fitted value(s) are linear in the observations, since the vector $\hat{\mathbf{Y}}$ of fitted values is obtained by multiplying a matrix with the vector $\mathbf{Y}$ of observations of the response variable. The matrix $\mathbf{H}$ plays an important role in the analysis of the properties of least squares regression. It is called the *hat matrix* or the *prediction matrix* since the formula $\hat{\mathbf{Y}} = \mathbf{H}\mathbf{Y}$ tells us how to transform the observations $\mathbf{Y}$ into the values $\hat{\mathbf{Y}}$ predicted by the model. We shall see below that the diagonal elements $h_{i,i}$ enter in an explicit way into the variance of the raw residuals, but for the time being we shall stress that the $h_{i,i}$'s measure the *influence* of the corresponding observation: a large value of $h_{i,i}$ (since the $h_{i,i}$ are never greater than 1, a large value means a value close to 1) indicates that the corresponding observation plays a crucial role in the computation of the resulting value of $\hat{\beta}$ and consequently, greatly influences the interpretation of the results. This notion of influential observation has to be carefully distinguished from the notion of outlying observation discussed below. Indeed, there is nothing wrong with an influential observation, it should not be disregarded, but should simply be looked at carefully in order to understand why some of the results of the regression are what they are.

We display the following formula for future reference.

$$\hat{\mathbf{Y}} = \mathbf{X}\hat{\beta} = \mathbf{H}\mathbf{Y} \qquad \hat{\epsilon} = \mathbf{Y} - \hat{\mathbf{Y}} = [\mathbf{I}_n - \mathbf{H}]\mathbf{Y}. \tag{3.19}$$

The components $\hat{\epsilon}_i$ of the vector $\hat{\epsilon}$ defined above are called the *raw residuals* since:

$$\hat{\epsilon}_i = \hat{y}_i - y_i. \tag{3.20}$$

Properties of the Residuals

The properties of the raw residuals $\hat{\epsilon}_i$ are summarized in the following list.

- the $\hat{\epsilon}_i$'s are mean zero, $\mathbb{E}\{\hat{\epsilon}\} = 0$;
- their variance/covariance matrix is given by the formula $\Sigma_{\hat{\epsilon}} = \sigma^2[\mathbf{I}_n - \mathbf{H}]$;
- in particular
  - the $\hat{\epsilon}_i$'s are correlated;
  - they do not have the same variance since $\sigma_{\hat{\epsilon}_i} = \sigma\sqrt{1 - h_{i,i}}$.

We use the notation $h_{i,j}$ for the (i,j)-th entry of the hat matrix $\mathbf{H}$. It is important to reflect on the meaning of these statements. Indeed, given the fact that the $\hat{\epsilon}_i$ come as candidates for realizations of the actual error terms ϵ_i, one could expect that they form a white noise. But we just learned that this is not the case. Indeed, there are at least two good reasons why the plot of the raw residuals should not look like a white noise ! First the variance of $\hat{\epsilon}_i$ changes with i, and second, the $\hat{\epsilon}_i$'s are correlated. Any plot of the raw residuals should show these facts.

Notice that even though the dependence of the $\hat{\epsilon}_i$ may look shocking at first, it should not be surprising since after all, the $\hat{\epsilon}_i$'s are computed from $\hat{\beta}$ instead of from the (deterministic) true value β which is not available, and also the estimator $\hat{\beta}$ is a function of all the observations y_j, and consequently of all the error terms ϵ_j. This is the source of this unexpected correlation between the residuals.

Estimator of the Variance σ^2

We first introduce the notation:

$$R_0^2 = \|\mathbf{Y} - \mathbf{X}\hat{\beta}\|^2 = \sum_{i=1}^{n} [y_i - \hat{\beta}_0 - \hat{\beta}_1 x_{i,1} - \cdots - \hat{\beta}_p x_{i,p}]^2 \tag{3.21}$$

for the minimum sum of squared residuals (i.e. the minimum value of the loss function $\mathcal{L}_2(\beta)$). In the notation used in our discussion of the simple least squares linear regression, this quantity was denoted by SSE and called the sum of squared errors. The variance parameter σ^2 is estimated by:

$$\begin{aligned}\hat{\sigma}^2 &= \frac{1}{n-p-1} R_0^2 \\ &= \frac{1}{n-p-1} \|\mathbf{Y} - \mathbf{X}\hat{\beta}\|^2 \\ &= \frac{1}{n-p-1} \sum_{i=1}^{n} [y_i - \hat{\beta}_0 - \hat{\beta}_1 x_{i,1} - \cdots - \hat{\beta}_p x_{i,p}]^2.\end{aligned}$$

As usual, the normalization of R_0^2 is done by division after subtracting the number of parameters ($p+1$ in our case) from the number of observations. This correction is included to make sure that $\hat{\sigma}^2$ is an unbiased estimator of the variance σ^2.

Standardized Residuals $\hat{\epsilon}'_i$

In order to resolve the issue of heteroskedasticity (i.e. fluctuations in the variance) we would like to divide the raw residuals by their standard deviations. In other words, we would like to use:

$$\frac{\hat{\epsilon}_i}{\sigma\sqrt{1-h_{i,i}}},$$

but since we do not know σ we instead use:

$$\hat{\epsilon}'_i = \frac{\hat{\epsilon}_i}{\hat{\sigma}\sqrt{1-h_{i,i}}} \tag{3.22}$$

The $\hat{\epsilon}'_i$ so defined are called the *standardized residuals.* Obviously, they are mean zero and have unit variance, but there is still a lot of correlation between them.

Studentized Residuals $\hat{\epsilon}^*_i$

In order to circumvent the problem caused by the serial correlation of the residuals considered so far, another type of residual was proposed. The $\hat{\epsilon}^*_i$ defined below are called the *studentized residuals.* For each $i = 1, \ldots, n$, we

- remove the observation (x_i, y_i);
- fit a linear model (by least squares) to the remaining $(n-1)$ observations;
- call $\hat{y}_{(i)}$ the prediction of the response for the value x_i of the regressor.

Notice that the value of the error term ϵ_i does not enter into the computation of the new *residual* $y_i - \hat{y}_{(i)}$, so we would hope that these new residuals are de-correlated. In any case, we set:

$$\hat{\epsilon}^*_i = \frac{y_i - \hat{y}_{(i)}}{\sqrt{\mathrm{var}\{y_i - \hat{y}_{(i)}\}}}. \tag{3.23}$$

These studentized residuals should look like white noise and a reasonable diagnostic for the fit of a linear model is to check how this materializes. As defined, the studentized residuals require enormous computer resources, since the computation of each single $\hat{y}^{(i)}$ requires a linear regression ! Fortunately, the set of all the studentized residuals can be computed from one single linear regression and a few updates. Indeed, simple arithmetic can be used to derive the following formula:

$$\hat{\epsilon}_i^* = \sqrt{\frac{n-p-1}{n-p-\hat{\epsilon}_i'^2}}\,\hat{\epsilon}_i'$$

linking the standardized residuals to the studentized ones. Despite the existence of a function `ls.diag`, which can be used with an object created by the function `lsfit`, there is no convenient way to produce the residual diagnostics from an object created by the function `lm`. For this reason, we provided a simple wrapper called `lm.diag`, which does just that. Its use is illustrated in Subsection 3.5.2 below.

More Residual Diagnostics

A standard way to gauge the goodness of fit of a regression model is to use graphical diagnostics. For example if a plot of the standardized or studentized residuals (whether it is a sequential plot, or a plot against the fitted values or the observations) shows values outside the range $[-2, +2]$, one should suspect that something is wrong. Either the model is not appropriate for the data, or if the model is reasonable, the observations responsible for these extreme residual values do not belong. These observations are usually called *outliers*. It is a good practice to try to identify the potential outliers and if one has faith in the model, to re-fit a regression after the outlying observations have been removed.

Revisiting the Utility Indexes Analysis

We use the utility data to illustrate how to produce the suites of regression diagnostics in `S-Plus`. They are created with the commands

```
> Ue.diag <- lm.diag(Ue)
> Ud.diag <- lm.diag(Ud)
```

Figure 3.9 gives the plots of the influence measures (i.e. the diagonal entries of the hat matrix) in the case of the simple regression of `UtilLR` against `EnronLR` and `DukeLR`, respectively. These numerical vectors are extracted from the diagnostic objects by means of the extension $\cdots$ `$hat`. In preparing Figure 3.9, we did not rely on the usual `tsplot` function to produce the plots of the vectors `Ue.diag$hat` and `Ud.diag$hat`. Instead, we used time series plots in order to have the dates appear on the horizontal axes. We shall learn how to do that in `S-Plus` when we introduce the `timeSeries` objects in Chapter 5. It is clear from this plot that in both cases, the Winter season and the period spanning the end of Summer and early

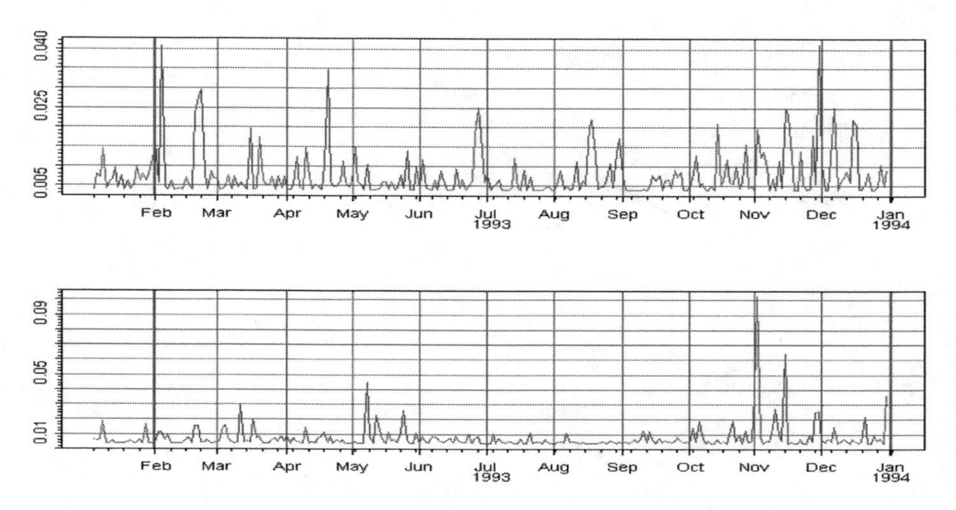

Fig. 3.9. Sequential plot of the $h_{i,i}$ in the case of the simple regression of utility index daily log-returns against ENRON's daily log-returns (top) and against DUKE's daily log-returns (bottom).

Fall are most influential. Notice nevertheless that the scales of the vertical axes are not the same, and taking this fact into account, one sees that a few days in November are extremely influential in the regression of the utility index log-returns against DUKE's log-returns. As we shall see in the later part of the book, departure from normality can be an indication of the existence of significant dependencies among the residuals, so it is always a good idea to check the Q-Q plots of the standardized and the studentized residuals against the quantiles of the normal distribution. We give these Q-Q plots in Figure 3.10. No significant departure from normality seems to be present. The numerical vectors containing the standardized residuals and the studentized residuals can be extracted from the `S-Plus` diagnostic objects by means of the extensions $\cdots$ `$stdres` and $\cdots$ `$studres` respectively. The Q-Q plots of Figure 3.10 were created with the commands:

```
> qqnorm(Ued.diag$stdres)
> qqnorm(Ued.diag$studres)
```

3.5.3 First Extensions

The formalism of linear models is so general that the theory can be applied to many diverse situations. We consider two such applications in this subsection. Many more will be discussed later in the chapter. Since the applications considered in this section will not be used in the sequel, this section can be skipped in a first reading.

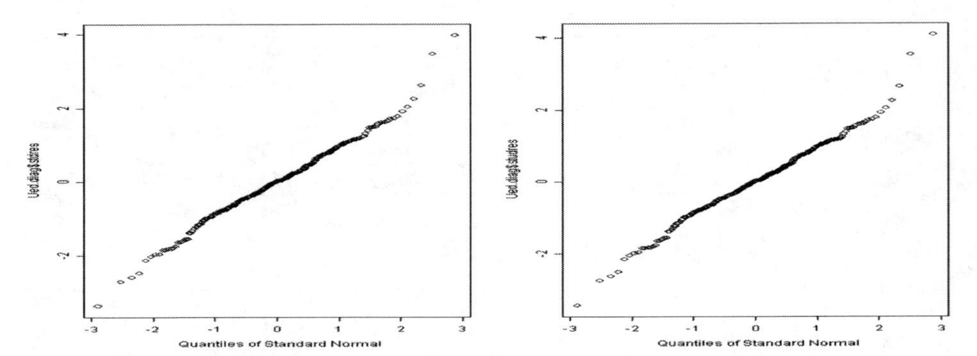

Fig. 3.10. Q-Q plots of the standardized residuals (left) and the studentized residuals (right) of the multiple least squares regression of the utility daily log-returns against both ENRON and DUKE daily log-returns.

Weighted Regression

Each diagonal entry of the hat matrix $\mathbf{H}$ gives an indication of the influence of the corresponding measurement. It is sometime desirable to decrease (resp. increase) the influence of a given measurement on the final regression result. This can be done by means of a minor change in the way each individual measurement actually contributes to the value of the loss function. Let us consider for example the weighted sum of squares:

$$\mathcal{L}_2^{(w)} = \sum_{i=1}^{n} w_i |y_i - \varphi(x_i)|^2. \tag{3.24}$$

for some predetermined set of nonnegative weights w_i. In order to avoid large contributions to the overall loss, any function φ minimizing this loss function will try very hard to have a value $\varphi(x_i)$ very close to y_i for the indices i for which the weight w_i is large. On the other hand, if a weight w_i is smaller than the bulk of the other weights, then the fit may be loose without causing much damage to the overall value of the loss.

The computations involved in the minimization of the weighted loss function $\mathcal{L}_2^{(w)}$ are of the same complexity as the computation of the minimum of the corresponding unweighted least squares loss function $\mathcal{L}_2$. For this reason, it is possible in `S-Plus` to specify a set of weights and expect as result the regression function φ minimizing the weighted loss function $\mathcal{L}_2^{(w)}$.

Example. Let us imagine that one tries to explain the variable `UtilLR` by the variable `DukeLR` by giving more importance to the days when the ENRON index is larger than the DUKE index. One can use simple least squares linear regression using the variable `ENRON.index/DUKE.index` as a set of weights. This would be accomplished by the `S-Plus` command:

```
> TstUtil.lm <- lm(UtilLR ~ DukeLR,
              weights=ENRON.index/DUKE.index)
```

Remark. Because of the parallel that we drew between the least squares regression and its robust cousin the least absolute deviations regression, it would be natural to consider a weighted loss function of the form:

$$\mathcal{L}_1^{(w)} = \sum_{i=1}^{n} w_i |y_i - \varphi(x_i)|, \tag{3.25}$$

and by analogy with the weighted least squares regression to have the weighted least absolute deviations regression function be the result of the search for the function φ minimizing $\mathcal{L}_1^{(w)}$. Unfortunately, this minimization is quite involved and weighted least absolute deviations regression is rarely an option: the function `l1fit` does not offer the possibility to include weights in the regression.

Seemingly Unrelated Regressions

Let us assume that for each $j = 1, \ldots, J$ we fit a multiple linear regression model

$$\mathbf{Y}^{(j)} = \mathbf{X}^{(j)}\beta^{(j)} + \epsilon^{(j)}$$

to the sample

$$(\mathbf{x}_1^{(j)}, y_1^{(j)}), \ldots, (\mathbf{x}_n^{(j)}, y_n^{(j)})$$

where the explanatory variables $\mathbf{x}_i^{(j)}$ all have the same dimension $k = p+1$. In such a situation, one performs the J regressions in sequence, independently of each other. However, it happens in many financial applications that the explanatory variables are the same for all the regressions. In this case the model can be rewritten as:

$$\mathbf{Y}^{(j)} = \mathbf{X}\beta^{(j)} + \epsilon^{(j)}$$

and it can be ran in `S-Plus` by binding all the response vectors into a $n \times J$ response matrix. According to the theory developed earlier, each parameter vector $\beta^{(j)}$ can be estimated by ordinary least squares, and the results is given in matrix form by:

$$\widehat{\beta}^{(j)} = [\mathbf{X}^t\mathbf{X}]^{-1}\mathbf{X}^t\mathbf{Y}^{(j)}.$$

Now comes the interesting part of this subsection. Even though the response variables $Y_i^{(j)}$ and $Y_{i'}^{(j')}$ are uncorrelated if they come from two different observations (i.e. if $i \neq i'$), we now assume that, the responses associated to the same observation are correlated. In other words, we assume the existence of a variance/covariance matrix $\Gamma = [\gamma_{j,j'}]_{j,j'=1,\ldots,J}$ such that:

$$\text{cov}\{Y_i^{(j)}, Y_i^{(j')}\} = \gamma_{j,j'}, \qquad j,j' = 1, \ldots, J,\ i = 1, \ldots, n, \tag{3.26}$$

or in vector form, $\Sigma_{\mathbf{Y}_i} = \Gamma$ for all $i = 1, \dots, n$. As we already pointed out, this assumption is quite realistic in many financial applications. For example it holds when the response variables $Y_i^{(j)}$ are the log-returns of J stocks in the same economic sector, and when the explanatory vectors $\mathbf{x}_i$ are observations of vectors of economic factors driving the dynamics of the share prices of the public companies in the sector. A linear model specified this way is called a set of seemingly unrelated regressions, or SUR for short. The usual ordinary least squares approach is not appropriate because of the dependence among the response variables. In order to understand why, we rewrite the model in the standard form of a linear model with a scalar response variable. In order to do so, we bind the response vectors $\mathbf{Y}^{(j)}$ into a $(nJ) \times 1$ column vector $\widetilde{\mathbf{Y}}$, the noise vectors $\epsilon^{(j)}$ into a $(nJ) \times 1$ column vector $\widetilde{\epsilon}$, the parameters $\beta^{(j)}$ into a $(kJ) \times 1$ column vector $\widetilde{\beta}$, and we create the new design matrix $\widetilde{\mathbf{X}}$ as the $(nJ) \times (kJ)$ matrix with J copies of $\mathbf{X}$ on the diagonal:

$$\widetilde{\mathbf{Y}} = \begin{bmatrix} \mathbf{Y}^{(1)} \\ \vdots \\ \mathbf{Y}^{(J)} \end{bmatrix}, \quad \widetilde{\epsilon} = \begin{bmatrix} \epsilon^{(1)} \\ \vdots \\ \epsilon^{(J)} \end{bmatrix}, \quad \widetilde{\beta} = \begin{bmatrix} \beta^{(1)} \\ \vdots \\ \beta^{(k)} \end{bmatrix}, \quad \widetilde{\mathbf{X}} = \begin{bmatrix} \mathbf{X} & \mathbf{0} & \cdots & \mathbf{0} \\ \mathbf{0} & \mathbf{X} & \cdots & \mathbf{0} \\ \vdots & \vdots & & \vdots \\ \mathbf{0} & \mathbf{0} & \cdots & \mathbf{X} \end{bmatrix}.$$

With these notation, the SUR model reads:

$$\widetilde{\mathbf{Y}} = \widetilde{\mathbf{X}}\widetilde{\beta} + \widetilde{\epsilon}$$

but the ordinary least squares estimate

$$\widehat{\widetilde{\beta}}^{(j)} = [\widetilde{\mathbf{X}}^t\widetilde{\mathbf{X}}]^{-1}\widetilde{\mathbf{X}}^t\widetilde{\mathbf{Y}}$$

is presumably not the best estimate. Indeed, even when the noise terms are Gaussian, this estimate does not coincide with the maximum likelihood estimator. This is due to the fact that the above least squares estimate is based on the assumption that the variance/covariance matrix of the noise vector $\widetilde{\epsilon}$ is a multiple of the $(nJ) \times (nJ)$ identity matrix I_{nJ}, and this is not the case in the present situation. Indeed, this variance/covariance matrix is not diagonal as a simple computation shows. It is in fact equal to what is called the Kroenecker product of the matrix Γ and the $n \times n$ identity matrix I_n. We shall not pursue the analysis any further. We refer the interested reader to the Notes & Complements section at the end of the chapter for references.

3.5.4 **Testing the CAPM**

We now present the Capital Asset Pricing Model (CAPM for short) as an illustration of the linear regression techniques introduced in this chapter.

We consider a market economy containing N financial assets, and we denote by R_{jt} the log-return on the j-th asset on day t. For the purposes of the present discussion, we assume the existence of a market portfolio, and we denote by $R_t^{(m)}$ its log-return on day t. Finally, we also assume the existence of lending and borrowing at the same risk-free rate which we denote by r, and which we assume to be deterministic. The Sharpe-Lintner version of the CAPM states that the excess return (over the risk-free rate) of each asset j is, up to noise, a linear function of the excess return of the market portfolio. In other words, for each j:

$$\widetilde{R_{jt}} = \alpha_j + \beta_j \widetilde{R_t^{(m)}} + \epsilon_{jt}$$

where the noise sequence $\{\epsilon_{jt}\}_t$ is uncorrelated with the market portfolio return. We use a tilde to denote the returns in excess of the risk-free rate:

$$\widetilde{R_{jt}} = R_{jt} - r \quad j = 1, 2, \ldots \qquad \text{and} \qquad \widetilde{R_t^{(m)}} = R_t^{(m)} - r. \tag{3.27}$$

One of the claims of the CAPM theory is that, if one uses excess returns, the intercept α_j appearing in the linear regression equation (3.27) should be zero. In other words, the regression lines should all go through the origin, whatever the choice of the asset. Notice that the same model can be used for small portfolio returns R_{jt} instead of individual stock returns.

For each stock, a least squares regression will provide estimates for the slope β_j and the intercept α_j. The estimate $\hat{\beta}_j$ is given by a normalized form of the covariance between the j-th asset (or the j-th portfolio) and the market portfolio: this is known as the investment beta for the j-th asset. It measures the sensitivity of the return to variations in the returns of the market portfolio. So assets (or portfolios) with a $\hat{\beta}_j$ greater than one are regarded as risky, while those with $\hat{\beta}_j$ smaller than one are much less sensitive to market fluctuations.

The validity of CAPM depends upon

- the existence of a significant linear relationship (so we shall look at the R^2);
- a zero intercept (so we shall test the hypothesis that $\alpha_j = 0$);
- uncorrelated normally distributed error terms (which we shall check by means of a residual analysis);
- all of this being stable (which we shall check by testing the constancy of the beta's over time).

Empirical Tests

In order to test the CAPM, we consider the daily returns on five stocks of one of the utility subindexes of the Dow Jones Industrial Average. We use the S&P 500 index as a proxy for the market portfolio, and we use the yield on the 13 weeks T-bill (see Section 3.8 for the definition of this instrument) as a proxy for the risk-free rate of borrowing from which the excess returns are computed. In the late nineties, gas and electricity trading became an important source of revenues. Deregulation raised high expectations, and speculative trading overshadowed hedging and risk management. These companies experienced growth throughout this period. Things changed dramatically in 2000. The California crisis and Enron's bankruptcy ignited a sudden reversal in trading activity, and a spectacular downfall for the sector. So we divided the data into the period starting 01/01/1995 and ending 01/01/1999, and the period starting 01/01/1999 and ending 01/01/2003. Figure 3.11 shows the results of simple linear regressions of the American Electric Power (AEP) weekly excess returns against the market excess returns over the first period on the left, and over the second period on the right. In each case we plotted both the least squares and least absolute deviations regression lines. The beta of the second period is much bigger than the beta of the first period, which is consistent with our interpretation of a risky stock in terms of the size of its beta. Only looking at this plot may not show this fact. Indeed it is partially masked by the fact that the scales on the vertical axes are not the same. See Tables 3.3 and 3.4 for the exact values of the least squares estimates of the betas. In the case of the first period, the shape of the cloud of points is different, and the robust estimate of β_{AEP} is greater.

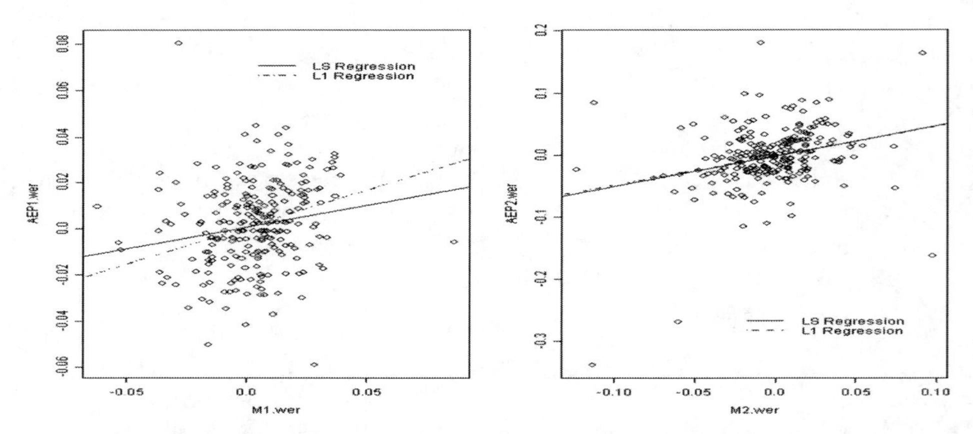

Fig. 3.11. Least squares (solid line) and least absolute deviations (dashed line) regressions of the American Electric Power (AEP) weekly excess returns against the market excess returns for the period starting 01/01/1995 and ending 01/01/1999 (left) and for the period starting 01/01/1999 and ending 01/01/2003 (right).

Figure 3.12 shows the results of the same linear regression analysis in the case of Texas Utilities (TXU). The results are qualitatively the same.

Next we test the null hypothesis of a zero intercept in the case of five of the largest electric company in the Dow Jones Utility Index. We reproduce the intercept estimates and the p-values of the t-test in Tables 3.3 and 3.4 for the two periods. In all cases, the test statistic could not reject the null hypothesis of a zero intercept. However, looking carefully at the estimates of the betas shows clearly that these betas change from one period to the next. But the academic literature is populated with many papers arguing that the model should be rejected on the basis of empirical

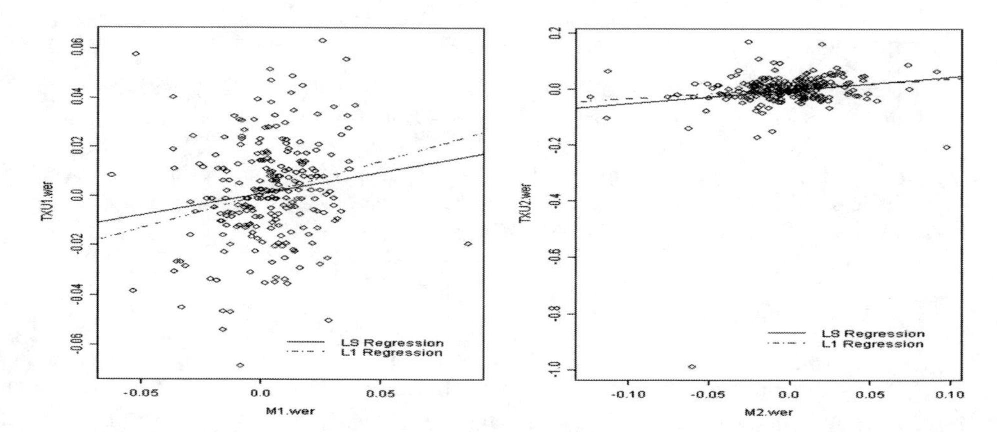

Fig. 3.12. Least squares (solid line) and least absolute deviations (dashed line) regressions of the Texas Utilities (TXU) weekly excess returns against the market excess returns for the period starting 01/01/1995 and ending 01/01/1999 (left) and for the period starting 01/01/1999 and ending 01/01/2003 (right).

	AEP Estimate(p-value)	**DUKE** Estimate(p-value)	**PCG** Estimate(p-value)	**SO** Estimate(p-value)	**TXU** Estimate(p-value)
Intercept	0.0009(0.4604)	0.0015(0.2826)	0.0011(0.5233)	0.0008(0.5691)	0.0011(0.4286)
Slope	0.1861(0.0050)	0.1563(0.0326)	0.0127(0.8893)	0.2520(0.0010)	0.1762(0.0176)

Table 3.3. Intercept and slope estimates, and corresponding p-values (in parentheses) for the least squares regression tests of the CAPM for a set of five electric companies over the period starting 01/01/1995 and ending 01/01/1999.

tests of this type. See the Notes and Complements for further discussion of the issue. In any case, our numerical results show that the stability of the betas over time (and consequently the choice of the periods over which one should estimate them) can become a serious issue. In order to avoid this difficult challenge, and in order to satisfy the pundits debating the validity of the model, we generalize the CAPM to time varying betas in Chapter 6, and we apply the filtering theory developed in that chapter to estimate these *changing betas*.

3.6 Polynomial Regression

After a short excursion into the realm of multivariate regression, we turn our attention back to simple regression (i.e. one real-valued explanatory variable x) and we force the regression function φ to be a polynomial. In other words, we restrict ourselves to regression functions of the form:

	AEP Estimate(p-value)	DUKE Estimate(p-value)	PCG Estimate(p-value)	SO Estimate(p-value)	TXU Estimate(p-value)
Intercept	-0.0009(0.7621)	-0.0010(0.7307)	-0.0029(0.5813)	0.0021(0.3492)	-0.0027(0.5840)
Slope	0.4982(0.0000)	0.1563(0.0000)	0.4294(0.0180)	-0.0184(0.8074)	0.4834(0.0052)

Table 3.4. Intercept and slope estimates, and corresponding p-values (in parentheses) for the least squares regression tests of the CAPM for a set of five electric companies over the period starting 01/01/1999 and ending 01/01/2003.

$$\varphi(x) = \beta_0 + \beta_1 x + \beta_2 x^2 + \cdots + \beta_p x^p \tag{3.28}$$

for some integer $p \geq 1$ and unknown parameters β_0, β_1, ..., β_p. This type of regression generalizes the simple linear regression which we saw earlier, since the latter corresponds to the case $p = 1$. As such it appears as a very attractive way to resolve some of the shortcomings we noticed. Unfortunately, polynomial regression can be very poor, especially when it comes to prediction of the response for values of the explanatory variables outside the range of the data. So our advice is to avoid its use unless one has VERY GOOD REASONS to believe that the model (i.e. the true regression function φ) is indeed a polynomial.

3.6.1 Polynomial Regression as a Linear Model

Even though a polynomial of degree p is a nonlinear function of the variable x when $p > 1$, it can be regarded as a linear function (affine to be specific) of the variables x, x^2, ..., and x^p. So, we could fit a linear model to the data as long as the observations x_1, x_2, ..., x_n on the single univariate explanatory variable x are replaced by n observations of x and its powers which we collect together in a design matrix:

$$\mathbf{X} = \begin{bmatrix} 1 & x_1 & x_1^2 & \cdots & x_1^p \\ \vdots & \vdots & \vdots & & \vdots \\ 1 & x_n & x_n^2 & \cdots & x_n^p \end{bmatrix}$$

But we should not have to do that, `S-Plus` should do it for us. The `S-Plus` command used for polynomial regression is:

```
lm(response ~ poly(regressor,degree))
```

to which we may want to add the `data.frame` by setting the parameter `data`.

3.6.2 Example of `S-Plus` Commands

We use the data set `FRWRD` containing the values in US $ of the 36 Henry Hub natural gas forward contracts traded on March 23rd 1998. We use these prices as our response variable, and we use the index of the delivery month of the forward contract as explanatory variable. For the sake of simplicity we use the integers $1, 2, \ldots, 36$. This example may not be representative of most regression problems, for, in general, the explanatory variable does not take values in a regular deterministic grid. We chose it for the sake of illustration: natural gas forward curves have a strong seasonal component which distinguishes them from most of the other commodity forward curves. Figure 3.13 gives the results of several polynomial regressions of the forward prices against the month indexes. We used degrees 3, 6 and 8 respectively. The plots were produced with the `S-Plus` commands:

```
> plot(1:36,FRWRD,main="Polynomial Gas Forward Curves")
> lines(1:36,fitted(lm(FRWRD~poly(1:36,3))))
> lines(1:36,fitted(lm(FRWRD~poly(1:36,6))),lty=3)
> lines(1:36,fitted(lm(FRWRD~poly(1:36,8))),lty=6)
> legend(locator(1),c("Degree=3","Degree=6","Degree=8"),
                                    lty=c(1,3,6),bty="n")
```

The estimates fitted to the response variable are extracted from the `lm` object by the function `fitted`. The points corresponding to the fitted values have been joined by straight line segments by the `S-Plus` function `lines`. This produces a continuous curve and gives the impression that the plot of the polynomial is actually given, even though we only plotted broken lines between the fitted points. More on the properties of this function `lines` later at the end of this chapter. None of the polynomial fits

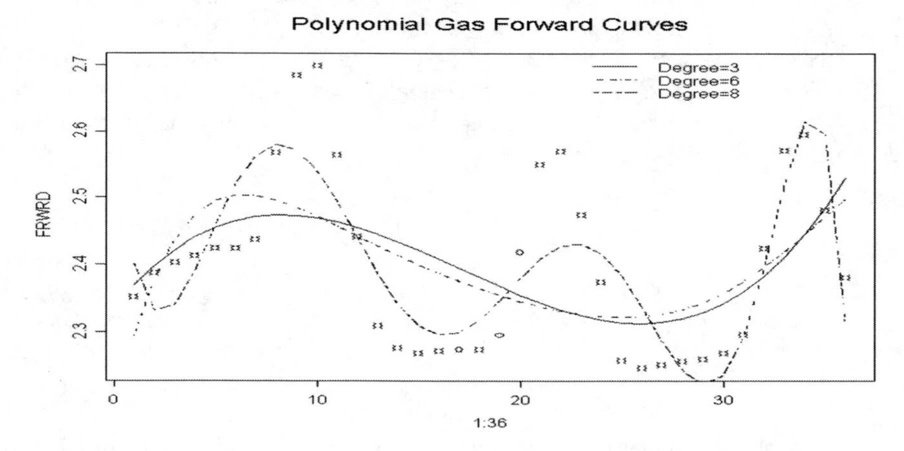

Fig. 3.13. Scatterplot of the prices on March 23rd, 1998 of the 36 traded natural gas forward contracts, together with the results of three different polynomial regressions.

given in Figure 3.13 is very good. Indeed, the forward prices do not seem to be a polynomial function of the month of maturity, and it would take a much higher degree to get a satisfactory fit throughout the 36 months period.

3.6.3 Important Remark

Polynomial regression, as we just introduced it, is a particular case of the estimation of the regression function φ when the latter is known to belong to a specific vector space. In the present situation the vector space is the vector space of polynomials of degree at most p. This space has dimension $p+1$. Indeed, the special polynomials $\mathbf{1}$, x, x^2, ..., x^p form a basis for this space, since any polynomial of degree at most p can be decomposed in a unique way as a linear combination of these particular $(p+1)$ polynomials. So any element of this vector space is entirely determined by $(p+1)$ numbers, the coefficients of the decomposition in this basis. This is why polynomial regression is a particular case of parametric statistics, and this is how we recast this seemingly nonlinear simple regression into the framework of a linear multivariate regression. But vector spaces have many different bases. In fact, one of the very nice features of vector algebra is the possibility for changing basis. Indeed, re-expressing a problem in a different basis may sometimes make it easier to understand and to handle.

Since what matters to us is the function φ and not so much the way in which it is parameterized, or the specific basis in which one chooses to decompose it, it should not matter how `S-Plus` handles the linear model set up to perform a polynomial regression. In fact, even though we like to think of the particular basis $\{\mathbf{1}, x, x^2, \ldots, x^p\}$ when we introduce a polynomial of degree at most p, there is no reason why `S-Plus` should work with this particular basis. In other words, there is no reason why the `S-Plus` function `poly` could not create the design matrix $\mathbf{X}$ by expressing the polynomial candidate for the regression function in a different basis of the same vector space of polynomials. In fact, this is exactly what it does. This is `S-Plus` internal politics which we should not have to be concerned with. More on this subject later in the appendix at the end of this chapter when we discuss `S-Plus`' idiosyncrasies.

3.6.4 Prediction with Polynomial Regression

Whether it is for speculative reasons or for hedging future price exposures, owners and operators of gas fired power plants are often involved in long term contracts with maturities going far beyond the longest maturity appearing in the publicly available forward curve data. They are not the only ones to do that, and some amateurs can be burned at this game. The California electricity crisis is full of instances of such mishaps. In any case, the prediction of the forward prices for these maturities is a challenge, and we show why one should not use polynomial regression in the case of natural gas. Having fitted a polynomial to the existing curve, it is straightforward to compute the value of this fitted polynomial for any time to maturity. A command of the form `sum(COEFS*(48^(0:8)))` will do just that for a time to maturity of 48 months if the coefficients of the polynomial are stored in the components of the vector `COEFS`. Indeed we should have:

$$\texttt{sum(COEFS} * (48\hat{}(0:8))) = \texttt{COEFS}[1] * 48^0 + \texttt{COEFS}[2] * 48^1 + \cdots + \texttt{COEFS}[9] * 48^8$$

which is exactly the desired evaluation of the value of the polynomial. Checking the prediction for times to maturity equal to 4 and 6 years gives:

```
> FPOLY <- lm(FRWRD~poly(1:36,8))
> COEFS <- poly.transform(poly(1:36,8),coef(FPOLY))
> PRED <- sum(COEFS*(48^(0:8)))
> PRED
[1] -271.0561
> PRED <- sum(COEFS*(72^(0:8)))
> PRED
[1] -42482.44
```

This is obviously not right, these numbers should not be negative, and not be so large. In order to illustrate how bad the situation is, we compute the values of the fitted polynomial for all the monthly maturities ranging from 0 to 48 months, and we superimpose these values on the scatterplot of the values of the forward prices for the first 36 months.

```
> TIME <- 1:48
> FPRED <- rep(0,48)
> for (I in 1:48) FPRED[I] <- sum(COEFS*(I^(0:8)))
> plot(TIME[1:36],FRWRD,xlim=c(-1,50),ylim=c(-300,5),
        main="Polynomial Gas Forward Price Prediction")
> abline(h=0,v=0)
> lines(TIME,FPRED)
```

The results are given in Figure 3.14. The scale of the values for the times to maturity

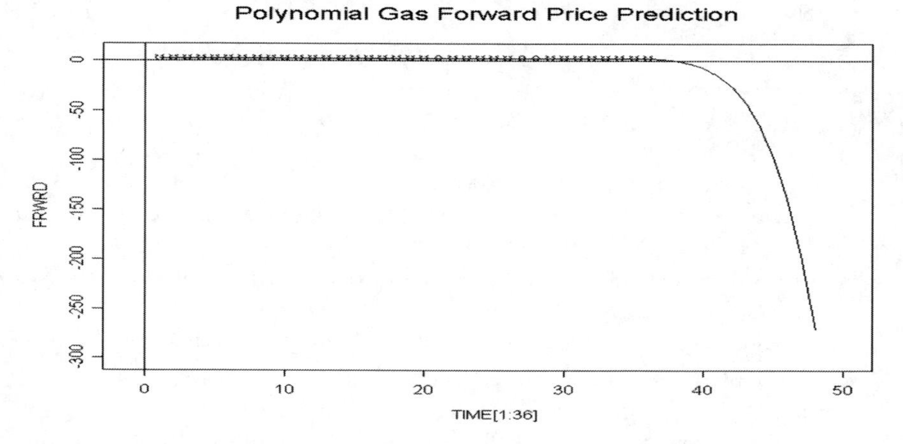

Fig. 3.14. Prediction of the price of a 48 month natural gas forward contract using the best of the polynomial regressions done on March 23rd, 1998.

larger than 3 years is so large that the original part of the forward curve is dwarfed to the point of looking like a horizontal straight line. We use this example to emphasize one more time how inappropriate it can be to try to predict the response for values of the regressor located too far outside the range of the data.

3.6.5 Choice of the Degree p

The choice of degree p is a particular case of the delicate problem of the choice of the dimension of a model. A larger degree (in general a large number of explanatory variables) leads to a smaller sum of squares errors and a larger R^2, which is desirable. The case of polynomial regression is a good example with which to illustrate this fact. If the sample is of size n, for most data sets, it is possible to find a polynomial of degree $n+1$ going through all the points and, consequently, providing a ZERO sum of squares !!! But if the data is noisy (i.e. if the variance σ^2 of the noise is not 0) the fit provided by this polynomial of degree $n+1$ is very unsatisfactory: we are fitting the noise, and this is not desirable.

In fact, too large of a degree will produce absurd predictions, especially if the model is used to predict values of the response variable for values of the regressor outside the range of the data. The general principle is:

BE PARSIMONIOUS.

Criteria (mostly of an intuitive nature) have been designed to help with the choice of dimension of a linear model (and in particular with the degree of a polynomial regression). The most popular of them seem to be based on principles of information theory and entropy measures. `S-Plus` provides several of them, but to our astonishment, the most popular seems to be the (in)famous AIC which stands for Akaike Information Criterion. Its use is widespread despite well-documented flaws. We will reluctantly conform to the common practice and shall mention its use throughout.

3.7 NONLINEAR REGRESSION

We have already encountered several examples of simple regression for which the regression function $\varphi(x)$ was a nonlinear function of the explanatory variable x. However, in all these cases, the estimation of φ was based on the solution of a linear model. We now consider examples of models which cannot be reduced to linear models. We shall apply the techniques developed in this section to the constructions of yield curves used by the various Central Banks all over the world.

We present the main ideas of nonlinear regression on an example, and we use the same type of illustration as in our discussion on polynomial regression: as before we use the analysis of commodity forward curves as a test bed. The explanatory variable is again the index of the month of delivery for a forward contract, and for the sake of simplicity we use the integers $1, 2, \ldots, 18$ which we put in a vector `x`. The response variable y is now the forward price of a crude oil contract. The main difference with the natural gas forward curve considered earlier is the lack of seasonality. Typical crude oil forward curves can be either increasing (we say that they are in backwardation) or decreasing (in which case we say that they are in contango). But in any case, they are almost always monotone with significant curvature.

A First Model

We propose to fit the forward quotes with a function of the form:

$$y = \varphi_\theta(x) = F_\infty \frac{x+K}{x+1}, \tag{3.29}$$

where the constant F_∞ has the interpretation of an asymptotic forward price while K is a positive constant. Economic theory justifies the existence of a limiting price F_∞ representing the cost of production. We use the notation $\theta = (F_\infty, K)$ for the parametrization of the regression function φ_θ. We choose the parametric form (3.29) because the forward curve is in contango when $K < 1$ and in backwardation when $K > 1$, and obviously flat when $K = 1$. Least squares regression can be used to determine the values of the parameter $\theta = (F_\infty, K)$ which minimize the sum of squares, giving:

$$\hat{\theta} = \arg\inf_\theta \sum_i |y_i - \varphi_\theta(x_i)|^2.$$

Obviously, these functions φ_θ are nonlinear. But they form a specific family parameterized by the two-dimensional parameter θ, and given the observations (x_i, y_i) contained in the data, one can try to use a minimization procedure to look for the optimal θ. This can be done using the `S-Plus` function `nls` whose call is similar to a call to `lm` in the sense that a formula and possibly a reference to the data frame used are required. In the present situation the formula will have to be nonlinear. It should read:

```
y ~ FINF*(x+K)/(x+1)
```

if we use the `S` objects `FINF` and `K` for the parameters F_∞ and K, respectively. We perform the nonlinear regression with the command:

```
> Ffit <- nls(y ~ FINF*(x+K)/(x+1), start=c(FINF=17,K=.1))
```

Should the explanatory and response variables be columns of a data frame, the name of this data frame should be passed to the function `nls` by setting the parameter `data`. The function `nls` relies on an iterative optimization routine, and this makes it somewhat unpredictable. So, it is always a good idea to initialize the optimization by providing reasonable values for the parameters. The initial values for the parameters are passed with the argument `start`. We chose to initialize F_∞ to 17 because looking at the scatter plot of (x, y), it appears that this value is in the range of the asymptotic value of the forward price. Also, since the forward curve seems to be in contango, we start K with a value smaller than 1. The numerical results produced by the command above can be viewed using the command `summary`

```
> summary(Ffit)
    Formula: y ~ (FINF * (x + K))/(x + 1)
 Parameters:
        Value Std. Error  t value
FINF 17.631500  0.0338120 521.4560
   K  0.790918  0.0139659  56.6321
 Residual standard error: 0.143233 on 32 degrees of freedom
 Correlation of Parameter Estimates:
    FINF
K -0.672
```

or they can be plotted with the commands:

```
> plot(x,y,main="Crude Oil Forward Curve")
> lines(x, fitted(Ffit))
```

As before, we use the function `fitted` to extract the fitted values from the `nls` object `Ffit`. The results are given in the left pane of Figure 3.15.

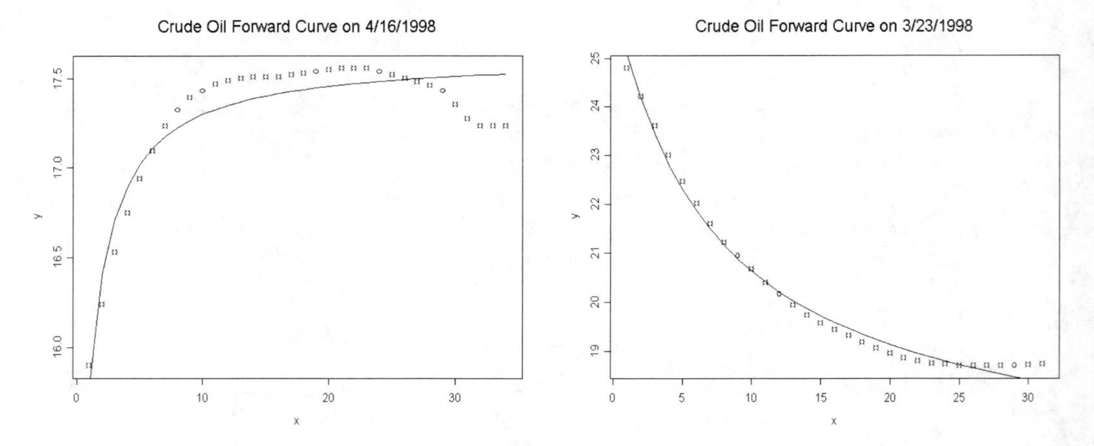

Fig. 3.15. Crude oil forward curves produced by nonlinear regression.

Transformation of the Variables

As we are about to see in the next subsection, the above model will be abandoned because it cannot give a reasonable account of some of the forward curve shapes which appear on a regular basis. We kept it anyway because of its simplicity and because of the following thought-provoking remark. Notice that:

$$\varphi_\theta(x) = F_\infty \frac{x+K}{x+1} = F_\infty \left(1 + (K-1)\frac{1}{x+1}\right),$$

so that if we set $z = 1/(x+1)$, then the model (3.29) can be rewritten as:

$$y = \beta_0 + \beta_1 z$$

if we set $\beta_0 = F_\infty$ and $\beta_1 = F_\infty(K-1)$. So it should be possible to obtain the results of the nonlinear regression used earlier with a simple linear regression of the response variable y against the explanatory variable z, and a simple transformation of the estimates of the slope and intercept to get back to estimates of the original parameters F_∞ and K. Recall that maximum likelihood estimates of transformed parameters can be obtained by transforming the maximum likelihood estimates of the parameters, so our strategy is consistent when our nonlinear least squares estimates are maximum likelihood estimates, and this typically is the case when the noise terms form a Gaussian white noise. This transformation idea can be implemented in `S-Plus` in the following way:

```
> z <- 1/(x+1)
> LFit <- lm(y~z)
> LFINF <- coef(LFit)[1]
> LFINF
  17.63148
> LK <-1+ coef(LFit)[2]/LFINF
> LK
  0.7909179
```

So the linear regression provides exactly the same estimates for the parameters F_∞ and K. This is a beautiful example of a situation where a clever transformation of one of the variables reduces the complexity of the problem. However, it is important to keep in mind that recovering the same values for the estimates is not the whole story. Indeed, both the function `nls` and the function `lm` provide a suite of test statistics, confidence intervals and regions, and regression diagnostic tests, and the correspondence between these goodies is not so clear any more. Let us give an example for the sake of illustration. The function `lm` returns the correlation coefficient between the estimates of the slope β_1 and the intercept β_0. However, the significance of this number for the correlation between the parameters of interest F_∞ and K, is anyone's guess, because these parameters are nonlinear functions of β_0 and β_1.

A Second Model

The right pane of Figure 3.15 gives an example of crude oil forward curve in backwardation, also produced by nonlinear regression. The data are from April 16, 1998. We originally tried to fit a curve from the parametric family of functions φ_θ given by (3.29) but the results were very poor. Because it can only play with two parameters, the fitting procedure could not match the curvature of the forward curve on that day. So we decided to introduce a third parameter to give the fitting procedure more leeway. Implementing the nonlinear regression from the parametric family:

$$y = \varphi_\theta(x) = F_\infty \frac{x + K_1}{x + K_2}$$

with the command:

```
> Ffit<-nls(y~FINF*(x+K1)/(x+K2),start=c(FINF=17,K1=2,K2=1))
```

we obtained the results reproduced in the right pane of Figure 3.15. This confirms the general philosophy that more parameters produce a better fit. But as we explained earlier, following this trend too systematically can lead to disaster, especially in the presence of random noise (which is not obvious in the present situation).

Nonlinear regression is a very touchy business, and it should not be practiced without a license !

3.8 TERM STRUCTURE OF INTEREST RATES: A CRASH COURSE

The size and level of sophistication of the market of fixed income instruments increased dramatically over the last 20 years, and it has become a prime test bed for financial institutions and academic research. The fundamental object to model is the term structure of interest rates. We approach it via the prices of treasury bond issues. Models for these prices are crucial for pricing the important swap contracts and liquid derivatives such as caplets, swaptions, etc., quantifying and managing financial risk, and setting monetary policy. We mostly restrict ourselves to Treasury issues to avoid having to deal with the possibility of default. The highly publicized defaults of counties (such as the bankruptcy of Orange County in 1994), of sovereigns (like Russia defaulting on its bonds in 1998) and the ensuing ripple effects on worldwide markets have brought the issue of credit risk to the forefront. However, a discussion of credit risk in this context would be beyond the scope of this book.

In order to be in a position to tackle some of the fundamental statistical issues of the bond markets, we need a crash course on the mechanics of interest rates and the fixed income securities. However, in order to make our life easier, we assume that all the bonds used are default free, that there have no embedded options, no call or convertibility features, and we ignore the effects of taxes and transaction costs.

Zero Coupon Bonds

We first introduce the *time value of money* by valuing the simplest possible fixed income instrument. It is a financial instrument providing cash flow with a single payment of a fixed amount (the principal X) at a given date in the future. This date is called the maturity date. If the time to maturity is exactly n years, the present value of this instrument is:

$$P_X(n) = \frac{1}{(1+r)^n} X. \tag{3.30}$$

So this formula gives the present value of a nominal amount X due in n years time. Such an instrument is called a *discount bond* or a *zero coupon bond.* The positive number r is referred to as the (yearly) *discount rate* or *spot interest rate* for time to maturity n, since it is the interest rate which is applicable today (hence the terminology *spot*) on an n-year loan. Formula (3.30) gives a one-to-one correspondence between bond prices and interest rates. More generally, at any given time t, we denote by $P(t, t+m)$ the price of a zero coupon bond with unit principal and time to maturity m, or maturity date $T = t+m$. This price can be expressed in terms of an interest by the formula:

$$P(t, t+m) = \frac{1}{(1 + r(t, t+m))^m}, \tag{3.31}$$

where $r(t, t+m)$ is the yearly spot interest rate prevailing at time t for time of maturity $T = t+m$. We assumed implicitly that the time to maturity $\tau = T - t$ is a whole number of years. This definition can be rewritten in the form:

$$\log(1 + r(t, t+\tau)) = -\frac{1}{\tau} \log P(t, t+\tau)$$

and considering the fact that $\log(1+x) \sim x$ when x is small, the same definition gives the approximate identity:

$$r(t, t+\tau) \sim -\frac{1}{\tau} \log P(t, t+\tau)$$

which becomes an exact equality if we use continuous compounding. We use the Greek letter τ for the time to maturity $\tau = T - t$. This formula justifies the terminology discount rate for r. Considering payments occurring in m years time, the spot rate $r(t, t+\tau)$ is the single rate of return used to discount all the cash flows for the discrete period from time t to time $t+m$. As such, it appears as some sort of composite of interest rates applicable over shorter periods. Moreover, this formula offers a natural generalization to continuous time models with continuous compounding of the interest. In this case, it reads:

$$P(t, T) = e^{-(T-t)r(t,T)}. \tag{3.32}$$

Coupon Bearing Bonds

Treasury Bills are the perfect example of zero coupon bonds. They are issued and sold at auctions on a regular basis. Their maturities are shorter than one year, and they may not be very useful when it comes to understanding bonds with much longer maturities. Most Treasury bonds carry coupons. In general, what is called a bond (or a coupon bearing bond), is a regular stream of future cash flows. To be more specific, a *coupon bond* is a series of payments amounting to C_1, C_2, ..., C_m, at times T_1, T_2, ..., T_m, and a nominal payment X at the maturity date T_m. X is also called the face value, or principal value of the bond. The bond price at time t should be given by the formula:

$$P(t) = \sum_{j=1}^{m} C_j P(t, T_j) + X P(t, T_m). \tag{3.33}$$

This all purpose formula can be specialized advantageously, for in most cases, the payments C_j's are coupon payments made at regular time intervals. Coupon payments C_j are most often quoted as a percentage c of the face value X of the bond. In other words, $C_j = cX$. This percentage is given as an annual rate, even though payments are usually made every six months in the US, or different frequencies depending upon the country. It is convenient to introduce a special notation, say n_y, for the number of coupon payments per year. For example, $n_y = 2$ for coupons paid semi-annually. If we denote by $r_1, r_2, \ldots, r_m$ the interest rates for the m periods ending with coupon payment dates $T_1, T_2, \ldots, T_m$, then the present value of the bond cash flow is given by the formula:

$$\begin{aligned} P &= \frac{C_1}{1 + r_1/n_y} + \frac{C_2}{(1 + r_2/n_y)^2} + \cdots + \frac{C_m}{(1 + r_m/n_y)^m} \\ &= \frac{cX}{n_y(1 + r_1/n_y)} + \frac{cX}{n_y(1 + r_2/n_y)^2} + \cdots + \frac{cX + X}{n_y(1 + r_m/n_y)^m}. \end{aligned} \tag{3.34}$$

Note that we divided the rates r_n by the frequency n_y because the rates are usually quoted in years. Formulae (3.33) and (3.34) are often referred to as the *bond price equations.* An important consequence of these formulae is the fact that on any given day, the value of a bond is entirely determined by the discount curve (i.e. the available sequence of discrete observations of the function $T \hookrightarrow P(t, T)$ on that day).

Remarks

1. Reference to the present date t will often be dropped from the notation when no confusion is possible. Moreover, instead of working with the absolute dates T_1, $T_2, \ldots, T_m$, which can represent coupon payment dates as well as maturity dates of various bonds, it will be often more convenient to work with the times to maturities, which we denote by $\tau_1 = T_1 - t$, $\tau_2 = T_2 - t$, $\ldots$, $\tau_m = T_m - t$. We will use whatever notation is more convenient for the discussion at hand.

2. Unfortunately for us, bond prices are not quoted as a single number. Instead, they are given by a *bid-ask* interval. We ignore the existence of this bid-ask spread in most of the discussion that follows, collapsing this interval to a single value by considering its midpoint only. We shall re-instate the bid-ask spread in the next section when we discuss the actual statistical estimation procedures.

3. Formula (3.33) shows that a coupon bearing bond can be viewed as a composite instrument comprising a zero coupon bond with the same maturity T_m and face value $(1 + c)X/n_y$, and a set of zero coupon bonds whose maturity dates are the coupon payment dates T_j for $1 \leq j < m$ and face value cX/n_y. This remark is much more than a mere mathematical curiosity. Indeed, the principal and the interest components of some US Treasury bonds have been traded separately under the

Treasury STRIPS (Separate Trading of Registered Interest and Principal Securities) program since 1985.

Constructing the Term Structure by Linear Regression?

In principle, formula (3.33) above can be used to recover the prices of the zero coupon bonds in terms of the coupon bonds quoted on the market. Indeed, let us assume for example that on a given day t, we have access to the prices $P_i(t)$ of n instruments whose future cash flows are given by payments $C_{i,j}$ at the times $T_1 < T_2 < \cdots < T_m$. Notice that we assume that the payments are made at the same time T_j (instead of the auction anniversary dates). In this case, formula (3.33) can be rewritten in the form:

$$P_i(t) = \sum_{j=1}^{m} C_{i,j} P(t, T_j),$$

which shows that the vector $\mathbf{P}(t, \cdot)$ of the prices of the zero coupon bonds can be recovered from the vector $\mathbf{P}(t)$ of the quoted coupon bonds $P_i(t)$ when the matrix $C = [C_{i,j}]_{i,j}$ is invertible, in which case we have:

$$\mathbf{P}(t, \cdot) = C^{-1}\mathbf{P}(t).$$

Once the zero coupons are determined for $T = T_j$, one produces a full term structure $T \hookrightarrow P(t, T)$ by mere linear interpolation. Unfortunately, the coupon payments of the coupon bearing bonds priced on the market do not take place on the same dates, and the resulting matrix C is practically never invertible. A way to overcome this problem is to assume that the prices observed on the market are in fact noisy perturbations, and to assume that in fact:

$$P_i(t) = \sum_{j=1}^{m} C_{i,j} P(t, T_j) + \epsilon_i, \qquad i = 1, \ldots, n$$

and to extract the coefficients $P(t, T_j)$ by an ordinary least squares multiple regression. This procedure is hardly ever used in practice and we shall not discuss it any further.

Clean Prices & Duration

Formulae (3.33) and (3.34) implicitly assumed that t was the time of a coupon payment, and consequently, that the time to maturity was an integer multiple of the time separating two successive coupon payments. Because of the nature of the coupon payments occurring at isolated times, the prices given by the bond pricing formula (3.33) are discontinuous, in the sense that they jump at the times the coupons are paid. This is regarded as an undesirable feature, and systematic price corrections are routinely implemented to remedy the jumps. Since the bond price jumps by the amount cX/n_y at the times T_j of the coupon payments, the most natural way to smooth the discontinuities is to adjust the bond price for the *accrued interest* earned by the bond holder since the time of the last coupon payment. This notion of accrued interest is quantified in the following way. If the last coupon payment (before the present time t) was made on date T_n, then the accrued interest is defined as the quantity:

$$AI(T_n, t) = \frac{t - T_n}{T_{n+1} - T_n} \frac{cX}{n_y}, \tag{3.35}$$

and the *clean price* of the bond is defined by the requirement that the transaction price be equal to the clean price plus the accrued interest. In other words, if $T_n \leq t < T_{n+1}$, the clean price $CP(t, T_m)$ is defined as:

$$CP(t, T_m) = P_{X,C}(t, T_m) - AI(t, T_n)$$

where $P_{X,C}(t, T_m)$ is the transaction price given by (3.33) with the summation starting with $j = n + 1$. Notice that in all cases, the price of a bond appears as a multiple of its nominal value X. Since the role of the latter is merely a multiplicative factor, we shall assume, without any loss of generality that $X = 1$ from now on.

The maturity of a zero coupon bond measures the length of time the bond holder has invested his money, but it is desirable to have an analog for the case of coupon bearing bonds. Since such a bond can be viewed as a sequence of individual payments, a natural proxy could be the expected maturity of all these payments. This is the concept of duration proposed by Macaulay. For the sake of simplicity, we give its definition when the maturity is equal to an integer multiple of the length of time between two consecutive coupon payments. The duration at time t of a bond with annual coupon payment C, nominal value $X = 1$, and time to maturity m is given by the formula:

$$D_C(t, m) = \frac{C}{1 + Y_C(t, m)} + 2\frac{C}{(1 + Y_C(t, m))^2} + \cdots + m\frac{1 + C}{(1 + Y_C(t, m))^m}, \tag{3.36}$$

where $Y_C(t, m)$ is the yield of the bond, i.e. the number such that we can write the price $P_C(t, m)$ of the bond in the form:

$$P_C(t, m) = \frac{1}{(1 + Y_C(t, m))^m}. \tag{3.37}$$

The following two properties of the Macaulay duration fit well with the intuition behind the above definition. The duration of a bond is always smaller than its actual maturity. Moreover, the duration of a zero coupon bond (corresponding to the case $C = 0$) is equal to the actual maturity of the bond.

The concept of duration plays an important role in the immunization of fixed income portfolios, but a discussion of the details would take us far beyond the scope of this book. We shall limit the use of the duration to the weighting of various bond prices in the least squares term structure estimation procedures which we discuss below.

The Three Different Forms of Term Structure

The above discussion was aimed at identifying the one-to-one correspondence between the prices of discount bonds and some interest rates. Recall that we assume that $X = 1$. We also hinted at the fact that, when the compounding frequency was increasing without bound, the interest compounding formula

$$P(t, T = t + m) = \frac{1}{(1 + r(t, t + m)/n)^{nm}}$$

converged toward its continuously compounded analog

$$P(t, T = t + m) = e^{-Y(t,t+m)(T-t)}.$$

This formula can easily be inverted to give:

$$Y(t, T) = -\frac{1}{T - t} \log P(t, T).$$

The continuously compounded interest rate $Y(t, T)$ prevailing at time t for the maturity T is sometimes called the yield, and the curve $t \hookrightarrow Y(t, T)$ is called the yield curve. Practitioners use still a third way to capture the term structure of interest rates. It requires the notion of instantaneous forward rate which we shall only define in the continuous case. The instantaneous forward rate $f(t, T)$ prevailing at time t for the date of maturity T is defined by:

$$f(t, T) = -\frac{\frac{d}{dT} P(t, T)}{P(t, T)} = -\frac{d}{dT} \log P(t, T). \tag{3.38}$$

This definition implies that:

$$P(t, t + \tau) = e^{-\int_0^\tau f(t,t+u)du} \tag{3.39}$$

and in terms of the spot rate or yield:

$$Y(t, t + \tau) = -\frac{1}{\tau} \int_0^\tau f(t, t + u)du. \tag{3.40}$$

Even though we shall not use this fact, it is easy to see that this relation can be inverted to express the forward rates as a function of the spot rates:

$$f(t, T) = r(t, T) + (T - t)r'(t, T). \tag{3.41}$$

So on any given day t, the term structure of interest rate can be given by any one of the following three curves:

$$x \hookrightarrow P(t, t + x)$$
$$x \hookrightarrow Y(t, t + x)$$
$$x \hookrightarrow f(t, t + x)$$

We shall take advantage of this convenience when it comes to estimating the term structure from available data.

3.9 Parametric Yield Curve Estimation

This section reviews the methods of yield curve estimation used by some of the central banks which report to the Bank for International Settlements (BIS for short). Apart from the U.S. and Japan, most central banks use parametric estimation methods to infer smooth curves from daily quotes of prices of bonds and other liquid interest rate derivatives. Caplets and swaptions are most frequently used, but for the sake of simplicity, we shall limit ourselves to bond price data.

We postpone the discussion of nonparametric methods to the next chapter. The use of parametric estimation methods is justified by the principal component analysis performed in Chapter 2. There, we showed that the effective dimension of the space of yield curves is low, and consequently, a small number of parameters should be enough to describe the elements of this space. Moreover, another advantage of the parametric approach is the fact that one can estimate the term structure of interest rates by choosing to estimate first the yield curves, or the forward curves, or even the zero coupon curves as functions of the maturity. Indeed, which one of these quantities is estimated first is irrelevant: once the choice of a parametric family of curves and of their parametrization has been made, the parameters estimated from the observations, together with the functional form of the curves, can be used to derive estimates of the other sets of curves. We shall most often parameterize the set of forward rate curves, and derive formulae for the other curves (yields curves and discount bond curves) by means of the relationships made explicit earlier in formulae (3.38), (3.40) and (3.41).

On any given day, say t, one uses the available values of the discount factors to produce a curve $x \hookrightarrow f(t,x)$ for the instantaneous forward rates as functions of the time to maturity x. For the sake of notational convenience, we shall drop the reference to the present t in most of our discussions below. In this section, we limit ourselves to the fitting of a parametric family of curves to the data. We shall revisit this problem in Section 4.3 of Chapter 4 when we discuss nonparametric smoothing techniques based on splines.

3.9.1 Estimation Procedures

In this section we restrict ourselves to the two most commonly used curve families: the Nelson-Siegel and the Swensson families. We refer to Problem 3.6 for an illustration of the use of a third exponential family. For the sake of simpler notation, we drop the current date t from the notation.

The Nelson-Siegel Family

This family is parameterized by a 4-dimensional parameter $\theta = (\theta_1, \theta_2, \theta_3, \theta_4)$. It is defined by:

$$f_{NS}(x,\theta) = \theta_1 + (\theta_2 + \theta_3 x)e^{-x/\theta_4} \tag{3.42}$$

where θ_4 is assumed to be strictly positive, and as a consequence, the parameter θ_1, which is also assumed to be strictly positive, gives the asymptotic value of the forward rate. The value $\theta_1 + \theta_2$ gives the forward rate today, i.e. the starting value of the forward curve. Since this value has the interpretation of the instantaneous (short) interest rate r_t prevailing at time t, it is also required to be positive. The remaining parameters θ_3 and θ_4 are responsible for the so-called *hump*. This hump does exist when $\theta_3 > 0$, however, it is a dip when $\theta_3 < 0$. The magnitude of this hump/dip is a function of the size of the absolute value of θ_3, while θ_3 and θ_4 govern the location, along the maturity axis, of this hump/dip. Once the four parameters have been estimated, a formula for the zero-coupon yield can be obtained by plain integration from formula (3.40). We get:

$$Y_{NS}(x,\theta) = \theta_1 + (1 - e^{-x/\theta_4}) - \theta_3\theta_4 e^{-x/\theta_4}. \tag{3.43}$$

A formula for the discount factor can be deduced by injecting the yield given by this formula into $P_{NS}(x) = e^{-xY_{NS}(x)}$. The Nelson-Siegel family and these formulae are used in countries such as Finland and Italy to produce yield curves.

The Swensson Family

To improve the flexibility of the curves and the fit, Swensson proposed a natural extension to the Nelson-Siegel's family by adding an extra exponential term which can produce a second hump/dip. This extra flexibility comes at the cost of two extra parameters which have to be estimated. The Swensson family is generated by mixtures of exponential functions of the Nelson-Siegel type. To be specific, the Swensson family is parameterized by a 6-dimensional parameter θ, and defined by:

$$f_S(x,\theta) = \theta_1 + (\theta_2 + \theta_3 x)e^{-x/\theta_4} + \theta_5 x e^{-x/\theta_6}. \tag{3.44}$$

As before, once the parameters are estimated, the yield curve can be estimated by plain integration of (3.44). We get:

$$\begin{aligned} Y_S(x,\theta) = \theta_1 - \frac{\theta_2\theta_4}{x}(1 - e^{-x/\theta_4}) + \theta_3\theta_4\left[\frac{\theta_4}{x}(1 - e^{-x/\theta_4}) - e^{-x/\theta_4}\right] \\ + \theta_5\theta_6\left[\frac{\theta_6}{x}(1 - e^{-x/\theta_6}) - e^{-x/\theta_6}\right]. \end{aligned} \tag{3.45}$$

The Swensson family is used by the Central Banks of many countries including Canada, Germany, France and the UK.

3.9.2 Practical Implementation

Enough talk, let's see how all these ideas work in practice.

Description of the Available Data

On any given day t, financial data services provide, for a certain number of bond issues, the times to maturity $x_j = T_j - t$, the coupon payments and their frequencies, and various pre-computed quantities. We used data from `Data Stream`. For the purposes of illustration, we chose to collect data on German bonds. These instruments are very liquid and according to BIS, the Deutsche Bundesbank uses the Swensson's extension of the Nelson-Siegel family to produce yield curves. As an added bonus, the coupons on the instruments we chose are paid annually, which makes numerical computations easier.

The Actual Fitting Procedure

Let B_j be the bond prices available on a given day t. Let $B_j(\theta)$ be the prices one would get using formula (3.33), with zero coupon values computed from formula (3.39) when the forward curve is given by the element of the parametric family of forward curves determined by the parameter θ. Then our estimate of the term structure of interest rates is given by the zero-coupon / yield / forward curve corresponding to the (vector) parameter $\hat{\theta}^*$ which minimizes the quadratic loss function:

$$\mathcal{L}(\theta) = \sum_j w_j |B_j - B_j(\theta)|^2 \tag{3.46}$$

for a given set of weights w_j which are usually chosen as functions of the duration (3.36) and the yields to maturity. The dependence of the loss function upon the parameters θ appears to be complex and extremely nonlinear. In any case, this least squares estimation procedure is very much in the spirit of the nonlinear regression discussed in Section 3.7. As explained earlier, fitting the parameters depends upon delicate optimization procedures which can be very unstable and computer intensive. For this reason we will give only a limited sample of results below.

Remarks

1. Many Central Banks do not use the full spectrum of available times to maturity. Indeed, the prices of many short-term bonds are very often influenced by liquidity problems. For this reason, they are often excluded from the computation of the parameters. For example each of the Bank of Canada, the Bank of England and the Deutsche Bundesbank consider only those bonds with a remaining time-to-maturity above three months. The French Central Bank also filters out the short term instruments.
2. Even though it appears less general, the Nelson-Siegel family is often preferred to its Swensson relative. Being of a smaller dimension, the model is more robust and less unstable. This is especially true for countries with a relatively small number of issues. Finland is one of them. Spain and Italy are other countries using the original Nelson-Siegel family for stability reasons.

3. The bid-ask spread is another form of illiquidity. Most central banks choose the mid-point of the bid-ask interval for the value of B_j. The Banque de France does this for most of the bonds, but it also uses the last quote for some of them. Suspicious that the influence of the bid-ask spread could overwhelm the estimation procedure, the Finnish Central Bank uses a loss function which is equal to the sum of squares of errors where the individual errors are defined as the distance from $B(j, \theta)$ to the bid-ask interval (this error being obviously 0 when $B(j, \theta)$ is inside this interval).
4. It is fair to assume that most Central Banks use accrued interests and clean prices to fit a curve to the bond prices. This practice is advocated in official documents of the Bank of England and the US Treasury.
5. Some of the countries relying on the Swensson family, first fit a Nelson-Siegel family to their data. Once this 4-dimensional optimization problem is solved, they use the argument they found, together with two other values for θ_5 and θ_6 (often 0 and 1), as initial values for the minimization of the loss function for the Swensson family. Even then, these banks opt for the Swensson family, only when the final θ_5 is significantly different from 0 and θ_6 is not too large! This mixed procedure is used by Belgium, Canada and France.

3.9.3 `S-Plus` Experiments

The following `S-Plus` code can be used to compute values of the forward rate function in the Nelson-Siegel model.

```
fns <- function(x,THETA)
{
  FORWARD<-THETA[1]+(THETA[2]+THETA[3]*x)*exp(-x/THETA[4])
  FORWARD
}
```

Correspondingly, the yield curve can be computed with the help of the function `yns` defined by:

```
yns <- function(x,THETA)
{
  TT <- THETA[3] * THETA[4]
  TTT <- THETA[4] * (THETA[2] + TT)
  EX <- exp( - x/THETA[4])
  YIELD <- THETA[1] + (TTT * (1 - EX))/x - TT * EX
  YIELD
}
```

These two functions were used to produce the graphs of Figure 3.16 from the commands:

```
> THETA <- c(.07,-.03,.1,2.5)
> XX <- seq(from=0,to=30,by=1/12)
> FORW <- fns(XX,THETA)
```

```
> YIELD <- yns(XX,THETA)
> par(mfrow=c(1,2))
> plot(XX,FORW,type="l",ylim=c(0,.25))
> title("Example of a Nelson Siegel Forward Curve")
> plot(XX,YIELD,type="l",ylim=c(0,.25))
> title("Example of a Nelson Siegel Yield Curve")
> par(mfrow=c(1,1))
```

Finally, the price $B_{NS}(\theta)$ of a coupon bond can be computed from its coupon rate

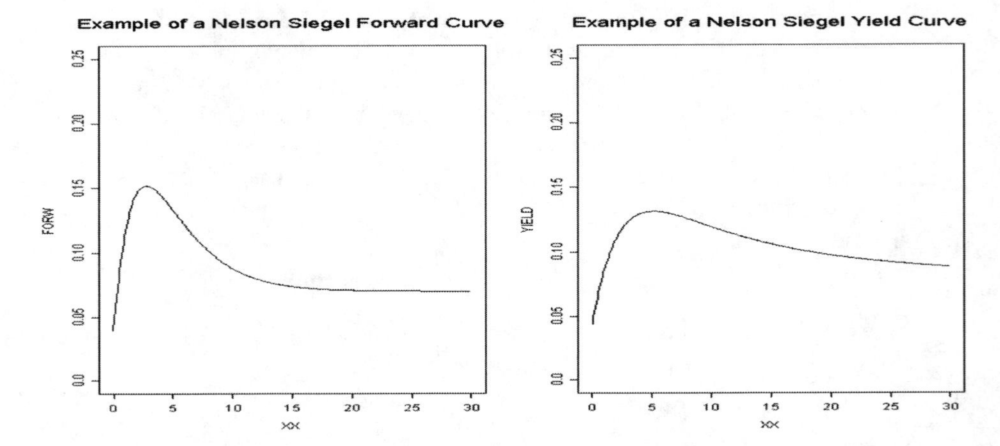

Fig. 3.16. Example of a forward curve (left) and a yield curve from the Nelson-Siegel family with parameters $\theta_1 = .07$, $\theta_2 = -.03$, $\theta_3 = .1$ and $\theta_4 = 2.5$.

COUPON, the accrued interest AI, the time to maturity LIFE given in years, and the parameters THETA of the Nelson-Siegel family, using the function bns, whose code implements the various formulae derived in this subsection.

```
bns <- function(COUPON,AI,LIFE,X=100,THETA=c(0.06,0,0,1))
{
    LL <- floor(1 + LIFE)
    DISCOUNT <- seq(to = LIFE, by = 1, length = LL)
    TT <- THETA[3] * THETA[4]
    TTT <- THETA[4] * (THETA[2] + TT)
    DISCOUNT <- exp( - THETA[1] * DISCOUNT +
            (TTT + TT * DISCOUNT) * exp(-DISCOUNT/THETA[4]))
    CF <- rep((COUPON * X)/100, LL)
    CF[LL] <- CF[LL] + X
    PRICE <- sum(CF * DISCOUNT) - AI
    PRICE
}
```

In order to illustrate the use of this function we chose the example of the German bond quotes available on May 17, 2000. After some minor reformatting and editing to remove the incomplete records, we imported the data into the data frame `GermanB041700`. Figure 3.17 shows the variables we kept for our analysis. We

	1	2	3	4	5	6	7	8
	Issue	Coupon	Maturity	Price	Intrst.Yield	Redemp.Yield	Accrud.Intrst	Life
1	1992	8.00	2002	105.28	7.60	5.20	7.67	2.04
2	1993	6.75	2003	104.37	6.47	5.16	6.47	3.04
3	1999	3.75	2009	90.43	4.15	5.13	1.07	8.72

Fig. 3.17. German bond quotes imported into `S-Plus` for the purpose of testing the nonlinear fit of the Nelson-Siegel family.

compute and minimize the loss (3.46) in the case of the Nelson-Siegel parametrization. We shall denote it by $\mathcal{L}_{NS}(\theta)$. It is given by the formula:

$$\mathcal{L}_{NS}(\theta) = \sum_j w_j |B_j - B_{NS}(j, \theta)|^2. \tag{3.47}$$

The parameters `THETA[j]` are obtained by minimizing the sum of square deviations (3.47). Since we do not know what kind of weights (if any) are used by the German Central Bank, we set $w_j = 1$. We use the `S-Plus` function `ms` to perform the minimization of the sum of square errors.

```
> GB.fit<-ms(~(Price-bns(Coupon,Accrud.Intrst,
                          Life,THETA))^2, data=GermanB041700)
> GB.fit
value: 8222.603 parameters:
     THETA1        THETA2      THETA3          THETA4
     0.06175748   6.12743 7.636384     1.45234
     formula:   ~    (Price - bns(Coupon,Accrud.Intrst,
                                                Life, THETA))^2
     47 observations
     call: ms(formula =   ~ (Price-bns(Coupon,Accrud.Intrst,
                         Life,THETA))^{}2, data=GermanB041700) }
```

3.9.4 Concluding Remarks

The results reported above show an extreme variability in the estimates at the *short end of the curve.* This confirms the widely-admitted fact that the term structure of interest rates is more difficult to estimate for short maturities. This is one of the reasons why many central banks do not provide estimates of the term structure for the left hand of the maturity spectrum.

All in all it seems clear that the various estimates are stable and reliable in the maturity range from one to ten years.

APPENDIX: CAUTIONARY NOTES ON S-Plus IDIOSYNCRACIES

The specific features of S-Plus which we describe in this appendix can be sources of headaches for first time users.

S-Plus & Polynomial Regression

We use a controlled experiment to illustrate an important point made in the text.

The Experiment

We generate a set of observations (x_i, y_i) for which the values of x are uniformly distributed between 0 and 100 and the values of y are, up to a noise which is normally distributed with variance one, given by the polynomial function:

$$y = \varphi(x) = 50 - 43x + 31x^2 - 2x^3.$$

The S commands to do that are:

```
> x <- runif(100,0,100)
> y <- 50 - 43*x + 31*x^2 -2*x^3 + rnorm(100)
```

Given the construction of x and y, we expect that a polynomial regression (using a polynomial of degree 3) of the variable y against the variable x will recover the exact coefficients we started with. Well, let's see:

```
> xylm <- lm(y ~ poly(x,3))
> xylm
Call:
lm(formula = y ~ poly(x, 3))
Coefficients:
 (Intercept) poly(x, 3)1 poly(x, 3)2 poly(x, 3)3
   -465738.7    -4798444    -2052599   -358566.4
Degrees of freedom: 100 total; 96 residual
Residual standard error: 1.012171
```

Obviously, the coefficients given in the S-Plus summary are not what we expected!

So What is Wrong?

It turns out that the estimated coefficients printed out by the program are the coefficients obtained in the decomposition of the function φ in a basis of 4 orthogonal polynomials which are **NOT** the polynomials $1 = x^0$, x, x^2 and x^3. As we explained in the section on polynomial regression, the internal manipulations of the program can be done in any basis. It turns out that, for reasons of numerical efficiency and stability, S-Plus chose to work with a basis of orthogonal polynomials different from the natural basis $\{1, x, x^2, x^3\}$. So how can we get back to our favorite basis, or at least to the basis in which we originally expressed the model? The answer is given by the function poly.transform which recovers the coefficients in the decomposition in the standard polynomial basis. Using it we get:

```
> poly.transform(poly(x,3),coef(xylm))
   x^0        x^1       x^2        x^3
50.39887 -43.02942 31.00065 -2.000004
```

which is what we expected!

Is There Something Wrong with the Function lines?

The function lines is a convenient graphical tool which can be used to produce a continuous curve by joining points on a graph. But as we are about to see, a careless call to this function can produce unexpected results. We use the ethanol data set from the S-Plus distribution to make our point. These data, though non-financial, are very well suited for the illustration we have in mind. They contain 88 measurements of the variable NOx giving the concentration of Nitric Oxide in a series of tests of a single-cylinder automobile engine. The regressor variables used to predict the response are the compression ratio C and the equivalence ratio E. We use the following commands to produce a scatterplot together with a smooth curve giving the general trend in the data. We shall come back to the function loess.smooth in the next chapter on nonparametric regression. It does not play a significant role here.

```
> attach(ethanol)
> dim(ethanol)
88  3
> names(ethanol)
"NOx" "C"   "E"
> plot(E,NOx)
> lines(loess.smooth(E,NOx))
```

The result is shown in Figure 3.18. Now, from the appearance of the result, one might think that a similar result could be obtained with a polynomial regression of high enough degree. We could do that by first computing the fitted values and then by joining these fitted values by straight line segments. After all, we have used this trick several times already. Using the commands:

```
> plot(E,NOx)
> lines(E, fitted(lm(NOx ~ poly(E,8))))
```

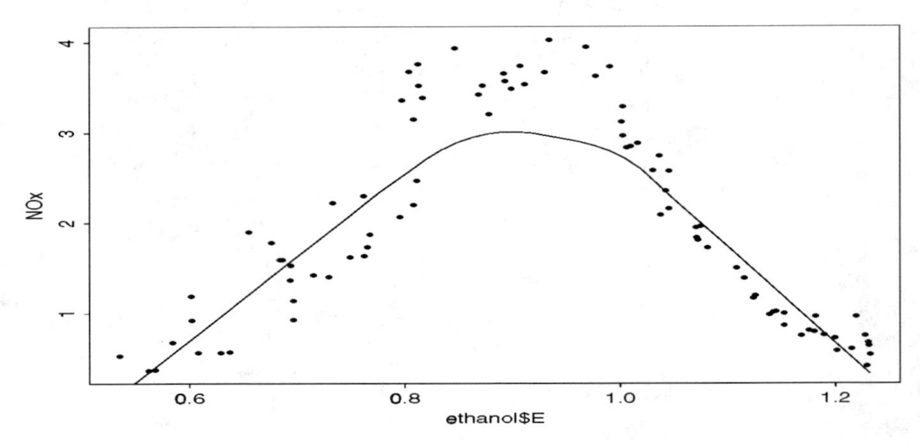

Fig. 3.18. Scatterplot of the ethanol data set.

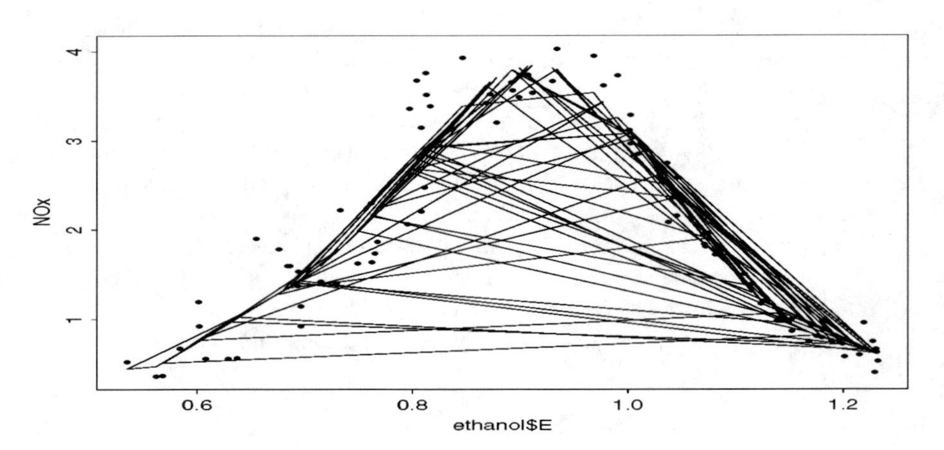

Fig. 3.19. Example of the WRONG use of the function `lines`.

we get the results reproduced in Figure 3.19. Obviously, something went wrong. The results are nothing near what we expected! To explain this apparent anomaly, we notice that, in our previous calls to the function `lines`, the values of the first components of the points to be joined were always in increasing order. On the other hand, if we look at the values of the current explanatory variable `E` we see that this is not the case.

```
> E
  1     2     3     4     5     6     7     8     9
0.907 0.761 1.108 1.016 1.189 1.001 1.231 1.123 1.042
.....................................................
.....................................................
```

```
  79    80   81    82    83    84    85    86    87    88
1.18 0.795 0.99 1.201 0.629 0.608 0.584 0.562 0.535 0.655
```

Now that we understand what is going wrong, can we find a solution to this problem? The answer is YES of course! We simply need to re-order the points so that the `E` -values are in increasing order. We can use the `S-Plus` function `order` to find the order of the entries of the vector `E`, and we can use this order to subscript the vectors containing the coordinates of the points. The following commands show how one can do all this.

```
> RKS <- order(E)
> EE <- E[RKS]
> NN <- NOx[RKS]
> par(mfrow=c(2,1))
> plot(E,NOx)
> plot(EE,NN)
> par(mfrow=c(1,1))
> plot(EE,NN)
> lines(EE, fitted(lm(NN ~ poly(EE,8))))
```

The results are reproduced in Figure 3.20, and this time, our expectations are met.

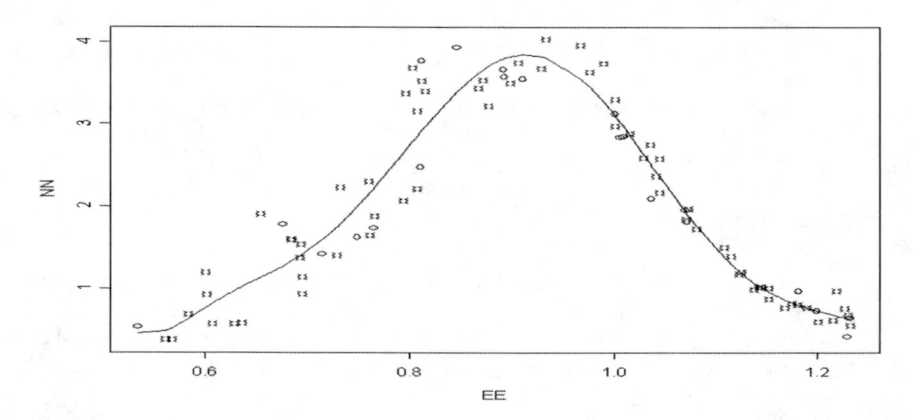

Fig. 3.20. Correct use of the function `lines` on the polynomial regression of degree 8 of the ethanol data.

PROBLEMS

Ⓔ **Problem 3.1** *The purpose of this problem is to analyze the data contained in the (data) script file* `strength.asc`. *Open and run the script to create the data set* `STRENGTH`. *The first*

column gives the fracture strength `X` *(as a percentage of ultimate tensile strength) and the second column gives the attenuation* `Y` *(the decrease in amplitude of the stress wave in neper/cm) in an experiment on fiberglass re-inforced polyester composites.*
1. Is a linear regression model reasonable to describe the data?
2. Perform a linear least squares regression of `attenuation` *against* `strength`, *determine the estimated intercept, the estimated slope and estimate the variance.*
3. Plot the fitted values of `attenuation` *against the corresponding actual values of* `attenuation`, *compute the coefficient of determination (the famous R^2), and assess the relevance of the linear model. Compute the predicted values of the variable* `attenuation` *for the values $x = 20$, $x = 50$ and $x = 75$ of* `strength`.
4. Compute the raw residuals and plot them against the fitted values of the `attenuation` *together with a* `qqnorm` *plot of these raw residuals. Comment on the results.*
5. Same question as in 2. and 3. (with the exception of the computation of R^2) for the least absolute deviations regression instead of the least squares regression.

Ⓔ **Problem 3.2** *The purpose of this problem is to perform a regression analysis of the 35 Scottish hill races data contained in the (data) script file* `hills.asc`.
1. Produce the scatterplot of the variable `time` *against the variable* `climb`, *and superimpose the least squares regression line of* `time` *against the variable* `climb`. *Repeat the same thing for* `time` *against* `dist`. *For each of these regressions, compute the R^2, and compare their values. Compute the raw residuals and produce their normal Q-Q plot. Comment.*
2. Redo the same thing using the least absolute deviations method instead, and compare the results.
3. Using the result of the regression of `time` *against* `dist`, *compute both in the case of the least squares regression and in the case of the absolute deviations regression, the predicted value of what the record time for a marathon should be? (use $x = 26.2$ for the distance of a marathon.*
4. For this question, we suppose that one of the race information has been entered incorrectly. For example, we create a new data set, say `Thills`, *with an erroneous information for the run* `Lairig Ghru`. *We are going to use:*

```
Lairig Ghru 28.0 2100 92.667
```

instead of

```
Lairig Ghru 28.0 2100 192.667
```

Create a new vector `Tdist` *identical to* `dist` *but we change the 11-th entry of the vector* `time` *and create a new vector* `Ttime` *with this new value 92.667. On the scatterplot of* `time` *against* `dist`, *superimpose the least squares regression line of* `time` *against* `dist`, *and of* `Ttime` *against* `Tdist`. *Create a new scatterplot of* `time` *against* `dist`, *and superimpose on this new scatterplot the least absolute deviations regression line of* `time` *against* `dist`, *and of* `Ttime` *against* `Tdist`. *Explain what you see, compare the regression coefficients, and explain.*
5. Perform the multiple least squares regression of `time` *against* `dist`, *and* `climb` *and compute the R^2, and compare this value of the coefficient of determination to the values of R^2 found for the simple least squares regressions of* `time` *against* `dist`, *and* `time` *against* `climb`. *Explain the results. Compute the raw residuals and produce their normal Q-Q plot. As in the case of R^2, ompare the result to the Q-Q plots of the simple least squares regressions.*
6. Perform the simple weighted *linear regression of the* `time` *against the* `climb` *using* `dist` *as set of weights, and compare the result to the regression of question 1.*

Ⓔ **Problem 3.3** *The purpose of this problem is to perform a first analysis of a classical data set contained in the script file* `basketball.asc`. *The intended analysis is very much in the spirit of the previous problem.*

1. We first use simple regression techniques.

1.a. Use least squares simple linear regression to explain the number of points scored by a player as given by the variable `points` *in terms of the variable* `height`.

1.b. Similarly, use least squares simple linear regression to explain `points` *in terms of the variable* `minutes` *giving the the time played per game.*

1.c. Use least squares simple linear regression to explain `points` *in terms of the values of* `age` *obviously giving the age of the player.*

1.d. Use regression diagnostics to compare these regressions and rank the variables `height`, `minutes` *and* `age` *according to their ability to predict the variable* `points`.

2. Use a least squares multiple regression to explain the values of the variable `points` *in terms of* `height` *and* `minutes`. *Similarly use a least squares multiple regression to explain* `points` *in terms of* `height`, `minutes` *and* `age`. *Does the inclusion of* `age` *in the regression improve the performance of the model?*

3. Redo question 1 with least absolute deviations regression instead of least squares regression. Compare the results and comment on the (possible) differences.

Ⓔ **Problem 3.4** *Import the data set* `MID1` *contained in the file* `poly31l12.asc` *in* `S-Plus` *format, and denote by the* 1st *and* 2nd *columns by X and Y respectively. The goal is to compare the least squares and least absolute deviations polynomial regressions of degree* 3 *to explain the values of Y from the corresponding values of X. In other words, we want to compare the real numbers β_0, β_1, β_2 and β_3, which minimize the criterion:*

$$C_2(\beta_0, \beta_1, \beta_2, \beta_3) = \sum_{j=1}^{134} |Y[j] - \beta_0 - \beta_1 X[j] - \beta_2 X[j]^2 - \beta_3 X[j]^3|^2$$

and the fit of the corresponding polynomial to those minimizing the criterion:

$$C_1(\beta_0, \beta_1, \beta_2, \beta_3) = \sum_{j=1}^{134} |Y[j] - \beta_0 - \beta_1 X[j] - \beta_2 X[j]^2 - \beta_3 X[j]^3|$$

and the corresponding polynomial fit. We use the `S-Plus` *functions* `lsfit` *and* `l1fit` *to implement the steps detailed in the text, and we shall not use the function* `lm`.

1. Give the commands needed to perform these two regressions, run these commands and give the resulting coefficients β_0, β_1, β_2 and β_3 and the optimal values of the criteria in both cases.

2. Construct the design matrix and compute the fitted values in both cases.

3. Give, on the same plot, the scatter plot of the values of `X` *and* `Y`, *the graph of the least squares polynomial regression and the graph of the absolute deviations polynomial. Say which of these two regressions looks better to you and explain why.*

Ⓔ **Problem 3.5** *The purpose of this problem is to analyze a classical example of nonlinear regression which can be found in most textbooks discussing the subject.*

The data, which are not of a financial nature, are contained in the data frame `Puromycin` *included in the* `S-Plus` *distribution. They contain three variables pertaining to a specific set of biomedical experiments on cells which were either* `treated` *or* `untreated` *with the drug Puromycin (this information is given by the third variable* `state`*). The response variable y is the initial velocity* `vel` *of the reaction while the regressor variable x is the enzyme* *concentration given in the first column* `conc`. *It is expected that the regression function $\varphi(x)$ will be given by the so-called Michaelis-Menten relationship:*

$$y = \varphi(x) = V_a \frac{x}{x + K},$$

where the constant V_a has the interpretation of an asymptotic velocity while K is a constant.
1. Attach the data frame `Puromycin` *to your* `S-Plus` *session and extract the rows corresponding to the treated cases. Then perform a nonlinear regression of the velocity y against the enzyme concentration x, starting with initial values $V_a = 200$ and $K = .1$ for the parameters.*
2. Give the values of the estimates obtained for the parameters V_a and K, and plot, on the same graph, the original data points of the "treated" sub-sample together with a broken line going through the points fitted by the nonlinear regression.

Ⓔ Ⓣ **Problem 3.6** *In this problem we introduce a new exponential family (called the generalized Vasicek family) and we use it to estimate the term structure of interest rates. For each value of the 4-dimensional parameter $\theta = (\theta_1, \theta_2, \theta_3, \theta_4)$ we define the function $f_{GV}(x, \theta)$ by:*

$$f_{GV}(x, \theta) = \theta_1 - \theta_2\theta_4 \frac{1 - e^{-x/\theta_4}}{x} + \theta_3\theta_4 \frac{(1 - e^{-x/\theta_4})^2}{4x} \tag{3.48}$$

where θ_4 is assumed to be strictly positive.
1. Mimic what was done in the text in the case of the Nelson-Siegel family and comment on the meanings and roles of the parameters θ_i's. Write an `S-Plus` *function* `fgv` *to compute the values of this function and plot the graphs of three of these functions for three values of the parameter θ which you will choose to illustrate your comments on the significance of the parameters.*
2. Derive an analytic formula for the price of the zero coupon bonds when the term structure of interest rates is given by an instantaneous forward curve (3.48). Follow what was done in the text for the function `bns` *and write an* `S-Plus` *function* `bgv` *to compute the values of the zero coupon bonds derived from the instantaneous forward interest rates given by the function* `fgv`. *Plot the zero coupon curves corresponding to the three values of θ chosen above in question 1.*
3. Derive an analytic formula for the yield curve, and write an `S-Plus` *function* `ygv` *to compute the yield curve, and plot the yield curves corresponding to the three values of θ chosen above in question 1.*
4. Using this new function family, estimate the term structure of interest rates (as given by the zero coupon curve, the forward curve and the yield curve) for the German bond data used in the text.

NOTES & COMPLEMENTS

The least squares method is a classic of statistical theory and practice. Its sensitivity to extreme features of the data has been well documented and the need for robust alternatives is widely accepted. The least absolute deviations method is the simplest example of a robust method of statistical estimation. Understanding the sensitivity of the estimation procedures to extreme measurements or outliers is a very important part of statistical inference, and an industry was developed on the basis of this. We refer the interested reader to the books of P. Huber [46] and Rousseeuw and Leroy [76] and the references therein. Realizing the importance of robust methods of estimation, `S-Plus` proposes a library of functions which can be used with the commercial distribution.

Complements to the tools of multivariate regression analysis and the theory of linear models, as introduced in this chapter, can be found in most multivariate analysis statistical textbooks. We refer the interested reader to the classical books of Montgomery and Peck [64], of Mardia, Kent and Bibby [61], or Chapter 14 of Rice's book [73] for an elementary exposition. Examples of `S-Plus` analyses with the function `lm` are given in [85].

Complements on seemingly unrelated regressions (SUR) can be found for example in the introductory econometric textbook of Ruud [71]. The CAPM was developed by Sharpe and Lintner after the pioneering work of Markowitz on the solution of the mean-variance optimal portfolio selection problem. An exposé of this theory can be found in any textbook on financial econometrics. We refer the interested reader to, for example, [41] or [16]. Most empirical tests of the model rely on market indexes as proxies for the market portfolio. The composition and the weights used in the computations of the market indexes have many features which vary over time (membership, capitalization of the members, ...) , and such features have been argued to be the main reason for the empirical rejection of CAPM. See for example the enlightening discussion by Roll in [74].

Our presentation of nonlinear regression is pretty elementary and it remains at an intuitive level. Nonlinear regression can be very technical, and the reader interested in mathematical developments, and especially the differential geometric aspects of likelihood maximization, is referred to Amari's book [2] or to the monograph [4] written in French by Antoniadis, Berruyer and Carmona. Nonlinear regression with `S` is explained in detail in Chapter 10 of Chambers' book [23], from which we borrow the classic example treated in Problem 3.5.

The Bank of International Settlements (BIS for short) provides information on the methodologies used by the major central banks to estimate the term structure of interest rates in their respective countries. It also provides data and samples of curve estimates upon request. The Nelson-Siegel family was introduced by Nelson and Siegel in [66] and the Swensson's generalization was proposed by Swensson in [82]. The use of cubic splines was proposed by Vasicek and Fong in [84]. We learned of the estimation of the term structure of interest rates using the Vasicek exponential family proposed in Problem 3.6 from Nicole El Karoui in a private conversation. The methods reviewed in this chapter are used by Central Banks for econometric analysis and policy making. Fixed income desks of investment banks are more secretive about their practice !

4

LOCAL & NONPARAMETRIC REGRESSION

This chapter is devoted to a class of regression procedures based on a new paradigm. Instead of searching for regression functions in a space whose elements are determined by a (small) finite number of parameters, we derive the values of the regression function from local properties of the data. As we shall see, the resulting functions are given by computational algorithms instead of formulae in closed forms. As before, we emphasize practical implementations over theoretical issues, and we demonstrate the properties of these regression techniques on financial applications: we revisit the construction of yield curves, and we propose an alternative to the Black-Scholes formula for the pricing of liquid options.

4.1 REVIEW OF THE REGRESSION SETUP

Although we have already worked in the regression framework for an entire chapter, we thought it would be useful to review once more the general setup of a regression problem, together with the notation used to formalize it. This will give us a chance to stress the main differences between the parametric point of view of Chapter 3 and the nonparametric approach of this chapter.

The general setup is the same as in our discussion of multiple linear regression. We have a sample of n observations

$$(\mathbf{x}_1, y_1), \ldots\ldots, (\mathbf{x}_n, y_n)$$

where for each $i = 1, 2, \ldots, n$, $\mathbf{x}_i$ is a vector of p numerical components which we arrange in a row $\mathbf{x}_i = [x_{i,1}, x_{i,2}, \ldots, x_{i,p}]$, and y_i is a real number. The components of the $\mathbf{x}_i$'s are the observed values of the p explanatory scalar variables, while the y_i's are the observed values of the corresponding response variable.

The $\mathbf{x}_i$'s can be deterministic: this happens when they are chosen by design. In this case, we assume that the y_i's are noisy observations of the values of a deterministic function $\mathbf{x} \hookrightarrow \varphi(\mathbf{x})$ which is to be estimated from the observations. But the $\mathbf{x}_i$'s can also be realizations of random vectors $\mathbf{X}_i$ (which are usually assumed to be independent), in which case the y_i's are interpreted as realizations of random variables Y_i so that the $(\mathbf{x}_1, y_1)$, ... , $(\mathbf{x}_n, y_n)$ appear as a sample of realizations of random couples $(\mathbf{X}_1, Y_1)$, ... , $(\mathbf{X}_n, Y_n)$. For the sake of notation, we will use the notation $(\mathbf{X}, Y)$ for a generic couple (random vector, random variable) with the same joint distribution. The statistical dependence of the Y-component upon the $\mathbf{X}$-components is determined by the knowledge of the (common) joint distribution of the couples $(\mathbf{X}_i, Y_i)$. This distribution *sits* in $(p+1)$ dimensions, and it can be a very complicated object. It is determined by the marginal distribution of the p-variate random vector $\mathbf{X}$ (which sits in p dimensions), and the conditional distribution of Y which gives, for each possible value of $\mathbf{X}$, say $\mathbf{x}$, the conditional distribution of Y given that $\mathbf{X} = \mathbf{x}$. Instead of considering the whole conditional distribution, one may want to consider first the conditional mean (i.e. the expected value of this conditional distribution) and this leads to the analysis of the function:

$$\mathbf{x} \hookrightarrow \varphi(\mathbf{x}) = \mathbb{E}\{Y|\mathbf{X} = \mathbf{x}\} \tag{4.1}$$

which is called the regression function of Y against $\mathbf{X}$. The graph of the function $\mathbf{x} \hookrightarrow y = \varphi(\mathbf{x})$ gives a convenient way of capturing the properties of such a regression. It is a one-dimensional curve when $p = 1$, a 2-dimensional surface when $p = 2$, and it becomes a hypersurface more difficult to visualize for larger values of the number p of explanatory variables.

In this chapter, except possibly for requiring that the function φ is (at least piecewise) smooth, nothing is assumed on the structure of φ (as opposed to the linearity, or polynomial character assumed in Chapter 3). In particular, we do not restrict the function φ to families of functions which can be characterized by a small number of parameters. This is what differentiates nonparametric regression from the parametric regression procedures seen in the previous chapter.

The univariate case $p = 1$ is considered first in what follows. In this case, one can view the search for the regression function φ, as a search for a graphical summary of the dependence between the two coordinates of the couples (x_1, y_1), ... , (x_n, y_n) which are usually visualized as a cloud of points in the plane. In this way, the graph of φ appears as a scatterplot smoother. We review the most frequently used nonparametric scatterplot smoothers in Section 4.3, and we illustrate their use in the construction of yield curves in Section 4.4. When we compare nonparametric smoothing to the parametric techniques introduced in Chapter 3, the main novelty is the notion of local averages. Indeed, in all the regression procedures considered so far, including polynomial regression, when computing the value $\varphi(x)$ of the regression function for a given value of x, each single observation (x_i, y_i) contributes to the result, whether or not x_i is close to x. This lack of localization in the x variable when averaging the y_i's is a source of many difficulties which the nonparametric smoothers are able to avoid.

This shortcoming due to the *lack of localization* in the x-variable is not universally shared by all parametric regression techniques. Natural splines regression is a case in point. Regression with natural splines is very popular in computer graphics. It is an integral part of parametric regression because it can be recast in the framework of the linear models of Chapter 3. However, because of their local character, natural splines share many of the properties of the nonparametric scatterplot smoothers, and we decided to include them in this chapter. We briefly explain how to use them in Section 4.2 below.

We illustrate the use of natural splines and nonparametric smoothers on the same `FRWRD` data set on which we have already tested polynomial regression in Chapter 3. Recall Figure 3.13. The last three sections of the chapter are devoted to the analysis of the multivariate case $p > 1$. We first introduce the kernel method which is recommended for small values of p. Then we discuss projection pursuit and other additive procedures designed to overcome the curse of dimensionality. For lack of time and space, we chose to ignore the tree-based regression methods, referring the interested reader to the references in the Notes & Complements at the end of the chapter.

4.2 Natural Splines as Local Smoothers

This section deals only with the univariate case $p = 1$. While still an integral part of parametric regression, natural splines offer a regression procedure satisfying the localization requirement formulated above. A spline of order $m+1$ is a function constructed from a subdivision of the range of x values. The points of this subdivision are called knots. This function is equal to a polynomial of degree $m + 1$ in between two consecutive points of the subdivision, and at each knot, all the left and right derivatives of order up to m match. This guarantees that the function is in fact m times continuously differentiable, discontinuities occurring possibly in the $(m + 1)$-th derivative, and only at the knots, but nowhere else. In most cases, this cannot be detected by the human eye on the plot of the graph of such a function. Given the knots, the splines of a given order form a finite dimensional vector space. Consequently, if we choose a basis for this linear space, each such spline function can be decomposed on this basis, and hence, it is entirely determined by a finite number of parameters, namely the coefficients in such a decomposition. It should now be clear that the discussion of polynomial regression given in Chapter 3 applies, and one can recast this type of regression within the class of linear models.

We refrain from discussing the details of the constructions of natural spline bases, restricting ourselves to a couple of illustrative examples. `S-Plus` has a fast implementation of a cubic natural splines regression. The typical form of the `S-Plus` command needed to run a natural splines regression is:

```
> ns.fit <- lm(response ~ ns(regressor, df=DF), data=DATA)
```

As in the case of polynomial regression, natural splines regression is performed by the generic `lm` method, after the data has been massaged by the specific function `ns`. Completely in analogy with the function `poly`, the function `ns` generates a design matrix from the basis of cubic splines associated with the specified sequence of knots and boundary conditions. In the case of natural cubic splines, the functions are taken to be linear (polynomial of degree 1) to the left of the smallest knot, and to the right of the largest knot. When the parameter `df` is supplied, the function `ns` chooses (`df` $-$ 1) knots at suitably chosen empirical quantiles of the x_i's. The knots can also be supplied with the optional parameter `knots` if the user so desires.

We illustrate the use of natural splines on the gas forward data used earlier to test polynomial regression. Recall Figure 3.13 in Chapter 3. The results are given in Figure 4.1. This plot was produced by the following `S-Plus` commands:

```
> plot(1:36,FRWRD,main="Natural Splines")
> lines(1:36,fitted(lm(FRWRD~ns(1:36,df=5))))
> lines(1:36,fitted(lm(FRWRD~ns(1:36,df=8))), lty=3)
> lines(1:36,fitted(lm(FRWRD~ns(1:36,df=15))), lty=6)
```

As seen from Figure 4.1, the natural splines regression is much more local than a plain polynomial regression. Indeed, with a large enough number of knots, it does a reasonable job of staying very close to the data throughout. Also, it is clear that the higher the number of knots, the tighter the fit, even at the price of losing some smoothness in the regression function.

Like all regression methods, natural splines can be used for prediction purposes. Indeed, the linear model used to fit a natural splines regression model, can also be used to predict the response for new values of the explanatory variable. But like in the case of polynomial regression, this application needs to be restricted to new values of the explanatory variable which are within the range of the data, for the same disastrous results can be expected outside this range.

4.3 NONPARAMETRIC SCATTERPLOT SMOOTHERS

Scatterplot smoothers can be viewed as nonparametric regression procedures for univariate explanatory variables, the idea being to represent a cloud of points (x_1, y_1), ..., (x_n, y_n) by the graph of a function $x \hookrightarrow y = \varphi(x)$. As explained earlier, the terminology nonparametric is justified by the fact that the function φ is not expected to be determined by a small number of parameters. In fact, except for a mild smoothness requirement, this regression function will not be restricted to any specific class.

We review some of the most commonly used scatterplot smoothers, and we give the precise definitions of those implemented in `S-Plus`. We restrict ourselves to the smoothers used in applications discussed in this book and the problem sets. For the sake of illustration, we compare their performance on data already used in our discussions of the polynomial and natural spline regressions.

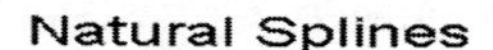

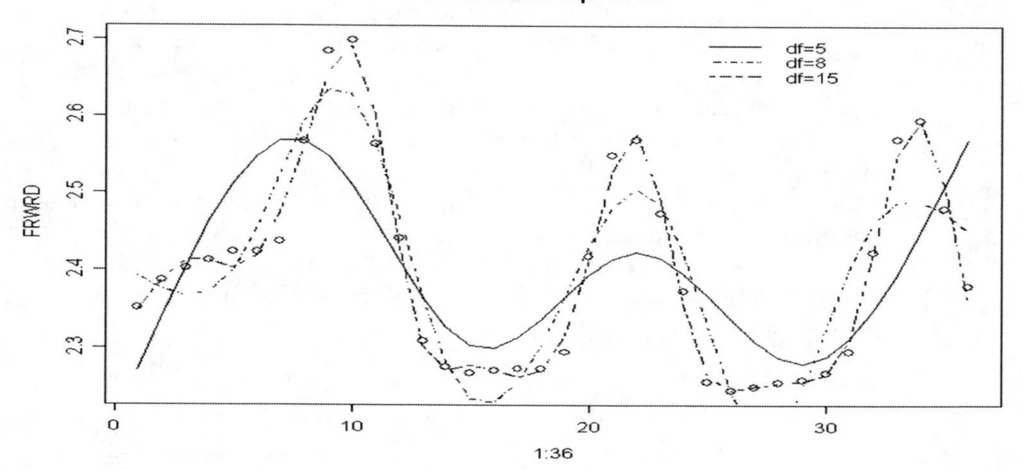

Fig. 4.1. Natural spline regressions with 5, 8 and 15 degrees of freedom, for the natural gas forward data which were used to illustrate polynomial regression in Chapter 3.

4.3.1 **Smoothing Splines**

We begin our review of scatterplot smoothers with smoothing splines. This choice was made to emphasize the similarities and differences with the natural splines discussed above. The idea behind the smoothing splines procedure is common to many applications in signal and image processing: it relies on regularization. To be specific, the scatterplot smoother φ is obtained by minimizing the objective function:

$$\mathcal{L}(\varphi) = \sum_{i=1}^{n} w_i |y_i - \varphi(x_i)|^2 + \lambda \int |\varphi^{(m)}(x)|^2 \, dx \tag{4.2}$$

for some constant $\lambda > 0$ called the smoothing parameter, for an integer m giving the order of derivation, and for a set of weights w_i which are most often taken to be unity. As usual, we use the notation $\varphi^{(m)}(x)$ to denote the m-th derivative of the function φ. The desired scatterplot smoother is the function $x \hookrightarrow \varphi(x)$ which minimizes the cost function $\mathcal{L}(\varphi)$, i.e. the argument of the minimization problem:

$$\varphi = \arg\min_f \mathcal{L}(f).$$

The goal of the first term in the objective function (4.2) is to guarantee the fit to the data, while the second term tries to ensure that the resulting scatterplot smoother is indeed smooth. The order of derivation m has to be chosen in advance, and the parameter λ balances the relative contributions, to the overall cost, of the lack of fit and the possible lack of smoothness. As defined, the minimizing function φ does not seem to have any thing to do with splines. But surprisingly enough, it turns out that the solution of the optimization problem (4.2) is in fact a spline of order $m+1$, i.e. an m times continuously differentiable function which is equal to a polynomial of degree $m+1$ on each subinterval of a subdivision of the range of the explanatory variable x. In particular, it has the same local properties as the regression by natural splines. `S-Plus` gives an implementation of this scatterplot smoother for $m=2$, in which case the resulting function φ is a cubic spline. The name of the `S-Plus` function is `smooth.spline`. The smoothing parameter λ appearing in formula (4.2) can be specified as the value of the (optional) argument `spar` of the function `smooth.spline`. Equivalently, the balance between the fit and smoothness terms of formula (4.2) can be controlled by setting the parameter `df` which stands for number of degrees of freedom, and which is essentially given by the integer part of the real number:

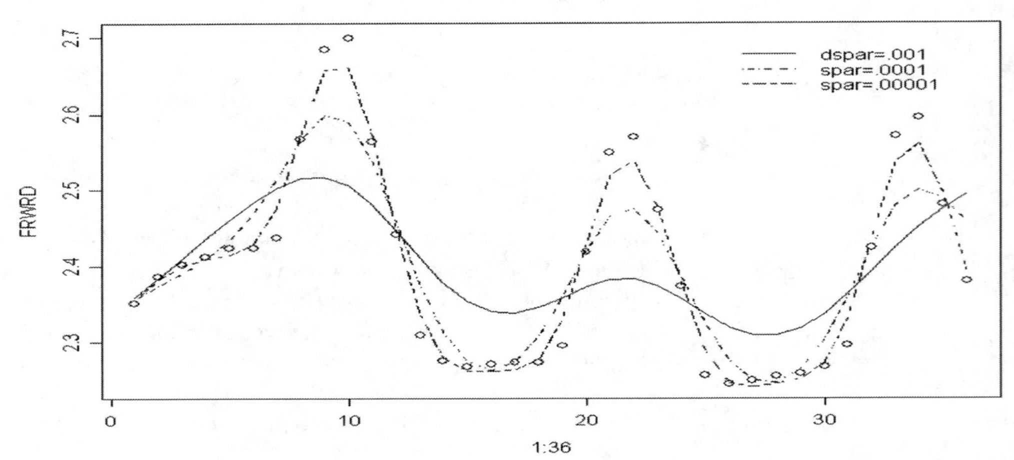

Fig. 4.2. Examples of smoothing splines with smoothing parameters $\lambda = .001$, $\lambda = .0001$ and $\lambda = .00001$.

$$n\frac{(\max x_j - \min x_j)^3}{\lambda}$$

(recall that n stands for the sample size). If neither of these parameters is specified, the function `smooth.spline` chooses λ by cross validation. The plots appearing in Figure 4.2 were created with the commands:

```
> plot(1:36,FRWRD,main="Smoothing splines")
> lines(smooth.spline(1:36,FRWRD,spar=.001))
> lines(smooth.spline(1:36,FRWRD,spar=.0001),lty=3)
> lines(smooth.spline(1:36,FRWRD,spar=.00001),lty=6)
```

The value of the smoothing parameter is set by the argument `spar`. Notice that the smaller this parameter, the closer the scatterplot smoother is to the points (since the fit contribution is more important) and that the larger this parameter, the smoother the resulting curve (since the smoothing contribution dominates the objective function).

4.3.2 **Locally Weighted Regression**

A quick review of the least squares paradigm for linear regression shows that the values of $\varphi(x)$ depend linearly on the observed responses y_i. Indeed, the optimal function for the least squares criterion happens to be of the form:

$$\varphi(x) = \sum_{i=1}^{n} w_i(x) y_i,$$

where each weight $w_i(x)$ depends upon all values x_j of the explanatory variable. As explained in the introduction, one of the goals of nonparametric smoothers is to keep this form of $\varphi(x)$ as a weighted average of the observed responses y_i, while at the same time choosing the weights $w_i(x)$ to emphasize the contributions of the y_i's corresponding to x_i's which are local neighbors of the value x at which the function φ is being computed.

A first implementation of this idea is given by the `S-Plus` function `loess`. This scatterplot smoother depends upon a parameter called `span` which gives the percentage of the total number of observations included in the neighborhood of each point. In practice, reasonable `span` values range from 0.3 to 0.5. Let us denote by k this number of points. So for each x at which one wishes to compute the function $\varphi(x)$, we denote by $N(x)$ the set of the k nearest values x_i of the explanatory variable, and by $d(x)$ the largest distance between x and the other points of $N(x)$. In other words:

$$d(x) = \max_{x_i \in N(x)} |x - x_i|.$$

To each $x_i \in N(x)$ we associate the weight

$$w_x(x_i) = W\left(\frac{|x - x_i|}{d(x)}\right) \tag{4.3}$$

where the weight function W is defined by:

$$W(u) = \begin{cases} (1 - u^3)^3 & \text{if } 0 \le u \le 1 \\ 0 & \text{otherwise} \end{cases}$$

and we choose $\varphi(x)$ to be the value of the weighted least squares regression line for the k points (x_j, y_j) for which $x_j \in N(x)$ and for the weights $w_x(x_j)$ defined in (4.3).

We illustrate the use of the `loess` function on the gas forward data on which we already tested several regression procedures. The results are given in the left pane of Figure 4.3. This plot was produced with the following `S-Plus` commands:

```
> plot(1:36,FRWRD,main="Loess Regression")
> lines(1:36, fitted(loess(FRWRD~seq(from=1,to=36,by=1))))
```

We did not set the value of the smoothing parameter `span` ourselves. Instead, we let the program use the default value. Obviously, the results are pretty bad. This example is used to show that using the default value for a parameter may not always be a good idea. Varying this parameter would have the usual effect, producing more or less smooth curves, as in the previous subsection.

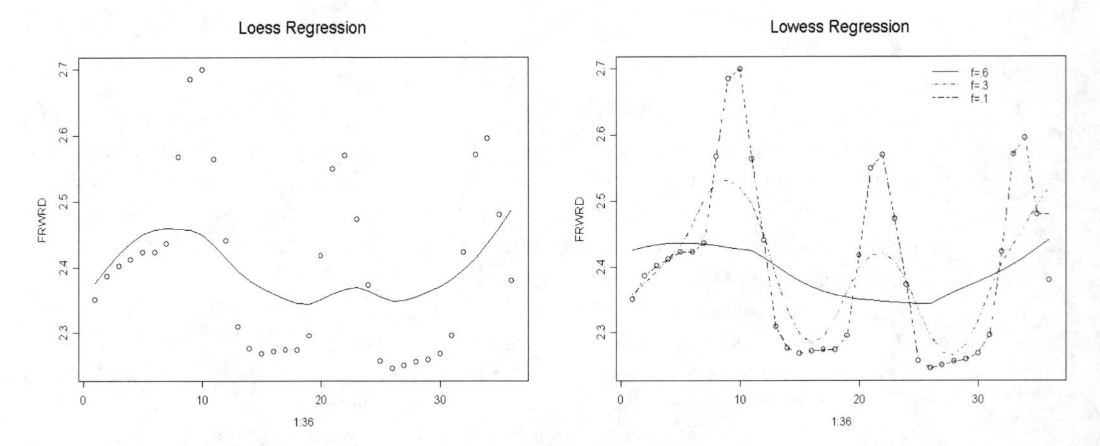

Fig. 4.3. `loess` regression using the default value of the parameter `span` (left), and robust smooth given by `lowess` (right) for the natural gas forward data.

4.3.3 A Robust Smoother

There is still another scatterplot smoother based on the same idea of local linear regression. It is a robust form of the weighted local linear regression given by the function `loess` described above. It is called `lowess`. Given a number of neighboring data points, this function performs locally a robust regression to determine the value of the function $\varphi(x)$. The use of this robust scatterplot smoother is illustrated in the right pane of Figure 4.3. We produced this plot with the following `S-Plus` commands:

```
> plot(1:36,FRWRD,main="Lowess Regression")
> lines(lowess(1:36,FRWRD,f=.6))
> lines(lowess(1:36,FRWRD,f=.3),lty=3)
> lines(lowess(1:36,FRWRD,f=.1),lty=6)
```

The optional parameter `f` gives the proportion of neighboring points used in the local robust regression. We used three values, `f`$= .6$, `f`$= .3$ and `f`$= .1$. The function `lowess` does not return an object of a specific class from which we extract the fitted values with the generic function `fitted`. It merely returns, in the old-fashioned way, a vector of x-values containing the values of the explanatory variable, and a vector of y-values containing the values fitted to the response variable. As before, we use the function `lines` to produce a continuous graph from discrete isolated points. Here, each of the values which are joined by straight lines to produce these three curves were obtained by averaging 12, 10 and 3 neighboring values of the response variable, respectively. This explains why these curves change from very smooth (and not fitting the data very well) to rather ragged, and why they give the impression of a broken line. We included the present discussion of the function `lowess` for the sake of completeness only. Indeed, since robust regressions are more computer intensive than least squares regressions, the `lowess` procedure can be very slow when the sample size n is large. For this reason, we do not recommend its use for large data sets.

4.3.4 The Super Smoother

The super smoother function `supsmu` is our last example of a scatterplot smoother before we start our long discussion of the kernel method. It is based on the same idea as the function `loess`, but its implementation has been optimized for computational speed. The main difference is the fact that the span (or equivalently the size k of the neighborhood $N(x_j)$) is now chosen by a form of local *cross validation* as a function of each x_j. Because of its computational speed, it is part of the projection pursuit algorithm which we present in detail later in this chapter. Calls to the function `supsmu`, and objects returned by these calls, have the structure described above in the case of the function `lowess`, so we refrain from illustrating them with examples.

4.3.5 The Kernel Smoother

As with the other scatterplot smoothers, the idea of the kernel smoother is to rely on the observed responses to neighboring values of x to predict the response $\varphi(x)$. The only difference is that, instead of relying on a limited number of observations y_i of the response, the local character of the averaging is realized by a weighted average of all the observed values y_i, the weights being decreased with the distance between x and the corresponding value x_i of the explanatory variable. To be more specific, the weights are computed by means of a *kernel* function $x \hookrightarrow K(x)$, and our good old enemy, the smoothing parameter. The latter is called *bandwidth* in the case of the kernel method, and it will be denoted by $b > 0$. By now, we should be familiar with the terminology and the notation associated with the kernel method. Indeed, we already introduced them in our discussion of the kernel density estimation method. We give a lucid account of the relationship between the applications of the kernel method to density estimation and to regression in the Appendix at the end of the chapter. The actual formula giving the kernel scatterplot smoother $\varphi(x)$ is:

$$\varphi(x) = \varphi_{b,K}(x) = \frac{\sum_{i=1}^{n} y_i K\left(\frac{x-x_i}{b}\right)}{\sum_{j=1}^{n} K\left(\frac{x-x_j}{b}\right)}. \tag{4.4}$$

Notice that the formula giving $\varphi(x)$ can be rewritten in the form:

$$\varphi(x) = \sum_{i=1}^{n} w_i(x) y_i \tag{4.5}$$

provided we define the weights $w_i(x)$ by the formula:

$$w_i(x) = \frac{K\left(\frac{x-x_i}{b}\right)}{\sum_{j=1}^{n} K\left(\frac{x-x_j}{b}\right)}. \tag{4.6}$$

Understanding the properties of these weights is crucial to understanding the very nature of kernel regression. These properties will be clear once we define what we mean by a kernel function. Recall that a nonnegative function $x \hookrightarrow K(x)$ is called a kernel function if it is integrable and if its integral is equal to 1, i.e. if it satisfies:

$$\int_{-\infty}^{+\infty} K(x)\, dx = 1,$$

in other words, if K is a probability density. The fact that the integral of $K(x)$ is equal to one is merely a normalization condition useful in applications to density estimation. It will not be of any consequence in the case of applications to regression since K always appear simultaneously in the numerator and the denominator: indeed, as one can easily see from formulae (4.5) and (4.6), multiplying K by a constant does not change the value of the regression function φ as defined in (4.4). But in order to be useful, the kernel $K(x)$ has to take relatively large values for small values of x, and relatively small values for large values of x. In fact, it is also often assumed that K is symmetric in the sense that:

$$K(-x) = K(x)$$

and that $K(x)$ decreases as x goes from 0 to $+\infty$. The above symmetry condition implies that:

$$\int_{-\infty}^{+\infty} xK(x)\, dx = 0 \tag{4.7}$$

which will be used in our discussion of the connection with kernel density estimation in the Appendix. Recall Figure 1.9 from Chapter 1 which gives the graphs of the four kernel functions used by the `S-Plus` density estimation method. They are also some of the most commonly used kernel functions when it comes to regression. Notice that the first three of them vanish outside a finite interval, while the fourth one

Kernel function	Formula
`box`	$K_{box}(x) = \begin{cases} 1, \text{ if } \lvert x\rvert \le .5 \\ 0 \;\; \text{otherwise} \end{cases}$
`triangle`	$K_{triangle}(x) = \begin{cases} 1 - \lvert x\rvert, \text{ if } \lvert x\rvert \le 1 \\ 0 \qquad\quad \text{otherwise} \end{cases}$
`parzen`	$K_{parzen}(x) = \begin{cases} (9/8) - (3/2)\lvert x\rvert + x^2/2, & \text{if } 1/2 \le \lvert x\rvert \le 3/2 \\ 3/4 - x^2, & \text{if } \lvert x\rvert \le 1/2 \\ 0 & \text{otherwise} \end{cases}$
`normal`	$K_{normal}(x) = \sqrt{2\pi}^{-1} e^{-x^2/2}$

Table 4.5. Table of the four kernel functions used by the `S-Plus` function `ksmooth`.

(theoretically) never vanishes. Nevertheless, since its computation involves evaluating exponentials, it will not come as a surprise that such a kernel can be (numerically) zero because of the evaluation of exponentials of large negative numbers: indeed for all practical purposes, there is no significant difference between e^{-60} and 0, and exponents which are that negative can appear very often! The `S-Plus` function `ksmooth` gives an implementation of the univariate kernel scatterplot smoother. The value of b is set by the parameter `bandwidth`. The kernel function K is determined by the choice of the parameter `kernel`. Four possible choices are available for the parameter `kernel`, `box` being the default. Explicit formulae are given in the table below. The option `triangle` gives the triangular function found in Figure 1.9, the `parzen` option gives a function proposed by Manny Parzen, who was one of the early proponents of the kernel method. Notice that this kernel replaces the cosine kernel used in density estimation. Finally the choice `normal` selects the Gaussian density function. We give the formulae for these functions in Table 4.5.

Except for the kernel function `box`, which leads to the crudest results, the other three kernels give essentially the same results in most applications. The situation is different when it comes to the choice of bandwidth parameter b. Indeed, the choice of the bandwidth is the *Achilles heel* of kernel regression. This choice can have an enormous impact, and the results can vary dramatically: small values of b give rough graphs which fit the data too closely, while too large a value of b produces a flatter graph. By experimenting with the choice of bandwidth, one can easily see that as b tends to ∞, the graph of the kernel smoother converges toward the horizontal straight line with intercept at the level of the mean $\overline{y}$ of the observed responses y_j. A rigorous proof of these empirical facts is given in full generality in Problem 4.3. As we explained earlier, this means that the regression is meaningless since the explanatory variable does not have any influence on the value of the prediction of the response variable. The following commands produce the plots in Figure 4.4 which illustrate this claim.

```
> plot(1:36,FRWRD,main="1-D Kernel Regression")
> lines(ksmooth(1:36,FRWRD,"normal",bandwidth=.01))
> lines(ksmooth(1:36,FRWRD,"normal",bandwidth=3),lty=3)
> lines(ksmooth(1:36,FRWRD,"normal",bandwidth=10),lty=5)
```

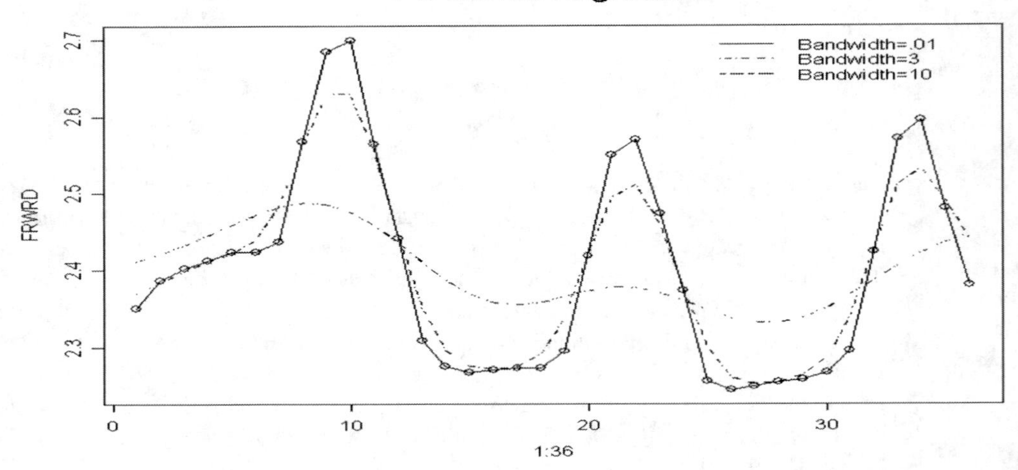

Fig. 4.4. Effect of the choice of bandwidth on the result of a kernel smoother.

More on the kernel scatterplot smoother later in Section 4.5 when we discuss the multivariate case and the kernel regression method.

4.4 More Yield Curve Estimation

Given our newly acquired knowledge of scatterplot smoothers and nonparametric curve estimation, we revisit the problem of estimation of the term structure of interest rates, as given for example by the instantaneous forward interest rate curves, which was originally tackled in Section 3.9 by means of parametric methods.

4.4.1 A First Estimation Method

The first procedure we present was called *iterative extraction* by its inventors, but it is known *on the street* as the *bootstrap method.* We warn the reader that this use of the word bootstrapping is more in line with the everyday use of the word bootstrapping than with the standard statistical terminology.

We assume that the data at hand consist of coupon bearing bonds with maturity dates $T_1 < T_2 < \cdots < T_m$ and prices $B_1 < B_2 < \cdots < B_m$. The so-called bootstrap method seeks a forward curve which is constant on each of the intervals $[T_j, T_{j+1})$. For the sake of simplicity, we assume that $t = 0$. In other words, we postulate that:

$$f(0,T) = f_j \qquad \text{for} \qquad T_j \leq T < T_{j+1} \qquad j = 1, \ldots, m-1$$

for a sequence $\{f_j\}_j$ of deterministic rates to be determined recursively by calibration to the observed bond prices. Let us assume momentarily that $f_1, \ldots, f_j$ have already been determined, and let us describe the procedure for identifying f_{j+1}. If we denote by X_{j+1} the principal of the $(j+1)$-th bond, by $\{t_{j+1,i}\}_i$ the sequence of coupon payment times, and by $C_{j+1,i} = c_{j+1}/n_y$ the corresponding payment amounts (recall that we use the notation c_j for the annual coupon rate, and n_y for the number of coupon payments per year), then the bond's price at time $t = 0$ can be obtained by discounting all the future cash flows associated with this bond:

$$B_{j+1} = \sum\nolimits_{t_{j+1,i} < T_j} P(0, t_{j+1,i}) \tfrac{c_{j+1} X_{j+1}}{n_y} \tag{4.8}$$
$$+P(0,T_j) \left(\sum\nolimits_{T_j < t_{j+1,i} \leq T_{j+1}} e^{-(t_{j+1,i} - T_j) f_{j+1}} \tfrac{c_{j+1} X_{j+1}}{n_y} + e^{-(t_{j+1,i} - T_j) f_{j+1}} X_{j+1} \right)$$

Notice that all the discount factors appearing in this formula are known since, for $T_k \leq t < T_{k+1}$ we have:

$$P(0,t) = \exp\left[-\sum_{h=1}^{k} (T_h - T_{h-1}) f_h - (t - T_k) f_{k+1} \right]$$

(recall formula (3.39) linking the price of the zero coupon bond to the instantaneous forward rate) and all the forward rates are known if $k \leq j$. Consequently, rewriting (4.8) as:

$$\frac{B_{j+1} - \sum_{t_{j+1,i} \leq T_j} P(0, t_{j+1,i}) \frac{c_{j+1} X_{j+1}}{n_y}}{P(0, T_j)}$$
$$= \frac{c_{j+1} X_{j+1}}{n_y} \sum_{T_j < t_{j+1,i} \leq T_{j+1}} e^{-(t_{j+1,i} - T_j) f_{j+1}} + e^{-(t_{j+1,i} - T_j) f_{j+1}},$$

and noticing that the left hand side can be computed, and that the unknown forward rate f_{j+1} appears only in the right hand side, this equation can be used to determine f_{j+1} from the previously evaluated values f_k for $k \leq j$.

Remark. Obviously, the forward curve produced by the bootstrapping method is discontinuous, since by construction, it jumps at all the input maturity dates. These jumps are the source of an artificial volatility which is due only to the method of estimation of the forward curve. This is the main shortcoming of this method of estimation. Several remedies have been proposed to alleviate this problem. The simplest one is to artificially increase the number of maturity dates T_j to interpolate between the observed bond (or swap) prices. Another proposal is to add a smoothness penalty which will force the estimated curve to avoid jumps. This last method is in the spirit of the smoothing spline estimation method which we discuss now.

4.4.2 A Direct Application of Smoothing Splines

For the purposes of this subsection we use the data contained in the `S-Plus data frame` `USBN041700`. These data comprise the quotes on April 17, 2000 of the outstanding US Treasury Notes and Bonds. Figure 4.5 gives the plot of the redemption yield as a function of the time to maturity, together with the plot of the smoothing spline. This plot was created with the following commands:

```
> plot(LIFE,INT.YIELD,main="Smoothing Spline .... Yields")
> lines(smooth.spline(LIFE,INT.YIELD))
```

The (smooth) yield curve plotted in Figure 4.5 is unusual because it has two humps,

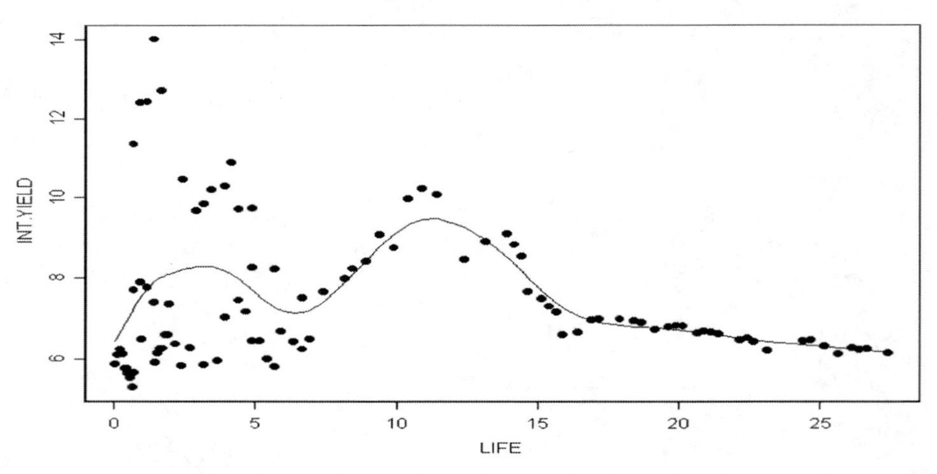

Fig. 4.5. Plot of the US Treasury Notes and Bonds redemption yields on April 17, 2000 together with the smoothing spline.

but it is still regarded as a reasonable yield curve.

4.4.3 US and Japanese Instantaneous Forward Rates

Even though we stated in Section 3.9 that the yield curves and forward rate curves published by the US Federal Reserve and the Bank of Japan were computed using smoothing (cubic) splines, they are not produced in the simplistic approach described above. The instantaneous forward rate curve produced on a given day t is the function $x \hookrightarrow \varphi(x)$ which minimizes the loss function:

$$\mathcal{L}_{JUS}(\varphi) = \sum_{i=1}^{n} w_i |P_i - P_i(\varphi)|^2 + \lambda \int |\varphi''(x)|^2 \, dx, \tag{4.9}$$

where $\varphi''(t)$ stands for the second derivative of $\varphi(t)$, the P_i's are the prices quoted for the outstanding bonds and notes available on day t, and the $P_i(\varphi)$'s are the prices one would get from pricing the bonds and notes on the forward curve given by φ using the fundamental pricing formula (3.33) in which the discount factors are computed from the forward curve given by φ. As in the parametric case, the weights w_i are chosen as functions of the duration of the i-th bond. See Section 3.9 for details.

The curve construction based on the minimization of the objective function $\mathcal{L}_{JUS}(\varphi)$ defined in (4.9) is reminiscent of smoothing splines regression. The difference lies in the fitting part $|P_i - P_i(\varphi)|^2$, which replaces the usual $|y_i - \varphi(x_i)|^2$. Instead of directly comparing the observation y_i to the values $\varphi(x_i)$ of the regression function, we compare the price P_i of a bond to the theoretical price $P_i(\varphi)$ that it would have if the function φ gives the true values of the instantaneous forward rate curve. This difference is enough to prevent us from directly using the `S-Plus` function `smooth.splines`, and it explains why the example we gave in Subsection 4.4.2 was for the construction of a yield curve $x \hookrightarrow \varphi(x)$ from observations y_i of the yields $\varphi(x_i)$.

4.5 Multivariate Kernel Regression

Multivariate kernel regression is a typical example of multivariate nonparametric nonlinear regression, but it can also be viewed as a high dimensional generalization of the procedure described in the subsection on the kernel scatterplot smoother, and especially the discussion of the function `ksmooth`. Indeed, most of what we said then, can be generalized to the case where the dimension p of the explanatory variables is not necessarily equal to 1. Indeed, formula (4.4) can be used in the form:

$$\varphi(\mathbf{x}) = \varphi_{b,K}(\mathbf{x}) = \frac{\sum_{j=1}^{n} y_j K\left(\frac{\mathbf{x}-\mathbf{x}_j}{b}\right)}{\sum_{j=1}^{n} K\left(\frac{\mathbf{x}-\mathbf{x}_j}{b}\right)}, \tag{4.10}$$

provided the function $\mathbf{x} \hookrightarrow K(\mathbf{x})$ is a kernel function in p dimensions, in the sense that it is a nonnegative function of p variables which integrates to one.

The simplest example of a p-dimensional kernel function is given by a function of the form:

$$K(\mathbf{x}) = k(\text{dist}(\mathbf{x}, 0)) \tag{4.11}$$

for some nonnegative and non-increasing function $k(\,\cdot\,)$ of one variable, and some choice of a notion of distance from the origin in p dimensions. Possible choices for this notion of distance include the usual Euclidean norms in $\mathbb{R}^p$:

$$\text{dist}(\mathbf{x}, 0) = \left(\sum_{j=1}^{p} x_j^2\right)^{1/2} \qquad \text{or} \qquad \text{dist}(\mathbf{x}, 0) = \left(\sum_{j=1}^{p} w_j x_j^2\right)^{1/2}$$

or non-Euclidean norms such as:

$$\text{dist}(\mathbf{x}, 0) = \sum_{j=1}^{p} |x_j| \qquad \text{or} \qquad \text{dist}(\mathbf{x}, 0) = \sup_{j=1,\dots,p} |x_j|.$$

These choices are popular because of their convenient scaling properties. With the exception of the Euclidean distance computed with different weights w_j for the different components x_j of the explanatory vector $\mathbf{x}$, all these kernel functions share the same shortcoming: all the components of the explanatory vector are treated equally, and this may be very inappropriate if the numerical values are on different scales. Indeed, in such a case, the value of the distance is influenced mostly (if not exclusively) by the variables having the largest values. We illustrate this point with a short discussion of an example which we will study in detail in the later part of this chapter. Let us imagine, for example, that the first explanatory variable is an annualized interest rate. Its values are typically of the order of a few percentage points. Let us also imagine that the second explanatory variable is a time to maturity. If for some strange reason this second variable is expressed in days instead of years, its values will be in the hundreds on a regular basis, and a distance of the type given above will ignore the small changes in interest rate, and report only on the differences in maturity. A change in unit in one of the variables can dramatically change the qualitative properties measured by these notions of distance, and consequently strongly affect the results of the kernel regression. This effect is highly undesirable. We discuss below alternative choices of kernel functions which can overcome this difficulty, as well as a standardization procedure which re-scales all the explanatory variables in an attempt to balance their relative contributions to the regression results.

Another very popular class of kernel functions is given by direct products (sometimes called tensor products) of one dimensional kernel functions. Indeed, if K_1, K_2, ..., K_p are one-dimensional kernel functions (possibly equal to each other), then the function:

$$K(\mathbf{x}) = K(x_1, x_2, \dots, x_p) = K_1(x_1)K_2(x_2)\cdots K_p(x_p) \qquad (4.12)$$

is obviously a p-dimensional kernel function. For these kernel functions, the weight multiplying the i-th response y_i is proportional to:

$$K\left(\frac{\mathbf{x}-\mathbf{x}_i}{b}\right) = K_1\left(\frac{x_1 - x_{i,1}}{b}\right) K_2\left(\frac{x_2 - x_{i,2}}{b}\right)\cdots K_p\left(\frac{x_p - x_{i,p}}{b}\right)$$

and from this expression one sees that there is no harm in choosing different values for the p occurrences of the bandwidth b in the right hand side. In other words, it is possible to choose p different bandwidths $b_1, b_2, \ldots, b_p$, one for each component of the explanatory variable. This feature of the direct product kernels makes them very attractive. In some sense, normalizing the scalar explanatory variables and using one single bandwidth amounts to the same thing as using different bandwidths for the components of the explanatory vector. See Subsection 4.5.2 for an example of standardization before running a kernel regression.

We now recast some of the most important properties of kernel regression as elementary remarks which also apply to the one dimensional case of the kernel scatterplot smoother `ksmooth` discussed earlier.

• The kernel regression estimate $\varphi(\mathbf{x})$ is a linear function of the observations. Indeed, the definition formula (4.10) can be rewritten in the form:

$$\varphi(\mathbf{x}) = \sum_{i=1}^{n} w_i(\mathbf{x}) y_i,$$

where the weights $w_i(\mathbf{x})$ are defined by:

$$w_i(\mathbf{x}) = \frac{K\left(\frac{\mathbf{x}-\mathbf{x}_i}{b}\right)}{\sum_{j=1}^{n} K\left(\frac{\mathbf{x}-\mathbf{x}_j}{b}\right)}.$$

Notice that these weights are nonnegative and they sum up to one. Because the kernel function is typically very small when its argument is large and relatively large when its argument is small, the weight $w_i(\mathbf{x})$ is (relatively) large when $\mathbf{x}$ is close (i.e. similar) to the observation $\mathbf{x}_i$ and small otherwise. This shows that the kernel regression function φ given by (4.10) is a weighted average of the observed values y_i of the response (and hence it is linear in the y_i's) with weights which give more importance to the responses from values of $\mathbf{x}_i$ close to the value $\mathbf{x}$ of the explanatory variables under consideration.

• The choice of bandwidth is a very touchy business. Many proposals have been made for an automatic (i.e. data driven) choice of this smoothing parameter. Whether one uses the results of difficult asymptotic analyses to implement bootstrap or cross validation procedures or simple rules of thumb, our advice is to be wise and to rely on experience to detect distortions due to a poor choice of bandwidth.

• As we explained earlier, it is tempting to use a separate bandwidth for each explanatory variable. This is especially the case when the kernel function is of the product type given in (4.12) and when the dynamic ranges of these variables are very different. For example, if a variable is expressed in a physical unit, changing the unit system may dramatically change the range of the actual values of the measurements, and small numbers can suddenly become very large as a consequence of the change of units. Accordingly, the influence of this variable on the computation of the kernel regression can increase dramatically. This undesirable effect is often overcome by normalizing the variables. See details in the discussions of the practical examples presented in Subsection 4.5.2 below.

• The sample observations of the explanatory vector form a cloud of points in the p-dimensional Euclidean space $\mathbb{R}^p$. The larger the dimension p, the further apart these points appear. Filling up space with points is more difficult in higher dimensions, and in any given neighborhood of a point $\mathbf{x} \in \mathbb{R}^p$, we are less likely to find points from the cloud of sample observations when p is large. This fact is known as Bellman's *curse of dimensionality.* When the number n of observations is not excessively large, the kernel regression has proven to be very powerful when the number of explanatory variables (i.e. the number p) is reasonably small, typically 2 or 3. How small this number should be obviously depends upon the sample size n, and the more observations we have, the larger the number of explanatory variables we may include. This form of Bellman's *curse of dimensionality* can easily be illustrated by heuristic arguments, but it can also be quantified by rigorous asymptotic results which show that n should grow exponentially with p. This is a serious hindrance.

4.5.1 Running the Kernel in S-Plus

When $p = 1$ the kernel regression is implemented by the scatterplot smoother `ksmooth` described in the Subsection 4.3.5. Unfortunately, there is no S-Plus function implementing the multivariate kernel regression. We propose to use the home-grown function `kreg` for the purposes of this book.

Since this function can be called quite often in a typical application, the core computations are done by a piece of C-code called by an S-wrapper. This C-code needs to be dynamically loaded each time, but this will be transparent to the user, at least if things are working according to plan. The output of the function `kreg` is a list containing the input variables and a variable `ypred`. A call to the function `kreg` of the form

```
> PRED <- kreg(X,Y,kernel=4,b=.4)
```

returns a variable `PRED$ypred` containing the values $\hat{y}_i = \varphi(\mathbf{x}_i)$ of the fitted values for the values of the regressor variable in the set of observations if the argument `xpred` is missing. When this argument is set to a vector of possible values of the explanatory variable/vector, as in the example:

```
> PRED <- kreg(X,Y,xpred=GRID,kernel=4,b=.4)
```

then `PRED$ypred` contains the values of this regression function φ for the specific values of the explanatory variables/vectors contained in the argument `xpred`. Note that the latter is equal to the observations `X` by default.

If in the two dimensional case, one may be interested in computing φ over a grid of points (for plotting purposes for example), one can generate such a surface plot by setting the parameter `PLOT` to `TRUE` by adding `PLOT=T` in the call to the function `kreg`. The surface is computed by default over a grid of 256×256 regularly spaced points between the minima and the maxima of the two explanatory variables. This grid can also be user specified.

4.5.2 An Example Involving the June 1998 S&P Futures Contract

This experiment is based on high frequency data on the S&P 500 index introduced and first manipulated in Chapter 5. For the sake of definiteness, we chose to work with the June 1998 futures contract for which we collected ALL the transaction records. Then for each of the 59 trading days between `March 15, 1998` and `June 8, 1998` we computed

- 6 indicators at 12:00 noon (for the morning transactions);
- the same 6 indicators for the afternoon transactions;

the goal of the experiment being to predict the values of the afternoon indicators from the knowledge of the morning ones. The 6 morning indicators are stored in a 59×6 matrix which we call `MORN.mat` and the corresponding afternoon values are stored in another `59 x 6` matrix which we call `AFT.mat`. These indicators were computed from the high frequency tick-by-tick data of all the quotes taking place in the morning and afternoon sessions, respectively. We shall explain later in Subsection 5.2.1 of Chapter 5 how these indicators were computed. For the time being, a quick explanation of what these indicators are will suffice. The first indicator is called `range`. It represents the difference between the highest quote and the lowest quote of the morning. The next indicator is called `nbticks`. It gives the number of transactions during the session we consider, whether it is the morning session or the afternoon one. The third indicator gives the standard deviation of the morning log-returns computed independently of the length of the time interval of the return, while the next two indicators give the volatility ratio `volratio` and the quantile slope `l2slope` which we will define rigorously in Chapter 5. Finally, the last indicator gives the mean separation time between two consecutive transactions. It is called `ticksep`. Printing the first five rows of the data matrix of our six morning indicators, we get:

```
> MORN.mat[1:5, ]
     range nbticks          vol  volratio l2slope ticksep
[1,]   4.8     199 9.68800e-08 0.0584073 298.328 7.12836
[2,]   6.5     199 1.01720e-07 0.0763445 293.569 6.36461
[3,]   5.1     200 6.59284e-08 0.0206114 299.409 7.00117
[4,]   4.5     200 7.13838e-08 0.0376023 317.831 6.59341
[5,]   8.8     207 1.06649e-07 0.0456482 334.903 6.62047
```

Obviously:

the 6 indicators **are not on the same scale**.

As a solution we propose to

standardize the explanatory variables.

As we already pointed out, this procedure is ubiquitous in nonlinear and nonparametric regression, and in classification. The following S-Plus commands do just that:

```
> MEANMORN <- apply(MORN.mat,2,mean)
> MEANMORN
   range nbticks         vol  volratio l2slope ticksep
 8.68474 203.389 1.46168e-07 0.0707302 309.809 5.74796
> MEANMORN.mat<-outer(rep(1,dim(MORN.mat)[1]),MEANMORN,"*")
> MEANMORN.mat
numeric matrix: 59 rows, 6 columns.
       range nbticks         vol  volratio l2slope ticksep
[1,] 8.68474 203.390 1.46168e-07 0.0707302 309.809 5.74796
[2,] 8.68474 203.390 1.46168e-07 0.0707302 309.809 5.74796
   ..............................................................
[58,] 8.68474 203.390 1.46168e-07 0.070730 309.809 5.747962
[59,] 8.68474 203.390 1.46168e-07 0.070730 309.809 5.747962
> VARMORN <- apply(MORN.mat,2,var)
> VARMORN
    range nbticks         vol    volratio l2slope  ticksep
  10.8161 15.7592 7.4550e-15 0.00280819 102.312 0.291711

> STDMORN <- sqrt(VARMORN)
> STDMORN.mat<-outer(rep(1,dim(MORN.mat)[1]),STDMORN,"*")
> NORMMORN.mat<-MORN.mat - MEANMORN.mat)/STDMORN.mat
```

The S-Plus function apply was designed to compute a given function on the rows or the columns of an array. So, the first command applies the function mean to each column (second dimension appearing as the second argument of the function apply). The second command creates a matrix with the same dimensions as MORN.mat (notice that we use the command dim to extract the dimensions of the matrix MORN.mat) by multiplying a vector of ones by the row of means, producing in this way, a matrix whose columns are repeating the means in question. We check that this is indeed the case by printing out the matrix MEANMORN.mat. Similarly, we create the matrix STDMORN.mat with constant columns equal to the standard deviation of the corresponding column of MORN.mat, and finally we standardize the original matrix MORN.mat by subtracting from each entry, the mean of the column and dividing by the standard deviation of the column. We check that we did exactly what we intended to do by computing the means and the variances of the columns of the matrix so obtained:

```
> apply(NORMMORN.mat,2,mean)
    range   nbticks       vol volratio   l2slope    ticksep
-1.05e-16 4.82e-16 2.59e-16 2.83e-17 3.46e-16  1.27e-15
> apply(NORMMORN.mat,2,var)
 range nbticks vol volratio l2slope ticksep
     1       1   1        1       1       1
```

and this shows that we succeeded in turning the explanatory variables into variables with empirical mean zero and empirical variance one. Let us now consider the problem of the prediction of the afternoon value of the volatility ratio at noon, i.e. when our information consists of the values of the 6 morning indicators. We shall see in Section 4.6 how to use the set of all 6 indicators as regressors, but because our sample size is only 59, we cannot hope to use kernel regression with more than 2 regressors (and even that may be a bit of a stretch). Based on the intuition developed with least squares linear regression, when it comes to the prediction of the values of `AFT.mat[,4]`, a sensible way to choose two explanatory variables out of the 6 morning indicators should be to find those morning indicators with the largest correlation with `AFT.mat[,4]`. However, basing our choice on this criterion alone could lead to disaster. We need to make sure that the correlation between the two regressors which we choose is as small as possible. Indeed, if they carry the same information, they will be equally correlated with `AFT.mat[,4]`, but the two of them together will not be more predictive than any single one of them taken separately. We do not dwell on this problem here. We simply refer to Problem 4.11 for an attempt to use Principal Component Analysis to find uncorrelated variables summarizing efficiently the information in the set of 6 variables. For the purposes of the present discussion, we restrict ourselves to choosing `MORN.mat[,4]` and `MORN.mat[,5]`, hoping that it is a reasonable choice. Recall that `MORN.mat[,4]` is the morning value of the volatility ratio whose afternoon value we try to predict, and that `MORN.mat[,5]` is some form of measure of the average rate at which the transactions occur during the morning.

The left pane of Figure 4.6 contains a three-dimensional scatterplot of the 59 observations of these three variables, while the right pane contains the two dimensional horizontal projection given by the 2-d scatterplot of the two explanatory variables. The three dimensional scatterplot is obtained by putting vertical bars over the points of the 2-D scatterplot of the explanatory variables, the heights of these bars being equal to the corresponding values of the response variable.

The two-dimensional scatterplot of the explanatory variables shows that the points are reasonably well spread throughout the rectangular region of the plot. Notice that the two plots of Figure 4.6 would look exactly the same if we had used the normalized indicators instead of the raw ones, the only changes being in the axis labels. At this stage, it is a good idea to check for the presence of isolated points far from the bulk of the data. Indeed as we have seen in our discussion of linear regression, the latter are very influential, and they often distort the results of the regression. We do not have such distant points in the present situation. The three dimensional plot shows that, except perhaps for the observations on the far right, linear regression could be a possibility. Consequently, we perform a (bivariate) linear least squares regression. The results are shown in Figure 4.7, and they do not confirm the early suspicion that a linear regression could do the trick. Because it is so difficult graphically to visualize why the regression plane would be pulled up or down and where, we perform a few kernel regressions to get a better sense of the response surface. The results are given in Figure 4.8. We used three different values of the bandwidth:

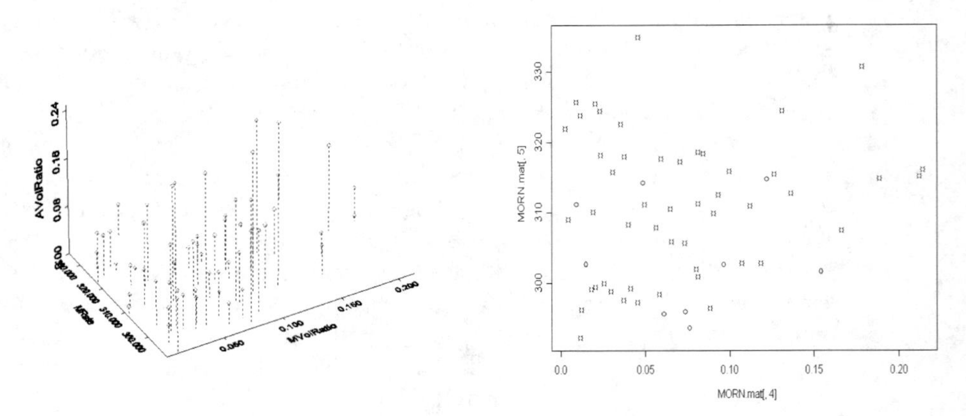

Fig. 4.6. Bar scatterplot of the Afternoon Volatility Ratio `AFT.mat[,4]` against the Morning Volatility Ratio `MORN.mat[,4]` and the Arrival Rate `MORN.mat[,5]`(left), and 2-D Scatterplot of `MORN.mat[,4]` and `MORN.mat[,5]` (right).

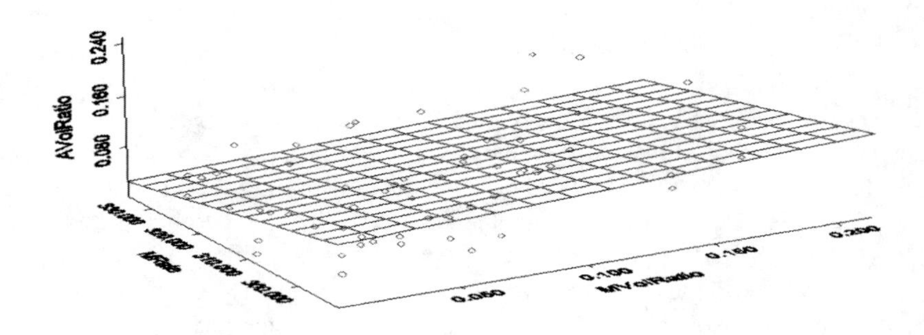

Fig. 4.7. Linear Regression of the afternoon volatility ratio `AFT.mat[,4]` against the morning volatility ratio `MORN.mat[,4]` and the morning rate of arrival `MORN.mat[,5]` of the transactions.

the first one is presumably too small because the surface is too rough, the last one is presumably too large because the surface is too smooth, and as Goldilocks would say, the middle one is *just right.*

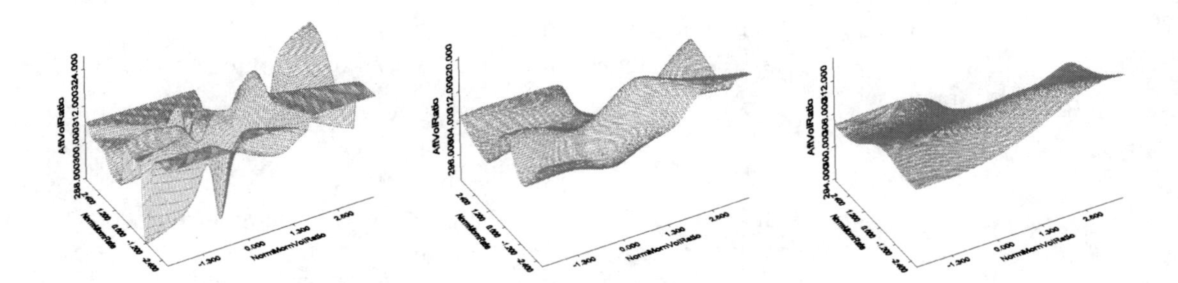

Fig. 4.8. Kernel regressions of the afternoon volatility ratio `AFT.mat[,4]` against the morning volatility ratio `MORN.mat[,4]` and the morning rate of arrival `MORN.mat[,5]` of the transactions in the original data matrix `MORN.mat`. We used a two dimensional Gaussian kernel with the bandwidths $b = 0.8$, $b = 2$ and $b = 4$ from left to right respectively.

4.6 PROJECTION PURSUIT REGRESSION

In their original proposal, the creators of the projection pursuit algorithm suggested writing the regression function $\varphi(\mathbf{x})$ of a model $y = \varphi(\mathbf{x}) + \epsilon$ in the form:

$$\varphi(\mathbf{x}) = \alpha + \sum_{j=1}^{m} \phi_j(\mathbf{a}_j \cdot \mathbf{x}). \tag{4.13}$$

Remember that we are working in the usual regression setting:

$$(\mathbf{x}_1, y_1), (\mathbf{x}_2, y_2), \ldots\ldots, (\mathbf{x}_n, y_n)$$

where the *explanatory variables* $\mathbf{x}_1, \mathbf{x}_2, \ldots, \mathbf{x}_n$ are p-dimensional. In other words, each $\mathbf{x}_i$ is a p-vector of the form $\mathbf{x}_i = (x_{i,1}, \ldots, x_{i,p})$ while the response variables $y_1, y_2, \ldots, y_n$ are univariate (i.e. scalar). As usual, we use bold face letters $\mathbf{x}$ to emphasize that we are dealing with a *multivariate* explanatory variable which we sometimes call an explanatory vector to emphasize that its dimension can be greater than one. The idea of the projection pursuit algorithm is to fight the *curse of dimensionality* inherent with large values of p, by replacing the p-dimensional explanatory vectors $\mathbf{x}_i$ by suitably chosen one-dimensional projections $\mathbf{a}_j \cdot \mathbf{x}_i$, hence the term *projection* in the name of the method. We now proceed to the estimation of the quantities α, and $\phi_1(\mathbf{a}_1 \cdot \mathbf{x}), \ldots, \phi_m(\mathbf{a}_m \cdot \mathbf{x})$ appearing in formula (4.13). The projection pursuit algorithm is based on an inductive procedure in which residuals are recomputed and fitted at each iteration, and for the purposes of the present discussion, one should think of the observed values y_i as the starting residuals, i.e. the residuals of order zero. Each time one of the terms in the sum appearing in the right hand side of (4.13) is estimated, the actual estimates are subtracted from the current values of the residuals, providing in this way a new set of residuals from which we proceed to estimate the next term in (4.13). This recursive fitting of the residuals justifies the term *pursuit* in the name of the method.

The constant α is naturally estimated by the mean of the observations of the response:

$$\hat{\alpha} = \overline{y} = \frac{1}{n}\sum_{i=1}^{n} y_i.$$

This sample mean is subtracted from the observations (i.e. the residuals of order zero) to get the residuals of order one. Next we proceed to the estimation of the first direction $\mathbf{a}_1$ and the first function ϕ_1. Because of computational considerations, it is important to choose a specific procedure capable of selecting the best function ϕ for each choice of direction given by the unit vector $\mathbf{a}$. `S-Plus` does just that by choosing, for each candidate $\mathbf{a}$, the function $\phi_{\mathbf{a}}$ given by the super-smoother algorithm (implemented as the function `supsmu`) obtained from the scatterplot of the projections of the data points $\{\mathbf{x}_i\}_{i=1,\ldots,n}$ on the direction $\mathbf{a}$, i.e. the values of the scalar products $\mathbf{a} \cdot \mathbf{x}_i$, and the corresponding values y_i of the response y (or the current residuals if we are in the middle of the recursive fitting procedure).

The creators of the projection pursuit algorithm proposed to fit recursively the residuals (starting from the values of the response variable) with terms of the form $\phi_{\mathbf{a}_j}(\mathbf{a}_j \cdot \mathbf{x})$ and to associate to each such optimal term a figure of merit, for example the proportion of the total variance of the response actually explained by such a term. In this way, the whole procedure would depend upon only one parameter. One could choose this *tolerance* parameter first (typically a small number) and one would then fit the response and successive residuals until the figure of merit (i.e. the proportion of the variance explained by $\phi_{\mathbf{a}_j}(\mathbf{a}_j \cdot \mathbf{x})$) dropped below the tolerance parameter. This would automatically take care of the choice of the order m of the model (4.13).

4.6.1 The `S-Plus` Function `ppreg`

Projection pursuit regression is implemented in `S-Plus` by the function `ppreg`. Had S-Plus implemented this algorithm in the way we described above, the proportion of the total variation in the response explained by the successive models (i.e. the sequential plot of $\cdots$`$esq`) would always be decreasing. Unfortunately, things are not that simple, and what we just described *is not exactly the algorithm implemented in* `S-Plus`.

When the arguments `m.min` and `m.max` are set and the function `ppreg` is run, the program does pretty much what we just described for m increasing from 1 to `m.max`, and the sequential plot of $\cdots$`$esq` would be decreasing if one could plot it at this stage. However, `S-Plus` does not stop after this *forward pass* over the data. It recomputes fitted models for $m =$ `m.max` $- 1$, $m =$ `m.max` $- 2, \ldots, m =$ `m.min` in this reverse order, and the actual output of the function `ppreg` consists of these results. For each value of m in this range, the program runs an optimization procedure to find simultaneously all the $\mathbf{a}_j$ appearing in the model (4.13). Unfortunately, minimization procedures are not always reliable, especially when the dimension is large and the function is not convex. Their results depend strongly on the initializations. Here is what S-Plus is doing in order to initialize this minimization.

Again, we proceed by induction, but we choose to go backward this time. For each value of m (varying from m.max $-$ 1 down to m.min) the program considers the unit direction vectors $\hat{\mathbf{a}}_1, \ldots, \hat{\mathbf{a}}_{m+1}$ found to be optimal in the model fitted previously with $m+1$ terms, and it computes the variances of the values fitted by the super-smoother algorithm supsmu:

$$\sigma_j^2 = \frac{1}{n}\sum_{i=1}^{n} \phi_{\hat{\mathbf{a}}_j}(\hat{\mathbf{a}}_j \cdot \mathbf{x}_i)^2.$$

Notice that we do not have to subtract the mean to compute the variance because we already subtracted the mean $\overline{y}$ of the response and after that the response (whether it is the actual response or the residual at a given stage of the recursion) is always centered. The minimization algorithm is then initialized with the m unit vectors $\hat{\mathbf{a}}_{j_1}, \ldots, \hat{\mathbf{a}}_{j_m}$ where the indices $j_1, \ldots, j_m$ are chosen in such a way that the variances $\sigma_{j_1}^2, \ldots, \sigma_{j_m}^2$ are the m largest among the $m+1$ variances $\sigma_1^2, \ldots, \sigma_{m+1}^2$. The rationale for this choice is simple. The relative size of the variance σ_j^2 is an indication of how important the contribution of $\hat{\mathbf{a}}_j$ (and of the corresponding function $\phi_{\hat{\mathbf{a}}_j}$) is.

In this way, for each value of m, S-Plus finds a model of the form:

$$\varphi(\mathbf{x}) = \overline{y} + \sum_{j=1}^{m} \phi_{\mathbf{a}_{m,j}}(\mathbf{a}_{m,j} \cdot \mathbf{x}). \tag{4.14}$$

However, the set $\{\mathbf{a}_{m,1}, \ldots, \mathbf{a}_{m,m}\}$ of unit vectors in the fitted model of order m, is not necessarily equal to the set of the first m vectors $\{\mathbf{a}_{m+1,1}, \ldots, \mathbf{a}_{m+1,m}\}$ of the set of unit vectors in the fitted model of order $m+1$. Even though it turns out that the plot of $\cdots$ $esq is very often decreasing (it can increase at times but this is not the rule in general) there is absolutely no reason why the proportion of the variation explained should be a decreasing function of m.

As a final remark, we mention the way in which S-Plus normalizes the functions ϕ_j appearing in the decomposition of the model (4.13). If one introduces the notation:

$$\beta_j = \sigma_j = \sqrt{\sum_{i=1}^{n} \phi_{\hat{\mathbf{a}_j}}(\hat{\mathbf{a}}_j \cdot \mathbf{x}_i)^2}, \qquad \varphi_j = \frac{1}{\sigma_j}\phi_{\hat{\mathbf{a}}_j}, \tag{4.15}$$

then obviously the model (4.13) can be rewritten in the form:

$$\varphi(x) = \overline{y} + \sum_{j=1}^{m} \beta_j \varphi_j(\mathbf{a}_j \cdot \mathbf{x}) \tag{4.16}$$

and in this form, the contribution of each of the functions φ_j can be viewed as having mean 0 and variance 1.

***Running the* `S-Plus` *Function* `ppreg`**

If in doubt, get help on the function `ppreg` by typing:

```
> help(ppreg)
```

A call to the projection pursuit function should be of the form:

```
> data.ppreg <- ppreg(x,y,min.term,max.term)
```

with possibly more options. See the help file for a complete list of all the options available. `x` should be a $n \times p$ - matrix containing the n observations on the p explanatory (predictor) variables, while `y` should be a $n \times 1$ vector containing the values of the corresponding observations of the response variable (i.e. the predictee). Notice that `y` could also be a matrix in the case where we want to predict several variables at the same time, but we shall not consider this possibility here. Such a command produces by default the sequential plot of the sequence $\{\text{esq}[m]\}_{m=1,\ldots,m.max}$ which gives essentially the *fraction of unexplained variance* (i.e. the residual sum of squares divided by the total sum of squares):

$$\begin{aligned} \texttt{esq}[m] &= \frac{RSS(m)}{TSS(m)} \\ &= \frac{\sum_{i=1}^{n} |y_i - \overline{y} - \sum_{k=1}^{m} \hat{\beta}_k \varphi_k(\hat{a}_k \cdot x_k)|^2}{\sum_{i=1}^{n} |y_i - \overline{y}|^2}. \end{aligned} \tag{4.17}$$

The value of m at which this function is minimum, or at least after which there is no significant decrease in the sequential plot of `esq`, gives a reasonable value for the number of terms to include in the projection pursuit. Several remarks are in order at this stage:

1. When the call to the function `ppreg` does not specify the optional variable `xpred` in the command line, then the residuals are returned in `ypred`. See illustrations of the use of this fact below.
2. If the variable `xpred` is specified in the command line, then the value of `ypred` returned by the function `ppreg` contains the fitted values, i.e. the predictions computed by the algorithm for the values of the explanatory vector $\mathbf{x}$ contained in `xpred`.

4.6.2 `ppreg` Prediction of the S&P Indicators

We revisit the analysis of the S&P indicators which was performed earlier in Subsection 4.5.2 with the help of kernel regression. This will lead to a first comparison of the performance of the two regression methods. We use the same notation, and since the introduction of the projection pursuit regression was motivated by problems with a large number of explanatory variables, we use for the prediction of any given afternoon indicator, the entire set of 6 morning indicators. Again, we choose to predict the volatility ratio, and for reasons already given in Subsection 4.5.2, we restrict ourselves to the analysis of the data set obtained by removing the two extreme days. The projection pursuit regression is performed using the following S-Plus commands:

```
> PPREG1 <- ppreg(MORN.mat[,1:6],AFT.mat[,4],
                                        min.term=1,max.term=12)
> tsplot(PPREG1$esq)
```

As before, we plot the output, PPREG1$esq, which gives the proportion of the variance still unexplained (as a function of the number of terms included in the sum). From the plot reproduced in Figure 4.9, it seems that 6 could be a reasonable choice.

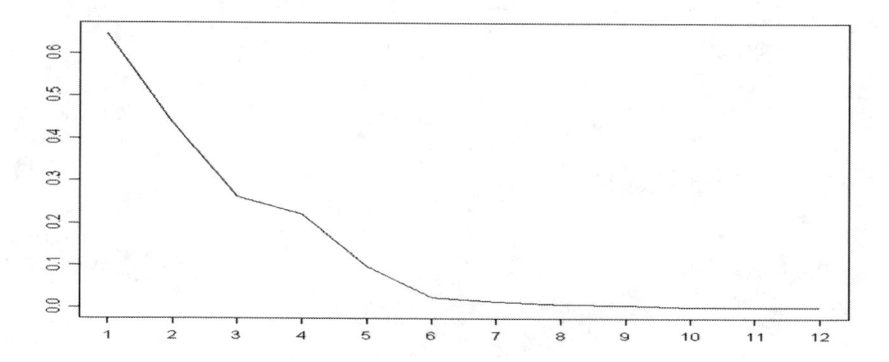

Fig. 4.9. Sequential plot of the esq output of the projection pursuit analysis of the six S&P 500 morning indicators.

So we re-run the projection pursuit algorithm with the command:

```
> PPREG1 <- ppreg(MORN.mat[,1:6],AFT.mat[,4],
                                        min.term=6,max.term=12)
```

Since the number of graphical tools is very limited when the number of explanatory variables is greater than 2 (see nevertheless the help file for a discussion of the tools provided to plot graphs for the functions ϕ_j) we decided to output the numerical figures of merit of this regression. We compute the relative sum of squares (analog of the $1 - R^2$ of the linear models) and the actual sum of the squares of the raw residuals.

```
> PPREG1$fl2
[1] 0.02345262
> sum(PPREG1$ypred*PPREG1$ypred)
[1] 0.005101791
```

These figures look very good. They should be compared to the corresponding figures which we would have obtained if we had applied other methods such as the kernel regression of the previous section or the neural network regression discussed in the Notes & Complements at the end of the chapter. For the sake of comparison with the results of the kernel method which we used earlier, we perform projection pursuit regression using only the two morning indicators used earlier.

```
> PPREG2 <- ppreg(MORN.mat[,4:5],AFT.mat[,4],
                                    min.term=1,max.term=12)
> tsplot(PPREG2$esq)
```

As before we plot the proportion of the variance remaining unexplained, and from the result reported in Figure 4.10, we decide that 7 is a reasonable choice for the order of the pursuit. We then rerun `ppreg` with `m.min=` 7, and we compute the

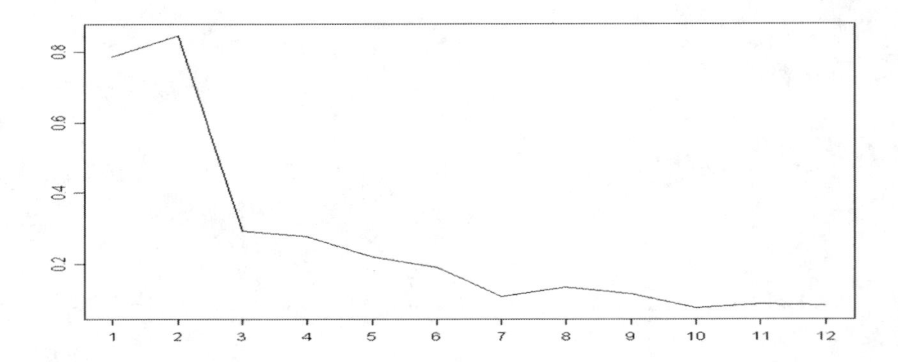

Fig. 4.10. Sequential plot of the `esp` when we use only two explanatory variables.

sum of the squares of the raw residuals.

```
> PPREG2 <- ppreg(MORN.mat[,4:5],AFT.mat[,4],
                                    min.term=1,max.term=12)
> sum(PPREG2$ypred*PPREG2$ypred)
[1] 0.1717635
```

This figure of merit is not as good as the number obtained earlier with the six explanatory variables, but it is much better than the one obtained with the kernel method. Indeed, computing the sum of square errors between the observations and the fitted values for the kernel regression with bandwidth $b = 2$ we find 5661.815. Even the bandwidth $b = .8$, which we rejected because we thought that the plot on the left pane of Figure 4.8 was too rough, leads to a dismal sum of square errors of 2524.696. In fact we need to lower the bandwidth to values in the range of $b = .1$ to get sums of square errors in the range of the score of the projection pursuit. But we rejected these small bandwidths because of the fear we had that the result of the kernel regression would merely be the result of *fitting the noise*.

Despite the apparent superiority of projection pursuit, we would not recommend to give up the kernel regression solely on the basis of this comparison with the sums of square errors. Indeed, if we compute the predictions of the `ppreg` regression function over the same grid as was used for the kernel regression, and we plot the prediction surface over this grid using the function `my.ppreg`, which we wrote to plot the results of projection pursuits from two explanatory variables (which is an oxymoron but which we use for pedagogical reasons):

```
> PPREG <- my.ppreg(MORN.mat[,4:5],AFT.mat[,4],
                                   min.term=7,max.term=12)
```

then we see from the plot reproduced in Figure 4.11 that the regression surface is not as smooth as the one obtained in Figure 4.8. Did we fit the noise or did the projection pursuit regression actually capture features that the kernel was oblivious to? The convoluted nature of the response surface evidenced by Figure 4.11 is presumably an indication that we got a smaller sum of square errors at the cost of noise fitting, and that predictions from this regression could be unstable and unreliable.

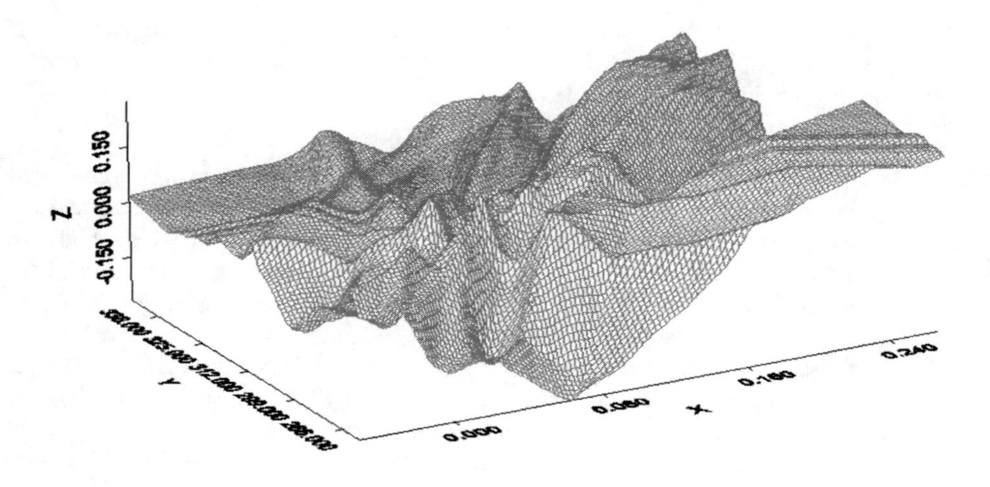

Fig. 4.11. Surface plot of the regression surface given by the projection pursuit algorithm in the case of the morning and afternoon S&P data.

Comparing the relative performance of the kernel and the projection pursuit regression methods is desirable. But one should not attempt to compare the results produced by following too closely the implementation prescriptions which we gave in this chapter. Indeed, the steps suggested for the implementation of the projection pursuit will very likely produce a good fit because the order of the model is based on a criterion involving the sum of square errors between the actual observations and the fitted values. However, using the projection pursuit regression model obtained in this way for the prediction of the response to new values of the explanatory variables may not be a very good idea, since we are not sure that we did not over-fit the model to get a smaller sum of square errors.

On the other hand, the prescriptions we used to choose the bandwidth of the kernel regression were not driven by the desire to get a small sum of square errors. So, we should not be surprised if the regression surfaces obtained in this way do not fit the data as well. Much smaller bandwidth values would be needed for that. The rationale for not choosing the bandwidth too small comes from the desire to use the regression model for prediction purposes.

In short, we suggested implementing the projection pursuit algorithm to have a reasonable fit to the data, while we gave recommendations for the choice of kernel bandwidth with the prediction of future response values in mind. These goals may be incompatible, and fair comparisons may be difficult. Quantifying how well a regression model fits the data is relatively easy: just look at the sum of square errors. Quantifying the predictive value of a regression model is more complicated: one possible way to do it is to fit a regression model to a training sample and quantify the prediction power of the method by using such a model on a different data set. This strategy is explained in the next paragraph, and we will use it in the section on option pricing and in numerous problems.

More on the Comparison of the two Methods

The question of the comparison of the performance of several nonparametric regression/prediction procedures is very delicate. Because they are based on statistical models with a solid theoretical foundation, the parametric regression methods (and especially the linear models) came with inferential tools for residual analysis. In the nonparametric world, the corresponding tools are only asymptotic (i.e. applicable when the sample size goes to ∞) and for this reason, they cannot be of much practical use for finite samples. As a consequence, practical *common sense recipes* are used instead. Since nonparametric methods can only be competitive when the sample size is relatively large, one often has the possibility of separating the data into two subsets, one used for *training* purposes, i.e. to fit a regression model, and the other one for *testing* purposes, i.e. to compare the figures of merit of several methods or several parameters, ... See the next section for an example. In financial applications, this is implemented in the form of *back testing*. In our specific example, after finding the kernel and the bandwidth, or the order and the parameters of the projection pursuit, using the June 1998 contract, we could use a different contract to test and compare the relative performances of the different methods or sets of parameter choices.

Confidence intervals offer a convenient way to quantify the accuracy of statistical estimates. In regression problems, confidence intervals are replaced by confidence bands, or sausages, containing the regression curves or surfaces. The computations needed to produce reliable confidence regions are usually very intensive. The most successful ones seem to have come out of *bootstrap*-like methodologies, but they are beyond the scope of this book.

4.7 Nonparametric Option Pricing

The goal of this section is to implement and compare several nonparametric methods of option pricing. We use real market data to compare the numerical performance of the various methods. After a couple of subsections devoted to introductory material on classical option pricing theory, the data are described in Section 4.7.3, and the details of the experiment are given in Section 4.7.4. The long Subsection 4.7.1 is not intended as a crash course on option pricing but merely as a convenient introduction to the mathematical theory of no-arbitrage option pricing, justifying the nonparametric approach used in this section.

4.7.1 Generalities on Option Pricing

For the sake of simplicity the following discussion is restricted to plain vanilla options on an underlying asset which can be thought of as a stock or an index. Moreover, to make our life easier, we use the terminology *option* when we actually mean *European call option.* The problem of pricing *European put options* can be addressed in exactly the same way, or can be reduced to the pricing of European call options because of a parity argument. We leave to the interested reader the easy task of adjusting the following discussion to this case. Also, we shall not discuss the pricing of options with American exercise or the so-called exotic options, such as barrier options.

An option is a contract which gives the buyer (of the option) the right (but not the obligation) to purchase at time T (called the expiration or maturity date of the option) one unit of the underlying asset or instrument at an agreed upon price K (called the strike price of the option). Such an option is a contingent claim and the contract unambiguously defines the payoff of the claim. In the situation at hand the payoff is a simple function of the price of the underlying asset at expiration. If we use the notation S_t to denote the price of this underlying asset at time t, then the payoff of our option is given by $f(S_T)$ where the function f is defined by $f(x) = \max\{x-K, 0\}$. Indeed, if at time of expiration the price S_T of the asset is smaller than K, then the option is worth nothing. Why would we buy one unit of the underlying for K if we can get it for less on the open market! Moreover, if the price S_T is greater than K, then it is clear that one should exercise the option, buy the unit of the underlying for K and re-sell immediately on the open market for a profit of $S_T - K$. The graph of the payoff function is given in Figure 4.12.

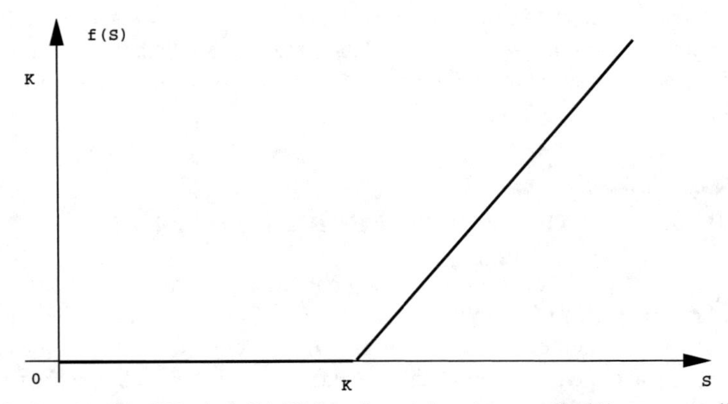

Fig. 4.12. Graph of the "Hockey Stick" function giving the payoff of European call options.

We shall use the notation $C_{T,K}(t, S)$ for the price of such an option at time t if the price (at this time) of the underlying asset is S.

Stochastic Models for Asset Prices

Obviously, the value S_T of the asset price at maturity cannot be predicted with certainty, and hence, it is reasonable to model it as the outcome of a random variable. Under this assumption the payoff is also a random variable, and the value of the option now depends on the probability that the price S_T at expiration is lower than the strike K, and on the various probabilities of the different ways this price could end up being greater than the strike. Because of this inherent randomness, it is reasonable to price the option by expectation, as an average over all the possible future scenarios weighted by their respective probabilities. At any given time t one can observe the current value of the underlying asset, say S, and consider the expectation of the payoff (which takes place precisely at time T since we are only considering options with an European exercise). It is quite clear that, because of the time value of money (from our discussion of the fixed income markets, we agree that one dollar later does not sound as valuable as one dollar now) this expectation should be discounted to give the current value of this expected payoff. Consequently, one is naturally lead to the following formula for the price of the option:

$$C_{K,T}(t, S) = e^{-r(T-t)}\mathbb{E}\{f_K(S_T)|S_t = S\}. \tag{4.18}$$

The positive number $r > 0$ appearing in the exponential in the right hand side stands for the interest rate. It is assumed to be constant over the life of the option. Finally the expectation is the expectation of the pay-off $f_K(S_T)$ conditioned by the knowledge that at time t the price of the underlying is S (i.e. $S_t = S$). This conditional expectation (like any other expectation) can be a sum when the random variables are discrete, but in general, it can be computed as an integral if one knows the density of the random variables in question. In the present situation, the expectation is in fact a conditional expectation, but this does not change the fact that the expectation can be computed as an integral: we just have to use the conditional density instead of the a-priori density. So, if we denote by $p(\,\cdot\,|S)$ the density of the random variable S_T of the asset price at time T knowing that $S_t = S$, then the expectation in (4.18) can be computed as:

$$\mathbb{E}\{f_K(S_T)|S_t = S\} = \int f_K(s')\,p(s'|S)\,ds'. \tag{4.19}$$

Intuitively, $p(s'|S)$ gives the probability that $S_T = s'$ given the fact that $S_t = S$. This probability density is often called the *objective density* or the *historical density* because it describes the statistics of the price of the asset at time T, as predicted by historical data. It depends upon many factors which need to be specified.

Aside on Dynamical Models

The static models of the statistical distribution of the underlying asset at a *fixed* given time are enough to price European call and put options, but they cannot suffice for the pricing of options whose settlements depend upon the entire trajectories of the underlying instruments. Even though we will not need the following material in this chapter, we include it as preparation for the forthcoming discussions in the last chapters of the book. Obviously, this paragraph can be skipped in a first reading.

One of the most popular dynamical models for stock prices is certainly the so-called geometric Brownian motion model proposed by Samuelson and used by Black, Scholes and Merton in their ground-breaking work on option pricing in the early seventies. The widespread use of their pricing formulae earned Merton and Scholes the Bank of Sweden Prize in Economics in Memory of Alfred Nobel, usually called the Nobel prize in economics, in 1997. Fisher Black passed away in 1995. According to this model, the dynamics of the underlying asset price are given by the *stochastic differential* equation:

$$dS_t = S_t[\mu dt + \sigma dW_t], \tag{4.20}$$

where the term dW_t is a white noise in continuous time (some sort of a mathematical monstrosity to which one has to give a rigorous meaning). The deterministic constant $\mu \in \mathbb{R}$ represents the rate of growth of the (logarithmic) return of the stock, while $\sigma > 0$ represents the volatility of the stock (log) return. Because the coefficients are constant, the stochastic differential equation (4.20) can be solved explicitly. The solution is given by:

$$S_t = S_t e^{[\mu - \frac{\sigma^2}{2}]t + \sigma W_t} \tag{4.21}$$

and the unexpected term $-t\sigma^2/2$ appearing in the exponential is called the Ito correction. This model is called the geometric Brownian motion model for stock prices. These technical facts are quoted here as preparation for Section 7.4, and Subsections 7.4.2 and 7.5.1 in Section 7.5.

We now come back to the derivation of a pricing formula for the option. The important fact is that, conditioned on being equal to S at time t, the asset price at time T is log-normally distributed with mean $\log S_t + (\mu - \sigma^2/2)\tau$ and variance $\sigma^2\tau$ where $\tau = T - t$. In other words:

$$\log S_T \sim N(\log S_t + (\mu - \sigma^2/2)\tau, \sigma^2\tau). \tag{4.22}$$

This fact is a consequence of the above choice for a dynamical model, but it can also be postulated independently. In any case, under these conditions, the expectation in (4.19) can be computed explicitly (see details below for a similar computation).

Risk Neutral Pricing

Even though the expectation of formula (4.18) is a very natural candidate for the price of the option, even in Samuelson's geometric Brownian motion model, it is not the correct one! Indeed, if options were priced in this way, market makers would have arbitrage opportunities making it possible to make a profit starting from nothing. So since we all know that there is no such a thing as a *free lunch,* something must be wrong with the pricing formula (4.18) and the distribution (4.22). In the absence of arbitrage, and for reasons well beyond the scope of the book, the option can indeed be priced by discounting an expectation, as long as the expectation is computed with respect to another log-normal distribution. This new density is called the *state price density,* and in the case of the Black-Scholes theory, it is obtained by merely replacing μ by r. In other words, in order to price an option on an underlying asset, we act as if the rate of growth of the (log) returns was the short interest rate. This justifies the name *risk neutral* probability for this new log-normal density.

Black-Scholes Formula

For the sake of completeness, we give the details of the computations leading to the Black-Scholes formula for the price of the option. This formula will be used in one of the nonparametric approaches presented below.

Using the explicit form of the density of the $N(0,1)$ random variable and formula (4.22), we derive the value of $C_{T,K}(t,S)$ as follows:

$$C_{T,K}(t,S) = \frac{e^{-r(T-t)}}{\sqrt{2\pi}} \int_{-\infty}^{+\infty} f\left(Se^{(r-\sigma^2/2)(T-t)}e^{\sigma\sqrt{T-t}z}\right) e^{-z^2/2}\,dz. \tag{4.23}$$

Now, using the fact that $f(x) = \max\{x - K, 0\} = (x - K)^+$ in the case of a European call option and using the notation:

$$d_1 = \frac{\log(S/K) + (r + \sigma^2/2)(T-t)}{\sigma\sqrt{T-t}} \quad \text{and} \quad d_2 = d_1 - \sigma\sqrt{T-t},$$

we get:

$$\begin{aligned}
C_{T,K}(t,S) &= \frac{e^{-r(T-t)}}{\sqrt{2\pi}} \int_{-\infty}^{+\infty} \left(Se^{(r-\sigma^2/2)(T-t)}e^{\sigma\sqrt{T-t}z} - K\right)^+ e^{-z^2/2}\,dz \\
&= \frac{1}{\sqrt{2\pi}} \int_{-\infty}^{+\infty} \left(Se^{-\sigma^2(T-t)/2+\sigma\sqrt{T-t}z} - Ke^{-r(T-t)}\right)^+ e^{-z^2/2}\,dz \\
&= \frac{1}{\sqrt{2\pi}} \int_{-d_2}^{+\infty} \left(Se^{-\sigma^2(T-t)/2+\sigma\sqrt{T-t}z} - Ke^{-r(T-t)}\right) e^{-z^2/2}\,dz \\
&= \frac{S}{\sqrt{2\pi}} \int_{-d_2}^{+\infty} e^{-\sigma^2(T-t)/2+\sigma\sqrt{T-t}z}e^{-z^2/2}\,dz \\
&\qquad - Ke^{-r(T-t)}\frac{1}{\sqrt{2\pi}} \int_{-d_2}^{+\infty} e^{-z^2/2}\,dz \\
&= S\Phi(d_1) - Ke^{-r(T-t)}\Phi(d_2),
\end{aligned}$$

where we performed the substitution $y = z + \sigma\sqrt{T-t}$ in the first of the two integrals. Recall that we use the notation Φ for the cumulative distribution function of the standard normal distribution. We summarize this computation in the following box:

> The price at time t of a European call with strike K and maturity T is given by the formula:
>
> $$C_{T,K}(t,S) = S\Phi(d_1) - Ke^{-r(T-t)}\Phi(d_2) \tag{4.24}$$
>
> if the price of the underlying risky asset is S at time t.

Notice that, instead of depending separately upon t and T, the conditional density and the actual price of the option depend only upon the difference $\tau = T - t$. The interpretation of this difference is very simple: it is the time to maturity of the option.

Figure 4.13 shows plots of the prices $C_{T,K}(t,S)$ (together with the prices $P_{T,K}(t,S)$ of the corresponding European put options) for various times to expiration $T-t$ when the strike price K, the volatility σ and the interest rate r are fixed. We notice that when regarded as functions of the underlying price S, the European call and put values appear as smoothed forms of the original hockey stick pay-off functions. Moreover, $C_{T,K}(t,S)$ vanishes near small values of S while $P_{T,K}(t,S)$ vanishes for large values of S.

Historical Volatility, Implied Volatility and the Smile Effect

Since the rate of growth μ of the stock was evacuated from the formula for the price of the contingent claims, the only parameter from the dynamics of the risky asset

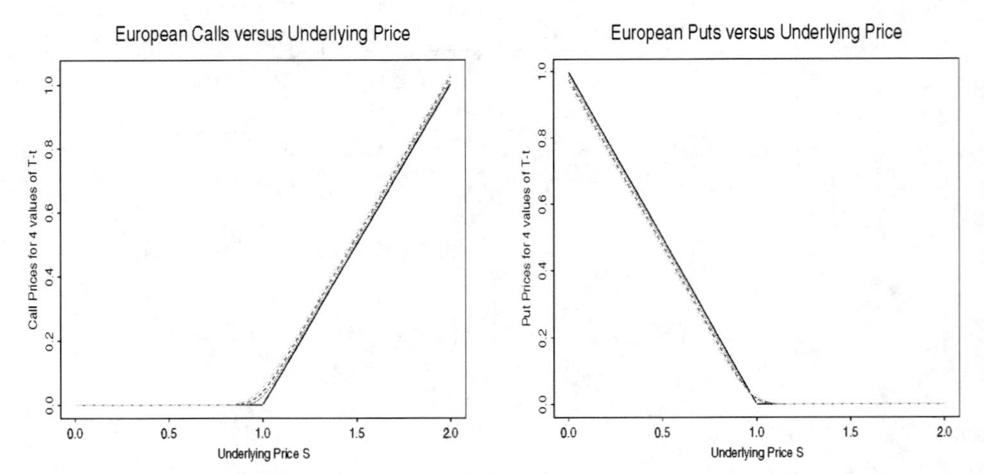

Fig. 4.13. Plots of the prices of European call (left) and European put (right) vanilla options as functions of the underlying price S when all the other parameters are held fixed. We used the values $K = 1$, $\sigma = 15\%$, $r = 10\%$ and $T - t = 0, 30/252 = 0.1190476, 60/252 = 0.2380952, 90/252 = 0.3571429$.

that is entering these formulae is the volatility σ and the problem we wish to discuss now is the estimation of its value.

In theory, the observation of a single trajectory (realization) of the price should be enough to completely determine the value of this parameter. This would be true if the price process S_t could be observed continuously! Unfortunately this cannot be the case in practice and we have to settle for an approximation. Given observations S_{t_j} of past values of the risky asset (usually the times t_j are of the form $t_j = t - j\delta t$), we use the fact that in the Black-Scholes model the random variables $\log(S_{t_j}/S_{t_{j+1}})$ are independent and normally distributed with mean $(\mu - \sigma^2/2)\delta t$ and variance $\sigma^2\delta t$. Consequently, the volatility can be estimated by the formula:

$$\hat{\sigma} = \frac{1}{(N-1)\sqrt{\delta t}} \sum_{j=0}^{N-1} [\log \frac{S_{t_j}}{S_{t_{j+1}}} - \overline{LS}]^2, \tag{4.25}$$

where the sample mean $\overline{LS}$ is defined by the formula:

$$\overline{LS} = \frac{1}{N} \sum_{j=0}^{N-1} \log \frac{S_{t_j}}{S_{t_{j+1}}}.$$

The volatility estimate provided by formula (4.25) is called the historical volatility. Even though $\overline{LS}$ provides (at least up to the multiplicative factor δt) an unbiased estimator of the growth rate μ, this fact is not used since the growth rate μ does not appear in the Black-Scholes formula of the arbitrage pricing of the claim.

Everything seems to be fine except for the fact that we are now about to encounter our first shocking evidence that the market does not have the good taste to follow the Black-Scholes model. This will be the first of a series of rude awakenings.

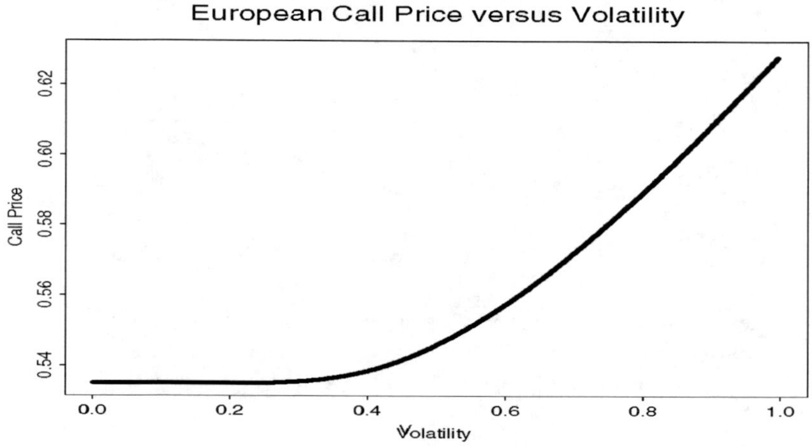

Fig. 4.14. Values of the price of a European call option as a function of the volatility when all the other parameters are held fixed. We used the values, $T - t = 90/252$, $r = 10\%$, $S = 1.5$ and $K = 1$.

A quick look at Figure 4.14, which gives the plot of $C = C_{T,K}(t, S)$ as a function of σ when all the other parameters (namely r, T, K and S) are held fixed, shows that C is an increasing function of σ, and consequently, that there is a one-to-one correspondence between the price of the option, and the volatility parameter σ. For each value of C, the unique value of σ which, once injected into the Black-Scholes formula, gives the option price C, is called the implied volatility of the option. This one-to-one correspondence is so entrenched in the practice of the markets that prices of options are most of the time quoted in percentage points (indicating a value for the *implied volatility*) rather than in dollars (for a value of C).

We now demonstrate the fact that the assumptions of the Black-Scholes model are in contradiction with market reality. We use quotes from January 6, 1993 (when the S&P index was at the $S = 435.6258$ level) on European calls on the S&P 500 index with maturity February 19, 1993, but with different strike prices. In general, when everything else is fixed, for a given set of strike prices K_j we have the corresponding call prices C_j quoted by the market. If the Black-Scholes model was in force we should expect that these quotes had been computed using formula (4.24) with a single volatility value for σ, the differences in the quotes being due only to the differences in the strikes (since the expiration is the same for all these options). Since given a strike price there is a one-to-one correspondence between option price and volatility, one should be able to compute the (implied) volatility used to price all these options. Moreover, plotting the values of these implied volatilities versus the

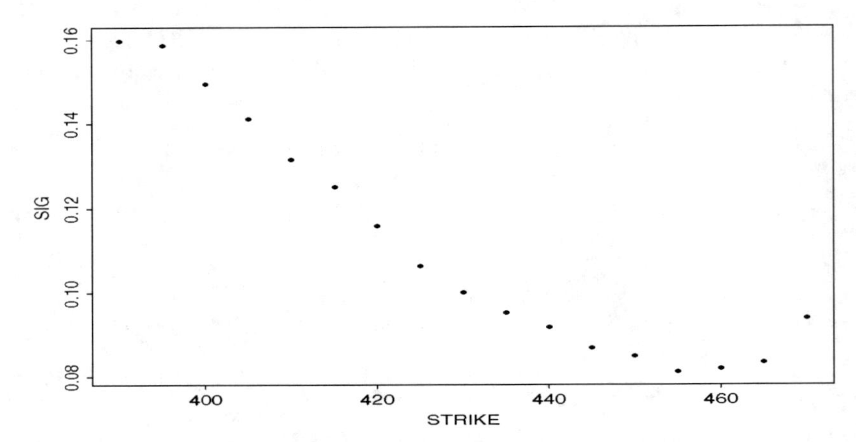

Fig. 4.15. Plot of the implied volatility values versus the strike prices of call options with the same maturity on the S&P 500 index. A convex (upward) curve is found instead of the horizontal line predicted by the Black-Scholes theory. This curve is called the smile.

corresponding strike prices one should expect a flat graph indicative of the uniqueness of the volatility used to price the options. Figure 4.15 shows that this is certainly not the case! The volatility plot forms a convex curve which has been called the volatility smile. This volatility smile is striking empirical evidence of the inadequacy of the Black-Scholes option pricing formula.

4.7.2 Nonparametric Pricing Alternatives

Because of the shortcomings of the Black-Scholes formula illustrated above, we resort to nonparametric techniques to price liquid options. We choose options on the S&P 500 because on any trading day, there are a large number of options traded with various times to maturity and strikes. We shall consider two competing approaches. The second one will rely partly on the Black-Scholes formula, but the first one is fully nonparametric.

Fully Nonparametric Pricing

In this approach, we do not make any assumption on the functional relationship between the measurable variables S (current price of the underlying asset), K (strike price of the option), τ (time to maturity), and r (short interest rate), which we consider as explanatory variables, and the price C of the corresponding (European call) option which we regard as the response variable. So if we bind together the four explanatory variables into a row $\mathbf{X} = [S, K, \tau, r]$ and if we set $Y = C$ for the response variable, then we are in the classical setting of (multiple) regression, and we can use any of the techniques seen so far to explain and predict the price Y from the knowledge of **X**. The goal of the first part of our experiment is to implement and compare the results of the kernel method and the projection pursuit regression in this setting.

Semi-Parametric Option Pricing

By definition of the implied volatility, the knowledge of the value of an option is equivalent to the knowledge of its implied volatility. Indeed one can go from one to the other by evaluation of Black-Scholes formula (4.24). So instead of pricing the option by computing its values in US dollars via this formula, we first derive the implied volatility, and then we evaluate Black-Scholes formula. Next we argue that this implied volatility is a specific function of a smaller number of variables, and we attempt to derive this functional dependence by a nonparametric regression, typically a kernel regression since after reducing the number of variables, the dimension is presumably not an issue any longer. This approach is called semi-parametric because it is a combination of a purely nonparametric approach (using the kernel to predict the implied volatility) and a parametric strategy (characterizing the implied volatility by a small number of parameters and computing the Black-Scholes formula.

4.7.3 Description of the Data

The data are contained in two text files named `trgsp.asc` and `tstsp.asc` from which one creates the `S-Plus` objects `TRGSP` and `TSTSP`. As you might guess from their names, these two data matrices are intended for training and testing purposes respectively. Each data file contains six columns, each row corresponding to an option. The meanings of these columns are as follows:

1. `SP` for the price of the index
2. `KK` for the strike price of the option
3. `TAU` for the time to expiration (in days)
4. `RR` for the spot interest rate
5. `ISIGMA` for the implied volatility
6. `CALL` for the price of the (European call) option

We chose to use repeated letters `KK` and `RR` instead of `K` and `R`, respectively, because `S-Plus` reserves some of the single letter names for protected objects. Also, you should not assume that the row numbers of the data frames have much to do with the dates at which these quotes were in force. The third column gives the *time to maturity* which is the difference $\tau = T - t$ between the time of maturity T and the time t at which the quote was given. The dates t and T are not given in the data sets, only their differences is. For the sake of definiteness, all the quotes are from 1993, and some of the options expire in 1994. In particular, using sequential plots (like those given by the function `tsplot`) would not make much sense in the present situation.

Warning: The values of `TAU` found in the third column are given in days. In most applications, they need to be expressed in years before they can be used in the formulae as values of $\tau = T - t$. This change of units should be done by dividing all the entries of this column by 252. This is the average number of trading days in one year.

4.7.4 The Actual Experiment

For the purposes of this experiment, we wrote two simple `S-Plus` functions, `bscall` which computes the price of a European call option from the Black-Scholes formula (4.24), and `isig` which computes the implied volatility by inverting the same Black-Scholes formula. Notice that both functions require that the parameter `TAU` which stands for the variable τ be given in years, while the third columns of `TRGSP` and `TSTSP` give the numbers of days until maturity. We choose the convention of 252 trading days per year to convert the number of days in years. In order to make sure that this function does what it is supposed to do, one can use it with the arguments `SP`, `KK`, `TAU`, `RR`, and `ISIGMA`, and check that one does indeed recover the last columns of the training `TRGSP` and testing `TSTSP` data matrices respectively.

In order to check that the sixth column is equal to the result of the function `bscall` when applied to the first five columns, we compute the range (i.e. the two-dimensional vector whose entries are the minimum and maximum of the original vector) of the difference and we check that its entries are equal to 0, at least up to small rounding errors due to the fact that the Black Scholes formula cannot be inverted exactly, and to the fact that the computation of the implied volatility is merely the result of a numerical approximation.

```
> BSTRG <- bscall(TAU=TRGSP[,3]/252,K=TRGSP[,2],S=TRGSP[,1],
                                    R=TRGSP[,4],SIG=TRGSP[,5])
> range(TRGSP[,6] - BSTRG)
[1] 0 0
> BSTST <- bscall(TAU=TSTSP[,3]/252,K=TSTSP[,2],S=TSTSP[,1],
                                    R=TSTSP[,4],SIG=TSTSP[,5])
> range(TSTSP[,6] - BSTST)
[1] 0 0
```

Fully Nonparametric Regression

In this first part, we use the function `kreg` to predict the option prices in the testing sample `TSTSP` using as explanatory variables the current price of the underlying index, the strike price, the time to maturity, and the current value of the short interest rate. *Obviously, we do not use the implied volatility, that would be cheating!* In other words, we use the training data in `TRGSP` to determine the regression function which we then use to predict the prices of the options contained in `TSTSP`. We also compute the raw sum of square errors which we call SSE_1 and a per-option error. We shall use them later to compare the performance of this four dimensional kernel regression with the other methods explained above.

Fully Fledged Kernel. Since the function kreg uses only one bandwidth, we need to standardize the explanatory variables before running a multidimensional kernel regression. Also, we take this opportunity to emphasize the modicum of care needed to standardize a testing sample from the results of the normalization of the training sample. In other words, we need to standardize the regressor variables using the means and the standard deviations of the observations of the regressor variables contained in the training sample. This is a typical source of error when one first enters the business of nonparametric *prediction*. We follow the steps outlined in Subsection 4.5.2. We first standardize the regressors in the training sample.

```
> X1 <- cbind(TRGSP[,1],TRGSP[,2],TRGSP[,3],TRGSP[,4])
> MEANX1 <- apply(X1,2,mean)
> MEANX1
[1] 441.91861365 439.84700000  50.33720000   0.03030124
> MEANX1.mat <- outer(rep(1,dim(X1)[1]),MEANX1,"*")
> VARX1 <- apply(X1,2,var)
> VARX1
[1] 4.275647e+01 5.053527e+02 6.411393e+02 6.558444e-07
> STDX1 <- sqrt(VARX1)
> STDX1.mat <- outer(rep(1,dim(X1)[1]),STDX1,"*")
> SX1 <- (X1-MEANX1.mat)/STDX1.mat
> apply(SX1,2,mean)
[1] -8.855560e-13  8.080203e-16 -1.350475e-16  9.482761e-13
> apply(SX1,2,var)
[1] 1 1 1 1
```

We now normalize the entries of the testing sample by using the same normalization as in the normalization of the training sample. Notice that, for the sake of convenience, the second and third commands use the fact that the training sample is larger than the testing sample.

```
> XPRED1 <- cbind(TSTSP[,1],TSTSP[,2],TSTSP[,3],TSTSP[,4])
> MEANX1 <- MEANX1.mat[1:dim(XPRED1)[1],]
> STDX1 <- STDX1.mat[1:dim(XPRED1)[1],]
> SXPRED1 <- (XPRED1-MEANX1)/STDX1
> apply(SXPRED1,2,mean)
[1]  0.555335926  0.075311219 -0.002338005  0.639329493
> apply(SXPRED1,2,var)
[1] 0.2279553 0.9240326 0.8392511 0.1225965
```

Notice that the columns of the standardized testing sample do not have mean zero and variance one. This is due to the fact that we have to use the means and standard deviations of the columns of the training sample if we want to pretend that we are going to predict the response for the individual elements of the testing sample, one by one, without assuming that we know the whole sample. Once this standardization is out of the way, we are ready to run the kernel regression. We use the Gaussian kernel function by setting the parameter kernel to 7 in the command below. The results would be qualitatively the same if we use another kernel function. The choice of bandwidth is even more delicate than before. Indeed, it is not possible to eye-ball this choice by looking at plots of the regression surfaces obtained for different values of the bandwidth. A possible approach is to use the default value proposed by the program. This value is derived from mathematical results aiming at the identification of optimal choices for the bandwidth. Such mathematical results have been proved in the limit of sample sizes going to infinity. They give specific prescriptions for the choice of the optimal bandwidth in the univariate case, the multivariate case being handle by a scaling correction. Another possibility would be to implement a form of cross-validation. However, for the sake of simplicity, we use the default value. Using this bandwidth we get:

```
> YPRED1.kreg <- kreg(SX1,TRGSP[,6],xpred=SXPRED1,
                                          kernel=7,b=.05)
> YPRED1 <- YPRED1.kreg$ypred
> SSE1 <- sum((YPRED1 - TSTSP[,6])*(YPRED1 - TSTSP[,6]))
> SSE1
[1] 256.4033
> sqrt(SSE1/length(YPRED1))
[1] 1.132262
```

Projection Pursuit. The implementation of projection pursuit is simpler because we do not have to standardize the explanatory variables. As before, we use the option data from TRGSP as a training sample to which we fit the regression model, which we then use to predict the prices of the options of the data set TSTSP. As before, we use as explanatory variables the current price of the underlying index, the strike price, the time to maturity, and the current value of the short interest rate. And as before, we compute the sum of square errors which we now call SSE_2, as a figure of merit.

```
> YPRED2.ppr <- ppreg(TRGSP[,1:4],TRGSP[,6],
                               min.term=1,max.term=10)
> tsplot(YPRED2$esq)
```

The plot of YPRED2$esq suggests using any integer between 2 and 9 for the number of terms to include in the pursuit. We choose 2 to be parsimonious, and we rerun the algorithm accordingly.

```
> XPRED2 <- as.matrix.data.frame(TSTSP[,1:4])
> YPRED2.ppreg <- ppreg(TRGSP[,1:4],TRGSP[,6],
                  min.term=2, max.term=10,xpred=XPRED2)
> YPRED2 <- YPRED2.ppreg$ypred
> SSE2 <- sum((YPRED2 - OTSTSP[,6])*(YPRED2 - OTSTSP[,6]))
> SSE2
[1] 42.95764
> sqrt(SSE2/length(YPRED2))
[1] 0.4634525
```

The above result shows a very significant improvment. It is a clear proof of the difficulties the kernel has when it has to handle high dimensions. Comparing the two fully nonparametric methods confirms the fact that the projection pursuit has a better control of the curse of dimensionality.

Semi-Parametric Estimation

The prediction methods used above are based on a brute force approach, using only the raw data, and no particular knowledge of the specifics of the actual problem at hand. We now inject some *finance* into the statistical mix, and as we are about to see, this is going to take us a long way.

The price of the option is now derived from a prediction of its implied volatility, by computing the function `bscall`. Here we follow the market practice which favors implied volatility over price. However, the thrust of the financial input that we mentioned above is in the way we predict the implied volatility. Instead of blindly mining the four dimensional training sample, we use a two dimensional explanatory vector made of financially meaningful variables: the time to maturity τ, and the moneyness $M = e^{r(T-t)}S/K$. As before, r is the short interest rate, S is the current price of the underlying asset, and K is the strike price of the option. This ratio M is called the *moneyness* of the option because it compares the index futures price $e^{r(T-t)}S$ to the strike K. For this reason, the option is said to be *in the money* when $M > 1$, *out of the money* when $M < 1$ and *at the money* when $M = 1$. Note that, in the computation of the moneyness M, the time to maturity $\tau = T - t$ has to be expressed in years because the spot interest rate r is quoted annually.

At this stage, it is not important which nonparametric regression we use to predict the implied volatility from these two explanatory variables. Indeed, in two dimensions, similar figures of merit can be achieved if we choose the smoothing parameters appropriately. We decided to use the kernel method for the sake of definiteness.
In order to complete the program described above, we first compute the moneyness for each of the options of the training and testing samples.

```
> TRGM <- exp(TRGSP[,4]*TRGSP[,3])*TRGSP[,1]/TRGSP[,2]
> TSTM <- exp(TSTSP[,4]*TSTSP[,3])*TSTSP[,1]/TSTSP[,2]
```

Next, we play the same *standardization* game to prepare for the kernel regression/prediction. Since this is not the first time that we are going through this exercise, we limit ourselves to giving the `S-Plus` commands.

```
> XX3 <- cbind(TRGSP[,3],TRGM)
> MEANXX3 <- apply(XX3,2,mean)
> MEANXX3
                TRGM
0.2025935 0.9897337
> MEANXX3.mat <- outer(rep(1,dim(XX3)[1]),MEANXX3,"*")
> VARXX3 <- apply(XX3,2,var)
```

```
> VARXX3
                         TRGM
0.01023078 0.002668094
> STDXX3 <- sqrt(VARXX3)
> STDXX3.mat <- outer(rep(1,dim(XX3)[1]),STDXX3,"*")
> SXX1 <- (XX3-MEANXX3.mat)/STDXX3.mat
> apply(SXX3,2,mean)
                                TRGM
2.104511e-014 1.794116e-014
> apply(SXX3,2,var)
    TRGM
 1      1
```

Now that we are done with the standardization of the training sample, we normalize the testing sample explanatory variables by using the coefficients computed from the training sample.

```
> XXPRED3 <- cbind(TSTSP[,3],TSTM)
> MEANXX3 <- MEANXX3.mat[1:dim(XXPRED3)[1],]
> STDXX3 <- STDXX3.mat[1:dim(XXPRED3)[1],]
> SXXPRED3 <- (XXPRED3-MEANXX3)/STDXX3
> apply(SXXPRED3,2,mean)
                        TSTM
-0.5131039 0.09892794
> apply(SXXPRED3,2,var)
                     TSTM
0.6833883 0.7110236
```

For the sake of illustration, we give the scatterplots of the normalized regressor variables in Figure 4.16. As expected, the points of the training sample cover the region we expect a standardized bivariate normal sample would cover. On the other hand, it appears that the options of the testing sample come in three very distinct subgroups according to the values of the time to maturity. This should not be too bad of a problem given the fact that, each point in the testing sample seems to be well surrounded by a large number of points of the training sample. That will certainly help the kernel regression to do its job. Finally, we compute semi-parametric predictions of the option prices of the testing sample (i.e. the entries in the column `CALL` in `TSTSP`) by first computing the kernel prediction of the implied volatility, and then by plugging these predictions into the Black-Scholes formula. We compute the sum of square errors as before, and we call it SSE_3. As before, we choose to work with the Gaussian kernel function, hoping that this will prevent divisions by 0 in our tests, and as before, the choice of the bandwidth is the crucial difficulty to overcome. Fortunately, the fact that we are working with a two dimensional regressor vector makes it possible to visualize the properties of the regression surface. So for the purposes of this project, we limit ourselves to the simplest method of all: we compare the plots of the regression surfaces for several values of the bandwidth, and we pick the bandwidth leading to the most reasonable implied volatility surface. This procedure relies heav-

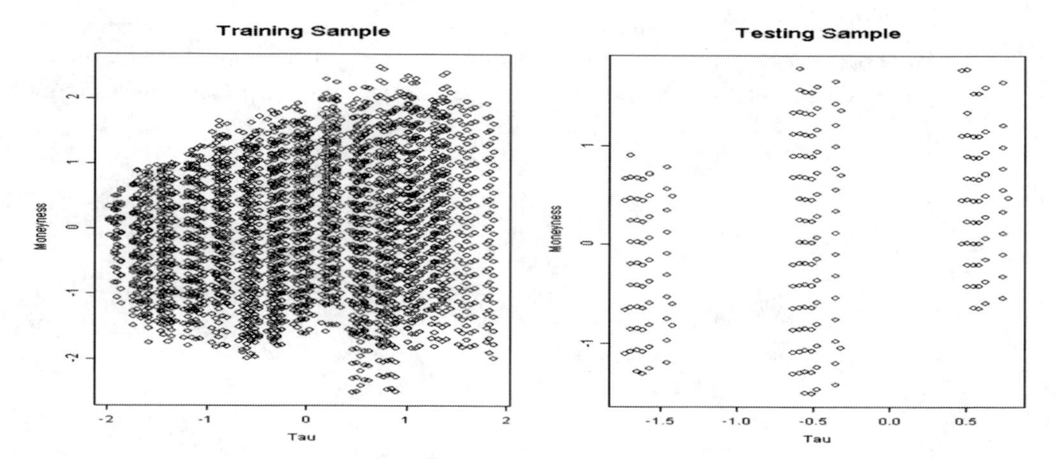

Fig. 4.16. Scatterplots of the standardized explanatory variables in the training sample (left) and the testing sample (right).

ily on our past experiences: it is more of an art form than a quantitative approach, however, it will be good enough here. To illustrate some of the features leading to our choice, we reproduce in Figure 4.17, the regression surfaces computed above a common grid of points in the (TAU, M)-plane, for three values of the bandwidth. The regression for the largest of the three bandwidths is too smooth and misses many of the features of the regression surface visible when the bandwidth is equal to 0.1, while the regression for the smallest of the two bandwidth is too rough, including effects which are presumably only due to noise. After extensive experimentation with

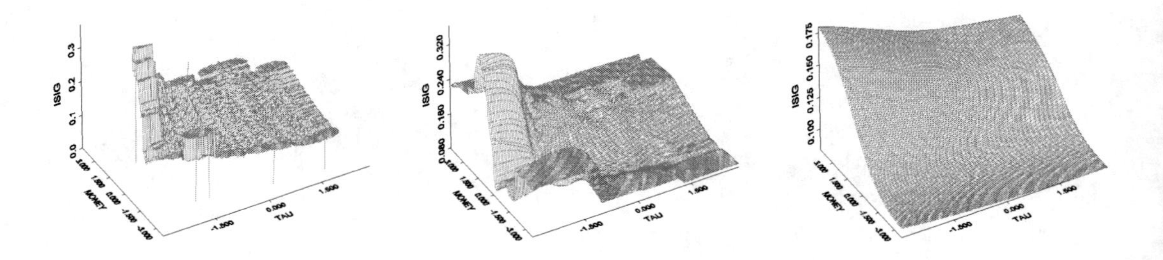

Fig. 4.17. Regression of the implied volatility on the time to expiration and the moneyness. From left to right, kernel regression surfaces for the bandwidths $b = 0.01$, $b = 0.1$ and $b = 1.0$.

the value of the bandwidth, we decided that $b = .033$ was a reasonable choice. According to the semi-parametric strategy outlined above, once the bandwidth has been chosen, we predict the implied volatilities of the testing sample by nonparametric regression, and we plug the predictions so-obtained into the Black-Scholes formula in order to obtain the semi-parametric predictions of the option prices. Finally, we compute the sum of square errors as before.

```
> YPRED3.kreg <- kreg(SXX1,TRGSP[,5],xpred=SXXPRED1,
                                          kernel=7,b=.033)
> ISIGMAPRED <- YPRED3.kreg$ypred
> YPRED3 <- bscall(TAU=TSTSP[,3],K=TSTSP[,2],S=TSTSP[,1],
                            R=TSTSP[,4], SIG=ISIGMAPRED)
> SSE3 <- sum((YPRED3 - TSTSP[,6])*(YPRED3-TSTSP[,6]))
> SSE3
[1] 6.470708
> sqrt(SSE3/length(YPRED3))
[1] 0.1798709
```

Next we compare the performance of the different methods used to predict the price of the (European call) options on the S&P 500, and we comment on the differences in the results of the various methods.

4.7.5 **Numerical Results**

We first compare the results of the three methods by comparing the sums of squares `SSE1`, `SSE2`, and `SSE3` obtained with our implementations (i.e. for the particular choices of the smoothing parameters which we made) of the three methods. Obviously, the semi-parametric method gives much better results.

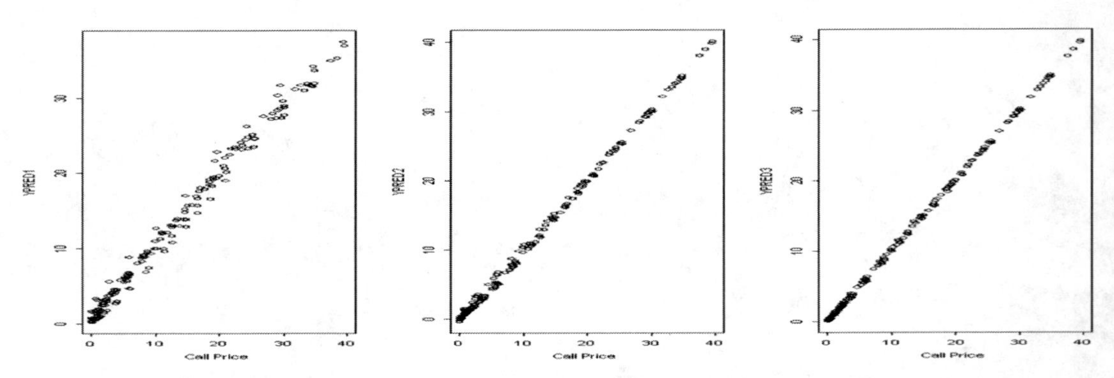

Fig. 4.18. Scatterplot of the predicted call prices against the actual prices. Left: 4- dimensional kernel regression. Center: projection pursuit regression. Right: semi-parametric regression based on the kernel prediction of the implied volatility and the Black-Scholes formula.

As a last attempt to compare the predictions given by the three regression procedures, we plot the predicted values against the actual prices of the call options in the testing sample. The three plots are given in Figure 4.18. They confirm the conclusions based on the numerical scores of merit compared earlier: the semi-parametric regression based on the 2-dimensional kernel gives the best results, and the 4-dimensional kernel is the worst because the points are scattered further away from the diagonal. Also, this plot shows that the kernel underestimates the highly priced options. It would be interesting to go back to the data and try to explain why.

As mentioned earlier, Problems 4.13 and 4.14 give examples of situations for which the conclusions are slightly different.

State Price Density

According to the discussion leading to (4.18), we have:

$$C_{T,K}(t,S) = e^{-r(T-t)}\mathbb{E}\{f_K(S_T)|S_t = S\} = e^{-r(T-t)}\int f_K(s')p(s')ds'$$

for some density $p(s')$, where we use the notation $f_K(x) = (x-K)^+ = max\{x - K, 0\}$ for the pay-off function of the European call option. The density $p(s')$ plays such an important role that it is given a special name: the state-price density. Its importance comes from the fact that it can be used to price any kind of European contingent claim with the same maturity. Indeed, if an option pays the amount $f(S_T)$ at maturity T, simple arbitrage arguments tell us that its price at time t should be given by the risk neutral expectation:

$$e^{-r(T-t)}\mathbb{E}\{f(S_T)\} = e^{-r(T-t)}\int_0^\infty f(s')p(s')\,ds'.$$

Using the fact that:

$$\frac{\partial f_K(x)}{\partial K} = \begin{cases} -1 \text{ if } K < x \\ 0 \quad \text{if } x < K \end{cases} \qquad \text{and} \qquad \frac{\partial^2 f_K(x)}{\partial K^2} = \delta_x(K)$$

where δ_x denotes the Dirac delta function at x, and allowing ourselves to interchange the derivatives and the integration, we get:

$$\begin{aligned}\frac{\partial^2 C_{K,T}(t,s)}{\partial K^2} &= e^{-r(T-t)}\frac{\partial^2}{\partial K^2}\int f_K(x)p(x)\,dx \\ &= e^{-r(T-t)}\int \frac{\partial^2 f_K(x)}{\partial K^2}p(x)\,dx \\ &= e^{-r(T-t)}\int \delta_x(K)p(x)\,dx \\ &= e^{-r(T-t)}p(K).\end{aligned}$$

This shows that all other variables being fixed momentarily, the state price density is proportional to the second derivative of the call price when viewed as a function of the strike price, more precisely:

$$p(K) = e^{r(T-t)} \frac{\partial^2 C_{K,T}(t,s)}{\partial K^2}. \qquad (4.26)$$

We propose to compute the state price density in each of the following cases:

1. $S = 445$, $r = 3\%$, $\tau = 21/252$,
2. $S = 445$, $r = 3\%$, $\tau = 45/252$.

In order to avoid to reproduce too large a number of `S-Plus` commands, we summarize some of the steps in words. First we choose a grid of values of the strike K and the underlying S over which we compute the density. We choose a regular grid `SEQK` of 1024 points between $K = 400$ and $K = 500$. In each case, we compute the moneyness with the prescribed values of S, r, and τ, and each of the values of K in the grid. We then standardize the constant τ and the vector of these values of the moneyness using the means and standard deviations of the training sample. We put the result in a matrix called `SSPD`. Once this is done, we compute the predictions of the implied volatilities and the corresponding call prices, and finally, we compute the second derivative, as approximated by the second difference (properly normalized by the size of the grid step which we called `DELTAK`).

```
> SPD.kreg <- kreg(SXX1,OTRGSP[,5],xpred=SSPD,
                                           kernel=7,b=.033)
> SPDSIGMA <- SPD.kreg$ypred
> SPDY <- bscall(TAU=rep(21/252,NP),K=seq(from=400,to=500,
        length=NP),S=rep(445,NP),R=rep(.03,NP),SIG=SPDSIGMA)
> DENS2 <- diff(SPDY,differences=2)/(DELTAK^2)
```

The plots are given in Figure 4.19. In each case, we plot of the state price density and we superimpose the theoretical state price density suggested by the Black-Scholes pricing paradigm (as used in our derivation of the Black-Scholes formula). Many interesting facts can be derived from these plots. We mention just a few to illustrate the interpretation of our results. Let us concentrate on the left plot of Figure 4.19. For this maturity of 21 days, the fact that the Black-Scholes density is below the empirical estimate on the left and right most parts of the graph, implies that the options in the money (both puts and calls) are underpriced by the Black-Scholes formula. Notice that this underpricing becomes an overpricing for the longer maturity of 45 days.

APPENDIX: KERNEL DENSITY ESTIMATION & KERNEL REGRESSION

This appendix represents an excursion away from the practical bent of this chapter. We revisit the topic of density estimation, elucidating its strong connections with the nonparametric regression methods. We first start with an informal motivation.

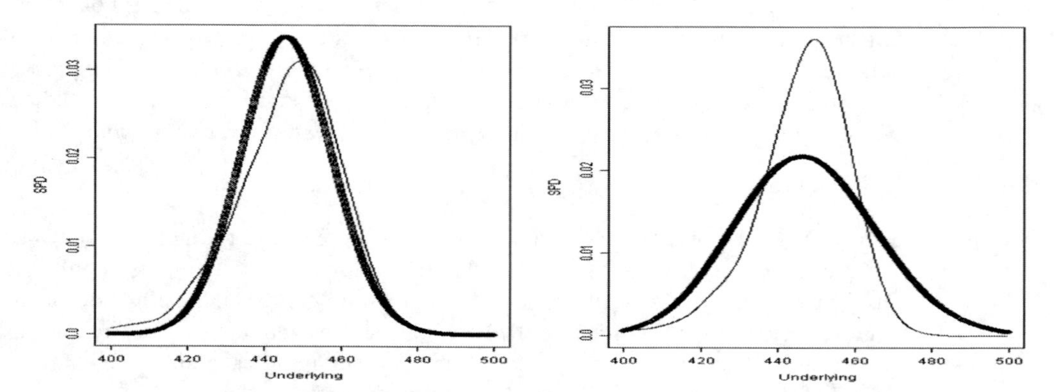

Fig. 4.19. Graphs of the estimates of the state price densities (thin line) together with the corresponding lognormal density of the Black-Scholes theory (thick line). In both cases, we assumed that the underlying was $S = 445$ and that the short rate was $r = 3\%$. The left pane is for options with 21 days to maturity, while the right pane is for options with 45 days to maturity.

The kernel regression is, as any other regression method, a way to estimate the value of the regression function:

$$\mathbf{x} \hookrightarrow \mathbb{E}\{Y|\mathbf{X} = \mathbf{x}\}.$$

Since the above right hand side is an expectation, it can be computed (at least in theory) as a sum when the distribution of Y is discrete, or as an integral when the distribution of Y has a density. In this case:

$$\mathbb{E}\{Y|\mathbf{X} = \mathbf{x}\} = \int y f_{\mathbf{X}=\mathbf{x}}(y)\, dy$$

if we denote by $f_{\mathbf{X}=\mathbf{x}}(y)$ the conditional density of Y given that $\mathbf{X} = \mathbf{x}$. By the definition of conditional probability, the conditional density appearing in this formula is given by the ratio:

$$f_{Y|\mathbf{X}=\mathbf{x}}(y) = \frac{f_{(\mathbf{X},Y)}(\mathbf{x}, y)}{f_{\mathbf{X}}(\mathbf{x})}$$

of the joint density $f_{(\mathbf{X},Y)}$ of $\mathbf{X}$ and Y to the (marginal) density $f_{\mathbf{X}}$ of $\mathbf{X}$. Consequently, the regression function can be rewritten as:

$$\mathbb{E}\{Y|\mathbf{X} = \mathbf{x}\} = \int y \frac{f_{(\mathbf{X},Y)}(\mathbf{x}, y)}{f_{\mathbf{X}}(\mathbf{x})} dy = \frac{1}{f_{\mathbf{X}}(\mathbf{x})} \int y f_{(\mathbf{X},Y)}(\mathbf{x}, y) dy \tag{4.27}$$

since the denominator (which does not depend upon y) can be pulled out of the integral. From these formulae it seems reasonable to expect that being able to estimate densities should lead to regression estimation procedures.

As we stated earlier, the purpose of this appendix is to shed some light on the intimate relationship between the kernel density estimation method presented in Chapters 1 and 2, and the kernel regression method presented in this chapter. The present discussion is an aside of conceptual interest, and it can be skipped in a first reading. To set the stage, we come back to the regression setting defined by a sample:

$$(x_1, y_1), \ldots\ldots, (x_n, y_n)$$

and for the sake of simplicity, we assume that the regressor variables are univariate (i.e. $p = 1$). It should be obvious that the following discussion also applies to the general case $p \geq 1$ without any change. We proceed to the estimation of the regression function by estimating the two densities entering the formula by the kernel method. First we estimate the density $f_X(x)$, and in order to do so we choose a kernel function K and a bandwidth $b > 0$. We get:

$$\hat{f_X}(x) = \frac{1}{nb} \sum_{j=1}^{n} K\left(\frac{x - x_j}{b}\right) \tag{4.28}$$

for the kernel estimate of the density $f_X(x)$. Next we work on the estimation of the joint density $f_{(X,Y)}(x, y)$ and we choose a tensor product kernel $\tilde{K}(x, y) = K(x)k(y)$ for some symmetric kernel k in the y variable, and a bandwidth b' which can be equal to or different from b, it will not matter in the end. In doing so we get:

$$\begin{aligned} \int y f_{(X,Y)}(x, y) dy &= \int y \frac{1}{nbb'} \sum_{j=1}^{n} K\left(\frac{x - x_j}{b}\right) k\left(\frac{y - y_j}{b'}\right) dy \\ &= \frac{1}{n^2 bb'} \sum_{j=1}^{n} K\left(\frac{x - x_j}{b}\right) \int y \, k\left(\frac{y - y_j}{b'}\right) dy \end{aligned} \tag{4.29}$$

and if we compute the integral appearing in the right hand side of (4.29) using the substitution $z = (y - y_j)/b'$ we get:

$$\begin{aligned} \int yk\left(\frac{y - y_j}{b'}\right) dy &= \int (b'z + y_j)k(z)b'dz \\ &= b'^2 \int zk(z)dz + b'y_j \int k(z)dz \\ &= b'y_j \end{aligned} \tag{4.30}$$

if we use the facts that $\int k(z)dz = 1$, by definition of a kernel, and $\int zk(z)dz = 0$, because of our symmetry assumption. Plugging formula (4.30) into formula (4.29) we get:

$$\int y f_{(X,Y)}(x, y) dy = \frac{1}{nb} \sum_{j=1}^{n} y_j K\left(\frac{x - x_j}{b}\right) \tag{4.31}$$

Finally, plugging formulae (4.28) and (4.31) into the expression (4.27) defining the regression function, we obtain:

$$\mathbb{E}\{Y|X = x\} = \frac{\sum_{j=1}^{n} y_j K\left(\frac{x-x_j}{b}\right)}{\sum_{j=1}^{n} K\left(\frac{x-x_j}{b}\right)} \tag{4.32}$$

which is exactly the formula giving the kernel regression estimate. In other words, the kernel regression is consistent with the notion of kernel density estimation. In fact, if one agrees to use formula (4.28) as a density estimator, then the definition of the regression function forces on us formula (4.32) as the only reasonable regression estimate !!

PROBLEMS

Ⓔ **Problem 4.1** *Open and run the (data) script file* `geyser.asc` *to create the data set* `GEYSER`. *It gives the characteristics of* 222 *eruptions of the "Old Faithful Geyser" during August* 1978 *and* 1979. *The first column gives the duration (in minutes) of the eruption and the second column gives the time until the following eruption (in minutes). Make sure that the* `S-Plus` *object you use has* 222 *rows and* 2 *columns.*
1. Perform a least squares linear regression of the second column on the first one and assess the significance (or lack thereof) of the regression. Give an interpretation to your results.
2. Perform several kernel regressions (say 5*) using the* `normal` *kernel and varying values for the bandwidth in order to see the two extreme regimes (very small bandwidths and very large bandwidths) discussed in the text. Choose the value of the bandwidth which you find most reasonable, and compare the resulting regression curve to the result of part 2.*

Ⓔ **Problem 4.2** *Attach the data set* `SHIP` *(or* `ship`*) included in the standard* `S-Plus` *distribution package, and make sure that the data are stored in a vector. Use the command:*

```
> SHIP <- as.vector(ship)
```

if needed. Create a vector `TIME` *containing the integers ranging from* 1 *to the number of entries in the vector* `SHIP`.
1. Compute the least squares regression line and the least absolute deviations regression line of `SHIP` *versus* `TIME` *and superimpose these two lines onto the scatterplot of* `TIME` *and* `SHIP`. *Compare the ways in which these lines account for the upward trend in the data and explain the differences.*
2. The purpose of this question is to fit a more general polynomial (i.e. of degree possibly greater than 1*) to the* `SHIP` *data. Perform polynomial regressions of degrees* 2, 4, 6, *and* 8 *successively, plot the results, and choose the value of the degree which seems the most reasonable.*
3. The purpose of this question is to use natural splines to smooth the data `SHIP`. *Vary the number of degrees of freedom (i.e. the parameter* `df`*). Use the values* 2, 6, 10, 14 *and* 18 *for the number of degrees of freedom (i.e. the parameter* `df`*) and for each of them fit a natural spline to the data. Explain how the smoothed curve changes with the value of the number of degrees of freedom and choose the one which seems the most reasonable.*
4. Finally we smooth the `SHIP` *data using a kernel smoother* `ksmooth` *with a* `normal` *kernel function. Use the values* 1, 5, 20, 50 *and* 125 *for the bandwidth, and for each of these values, superimpose the graph of the kernel scatterplot smoother on the actual scatterplot of the original data. Explain how the smoothed curve changes with the bandwidth.*

Ⓣ **Problem 4.3** *The mathematical results which you are asked to prove in this problem were evidenced numerically in the two problems above. Given a kernel function K and n couples (x_1, y_1), ..., (x_n, y_n) of real numbers, compute the limits:*

$$\lim_{b \searrow 0} \varphi_{b,K}(x) \qquad \text{and} \qquad \lim_{b \nearrow \infty} \varphi_{b,K}(x)$$

of the kernel smoother when the bandwidth goes to 0 or ∞.

You can choose one of the four kernel functions given in the text (pick your favorite) or you can try to give this proof for a general kernel function if you feel comfortable enough.

Ⓣ **Problem 4.4** *Given two functions f and g their convolution $f * g$ is the function defined by the formula:*

$$[f * g](x) = \int_{-\infty}^{+\infty} f(y)g(x-y)\, dy$$

1. Use a simple substitution to check that $f * g = g * f$.
2. Prove that:

$$K_{triangle} = K_{box} * K_{box}$$

and that:

$$K_{parzen} = K_{triangle} * K_{box}.$$

Recall the definitions of these kernels given in Table 4.5.
3. Use `S-Plus` *to plot the four kernel functions given in Table 4.5. Explain your work, and try to create the plots without using any mathematical formula.*

Ⓔ **Problem 4.5** *Open and run the (data) script file* `vineyard.asc` *to create the data set* `VINEYARD`.
1. Superimpose (on the same plot) the scatterplot of `nb` *and* `lugs90`, *the graph of the polynomial regression of* `lugs90` *on* `nb` *with a polynomial of degree 3 and the graph of the polynomial regression (still of degree 3) of the column of 50 measurements obtained by removing the first row and the last row from* `lugs90` *on the column of 50 values obtained by removing the first row and the last row from* `nb`.
2. Same thing with the natural spline regression with degree of freedom $df = 5$.
3. Same question with the smoothing spline regression.
4. Still redo the same thing, but with the kernel regression with a normal kernel and a bandwidth equal to 5.
5. Compare the results obtained in questions 2 and 3 to the results obtained in questions 4 and 5. Explain which features of the regression algorithms are responsible for the differences.
NB: *If you are not sure of what the* spline *regressions do, just compare the polynomial and the regression methods used in questions 2 and 5 respectively.*

Among the other regression methods seen in the text, which methods would give results of the type obtained in questions 2 and 3 and which methods would give results of the type obtained in question 4 and 5?

Ⓔ **Problem 4.6** *The purpose of this problem is to analyze the data set* `ROCK` *included in the script file* `rock.asc`, *to illustrate the inner workings of the* `S-Plus` *implementation of the projection pursuit regression algorithm. The data consist of 48 measurements on 4 cross-sections of each of 12 oil-bearing rocks. Our goal is to predict the permeability y from the other 3 measurements (the total area, the total perimeter and a measure of roundness of the pores in the rock cross section).*

1. Attach the data frame `rock` *and run the projection pursuit algorithm to perform such a prediction.*
2. Assess the quality of the fit by computing the sum of square residuals.
3. Produce the plots of the first three additive terms in the projection pursuit regression formula and give the 2-D scatterplots of the response against the fitted values and the residuals respectively.

Ⓔ Ⓢ Ⓣ **Problem 4.7** *The goal of this problem is to illustrate one of the most important features of the projection pursuit regression: its ability to detect interactions between the regressor variables. We first show that the function $f(x_1, x_2) = x_1x_2$ is easily written in the form used by the projection pursuit algorithm.*
1. Determine μ, β_1 and β_2 so that the identity:

$$x_1x_2 = \mu + \sum_{j=1}^{2} \beta_j\phi_j(\mathbf{a}_j^t\mathbf{x})$$

holds for all $\mathbf{x}^t = [x_1, x_2]$ with $\mathbf{a}_1^t = [1, 1]$, $\mathbf{a}_2^t = [1, -1]$ and $\phi_1(t) = t^2$ and $\phi_2(t) = -t^2$.
2. Create data vectors $X1$ and $X2$ of length $n = 512$ and with entries forming independent samples from the uniform distribution $U(-1, +1)$ over the interval $[-1, 1]$. Create the vector Y according to the formula:

$$Y = X1 * X2 + \epsilon,$$

where $\epsilon = \{\epsilon_j\}_{j=1,\dots,512}$ is a Gaussian white noise (i.e. a sequence of independent identically distributed normal random variables with mean zero) with standard deviation $\sigma = .2$. Give a scatterplot of Y against $X1$, and of Y against $X2$.
3. Run the projection pursuit algorithm to regress Y on $X1$ and $X2$ with a number of terms between `min.term=` *2 and* `max.term=` *3. Produce scatterplots to visualize the graphs of the functions $\phi_j(t)$ found by the algorithm (if in doubt, read the on line help file for the function* `ppreg`*). Does that fit with the results of the computations done in question 1. above? Finally, do a scatterplot of Y versus $\hat{Y}$ and another scatterplot of the residuals versus $\hat{Y}$ and comment.*
NB: recall that we use the notation ^{t} to denote the transpose of a vector or a matrix.

Ⓔ **Problem 4.8** *The purpose of this problem is to analyze some of the features of the basketball data introduced in Problem 3.3. The first two questions are an attempt to evaluate the predictive value of a variable (*`age` *in the present situation).*
1. Use projection pursuit to regress `points` *against* `height`, `minutes` *and* `age`. *Redo the regression without using the variable* `age` *and compare with the previous result.*
2. Similarly, use projection pursuit regression to explain the number of assists per game as given by the variable `assists` *in terms of the explanatory variables* `height`, `minutes` *and* `age`. *Compare with the result obtained without using the predictor variable* `age`.
3. Choose (at random if possible) 10 players, create a new data set, say `BTST` *(for Basketball TeST sample), with these 10 rows and another data set, say* `BTRG` *(for Basketball TRaininG sample) with the remaining 86 players. Keep only the columns corresponding to the variables* `height`, `minutes` *and* `points`. *Use projection pursuit regression to fit a model to the training sample* `BTRG` *and use the coefficients of the model to predict the values of the variable* `points` *for the values of the variables* `height` *and* `minutes` *of the 10 players used to create the test sample* `BTST`. *Compute and note the residual sum of squares for these 10 predictions*

Ⓔ **Problem 4.9** *Open and run the (data) script file* `mind.asc` *to create the data set* `MIND`*, and give the names* `X1`*,* `X2`*,* `X3` *and* `Y` *to its 4 columns. The goal is to regress the last variable* `Y` *against the first three columns using projection pursuit.*
1. Determine the best order for the model.
2. Run a model of the order determined in question 1 and give the (directional) unit vectors found by the program. Comment.
3. Plot the functions φ found by the program. Can you make a guess at the functional dependence of `Y` *upon* `X1`*.* `X2` *and* `X3`*?*

Ⓔ **Problem 4.10** *The goal of this problem is to mimic the volatility ratio prediction experiment described in the text in order to predict the afternoon value of the volatility indicator. The data needed for this problem are contained in the data matrices* `MORN.mat` *and* `AFT.mat` *used in the text.*
1. Give the two dimensional scatterplot of the morning volatility and arrival rate indicators. Plot a three-dimensional scatter plot of these two explanatory variables together with the afternoon volatility indicator, plot the result of the least squares linear regression, and explain the results by identifying overly influential data points.
2. Plot the regression surfaces obtained by kernel regression (use the Gaussian kernel and three bandwidths which you will choose carefully after standardizing the explanatory variables), and confirm the explanations given in question 1. above.
3. Remove the excessively influential measurements identified earlier, recompute the kernel regression on the remaining data, plot the new regression surface, and compute the residual sum of squares.
4. Perform a projection pursuit analysis of the reduced data set, plot the projection pursuit regression surface, compute the new residual sum of squares and compare them both with the results obtained in question 3. above.

Ⓔ **Problem 4.11** *The goal of this problem is to demonstrate the use of principal components analysis in the selection of a minimal informative set of explanatory variables. We use data sets* `MORN.mat` *and* `AFT.mat` *used in the text. Open and run the (data) script file* `indicators.asc` *to create these two data sets.*
1. Use the kernel method to regress the afternoon volatility ratio (third column of the data matrix `SAFT.mat`*) against the six variables of* `SMORN.mat`*, and compute the sum of square errors. (NB: explain your bandwidth choice).*
2. Perform a principal component analysis of the `SMORN` *data set and compute the vector of the daily values of the two most important components* `PC1` *and* `PC2`*. Use the kernel method to regress the afternoon volatility ratio against* `PC1` *and* `PC2`*, and compute the sum of square errors. Again, you will need to justify your choice of bandwidth. Compare with the sum of square errors found in question 1, the sum of square errors found in the experiment with the kernel method described in the text. Comment.*

Ⓔ **Problem 4.12** *This problem is a sequel to Problem 4.11. It uses the data contained in the matrices* `SMORN.mat` *and* `SAFT.mat`*. Open and run the (data) script file* `spindicators.asc` *to create these two matrices. They contain the six indicators computed for the September S&P 500 futures contract from the mornings and afternoons transactions of the period ranging from June 1, 1998 to September 5, 1998.*
1. Use the regression function computed for the kernel regression of the afternoon volatility ratio of the modified June contract data (fourth column of `SAFT.mat`*) against the morning volatility ratio and rate of arrival indicators of the same modified June contract data*

(fourth and fifth columns of SMORN.mat*) to predict the afternoon volatility ratios given in the fourth column of* SAFT.mat *from the morning volatility ratio and rate of arrival indicators of* SMORN.mat*. Compute the sum of square errors.*
In other words, use the June contract data as training sample and the September contract data as testing sample.
2. Redo the same thing using as explanatory variables the first two (linear combinations) principal components of the six indicators found in the PCA analysis of the June matrix data done in Problem 4.11. Comment.

Ⓔ **Problem 4.13** *The goal of this problem is to perform the option pricing analysis described in the text on a different data set. We consider a training sample of* $5,000$ *trades of European call options on the S&P 500 index which took place in 1993, and we consider a testing sample of* 500 *different trades which took place the same year. The data are contained in the script files* trgsp2.asc *and* tstsp2.asc*, respectively.*
1. Perform the analysis described in the text and compute the three figures of merit, especially the squared error per option.
2. Compare the performances of the three prediction procedures on the testing sample. Comment on the differences with the results reported in the text. Hint: when dealing with the semiparametric method, use the scatterplots of the explanatory variables as a guide.

Ⓔ **Problem 4.14** *Like Problem 4.13 above, the present problem aims at the implementation on a different data set, of the option pricing analysis described in the text. We are still considering European call options on the S&P 500 index, but the data are more recent since we are using quotes from the year 2000. Also, the samples are larger and more importantly, slected at random. The data are contained in the files* trgsp3.asc *and* tstsp3.asc*, respectively.*
1. Perform the analysis described in the text and compute the three figures of merit, especially the squared error per option.
2. Compare the performances of the three prediction procedures on the testing sample. Comment on the differences with the results reported in the text, and the results of the experiment conducted in Problem 4.13 above. As before, use the scatterplots of the explanatory variables as a guide to understand and possibly explain the differences.

Ⓔ **Problem 4.15** *This problem illustrates the use of temperature data, and especially the numbers of Heating degree Days and the numbers of Cooling Degree Days (HDD and CDD for short), to predict the price of a commodity. The idea of this part of the project has its origin in a claim found in the introductory article written by Geoffrey Considine for the Weather Resource Center of the Chicago Mercantile Exchange website.*

We first describe the data. They are contained in the text file corntemp.asc*. Once opened in* S-Plus*, run it as a script, and an* S-Plus *object (a data frame to be specific) named* CORNTEMP *will be created. It is a* 606×3 *- matrix of numbers. Each row corresponds to a month, starting from July 1948 (the first row), and ending December 1998 (the last row). The first column gives the official price a farmer could get for a bushel of corn in Iowa that month. This variable is called* MCorn *for Monthly Corn price. The second column gives the monthly average heating degree days in Des Moines as measured at the meteorological station of the airport. Heating Degree Days (HDD's) and Cooling Degree Days (CDD's) are discussed in details in Chapter 5. For the purposes of the present problem it is enough to know that the higher the temperature in the summer, the larger the number of CDD's and the smaller the number of HDD's, and the cooler the temperature, the larger the number of*

HDD's and the smaller the number of CDD's. The second column of the data matrix was computed by summing up the HDD's of the month, and dividing the total by the number of days in the month. This variable is called `MDMHDD` *for Monthly Des Moines HDD. The third column gives the monthly CDD averages for Des Moines. They were computed in the same manner. This variable is called* `MDMCDD`.

NB: *The data have been "cleaned" but there are still some "NA". Explain how you handle these entries.*

The goal of the problem is to predict the November and December prices of corn in Iowa from the summer temperatures in Des Moines as captured by the numbers of CDD's and HDD's.

1. For each year between 1948 *and* 1998,

- *Extract the November price of corn in Iowa;*
- *Choose and compute regressor variables involving the values of* `MCorn`, `MDMHDD` *and* `MDMCDD` *up to (and possibly including) July of the same year but not later. You are allowed to use up to* 4 *variables (but remember that, as we repeated over and over, the smaller this number the better).*

Once this is done, regress the November price of corn in Iowa on the regressor variables you chose according to the rules of the second item above. You are expected to perform three regressions, at least one of them being linear, and at least one of them being nonparametric. For each of these regressions, you need to give the list of the steps you take, and the parameters you use, so that your results can be reproduced. Also, in each case you will compute the proportion of the variance of the response variable explained by the regression.

2. Same question as above, replacing the November price of corn by the December price, and the July limit by August.

Ⓔ **Problem 4.16** *The data to be used for this problem are contained in the* `S-Plus` *script file* `pprice.asc`. *Running this script will create the data frame* `PPRICE` *for which each row corresponds to a particular date.*

◇ *The first column contains the values of a variable named* `GasSpot` *which gives the spot price of natural gas on that date.*

◇ *The second column contains the values of a variable named* `SDTemp` *which gives the average temperature in San Diego over the* 31 *days preceding the date in question.*

◇ *The third column contains the values of a variable named* `PPower` *which gives the average over the* 5 *days preceding the date in question, of the spot price of firm on peak electric power at the Palo Verde station.*

◇ *Finally, the fourth column contains the values of a variable named* `FPower` *which gives the average over the two weeks following the date in question, of the spot price of firm on peak electric power at the Palo Verde station.*

Form a data frame `TRG` *with the first* 250 *rows of* `PPRICE`. *The entries of* `TRG` *correspond to the period from 2/4/1999 to 2/3/2000. You shall also need to form a data frame* `TST` *with the last* 80 *rows of* `PRICE`. *The entries of* `TST` *correspond to the period from 7/13/2001 to 11/9/2001. We avoid the period in between because of the extreme volatility of the natural gas and power prices. This does not mean that we are not interested in studying periods of high volatility, quite the contrary. It is merely because the economic fundamentals were not the only driving force during this crisis period.*

The goal of the problem is to predict the values of the average price of (firm on peak) electric power over the next two weeks from past values of explanatory variables such as the weather (as quantified by the average temperature in San Diego), the price of natural gas, and possibly past values of the price of electricity at the same location. We use the data in `TRG` *as a training sample to fit a regression model, and we compute the predictions given by such a model for the data in the testing sample* `TST`.

Warning. *The variable* `PPower` *should not be used in the first 4 questions. Moreover, for all the predictions considered below, the figure of merit should be the square root of the mean squared error.*

1. Fit a least squares linear regression model for `FPower` *against* `GasSpot` *and* `SDTemp` *using the data in* `TRG`, *use this model to predict the values of* `FPower` *in* `TST` *from the corresponding values of the explanatory variables, and compute the figure of merit.*
2. Same question with least absolute deviations linear regression instead.
3. Same question using projection pursuit. Explain your work.
4. Same question using kernel regression. Again make sure that you explain all the steps you take, and justify your choice of kernel function and bandwidth.
5. Fit a least squares linear regression model for `FPower` *against* `GasSpot`, `SDTemp` *and* `PPower` *using the data in* `TRG`, *as before, use the fitted model to predict the values of* `FPower` *in* `TST` *from the corresponding values of the three explanatory variables, and compute the figure of merit.*
6. Same question with least absolute deviations linear regression instead.
7. Same question using projection pursuit.
8. Compare the numerical results obtained with the various methods, and explain why they could have been expected.

Ⓔ **Problem 4.17** *The data to be used in this problem are contained in the* `S-Plus` *script file* `crude.asc`. *Running this script will create the data frame* `CRUDE` *containing a numerical matrix with* 3325 *rows and* 12 *columns, and the numeric vector* `COma` *of length* 3325.

In both data structures, each row corresponds to a date, the first one being 4/18/1989, and the last one being 8/12/2002. Each row of the vector `COma` *contains the average of the crude oil spot price over the period of* 5 *days starting on (and including) the date indexing the row. Each row of the matrix* `CRUDE` *gives the prices of the* 12 *futures contracts of crude oil as traded the day before. Form a data frame* `TRGCRUDE` *and a vector* `TRGCOma` *with the first* 2500 *rows of* `CRUDE` *and* `COma` *respectively. You shall also need to form a data frame* `TSTCRUDE` *and a vector* `TSTCOma` *with the last* 825 *rows of* `CRUDE` *and* `COma` *respectively.*
The goal is to predict the values of the average spot price over the next five days from the prices of the crude oil futures contracts traded the day before, by fitting a regression model to the training data contained in the data sets `TRGxxx`, *and using the model to compute predictions for the values of the response in the testing sample* `TSTCOma`.

Warning. *For all the predictions considered in this problem, the figure of merit should be the square root of the mean squared error. It is very important that you explain your work in detail. In particular, explain your choices for the order of the models, the kernel functions, and the bandwidths you use.*

1. Fit a least squares linear regression model for `COma` *against the* 12 *explanatory variables given by the prices of the futures contracts the day before using the data in* `TRGCRUDE` *and* `TRGCOma`, *use this model to predict the values of* `COma` *in* `TSTCOma` *from the corresponding values of the explanatory variables, and compute the figure of merit.*
2. Fit a projection pursuit regression model for `COma` *against the* 12 *explanatory variables given by the prices of the futures contracts the day before using the data in* `TRGCRUDE` *and* `TRGCOma`, *use this model to predict the values of* `COma` *in* `TSTCOma` *from the corresponding values of the explanatory variables, and compute the figure of merit.*

3. Perform the PCA of the data in TRGCRUDE, *plot the first four loadings, give the proportions of the variance they explain, and compute the first two principal components.*
4. Fit a one dimensional kernel regression model for COma *against the first principal component using the data in* TRGCRUDE *and* TRGCOma, *use this model to predict the values of* COma *in* TSTCOma *from the corresponding values of the explanatory variables, and compute the figure of merit.*
5. Fit a two dimensional kernel regression model for COma *against the first two principal components using the data in* TRGCRUDE *and* TRGCOma, *use this model to predict the values of* COma *in* TSTCOma *from the corresponding values of the explanatory variables, and compute the figure of merit.*
6. Compare the numerical results obtained with the various methods, and explain why they could have been expected.

As stated, the next three problems require the use of neural networks. The reader can choose to tackle them using the information given in the Notes & Complements, or by ignoring the questions requiring the use of neural network regression.

Ⓔ **Problem 4.18** *The goal of this problem is to compare the results of the analysis of the basketball data obtained by projection pursuit to the results one can obtain using neural network regression instead.*
1. Produce numerical evidence that the variable age *does not add to the prediction of the variable* points *and/or the variable* assists *in the case of neural network regression.*
2. Use the same training sample $BTRG$ *to fit a neural network and use the coefficients (i.e. weights) of the model to predict the values of the variable* points *in the test sample* $BTST$. *Compute the residual sum of squares for these* 10 *predictions and use this value to say which of the projection pursuit or the neural network method did better (for this particular criterion and on this particular data set).*

Ⓔ **Problem 4.19** *The data set* SUBSP *to be used for this problem are contained in the* ascii *file* subsp.asc. *Read it in an* S-Plus *object of the same name, It is a single column containing* 1000 *rows giving the daily returns on a sub-index of the S&P 500 index (the Electronic Equipment subindex, to be specific) during the period beginning January 1993 and ending December 31, 1996. In other words, the first row gives the closing on January 4, 1993, the second row the closing on January 5, 1993 ...... and finally the last row contains the values of the subindex at the closing bell on Tuesday December 31, 1996. The markets are open approximately* 250 *days each calendar year so that* SUBSP *contains four years worth of daily quotes.*

The goal is to produce a forecasting for the future changes in the index. We shall use the first three years for training of our prediction system and we shall test them on the last year.
1. Construct a neural network using the first 750 *daily values of the sub-index to predict the* 5*-day move in the future (i.e. the variable which on day* I *has the value* SUBSP[I+5]-SUBSP[I]*) using only the past changes over the last day, the last week, the last three weeks and the last twelve weeks (i.e. the variables which on day* I *have the values* SUBSP[I]-SUBSP[I-1], SUBSP[I]-SUBSP[I-5], SUBSP[I]-SUBSP[I-15] *and* SUBSP[I]-SUBSP[I-60]*). Use this neural net to predict the next* 245 *values of the five day increment in the index (i.e. the same response variable* SUBSP[I+5]-SUBSP[I] *for* I=751,....,995*) and compute the sum of the squares of the errors made.*

2. Compute the same sum of square errors using now the latest value of the 5-day increment available (i.e. `SUBSP[I]-SUBSP[I-5]`*) instead of the neural net prediction and compare to the result of question 1. Any comments?*

Ⓔ **Problem 4.20** *Import the data set* `BLIND` *contained in the* `ascii` *file* `blind.asc` *in* `S-Plus` *and denote by* `X1`*,* `X2`*,* `U1`*,* `U2` *and* `Y` *the 5 columns.*
1. We consider the problem of the prediction of variable `Y` *from the variables* `X1` *and* `X2`*. Use the first 100 rows as training sample and compute the sum of square residuals for the prediction of the last 28 values of* `Y` *from the corresponding values of* `X1` *and* `X2`

- *using a projection pursuit regression (with a number of terms between 2 and 4);*
- *using a neural network regression (with two units in the hidden layer)*

Compare the two results.
2. The results obtained with the neural network regression should be UNUSUALLY good. Try to explain this fact by regressing, using the projection pursuit regression (with the same min & max numbers of term as above)

- `Y` *versus* `U1` *and* `U2`*;*
- `U1` *versus* `X1` *and* `X2`*;*
- `U2` *versus* `X1` *and* `X2`.

NOTES & COMPLEMENTS

The use of natural splines was proposed as a transition step from the set of full parametric regression procedures presented in the previous chapter into a set of local regression methods leading to the nonparametric procedures discussed in this chapter. Besides natural splines, `S-Plus` also offers an implementation of B-splines . See the classic text by de Boor [28] for details on the zoology of the various families of splines. We recommend using these splines as an investigation tool, possibly as a coding device in additive models, but we encourage the user to be extremely careful using the results of spline regressions for prediction purposes, too many pitfalls are to be avoided. The rationale behind the definition of the smoothing splines is completely different from the typical search for a regression function in a parametric family. The fact that the argument of the minimization of the penalty function used to defined smoothing splines is indeed a spline is highly nontrivial. It is a result of Kimberdoff and Whaba that the solution of this optimization problem is in fact a spline of order $m + 1$.

The scatterplot smoothers described in the chapter have been chosen to illustrate the notion of locality in regression. Their main role is to prepare for the introduction of the multivariate nonparametric methods which occupy the rest of the chapter.

The so-called bootstrap method of estimation of the instantaneous forward-rate curve was introduced by Fama and Biss in [35]. Details on the implementations used by the Japanese and US Central Banks to produce their yield curves can be found in the comprehensive document [7] made available by the Bank of International Settlements.

Multi-dimensional kernel regression is very appealing because of the simplicity and clarity of the principle on which it is based: the algorithm mines the data to find records of the explanatory (vector) variable which are similar to the point at which the regression is being computed, and it returns a weighted average of the responses according the weights being proportional to the measures of similarities. Rigorous mathematical results can be proven on the desirable properties of this regression method. In particular optimality of the kernel function and of the bandwidth choices can be proven, but these results are unfortunately asymptotic in nature, i.e. valid only in the limit $n \to \infty$ of large sample sizes. Despite the simplicity of the rationale behind the kernel method, the mathematical proofs remain very technical. They have to deal with a subtle balance between the errors due to the bias and the variance of the estimator. Indeed, sampling from the conditional distribution of the response for neighboring values of the explanatory variable introduces a bias which, contrary to the case of the parametric methods seen in the previous chapter, cannot be avoided in the present situation. The reader interested in the mathematical analysis of the kernel regression as well as other nonparametric regression methods is referred to the book of Haerdle [42] and Hastie, Tibshirani, and Friedman [44].

The projection pursuit algorithm was originally proposed by Friedman and Stuetze in 1981 in a short article in the Journal of the American Statistical Association. The implementation of the general idea of projection pursuit is not limited to regression problems. It has seen applications to density estimation, classification and pattern recognition problems. More recently, a variation on the same idea was re-discovered, and it gained a lot of attention in the signal processing and image analysis communities: this new flavor of projection pursuit was proposed by Mallat and Zhang. It was called *matching pursuit*, but the idea remains the same. We limited our discussion to projection pursuit regression as it is implemented in `S-Plus`.

Neural networks were very popular in the late eighties, especially in statistical pattern recognition circles. The `S-Plus` library `nnet` contributed by Venables and Ripley contains an implementation of the so-called feed-forward neural networks. See Section 11.4 of their book [85]. For the sake of simplicity they restrict their discussion to the case of one single hidden layer. Formally, such a regression/prediction procedure models the dependence of the response variable y upon the explanatory variables $\mathbf{x}$ by a function of the form:

$$y \approx \varphi_{weights}(\mathbf{x}) = \phi_{out}\left(\alpha + \sum_{j=1}^{p} w_{in,out,j}x_j + \sum_{j=1}^{nbunits} w_{2,j}\phi_{in}\left(\alpha_j + \sum_{\ell=1}^{p} w_{1,\ell}x_\ell\right)\right) \tag{4.33}$$

which is best understood in the light of the following comments:

- the numbers α_k play the roles played by the intercepts in linear regression. As in the case of linear regression, they can be subsumed by introducing a *dummy* input unit, which contains the number 1 irrespective of the observation, and adding weights for the outputs of this unit;
- the activation functions ϕ_{out} and ϕ_{in} are usually the same (typically the logistic function) but it happens quite often that the output activation ϕ_{out} is chosen to be linear;
- the p weights $w_{in,out,j}$ are used for the direct links between the input units and the output units;
- the p weights $w_{1,\ell}$ are used for the links between the input units and the units in the hidden layer while the weights $w_{2,j}$ are used for the links between the units in the hidden layer and the output unit.

A neural network of the type given by formula (4.33) is fit to the data $(\mathbf{x}_1, y_1), \dots, (\mathbf{x}_n, y_n)$ by choosing the number of units in the hidden layer, the type of activation functions and possible direct links and most importantly by choosing the weights. This choice is usually made by solving the optimization problem:

$$\arg \min_{weights} \sum_{i=1}^{n} |y_i - \varphi_{weights}(\mathbf{x}_i)|^2. \tag{4.34}$$

In other words, for each possible choice of a set of weights we compute the fitted values $\varphi_{weights}(\mathbf{x}_i)$ of all the observations in the training sample, and the figure of merit given in formula (4.34) for this set of weights is the sum of the square discrepancies with the actual observed values y_i's. The optimal set of weights should minimize this sum of square errors. Unfortunately, as experience shows, this minimization problem is very difficult because of the existence of many local minima and many of which are not satisfactory because they lead to poor predictions.

Because of the delicate optimization behind the fit of a neural net to data, implementations are difficult to come by. Venables and Ripley provide a function `nnet` to train a *feed-forward* one layer neural networks. A generic call to this function should look like:

```
> x.nnet <- nnet(X,Y,size=NB,Wts=W0,rang=R,linout=B,
                                        skip=BB,maxit=MAXIT)
```

where the parameters have the following meaning.

- `X` is the $n \times p$ design matrix, one row per observation in the training sample (whose size is denoted by n) and one column per explanatory variable;
- `Y` is the $n \times 1$ vector of the values of the response variable in the training sample;
- `size` gives the number of units in the hidden layer;
- `Wts` is the optional set of weights used to initialize the optimization;
- when the argument `Wts` is missing, the optimization procedure is initialized with random weights in the range $[-R, +R]$ where the number R is given as the parameter `rang`.
- `linout` is a boolean variable which, when set to `TRUE`, will force the output activation function ϕ_{out} to be linear or affine;
- `skip` is a boolean variable which when set to `TRUE` will force the existence of direct links from the input units to the output unit(s).

We found neural network regression difficult to use because of the difficulties in the choice of network and the search for parameters, and we chose not to discuss it in the text because of its poor performance compared to the other methods presented in the book.

The section on nonparametric option pricing and state price density estimation was inspired by Yacine Ait-Sahalia's Ph.D. thesis, and we use part of his data for illustration. The paper [48] is the first article we know of, where options are systematically priced from historical data using (modern) nonparametric regression procedures including neural networks. The more recent contribution [1] of Ait-Sahalia and Lo concentrated on the use of the kernel method. Their reason for choosing the kernel over the other nonparametric regression procedures comes from their desire to price more general contingent claims with (possibly) very complex payoff functions. Indeed, the kernel regression is more amenable to the estimation of the second derivative of the regression function with respect to the strike price. We decided to add the use projection pursuit for the sake of comparison.

Some very popular nonparametric regression methods have not been mentioned in the text. The *k-nearest neighbors* method is one of them. It is very similar in spirit to the kernel regression when one uses the box kernel. Indeed in both cases the value of the regression function at a given point is given by the average of the observed responses for a set of neighboring explanatory vectors. The difference is only in the definition of these neighboring points: we choose all the points within a certain distance (given by the value of the bandwidth) in the case of the kernel regression independently of their number, while we choose the k-nearest points in the case of the k nearest neighbors regression. The smoothing parameter is now the number k of neighbors involved in the averaging. Numerical results are very similar, and like in the case of the kernel method nonparametric density estimation can also be done by the $k$nearest neighbors method. However, adjusting implementations to allow for categorical explanatory variables appears to be easier with the k nearest neighbors method. The interested reader is referred to Silverman's book [79] for a detailed account of the k nearest neighbors method in the context of density estimation, and to Kohonen's book [54] for artificial intelligence applications.

Classification and clustering can be viewed as regression problems for which the response variable can only take finitely many values. All the nonparametric methods discussed above can be adapted to solve classification problems. However, because they offer intuitive interpretations, tree based methods remain the most appealing classification procedures. `S-Plus` offers a set of methods to manipulate trees for regression and classification, based on the fundamental work of Breiman, Friedman, Olshen, and Stone [12]. As in the case of regression, classification suffers from the curse of dimensionality, and principal component analysis is viewed as a reasonable technique for reducing the dimension of the explanatory vectors. Alternatives based on coding and computing efficiency arguments have been proposed, for example wavelet packets. The interested reader is referred to Wickerhauser's book [87] for details. Finally, we close with a reference to a recent attempt to bring ideas and techniques for data mining and machine learning that were developed in the artificial intelligence community under the umbrella of mathematical statistics. These techniques go under the name of boosting and bagging and they are intended to combine several classification or regression algorithms into a single, better performing one. This might seem like an impossible dream, but it does happen in some cases, and theoretical arguments can be given to justify such seemingly unrealistic expectations. See, for example, the recent book of Witten and Frank [88], and the graduate text by Hastie, Tibshirani and Friedman [44].

Part III

TIME SERIES & STATE SPACE MODELS

5

TIME SERIES MODELS: AR, MA, ARMA, & ALL THAT

Time series are ubiquitous in everyday manipulations of financial data. They are especially well suited to the nature of financial markets, and models and methods have been developed to capture time dependencies and produce forecasts. This is the main reason for their popularity. This chapter is devoted to a general introduction to the linear theory of time series, restricted to the univariate case. Later in the book, we will consider the multivariate case, and we will recast the analysis of time series data in the framework of state space models in order to consider and analyze nonlinear models.

5.1 NOTATION AND FIRST DEFINITIONS

The goal of time series analysis is to analyze data containing finite sequences of measurements, each coming with a time stamp, these time stamps being ordered in a natural fashion. The purpose of the analysis is to quantify the dependencies across time, and to take advantage of these correlations to explain the observations at hand, and to infer properties of the unobserved values of the series.

We have already encountered many instances of time series (recall, for example, the coffee futures data as plotted in Figure 2.4). However, in each case, we transformed the data to reduce the serial correlation to a minimum, and we used statistical techniques which are oblivious to the order of the data: this was our way of getting rid of the serial dependence in the data. It is now time to investigate the various ways one can model this dependence and take advantage of the properties of these models.

5.1.1 Notation

Most statistical problems deal with data in the form

$$x_0, x_1, \ldots, x_n. \tag{5.1}$$

In the regression applications considered so far, the order in which the observations were collected did not play any role. We are now interested in applications for which the order of the x_i's plays a crucial role in the interpretation of the data, as well as in the definition of the inferential problems we consider.

In most applications, the label n of the observation x_n corresponds to a time stamp, say t_n, giving the time at which the measurement was taken. As always, it is convenient to view the observations (5.1) as realizations of random variables $X_0, X_1, \ldots, X_n$ which we shall sometimes denote $X_{t_0}, X_{t_1}, \ldots, X_{t_n}$. These $n+1$ random variables will quite often be regarded as a subset of a (possibly infinite) sequence $\{X_t\}$ of random variables.

The x_n's (and hence the X_n's) can be scalars (in which case we talk about univariate time series) or vectors (in which case we talk about multivariate time series). As before, we try to use regular fonts for scalars and bold-face fonts for vectors.

Most of this chapter is devoted to the analysis of time series models. A *model* is a set of prescriptions for the joint distributions of the random variables (or random vectors in the multivariate case)

$$X_{i_1}, X_{i_2}, \ldots, X_{i_n}$$

for all possible choices of the finite ordered sequence $i_1 < i_2 < \cdots < i_n$ of time stamps. These joint distributions are completely determined by the model in some cases, while in other cases, only partial information is provided by the prescriptions of the model.

5.1.2 **Regular Time Series and Signals**

Regular time series are sets of measurements taken at regular time intervals. In other words, the time stamps of the sequence $\{t_j\}_{j=0,1,\ldots,n}$ are of the form $t_j = t_0 + j\Delta t$ for $j = 0, 1, \ldots, n$. Such a sequence of times is determined by its start t_0, its length $n+1$, and the time interval Δt between two successive times. Note that, instead of giving the sampling interval Δt, one can equivalently give the sampling frequency, or the time of the final measurement. Once the time sequence has been defined, one can then give the sequence of corresponding measurements separately.

Figure 5.1 gives an example of such a regular time series. It is a speech signal which we made up by recording the short sentence "`how are you`", digitizing the sound file, and collecting the resulting numerical values in an `S-Plus` numerical vector which we called `HOWAREYOU`. In part because of their frequent occurrence in applications to signal analysis (as traditionally performed by electrical engineers), regular time series are often called signals. We created such a `signalSeries` object `HOWAREYOU.ss`, and we produced the plot of Figure 5.1 with the `S-Plus` commands:

```
> HOWAREYOU.ss<-signalSeries(data=HOWAREYOU,from=1,by=1)
> plot(HOWAREYOU.ss, main="Speech Signal 'HOW ARE YOU'")
```

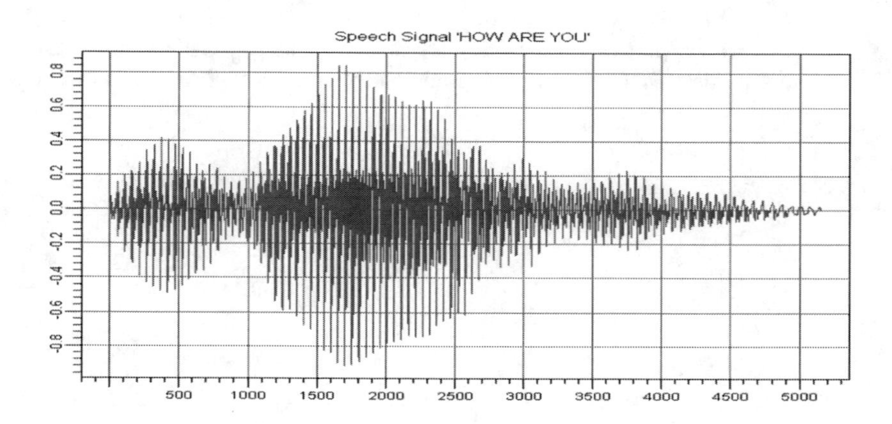

Fig. 5.1. Plot of the sound "How Are You" digitized at 8000Hz

The time stamps used to label the elements of the signal are simply successive integers starting from one. This should be contrasted with what comes next.

5.1.3 Calendar and Irregular Time Series

Most of the financial time series do not have the good taste to be regular in the sense given above. They differ from the signal series discussed above in several ways, starting from the fact that the time stamps are given by dates, thus the name calendar time series. Even though calendar time series are particular cases of a larger class of irregular time series, they will be the only ones considered here. Oftentimes, these data are daily, and gaps due to weekends and holidays create irregularities. Figure 5.2 gives the daily closing prices of the S&P 500 index on the New York Stock Exchange (NYSE for short) for the period ranging from January 4, 1960 to June 6, 2003. Here is a (very) small subset of the data used to produce the plot.

```
     Date     Open     High      Low    Close
    .....    .....    .....    .....    .....
17-Sep-01  1092.54  1092.54  1037.46  1038.77
10-Sep-01  1085.78  1096.94  1073.15  1092.54
 7-Sep-01  1106.40  1106.40  1082.12  1085.78
 6-Sep-01  1131.74  1131.74  1105.83  1106.40
 5-Sep-01  1132.94  1135.52  1114.86  1131.74
 4-Sep-01  1133.58  1155.40  1129.06  1132.94
31-Aug-01  1129.03  1141.83  1126.38  1133.58
30-Aug-01  1148.60  1151.75  1124.87  1129.03
29-Aug-01  1161.51  1166.97  1147.38  1148.56
28-Aug-01  1179.21  1179.66  1161.17  1161.51
    .....    .....    .....    .....    .....
```

As we can see, the time stamps can be very irregularly spaced. Nevertheless, the scale of the plot of Figure 5.2 does not allow us to see the gaps due to weekends and holidays. These data were downloaded from a data service on the internet, and

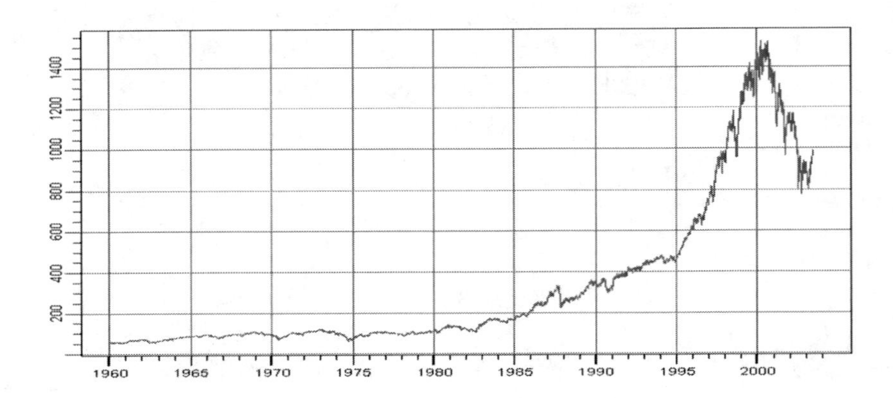

Fig. 5.2. Daily closes of the S&P 500 index from January 4, 1960 to June 6, 2003.

they came as an `Excel` spreadsheet. `S-Plus` offers special menu items to import such files into data frames. The following command would read the data if they were included in a text file named `FILENAME`.

```
> TMP <- read.table(FILENAME,
            col.names=c("Date","Open","High","Low","Close"))
```

We can create an object of class `timeSeries` with the command:

```
> SP.ts <- timeSeries(positions=as.character(TMP$Date),
                                          data=TMP[,2:5])
```

As in the case of `signalSeries`, the constructor `timeSeries` requires an argument `data` for the actual data forming the entries of the time series. The main difference comes from the time stamps. They need to be an ordered vector of class `timeDate` put in the slot `positions`.

Remark. This is the only example in the book for which we warn the reader *not to try it at home.* Indeed this example cannot work in its simple form. `S-Plus` reads a date like "17-Sep-91" and interprets it as being September 17, 91, not knowing that the year is in fact 1991. This would not be a major problem if all the dates were prior to January 1, 2000. Unfortunately, dates like "17-Sep-01" are interpreted as September 17, 1, and the list of time stamps is not ordered as it should be. `S-Plus` offers many powerful tools to read and manipulate dates. We discuss some of them in what follows.

In the financial industry, one of the first difficulties faced by the data analyst is the fact that financial data providers use different formats. In what follows, we give several examples to illustrate how the `timeSeries` objects and methods of `S-Plus` can be used to resolve these issues.

5.1.4 Example of Daily S&P 500 Futures Contracts

We consider another example of daily data, but instead of considering the spot level of the S&P 500, we now consider trades of futures contracts. The following is a small snapshot of a huge text file containing all the transactions on several futures contracts on the S&P 500 index. For the purposes of illustration, we shall concentrate on the contracts with March 1990 and June 1990 deliveries.

"contract"	"date"	"open"	"high"	"low'	"close"	"volume"	"op. int."
...	...	...	...	...	...	...	...
SPIMMDMAR1990	19900102	35632	36285	35550	36250	32475	105621
SPIMMDMAR1990	19900103	36330	36480	36130	36170	39916	105464
SPIMMDMAR1990	19900104	36140	36260	35610	35985	45267	104880
SPIMMDMAR1990	19900105	35920	35960	35430	35460	42288	103532
SPIMMDMAR1990	19900108	35407	35840	35360	35755	40983	105653
SPIMMDMAR1990	19900109	35775	35810	35190	35220	44577	104660
...	...	...	...	...	...	...	...
SPIMMDMAR1990	19900313	33730	33880	33490	33560	13940	41689
SPIMMDMAR1990	19900314	33605	33825	33470	33740	10542	35661
SPIMMDMAR1990	19900315	33742	33950	33675	33830	12110	32426
SPIMMDJUN1990	19900102	36685	36685	35970	36660	575	2986
SPIMMDJUN1990	19900103	36737	36890	36560	36590	1534	3593
SPIMMDJUN1990	19900104	36555	36670	36030	36410	603	3055
...	...	...	...	...	...	...	...

Reading Data in an `S-Plus` *Time Series Object*

We want to create a time series for each of the futures contracts. Assuming that the data visualized above are stored in a text file whose pathname is given by the character string `FILENAME`, we can use the same commands as before to read these data in a table named `FILE`, extract the dates corresponding to the March 1990 contract in an `S-Plus` object `MDATE`, specify the input format of the time stamps, and finally create the actual time series with these time stamps, and the data given by the five columns of `FILENAME` giving for each day, the opening price, the high and the low of the day, the closing price, the volume and the open interest on that day.

```
> FILE <- read.table(FILENAME, col.names = c("code",
    "delivery","open","high","low","close","volume","oi"))
> MDATE <- FILE[FILE$code == "SPIMMDMAR1990", 2]
> options(time.in.format = "\%4Y\%2m\%2d")
> DSPH90.ts <- timeSeries(positions = as.character(MDATE),
              data =FILE[FILE$code =="SPIMMDMAR1990", 3:8])
```

The string of characters specifying the option `time.in.format` implies that, while reading the `date` as a string of characters, `S-Plus` should understand the first four characters as the year, the next two characters as the month, and the last two characters as the day. This is a typical example of the use of the constructor `timeSeries` with its arguments `positions` and `data`. To read the next contract we use the commands

```
> JDATE <- FILE[FILE$code == "SPIMMDJUN1990", 2]
> DSPM90.ts <- timeSeries(positions = as.character(JDATE),
            data = FILE[FILE$code =="SPIMMDJUN1990", 3:8])
```

The `timeSeries` objects created in this way are multivariate time series in the sense that the x_j's are 5-dimensional vectors. It is easy to see that the two contracts overlap: there are large time intervals during which both contracts are actively traded. Our next task is to roll these futures contracts over each other to create a single time series by keeping, on each given day, the value of one contract only. Typically we choose the contract closest to delivery, except that we jump from one contract to the next when we get too close to expiration, say within one month of it, just to avoid artifacts due to delivery proximity. The following commands can be used to do this.

```
DAYCUT <- timeDate("02/15/1990")
TMP1 <- DSPH90.ts[positions(DSPH90.ts) <= DAYCUT)
TMP2 <- DSPM90.ts[positions(DSPM90.ts) > DAYCUT)
DSP.ts <- concat(TMP1,TMP2)
```

We chose by convention to switch from the March contract to the June contract on February 15, and we used the constructor `timeDate` to create this cut-off date. Then, we extracted the part of the March contract prior to this date by subscripting the time series `DSPH90.ts`, and we extracted the part of the June contract posterior to that date. We then concatenated these two `timeSeries` objects into a new one with the method `concat`. By choosing a sequence of day cuts at which we *jump* from one contract to the next, and repeating this procedure (in a more sophisticated program, though in the spirit of the above set of commands) over all the contracts, we create a `timeSeries` object, which we call `DSP.ts`, for daily S&P. The command:

```
> DSP.ts[1:7]
```

displays the first seven entries of the series. The result looks like:

Positions	open	high	low	close	volume	oi
01/02/1990	35632	36285	35550	36250	32475	105621
01/03/1990	36330	36480	36130	36170	39916	105464
01/04/1990	36140	36260	35610	35985	45267	104880
01/05/1990	35920	35960	35430	35460	42288	103532
01/08/1990	35407	35840	35360	35755	40983	105653
01/09/1990	35775	35810	35190	35220	44577	104660
01/10/1990	35175	35320	34670	35130	56386	105042

Plotting a Time Series in `S-Plus`

The generic method `plot` can be used with `timeSeries` objects. For example the commands:

```
> plot(DSP.ts[,6])
> title("S&P500 Daily Open Interest")
```

produce the plot of the open interest reproduced in Figure 5.3. This plot shows the effects of the artificial roll-over method used to create a time series from separate contracts. This is a clear indication that we should not use this series for an analysis free of the artifacts introduced by our rolling method.

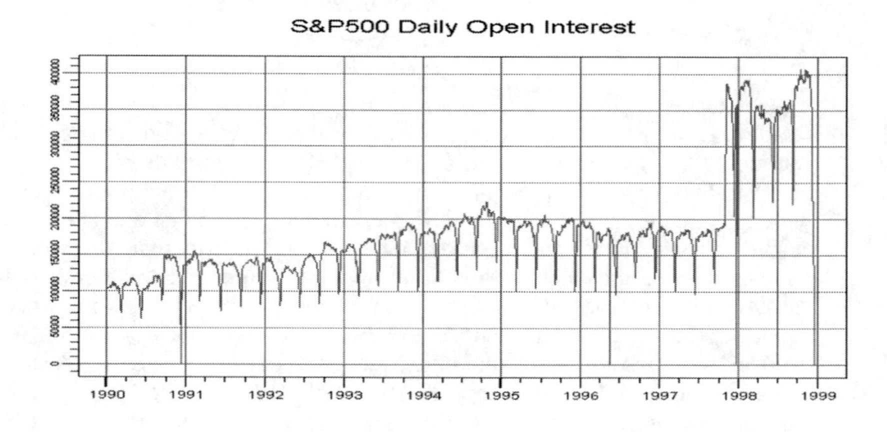

Fig. 5.3. Plot of the time series of daily open interest, as created by rolling over futures contracts on the S&P 500 index.

5.2 HIGH FREQUENCY DATA

In our quest for a broad survey of all the forms of financial data amenable to statistical analysis, we now turn to transaction-by-transaction data. They are the result of a different data collection process: a record is added to the data file each time a new transaction takes place. These data are also called trade-by-trade data, or even tick-by-tick data. They offer a unique insight into the study of trading processes and market microstructure. In this section, we study the special characteristics of high-frequency data, identifying the most striking differences between it and lower-frequency data, and introducing new tools and new methods tailored to the new challenges introduced by this new type of data.

Tick-by-tick data are available for liquid futures contracts, and in this section, we consider the example of futures contracts on the S&P 500 index for the sake of illustration. We shall use IBM stock tick-by-tick prices later in Chapter 7. Here is the way the data of the September 1998 contract look like.

"date"	"time"	"close"
19971008	14:53:38	1013.20
19971017	10:59:16	986.00
19971027	10:02:13	960.00
19971103	10:28:51	968.00
19971105	09:08:44	975.00
19971106	10:59:21	969.00
19971124	12:52:52	986.90
19971209	10:58:18	1015.00
19971210	09:22:05	1005.70
19971224	09:27:21	968.00
...	...	...

We notice that the time stamps can be very sparse. This is due to the fact that these trades occurred on days very far from the maturity of the contract: speculators are actively trading contracts closer to delivery! However, the situation changes dramatically when we look at the data later in the life of the contract. Indeed, one sees that not only are the transactions more frequent, but in fact, a large number of transactions can take place during the same minute. Given the fact that each row contains only one number besides the date and time, we shall assume that this number is the price at which the transaction was settled, not a bid or ask price. This information is not always given by the data provider, and the data analysis may be forced to make this sort of assumption. One of the unexpected surprises with high-frequency financial data is the fact that the notion of price is not clearly defined. Strangely enough, there are many reasons for that. The first one is clear from the data reproduced below: different values can be quoted with the same time stamp, so what is the price at that time?

"date"	"time"	"close"
...	...	...
19980804	11:05:00	1103.50
19980804	11:05:00	1103.00
19980804	11:06:00	1102.80
19980804	11:06:00	1102.60
19980804	11:06:00	1102.50
19980804	11:06:00	1102.40
19980804	11:06:00	1102.20
19980804	11:06:00	1102.00
...	...	...
19980804	11:06:00	1102.50
19980804	11:06:00	1102.40
19980804	11:06:00	1102.20
19980804	11:06:00	1102.00
19980804	11:07:00	1101.70
...	...	...

Whether one looks at this particular portion of the data set or not, it happens very often that, many seconds do not appear because there is no transaction at these times. Another idiosyncrasy of high-frequency data is the fact that the bid and ask prices do not make sense all the time. Indeed, when the frequency is high enough, the time interval between two transactions is so small that the price cannot move out of the bid-ask spread, muddying both the definition of the notion of price and of bid-ask spread at the same time. We avoid this issue by considering the settlement price as given above.

Reading High Frequency Data into `S-Plus`

Reading high-frequency data into `S-Plus` is not more difficult than reading any other type of data, as long as the format of the data is known. The only new difficulty is the sheer size of the data files: processes and computing times become longer, and memory is frequently an issue.

To continue the discussion of the futures contracts considered in this section, we notice that the time stamp is spread over the first two columns of the data file, the first column giving the date while the second column gives the time of the day at which the transaction took place. So we need to change the option `time.in.format` to accommodate this change.

Let us assume that we read the transaction data of the September 1998 contract in a `timeSeries` object which we called `SEP98.ts`. Looking at the beginning of the file we see:

```
                      Positions      1
10/08/1997 14:53:38.000 1013.2
10/17/1997 10:59:16.000  986.0
10/27/1997 10:02:13.000  960.0
11/03/1997 10:28:51.000  968.0
11/05/1997 09:08:44.000  975.0
   .....      .....      .....
08/19/1998 12:09:00.000 1105.7
08/19/1998 12:09:00.000 1105.8
08/19/1998 12:09:00.000 1105.7
08/19/1998 12:09:00.000 1105.6
08/19/1998 12:09:00.000 1105.5
08/19/1998 12:09:00.000 1105.6
08/19/1998 12:10:00.000 1105.7
```

Figure 5.4 gives the plot of a few days of all the transactions reported on the June 15 1998 futures contract on the S&P 500 index as traded on the Chicago Mercantile Exchange (CME for short). The vertical bands correspond to the time elapsed between the close of a given day and the opening of the next day. Obviously there is no data in this time interval. The graphing routine uses linear interpolation to provide values in these gaps. It is important to remember that the values appearing in these bands are artifacts of the plotting routine, they do not correspond to transaction prices.

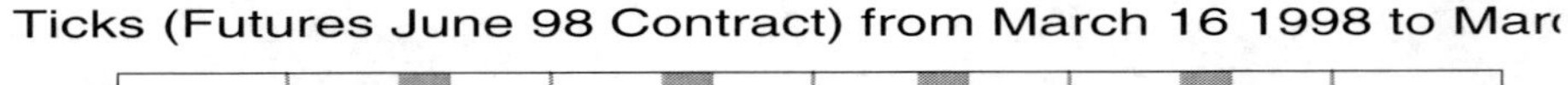

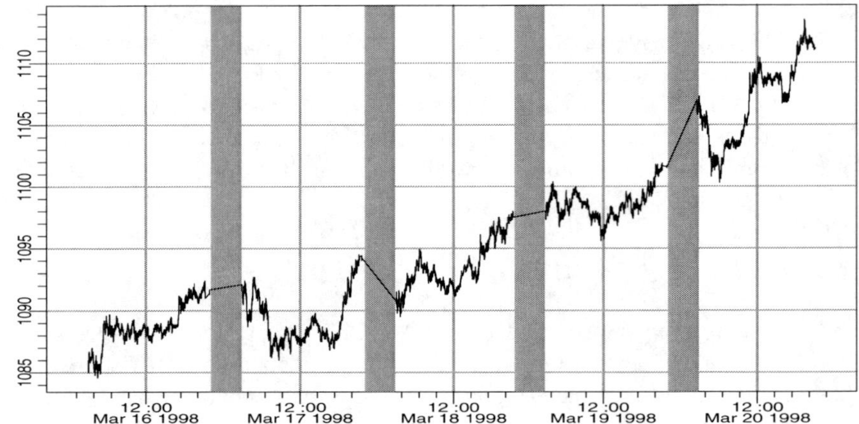

Fig. 5.4. Intra-day transactions of the June 15, 1998 S&P 500 futures contract recorded between the opening on March 16 and the close on March 20, 1998.

Remark on Quantized Ticks. Price changes from one transaction to the next are quoted in multiples of tick size. This tick size varies from one exchange to another. Typical values are (or used to be) one eighth and one sixteenth of a dollar. This practice is obsolete on a certain number of exchanges. For example, all New York Stock Exchange (NYSE) and New York Mercantile Exchange (NYMEX) stocks are traded in decimals since January 29, 2001. Nevertheless, practitioners should be aware of the fact that high-frequency data, and especially historical high-frequency raw data which has not been pre-processed, quite often take only discrete values: this can introduce numerical artifacts, and in particular spurious correlation. We illustrate this fact with the S&P 500 data considered in this subsection. We computed the fractional parts of the transaction prices (obtained by removing the integer parts to the actual prices), and we plotted their histogram in Figure 5.5. The quantification effect appears clearly.

5.2.1 `TimeDate` **Manipulations**

It is very easy to develop specific functions satisfying the needs of most time series data analysis. As an illustration, we give the code for two "home grown" functions which we wrote to extract the beginning of a given day, and noon of the same day. In other words, given a `timeDate`, the first function extracts the day, and returns a `timeDate` including hours, minutes and seconds of the beginning of that same day.

```
begday <- function(DAY)
{
    DAYDATE <- as.character(timeDate(as.character(DAY),
                                      format="%m/%d/%Y"))
```

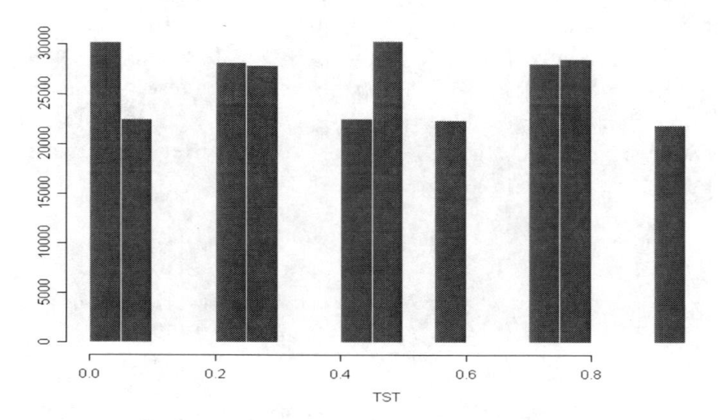

Fig. 5.5. Histogram of the fractional part of the intra-day transaction prices of the June 15, 1998 S&P 500 futures contract. The discrete nature of the data shows clearly.

```
        timeDate(paste(DAYDATE,"00:00:00.000",sep=" "),
                        format="%m/%d/%Y %02H:%02M:%02S.%03N")
    }
```

The second function returns the middle of the day.

```
    noon <- function(DAY)
    {
        DAYDATE <- as.character(timeDate(as.character(DAY),
                                              format="%m/%d/%Y")
        timeDate(paste(DAYDATE,"12:00:00.000",sep=" "),
                        format="%m/%d/%Y %02H:%02M:%02S.%03N")
    }
```

We used these functions to separate the morning trades from the afternoon trades, and compute the indicators studied in Chapter 4.

Minute by Minute Series

Many tools of time series analysis depend upon the assumption that the observations are regularly spaced in time. This is particularly true of all the model-fitting procedures based on the auto-covariance and correlation functions which we introduce later in this chapter. Moreover, most financial data analysis depends upon the computations of returns, and the latter are difficult to compute, interpret and compare when the prices are not given at regular time intervals.

As we just saw, high-frequency data are generically not regularly spaced. A natural solution to this problem is to create a regular time series from the data at hand, just by interpolation/extrapolation and resampling the extended data at regular time intervals. `S-Plus` has a very powerful function to do this with `timeSeries` objects. It is called `align`. We recommend checking the help file for this function to understand how it takes a frequency to create a grid of regularly spaced values obtained from the original series using a specific prescription (nearest measurement before, after, average of the two, ...). The following is an example of the kind of results we can get using this function. From the irregularly spaced data, we created a minute-by-minute time series.

```
07/17/1998 08:56:00.000 1191.6
07/17/1998 08:57:00.000 1191.4
07/17/1998 08:58:00.000 1191.2
07/17/1998 08:59:00.000 1191.3
07/17/1998 09:00:00.000 1191.8
07/17/1998 09:01:00.000 1192.0
07/17/1998 09:02:00.000 1191.9
07/17/1998 09:03:00.000 1191.8
07/17/1998 09:04:00.000 1192.4
07/17/1998 09:05:00.000 1192.4
```

Extracting One Day

The following commands show how one can easily extract one day, August 12, 1998 in the present example, and plot the corresponding sub-`timeSeries`.

```
> DAY <- timeDate("08/12/1998")
> SP081598 <- SEP98BM[positions(SEP98BM)>=DAY &
                         positions(SEP98BM)<DAY +1]
> plot(SP081598)
> title("S\&P500 Minute by Minute on 08/12/98") }}
```

The results are reproduced in Figure 5.6.

Extracting Several Days

One can extract several days in exactly the same way. We extract and plot three days worth of data for the sake of illustration. Notice once more the shaded bands between the close of one day and the opening of the next. Their role is to emphasize that there is no data in these time periods and that the (linear) plot supplied for the sake of continuity in these regions should be interpreted with a modicum of care.

```
> SPDAYS <- SEP98BMPTS[positions(SEP98BMPTS)>=DAY &
                          positions(SEP98BMPTS)<DAY+3]
> plot(SPDAYS)
> title("SP500 Minute by Minute from 08/12/98 to 08/14/98")
```

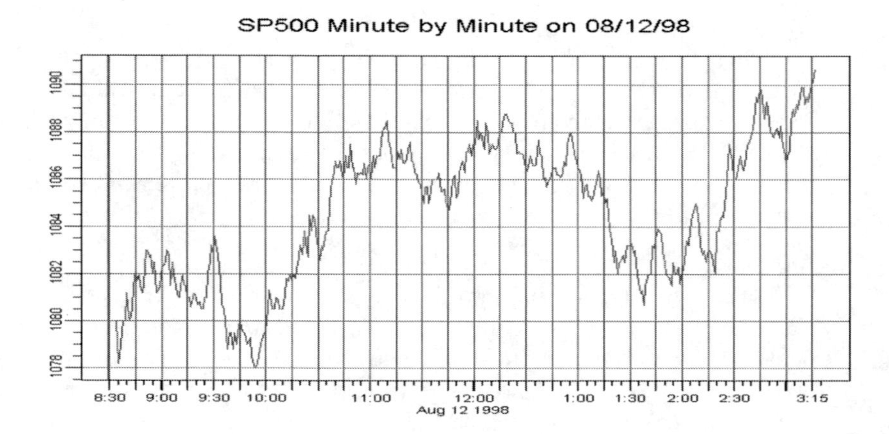

Fig. 5.6. Plot of one day's worth of the minute by minute time series.

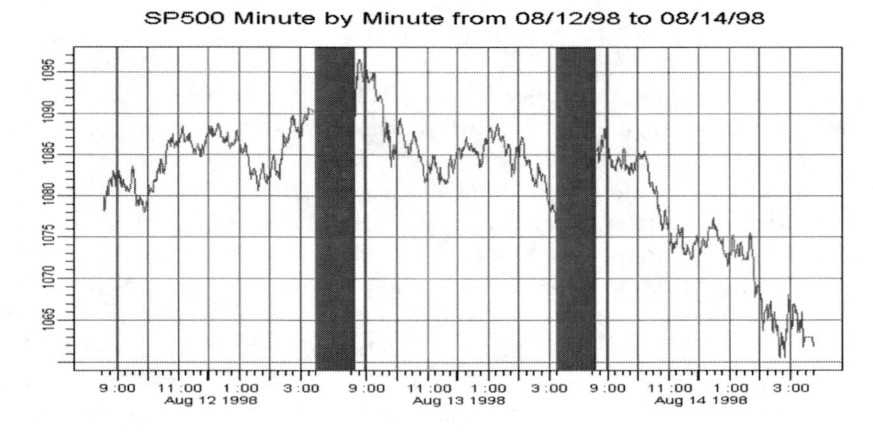

Fig. 5.7. Plot of three days' worth of the minute by minute time series.

Computing the Morning and Afternoon Indicators

Even though we do not aim at a thorough analysis of market microstructure, we present some of the statistics used by data analysts to capture the dynamics of high-frequency data. These dynamics are quantified by two types of indicators. The first type tries to capture patterns in the distributions of the actual times at which the transactions take place.

◇ ***Number of Transactions***. This is the first and the simplest possible indicator. It counts the number of transactions taking place during the period in question, giving a rough idea of the level of trading activity;

◇ ***Transaction Inter-arrival Time***. The mean of the lengths of the time intervals separating two successive transactions. Small values are signs of high trading activity.

◇ ***Transaction Rate***. This indicator is different from the classical rate at which arrivals take place in an arrival process, like in classical queuing theory for example. The spirit is the same, but the definition is more involved. It is computed in the following way: we compute the empirical quantile function (remember that this is the inverse of the empirical cdf) and we measure how steep its graph is, by computing the slope of the least squares simple linear regression.

The second type of indicator is more traditional in the sense that it consists of statistics usually computed for regular financial time series. In order to compute them, we first need to create a minute by minute series (obviously the time period Δt does not have to be one minute. Five minutes and ten minutes are also popular choices in the less active markets). Here we show how to use the powerful function `align` to produce a minute-by-minute series.

```
> TMP <- SERIES[positions(SERIES) >= begday(DAY)
                        & positions(SERIES) <= noon(DAY)]
> TMP <- align(TMP, by = "minutes", how = "before")
```

This example assumes that `DAY` is a `timeDate`, and that we are trying to compute the minute by minute series of the original high-frequency data contained in the `timeSeries` `SERIES` for the morning of that day. The first command extracts the entries of the morning from the time series `SERIES` by keeping the entries whose positions are after the beginning of the day, as given by `begday(DAY)`, and before noon, as given by `noon(DAY)`. We then *feed* the time series `TMP` so created to the function `align`, requesting that a minute-by-minute series be created by choosing for each minute time-stamp the last value present in the original series `TMP`. Other options are possible for the parameter `how`. The reader is referred to the help file. Once a regular series is created, we can compute the *usual suspects,* starting with the log-returns. Notice that computing these statistics requires the alignment of the series for all the returns to be computed over a period with the same length.

◇ ***Range***. The maximum one-minute log-return minus the minimum one-minute log-return over the morning or afternoon period;

◇ ***Volatility***. The standard deviation over the morning or afternoon period, of the one-minute log-returns.

◇ ***Volatility Ratio***. This indicator tries to tell us, given a level of volatility, if the market is trending in one direction, or *moving sideways.* Mathematically, this indicator is defined as the ratio of the sum of the absolute values of the one-minute log-returns to the absolute value of the log-return over the entire period. This number is always between 0 and 1, and it is close to one when the price always moves in the same direction, while it is close to zero when the moves are alternating between up and down. This phenomenon cannot be captured by the volatility, which measures the amount of action, and not the direction in which the action takes place!

Working with regular time series is very convenient for many reasons, not the least being the fact that we can take advantage of all the tools we are about to present in the remainder of the book. But one should keep in mind the fact that aligning a time series comes at a cost: the clear loss of information in time periods with high activity, and the introduction of undesired artifacts in periods of low trading activity.

From now on, we try to avoid the technical difficulties inherent to the analysis of irregular time series, we consider only models for regular time series, and when we fit these models to real data, we act as if the time series under consideration were regular.

5.3 Time Dependent Statistics and Stationarity

Now that we know the type of data amenable to practical time series analysis, we turn to the discussion of the first properties of the theoretical models. Throughout this section we assume that we are given such a theoretical model $\{X_t\}_t$ for a time series.

5.3.1 **Statistical Moments**

The mean function μ_X is defined as the (deterministic) function of time given by:

$$t \hookrightarrow \mu_X(t) = \mathbb{E}\{X_t\}.$$

Similarly, we define the variance function var_X (resp. the standard deviation function σ_X) as the (deterministic) function of time given by:

$$\begin{aligned} t \hookrightarrow \text{var}_X(t) &= \text{var}\{X_t\} = \mathbb{E}\{(X_t - \mu_X(t))^2\}, \\ (\text{resp. } t \hookrightarrow \sigma_X(t) &= \sqrt{\text{var}_X(t)} = \mathbb{E}\{(X_t - \mu_X(t))^2\}^{1/2}). \end{aligned}$$

Even though these statistics can capture some of the time dependent features of the series, they do not carry any information on the way individual random variables entering the series depend upon each other. With this in mind, we consider statistical moments involving several X_t's simultaneously. The auto-covariance function γ_X is defined as the (deterministic) function of two instants given by:

$$\gamma_X(s,t) = \text{cov}\{X_s, X_t\} = \mathbb{E}\{(X_s - \mu_X(s))(X_t - \mu_X(t))\}.$$

As one would expect, the auto-correlation function ρ_X is defined as the (deterministic) function of two instants given by:

$$\rho_X(s,t) = \text{cor}\{X_s, X_t\} = \frac{\mathbb{E}\{(X_s - \mu_X(s))(X_t - \mu_X(t))\}}{\sigma_X(s)\sigma_X(t)}.$$

We shall also use the notion of partial auto-correlation function. Its definition is best understood in the framework of stationary time series once the notion of linear prediction has been introduced. So stay tuned if you want to know more about partial auto-correlation functions.

The above concepts are limited to moments of orders one and two. They have been regarded as sufficient for quite some time, and many time series analysis tools have been designed to study models only on the basis of their first two moment functions. There are several reasons for that: first, these moments characterize entirely the Gaussian models. Moreover, their attractiveness is increased by the fact that the least squares methods only depends upon these moments. Unfortunately they are not sufficient to handle nonlinearities for which the use of higher order moments is required. Despite the fact that the importance of non-Gaussian and nonlinear time series is increasingly recognized, we shall not discuss these matters any further in this chapter. See the Notes & Complements at the end of the chapter for references.

5.3.2 The Notion of Stationarity

Stationarity is a crucial property of stochastic processes and time series models. It will be a *sine qua non* condition for the implementation of most estimation and prediction algorithms. The notion of stationarity can be described in the following manner.

Mathematical Definitions

A time series model for $\{X_t\}_t$ is said to be stationary if all its statistics remain unchanged after any time shift, i.e. if they are the same as the statistics of $\{X_{t_0+t}\}_t$ for all possible choices of t_0. This notion of stationarity is sometime called strong stationarity. The first obvious consequence of stationarity is that the mean function of a stationary time series is constant. The same holds for the variance and the standard deviation functions. Moreover, the auto-covariance function (and consequently the auto-correlation function as well) is a function of the difference between its arguments. This means that:

$$\mu_X(t) = \mu_X, \quad \mathrm{var}_X(t) = \sigma_X^2, \quad \sigma_X(t) = \sigma_X, \quad \text{and} \quad \gamma_X(s,t) = \gamma_X(t-s), \tag{5.2}$$

for some constants μ_X and σ_X, and for some function of one variable for which we still use the notation γ_X.

There is a weaker notion of stationarity which will be useful in the sequel. A time series model is said to be weakly stationary if its mean function is constant, and its auto-covariance function is a function of the difference of its arguments. Obviously, a stationary series (in the strong sense given above) is necessarily weakly stationary as implied by formula (5.2). However, the converse is not true in general. In a nutshell, (strong) stationarity means that the moments of all orders are invariant under time shifts, while weak stationarity merely requires that the moments of order 1 and 2 are shift invariant.

Remark. Despite this fact, the two notions of stationarity coincide for Gaussian time series models, i.e. when all the finite dimensional marginal distributions are multivariate Gaussian. Indeed, a multivariate Gaussian distribution is entirely determined by its mean vector and its variance/covariance matrix, which in turn is determined by the covariances of the random variables taken two by two.

Linear Prediction and Partial Auto-Correlation Function

Building a model for time series data is a required step in the prediction of future values. So, once a model for $\{X_t\}_t$ has been chosen, in many practical applications the goal is to produce, for a given time t (which we shall refer to as the present time), predictions for the (future) values of the outcomes $x_{t+1}, x_{t+2}, \ldots$ of the random variables $X_{t+1}, X_{t+2}, \ldots$. Obviously, these predictions should be non-anticipative in the sense that they should be solely based on the (present and past) observations $x_t, x_{t-1}, \ldots$. In other words, we are not allowed to use a crystal ball to look into the future when it comes to give predictions!

Given the observations $X_0 = x_0$, $X_1 = x_1$, ..., $X_t = x_t$ up to the present time t, and a mean zero random variable **Z**, we shall use the notation $E_t^{(m)}(Z)$ for the best prediction of Z by linear combinations of the m values $x_s - \mu_X$ for $t-m+1 \leq s \leq t$. In other words, when viewed as a function of the random variables $X_{t-m+1} - \mu_X$, $X_{t-m+2} - \mu_X$, ..., $X_t - \mu_X$, $E_t^{(m)}(Z)$ is the linear combination $\alpha_1(X_{t-m+1} - \mu_X) + \alpha_2(X_{t-m+2} - \mu_X) + \cdots + \alpha_m(X_t - \mu_X)$ which minimizes the quadratic error:

$$\mathbb{E}\{\|Z - \alpha_1(X_{t-m+1} - \mu_X) + \alpha_2(X_{t-m+2} - \mu_X) + \cdots + \alpha_m(X_t - \mu_X)\|^2\}. \quad (5.3)$$

Notice that, in the Gaussian case (i.e. in the case of (jointly) Gaussian time series models), this prediction operator is given by the conditional expectation:

$$E_t^{(m)}(Z) = \mathbb{E}\{Z | X_t, X_{t-1}, \ldots, X_{t-m+1}\}.$$

The best linear predictor $E_t^{(m)}(Z)$ is a linear combination of the $X_{t-m+1} - \mu_X$, $X_{t-m+2} - \mu_X$, ..., $X_t - \mu_X$, so it belongs to the linear space generated by these random variables. Moreover, since $E_t^{(m)}(Z)$ minimizes the quadratic error (5.3), it minimizes the distance between Z and this linear space. Consequently, $E_t^{(m)}(Z)$ can be viewed as the orthogonal projection of the random variable Z onto the linear space generated by the random variables $X_{t-m+1} - \mu_X$, $X_{t-m+2} - \mu_X$, ..., $X_t - \mu_X$. In fact, as we shall see in the sequel, this interpretation as an orthogonal projection is a great help when it comes to guessing and proving the properties of the best linear prediction operator $E_t^{(m)}$. The first instance is the following: because of the properties of orthogonal projections, $Z - E_t^{(m)}(Z)$ is orthogonal to all of the $X_j - \mu_X$ and consequently:

$$\mathbb{E}\{(Z - E_t^{(m)}(Z))(X_j - \mu_X)\} = 0, \qquad j = t-m+1, t-m+2, \ldots, t-1, t.$$

The random variable $Z - E_t^{(m)}(Z)$ is often referred to as an innovation because it represents, in a minimal way, the information needed to produce the outcome of Z, which cannot be given by linear combinations of the past values $X_{t-m+1} - \mu_X$, $X_{t-m+2} - \mu_X$, ..., $X_t - \mu_X$.

We shall use the notation E_t without the superscript $^{(m)}$ when $m = t$, in other words, when we use the entire available past to construct the prediction. This special prediction operator will be used extensively in Chapter 6.

With the notion of best linear predictor at hand, it is now easy to define the partial auto-correlation coefficients. The k-th partial auto-correlation coefficient $\phi_{k,k}$ is defined as the last coefficient in the linear combination giving $E_t^{(k)}(X_{t+1})$. In other words, if:

$$E_t^{(k)}(X_{t+1} - \mu_X) = \alpha_k(X_t - \mu_X) + \alpha_1(X_{t-1} - \mu_X) + \cdots + \alpha_1(X_{t-k+1} - \mu_X)$$

then we set $\phi_{k,k} = \alpha_1$. In this way, one sees that the partial auto-correlation coefficient $\phi_{k,k}$ measures the correlation between X_{t+1} and X_{t-k+1} (or equivalently, between X_t and X_{t-k} because of stationarity) after adjustment for the intermediate lagged variables.

Because the best linear predictor is an orthogonal projection, our knowledge of Euclidean geometry tells us that it can be computed in terms of inner products. In this way one can prove that the partial auto-correlation coefficient $\phi_{k,k}$ is given by the formula:

$$\phi_{k,k} = \Gamma_{X,k}^{-1} \gamma_{X,k}, \tag{5.4}$$

where the $k \times k$ matrix $\Gamma_{X,k}$ and the k-dimensional vector $\gamma_{X,k}$ are defined by:

$$\Gamma_{X,k} = [\gamma_X(i-j)]_{i,j=1,\dots,k} \qquad \gamma_{X,k} = [\gamma_X(1), \gamma_X(2), \dots, \gamma_X(k)]^t. \tag{5.5}$$

This equivalent form of the definition of the partial auto-correlation function is preferable because it is immediately amenable to estimation. Indeed, if we assume momentarily that we know how to estimate the auto-covariance function γ_X from given data $x_0, \dots, x_n$, see the discussion leading to (5.8) below, then the sequence $\{\phi_{k,k}\}_k$ of partial auto-correlations will be estimated by the sequence $\{\hat{\phi}_{k,k}\}_k$ given by:

$$\hat{\phi}_{0,0} = 1 \qquad \text{and} \qquad \hat{\phi}_{k,k} = \hat{\Gamma}_{X,k}^{-1} \hat{\gamma}_{X,k}, \tag{5.6}$$

where the $k \times k$ matrix $\hat{\Gamma}_{X,k}$ and the k-dimensional vector $\hat{\gamma}_{X,k}$ are computed from the estimate $\hat{\gamma}_X$ of the auto-covariance function via the formulae (5.5). This empirical estimate $\hat{\gamma}_X$ will be defined in the next subsection.

The notion of partial auto-correlation function may seem obscure at this stage, but please, bear with me for a short while: soon we shall give an enlightening interpretation of the partial auto-correlation function in terms of auto-regressive models of increasing orders fitted to the time series.

Time Averages as Statistical Estimates

One of the main consequences of the stationarity of a time series is the fact that the *theoretical moments* introduced earlier can be computed (or at least estimated) by time averages. Indeed, when stationarity holds, the moment empirical estimates (see formulae (5.7) and (5.8) below) are time averages which converge when the number of terms used in the sum increases without bound. The existence of these limits is a good sign since it gives stability of the empirical estimates, but unfortunately the limits they converge to can still be random, and they can change with the particular realization of the time series. Fortunately, there are many situations in which these limits are not random. Time series with this property are called *ergodic*. Roughly speaking, ergodicity allows us to replace space averages such as expectations, co-variances, ..., by time averages. Ergodicity is a very subtle mathematical property and it is difficult to check that a given set of numbers $x_1, \dots, x_n$ is a sample realization from an ergodic time series model. For this reason, we shall take ergodicity for granted in the sense that, whenever we find that a time series model is stationary, we shall implicitly assume that it is also ergodic. This is a bold attitude, but it is justified by the fact that, without it, we wouldn't be able to do much.

In practice, if $x_1, \dots, x_n$ is a finite set of observations from a time series model $\{X_t\}$ which we assume to be stationary, then we use the *time average*

$$\widehat{\mu}_X = \overline{x} = \frac{1}{n} \sum_{i=1}^{n} x_i \tag{5.7}$$

to estimate the mean μ_X. As explained above, the stationarity of the series implies that:

$$\lim_{N \to \infty} \frac{1}{N} \sum_{i=1}^{N} X_i = \tilde{\mu}_X$$

almost surely, i.e. for all typical sequence of observations, where $\tilde{\mu}_X$ is a random variable. Moreover, ergodicity implies that the object $\tilde{\mu}_X$ (which was thought to be random) is in fact a deterministic number which is just the true value of the mean μ_X of the series. This is the justification for the use of the time average (5.7) as an estimate of the mean. Similarly, the auto-covariance function $\gamma_X(h)$ is estimated by time averages of the form:

$$\widehat{\gamma}_X(h) = \frac{1}{n} \sum_{i=1}^{n-|h|} (x_i - \overline{x})(x_{i+|h|} - \overline{x}), \tag{5.8}$$

which are defined both for positive and negative lags h as long as $-n < h < n$. It is important to notice that the larger the lag absolute value $|h|$, the smaller the number of terms in the above sum. If we were to compute a confidence interval for the estimate $\widehat{\gamma}_X(h)$ we would see that this interval grows very fast with $|h|$. As a consequence, it is wise to restrict the estimation of the auto-covariance function to lags which are small compared to the sample size n. In `S-Plus`, you can specify this maximum number of lags for which the auto-covariance function is estimated. If you don't, the program uses a multiple of the logarithm of n as a proxy. Indeed this number is usually much smaller than the sample size. Notice also that we divide by n while there are only $n - |h|$ terms in the summation. This departure from the standard definition of the empirical auto-covariance function becomes irrelevant for large samples, since when n is large, dividing by n or $n - |h|$ does not make much difference, especially with the limitation we imposed on the size of $|h|$. Moreover, dividing by n guarantees that the function $\widehat{\gamma_X}$ is nonnegative definite, a mathematical property of crucial importance for spectral theory. Since we shall not address spectral theory issues in these lectures, we do not go any further down this avenue.

As one could expect, the estimate of the sample auto-correlation function is defined by:

$$\widehat{\rho}_X(h) = \frac{\widehat{\gamma}_X(h)}{\widehat{\gamma}_X(0)}, \qquad -n < h < n. \tag{5.9}$$

The distribution theory of the estimators $\hat{\mu}_X$, $\widehat{\gamma}_X$ and $\widehat{\rho}_X$ is well understood in the case of Gaussian time series, and confidence intervals and tests can be derived for these estimates. In the general case of possibly non-Gaussian series, only approximate tests are available. We discuss some of them in our discussion of the white noise series below.

5.3.3 The Search for Stationarity

The success of time series analysis depends strongly upon the satisfaction of two somewhat independent conditions. First one needs to be able to massage the data into a stationary time series, and second one needs to model this stationary time series and estimate the parameters of the model.

In this subsection, we consider the first of these two points. Unfortunately, there is no universal recipe to turn a given time series into a stationary one, and experience will have to lead the analyst in this endeavor. For the record, we review several of the most commonly used procedures. We first discuss general strategies in an abstract setting, postponing the implementation of these ideas to the next subsection.

Statistical tests are required to quantify the extent to which a search for stationarity is successful. Tests are not very powerful when the alternative hypothesis is too general. More restrictive alternative hypotheses are used in the commonly used tests for stationarity. We discuss the basic form of these tests at the end of Subsection 5.4.7. They are known under the name of Dickey-Fuller tests, or unit-root tests.

Removing Trends and Seasonal Components

We first consider the issues connected with the analysis of the mean. We already mentioned that it is common practice to subtract the mean μ_X, and use the series $X_t - \mu_X$ instead of X_t. This is convenient when the mean function $t \hookrightarrow \mu_X(t)$ is constant, since in this case, μ_X can easily be estimated. However, we cannot expect to be able to do that in general. Fortunately, it happens quite often that the mean function depends upon time in a manner which can be identified. Let us assume for example that the data, say x_t, can be reasonably well described, by an equation of the form:

$$x_t = m(t) + p(t) + \tilde{x}_t, \tag{5.10}$$

where $m(t)$ is a deterministic monotone (increasing or decreasing) function of t, $p(t)$ is a deterministic periodic function of t, and $\tilde{x}_t$ is a mean-zero stationary time series. The function $t \hookrightarrow m(t)$ is called trend, and $t \hookrightarrow p(t)$ seasonal component. Several simple techniques can be used to identify the deterministic components $m(t)$ and $p(t)$. If we bundle together the trend $m(t)$ and the seasonal component $p(t)$ and rename their sum $\varphi(t)$, and if we rename the stationary time series $\tilde{x}_t$ by ϵ_t, equation (5.10) becomes

$$x_t = \varphi(t) + \epsilon_t$$

which is exactly the type of equation amenable to regression analysis. Indeed, regression techniques are the methods of choice for the identification of the components appearing in (5.10), at least as long as these techniques are adjusted for the fact that the noise term ϵ_t can exhibit a significant dependent structure, and the fact that the regression function $\varphi(t)$ is the sum of a monotone function and a periodic function. We did not discuss monotone regression. Nevertheless, see the discussion of the differentiation operator below.

Problem 5.12 gives an example of identification of a periodic function using nonlinear regression. Seasonal components are often identified using Fourier (also called spectral) analysis. These techniques have proven to be extremely successful in many engineering applications, and numerous books and software packages have been written to make them available. `S-Plus` contains a suite of methods for the spectral analysis of signals, but we shall not use them in this book.

Stabilization of the Variance

Setting $\tilde{X}_t = X_t - \mu_X(t)$, it is always possible to write a time series as the sum of a deterministic function plus a mean-zero time series since $X_t = \mu_X(t) + \tilde{X}_t$. If the function $\mu_X(t)$ can be estimated from the data, (see examples below) it can be subtracted from the original data, and our modeling efforts should concentrate on the mean-zero time series $\tilde{X}_t$. It happens frequently that the standard deviation varies significantly with time, and since the local variance is more difficult to estimate than the local mean function, this may be a serious hurdle.

Variance stabilization transformations are ways to correct for these variations when the variance function is an explicit function of the mean in the sense that $\text{var}_{\tilde{X}}(t) = \varphi(\mu_{\tilde{X}}(t))^2\sigma^2$ for some constant $\sigma > 0$, and a known function φ. In such a case, it is easy to show that the variance of the time series $Y_t = \psi(\tilde{X}_t)$ is essentially constant and equal to σ^2 if the function ψ is such that $\psi'(\mu) = 1/\varphi(\mu)$. The time series $\{Y_t\}_t$ is more likely to be stationary, and in any case, it is more amenable to analysis than the original series. One will also demand that the function

ψ be invertible in order to be able to return to $\tilde{X}_t$ (and X_t) after analysis of the time series $\{Y_t\}_t$.

Examples of variance stabilization transformations include the function $\psi(\mu) = \log(\mu)$ (which is often used in the analyses of time series of financial returns) and the function $\psi(\mu) = \sqrt(\mu)$. According to the discussion above, the transformation $\psi(\mu) = \log(\mu)$ is recommended when the variance varies like the square of the mean (i.e. $\varphi(\mu) \sim \mu$), while the transformation $\psi(\mu) = \sqrt(\mu)$ is recommended when the variance varies like the mean (i.e. $\varphi(\mu) \sim \sqrt(\mu)$) as we find in many situations involving the Poisson distribution.

The Use of Differentiation

If $m(t)$ is a constant function, a mere differentiation should make it disappear. This intuition from calculus can be implemented in the case of time series by introducing the analog of the differentiation operator. We shall use the suggestive notation ∇ for this operator. At the model level, ∇ is defined by:

$$\nabla X_t = X_t - X_{t-1} \tag{5.11}$$

and for a (finite) data set $x = (x_0, x_1, \dots, x_N)$, the (first) difference $y = \nabla x$ is given by $y = (x_1 - x_0, x_2 - x_2, \dots, x_N - x_{N-1})$. Notice that y_t is defined for $t = 1, 2, \dots, N$ while x_t is defined for $t = 0, 1, 2, \dots, N$! It is now time to give some respectability to a practice we have used several times already: taking differences to turn a non-stationary time series into a stationary one. Recall that the typical instance is the computation of the log-returns of a financial time series. So we elevate this example to the rank of definition: we shall say that a time series is integrated of order one, or that it has one unit-root, if its difference is stationary. We shall use the notation $I(1)$ for the class of these time series.

In the same way a simple difference can *kill* a constant term, two successive differences will get rid of a linear trend (a function m of the form $m(t) = at + b$). Iterating the definition (5.11) of the first difference operator, we find definition formulae for the higher order differential operators. For example, the second order difference operator is given by:

$$\begin{aligned}\nabla^2 X_t = \nabla[\nabla X]_t = [\nabla X]_t - [\nabla X]_{t-1} &= X_t - X_{t-1} - (X_{t-1} - X_{t-2}) \\ &= X_t - 2X_{t-1} + X_{t-2}.\end{aligned}$$

More generally, by successive applications of the difference operator one can remove any kind of polynomial trend! This remark makes it plain how useful the higher order difference operators can be. To keep up with the definition introduced above, we say that a time series is integrated of order p, or that it has p unit-roots, if its p-th order difference is stationary, and we denote the class of time series with this property by $I(p)$. In other words:

$$\{X_t\}_t \in I(p) \qquad \text{means that} \qquad \{\nabla^p X_t\}_t \quad \text{is stationary}$$

Obviously, the notation $I(0)$ will be used for stationary time series. We use the notation ∇ to conform with the conventions used in most of the textbooks on time series analysis, and to emphasize the analogy with the differentiation of functions of a (continuous) variable t.

5.3.4 The Example of the CO_2 Concentrations

We choose to illustrate this discussion with the classical example of the concentration of CO_2 at Mauna Loa, Hawaii. The measurements are monthly. They range from January 1958 to December 1975. These data are contained in a data set `co2` included in the `S-Plus` distribution. Figure 5.8 gives a plot of these values. This plot reveals a linear upward trend and a cyclic behavior whose period seems to be 12 months (none of these remarks should come as a surprise).

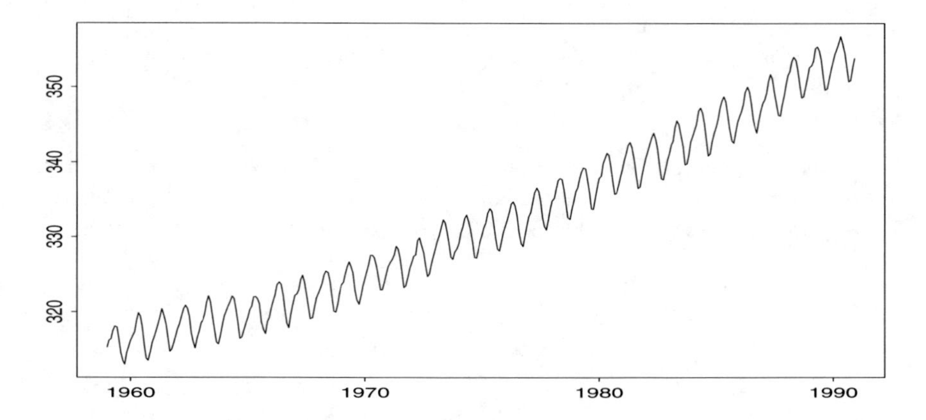

Fig. 5.8. Monthly concentrations of CO_2 at Mauna Loa, Hawaii from January 1958 to December 1975

The first step is to identify and remove the seasonal component. S-Plus provides a function to do just that. It is called `stl`. Unfortunately, it cannot be used with `timeSeries` objects, so we wrote a wrapper called `sstl` whose usefulness we proceed to illustrate. The following `S-Plus` commands were used to produce the plots contained in Figure 5.9.

```
> co2.stl <- sstl(co2.ts,"periodic")
> tsplot(co2.stl$sea)
> tsplot(co2.stl$rem)
```

The function `sstl` returns a list of two elements. The first one is identified with the extension $\cdots$`$sea`. It gives the seasonal component identified by the program. The second one is identified with the extension $\cdots$`$rem` and it gives the remainder term,

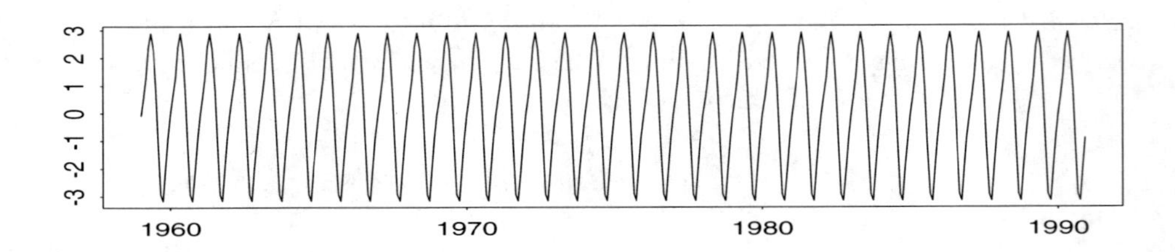

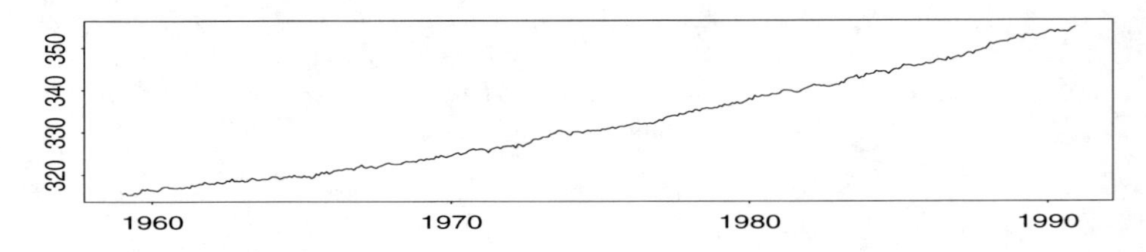

Fig. 5.9. The plot on the top gives the seasonal component as identified by the function `sstl`. The lower plot gives the remainder.

i.e. what is left once the seasonal component has been removed from the original series. Its plot is given at the bottom of Figure 5.9. It shows an obvious upward trend which could possibly be linear. To test this idea, we perform a simple least squares linear regression.

```
> plot(co2.stl$rem)
> l2line <- lsfit(positions(co2),co2.stl$rem)
> abline(l2line)
> tsplot(l2line$resid)
```

The results are shown in Figure 5.10. The top plot shows that the linear fit is reasonable, but the bottom plot shows that the residuals exhibit a definite "U-shape" behavior: they do not look like a stationary time series. We are not done yet. The linear regression was not enough to completely explain the lack of stationarity. Experience will teach us that this type of time series can be made *more stationary* by differentiation (differencing).

```
> co2d <- diff(l2line$resid)
> ts.plot(co2d)
```

The result of the differentiation is given in Figure 5.11.

It is now reasonable to assume that the series is stationary and one can go to the next step of the modeling effort. We shall come back to this example and continue its analysis after we introduce the auto-regressive and moving average models.

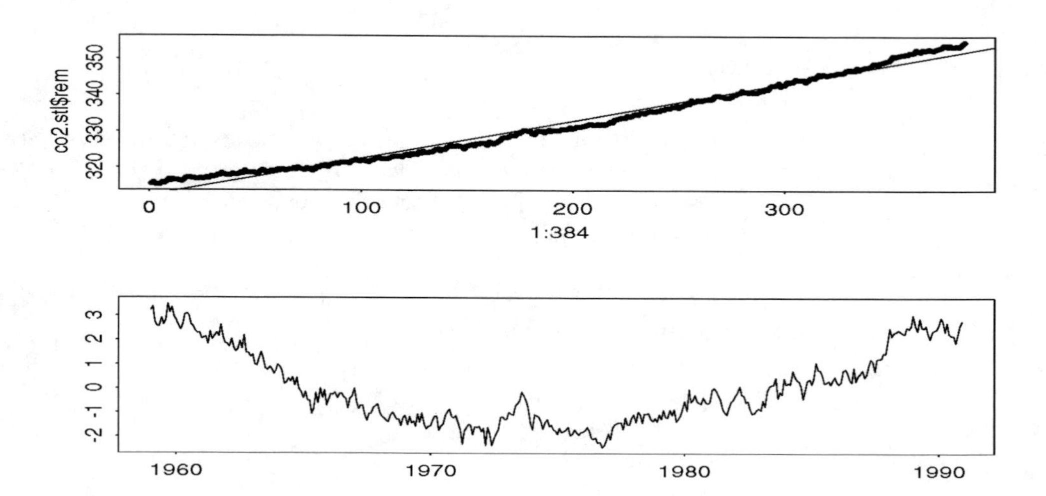

Fig. 5.10. Top: Residuals after removing the seasonal component and least squares regression. Bottom: Residuals after removing the linear trend from the residuals.

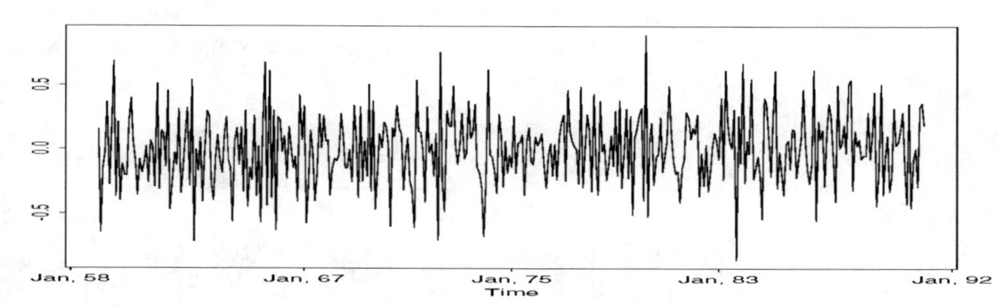

Fig. 5.11. CO_2 concentration remainder after applying the difference operator to the residuals of the linear regression.

It is not difficult to imagine that the same result could have been obtained by removing a quadratic trend instead of a linear one, or even by two successive differentiations.

5.4 First Examples of Models

We review the most commonly used examples of time series models.

5.4.1 White Noise

A finite sequence

$$w_0, w_1, \dots\dots, w_n$$

of real numbers is said to form a white noise if they are observations from a (possibly infinte) sequence of independent and identically distributed (i.i.d. for short) mean zero random variables, say $\{W_t\}_t$. Such a sequence is obviously stationary and $m_W = \mathbb{E}\{W_t\} = 0$ by definition. Also, if $s \neq t$ we have:

$$\text{cov}\{W_s, W_t\} = \mathbb{E}\{W_s W_t\} = 0$$

because of the independence assumption. Consequently:

$$\gamma_W(h) = \begin{cases} \sigma^2 & \text{if } h = 0 \\ 0 & \text{otherwise} \end{cases} \tag{5.12}$$

where σ^2 denotes the common variance of the random variables W_t.

Simulation

We review in this subsection the discussion already presented in the introductory session to `S-Plus`. We use the i.i.d. definition of a white noise to generate a normal white noise with variance $\sigma^2 = 4$ and plot it with the `S-` command `WN <- rnorm(1024,0,4)`. The result is reporduced in Figure 5.12. The top pane shows

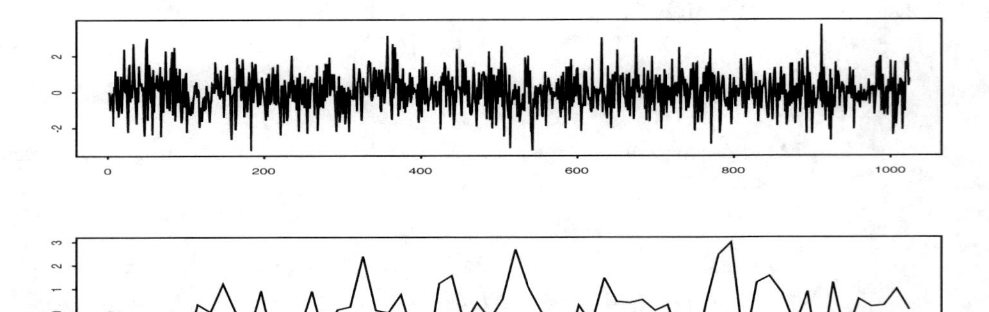

Fig. 5.12. Normal (i.e. Gaussian) white noise time series with variance $\sigma^2 = 4$.

the entire vector of length 1024, while the bottom pane zooms into a sub-vector of length 70. The result is a much smoother looking curve. This shows that the wiggly nature of the plot, as well as its roughness, depends strongly upon the scale used to look at the series.

Testing for White Noise

The need to check that the residuals of a fitted model form a white noise is going to be a nagging problem recurring throughout the remainder of the book. For the purposes of the present discussion, we assume that these residuals form a stationary series. We tackle the problem of the serial correlation with a graphical tool first (see first bullet point below), and then, with a quantitative test in the third bullet point. We also discuss the issue of identification of the marginal distribution of the individual terms in the second bullet point.

• The first available tool is graphical. In the same way Pavlov's dogs adjusted their behaviors, we should develop the reflex of plotting the auto-correlation function of any time series we bump into. The estimate which we can compute from the data makes sense when the time series is stationary, but its plot may be useful even when the time series is not. The following `S-Plus` commands can be used to compute and plot the auto-covariance function and the auto-correlation function of the white noise time series created earlier.

```
> WNacov <- acf(WN,40,"covariance")
> WNacor <- acf(WN,40,"correlation")
```

The results are given in Figure 5.13. The looks of the two plots are identical. After all, the values of these two functions differ only by a scaling factor: the common variance of all the entries of the series. In other words, these two plots only differ via the vertical axis labels. Indeed the auto-correlation function is normalized to start from the value 1 for lag 0. Notice also the presence of a band around the horizontal axis of the auto-correlation function. It gives approximate 95% confidence limits. These limits should not be taken too seriously. The confidence interval is only approximate: it is given as a tool to identify the values which are not significantly different from 0. Let us emphasize once more that, precisely because of the definition (5.8) of the estimated auto-covariance function (and auto-correlation function as well) it is not reasonable to compute the estimates for too large a value of the lag. Indeed, for the estimate to be useful, it needs to be based on a summation containing as many terms as possible. And since this number of terms is limited by the value of the lag, a compromise has to be reached. In this example we chose to compute and plot the acf coefficients for 40 lags. The maximum number of lags for which these functions are computed is usually chosen as a multiple of the logarithm of the length of the time series. Since the auto-correlation function of a white noise is non-zero only for lag zero, the confidence band should contain all the values except for the first one. This visual test is useful in detecting obvious correlations.

• Plotting a histogram (and a density estimate) of the entries of the series should be done to check for normality. In fact as we saw in the first chapter, a Q-Q plot is preferable. Such a plot could give an indication that the marginal distribution is Gaussian, and hence that the series entries could actually be independent (which is, as we know, much stronger than uncorrelated). Even though this graphical test could be complemented by a goodness of fit test (such as a χ^2 or a Kolmogorov-

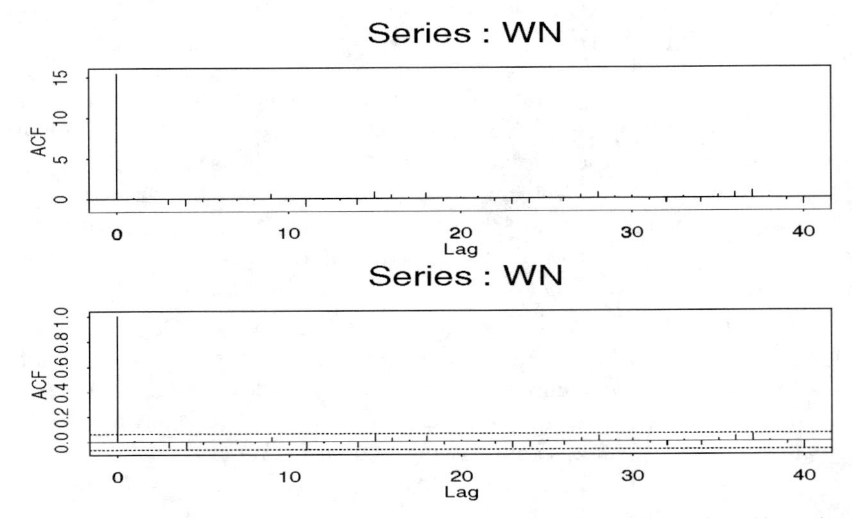

Fig. 5.13. Autocovariance (top) and autocorrelation (bottom) functions of a white noise time series.

Smirnoff test), it will only give information on the marginal distribution: it is very difficult to test for *joint normality* of all the marginal distributions ! If the normality of the marginal distributions were to be rejected, the apparatus introduced in the first chapter would be needed to fit GPD's.

• There are several powerful tests for the white noise hypothesis. The most common ones come under the name of *portmanteau* tests. They are based on the fact that, under the null hypothesis of a Gaussian white noise, appropriately weighted sums of the squares of the estimates of the auto-correlation function should follow a χ^2 distribution. `S-Plus` has one such test, but unfortunately, it is buried in the residual analysis of the fit to an ARIMA model. See Subsection 5.5.2 below for details.

Warning: the Different Forms of White Noise

There are many instances in which the terminology white noise is used even though the i.i.d. assumption is violated. In these instances, one merely demands that (5.12) holds to justify the term white noise. The reason for this confusing practice is that this weaker assumption is enough to justify most of the *least squares procedures*. In order to better understand what is at stake here, let us introduce clear definitions for the various forms of white noise. A time series $\{W_t\}_t$ is said to be a white noise if it satisfies the properties given in the two bullets which follow:

- the W_t are mean-zero i.e. $\mathbb{E}\{W_t\} = 0$ for all t;
- they have the same variance and they are uncorrelated:
  - in the strong sense, i.e. when all the W_t's are independent;
  - in the weak sense, i.e. when $\mathbb{E}\{W_t W_s\} = 0$ whenever $s \neq t$.

In any case, whether or not the independence holds, we shall use the notation $W \sim WN(0, \sigma^2)$ for any white noise with variance σ^2.

In this chapter, and in most of the remainder of the book, all the white noise time series will be assumed to be white noise series in the strong sense, i.e. with independent terms. We shall revisit this convention later in the last chapter when we discuss the so-called ARCH models. Hopefully, at that time, we shall be able to shed enough light on this problem to clear up some of these ambiguities.

For a statistical analysis, reaching a white noise (in the strong sense) is the end of the *modeling road*. Indeed, the statistician massages the data to extract significant component, after significant component, ... until she gets residuals forming a white noise. Then she is done, unless of course these residuals still contain some structure which can be identified and extracted. This is only possible if this residual white noise is a white noise in the weak sense, not in the strong sense. But of course, one should not expect this task to be easy: after all, for all of its tools, linear analysis cannot go beyond a weak white noise. Extracting substance from its *guts* will require skill and finely-sharpened tools. We will give examples in our analysis of the ARCH/GARCH and stochastic volatility models later in Chapter 7.

5.4.2 **Random Walk**

In our introductory session to `S-Plus`, a random walk was defined as an integral of a white noise, and we used the `S` function `cumsum` to construct a sample of a random walk from a sample of a white noise. At the level of the models, we say that $\{X_n\}_n$ is a random walk if there exists a white noise $\{W_t\}_t$ such that:

$$X_n = X_0 + W_1 + W_2 + \cdots + W_n \;, \tag{5.13}$$

in other words, if the whole sequence $\{X_n\}_{n\geq 0}$ is determined by X_0 and the induction formula $X_{n+1} = X_n + W_{n+1}$. Usually, X_0 is assumed to be independent of the entire white noise sequence $\{W_t\}_t$. Notice that $\mu_S(n) = \mathbb{E}\{X_n\} = \mathbb{E}\{X_0\}$ for all n. However, even though the mean function is constant, the random walk is not stationary. Indeed:

$$\begin{aligned}
\text{var}\{X_n\} &= \text{var}\{X_0\} + \text{var}\{W_1\} + \cdots + \text{var}\{W_n\} \\
&\quad + 2\text{cov}\{X_0, W_1\} + \cdots + \text{cov}\{X_0, W_n\} + \text{cov}\{W_1, W_2\} + \cdots \\
&\quad + \text{cov}\{W_1, W_n\} + \cdots\cdots \\
&= \text{var}\{X_0\} + n\sigma^2,
\end{aligned}$$

which is obviously changing with n. Notice that the independence of the terms appearing in (5.13) guaranteed the fact that the variance of the sum is equal to the sum of the variances. Keep in mind that this is not true in general. Even though the random walk is not stationary, its first difference is. Indeed $X_t - X_{t-1} = W_t$ is a white noise (and hence is stationary). So $\{X_t\}_t \sim I(1)$. These processes are also called *root one* processes for reasons we shall discuss below.

Random Walk with Drift

It happens quite often that the log-returns of market indexes have a small positive mean over significantly long periods of time. This remark is consistent with our discussion of the Samuelson's model for stock prices and stock indexes given in Section 4.7, and it justifies the introduction of the model

$$X_{n+1} = \mu + X_n + W_{n+1} \tag{5.14}$$

dubbed random walk with drift, the mean $\mu = \mathbb{E}\{X_{n+1} - X_n\}$ being called the drift. Summing both sides of (5.14) for different values of n, we obtain the analog of (5.13)

$$X_n = X_0 + n\mu + W_1 + W_2 + \cdots + W_n\ , \tag{5.15}$$

which shows that a random walk with drift is equal to a pure random walk (i.e. without a drift) plus a linear function with slope equal to the drift μ.

Random walk and random walk with drift models are not stationary. A quick look at their acf's show that they decay very slowly. This behavior is very different from the fast decay of the acf's of the stationary models which we study in this book.

5.4.3 **Auto Regressive Time Series**

The random walk model satisfies the induction equation:

$$X_t = X_{t-1} + W_t$$

giving the value of the series at time t in terms of the preceding value at time $t-1$ and a noise term. This shows that X_t is a good candidate for a (least squares) regression on its past value X_{t-1}. We now define a large class of time series with this property.

A mean-zero time series $X = \{X_t\}_t$ is said to be auto-regressive of order p (with respect to a white noise $W = \{W_t\}_t$) if:

$$X_t = \phi_1 X_{t-1} + \phi_2 X_{t-2} + \cdots + \phi_p X_{t-p} + W_t \tag{5.16}$$

for some set of real numbers $\phi_1, \phi_2, \ldots, \phi_p$. More generally, we say that $X = \{X_t\}_t$ is auto-regressive of order p if there exists a number μ_X (which will necessarily be the common mean of the random variables X_t) such that the series $\{(X_t - \mu_X)\}_t$ is auto-regressive of order p in the sense given above. In any case, we use the notation $X \sim AR(p)$. AR models are very important because of their simplicity, and because of the efficient fitting algorithms which have been developed. We shall give several example of their usefulness.

Identification of the Coefficients

We explain how to estimate the coefficients of a stationary autoregressive model when the order of the model is known. See the next subsection for a discussion of several possible ways to find the order of the model. So we momentarily assume that we know the order of the autoregressive model, and we present a general strategy on a specific example: for the sake of definiteness, we assume that the order of the autoregressive series is equal to 2, and we try to estimate the coefficients of the model. This model is of the form:

$$X_t = \phi_1 X_{t-1} + \phi_2 X_{t-2} + W_t \tag{5.17}$$

and our goal is to estimate the values of ϕ_1 and ϕ_2 and possibly of the variance of the noise W_t. Multiplying both sides of this definition by X_t and taking expectations we get:

$$\begin{aligned}\mathbb{E}\{X_t^2\} &= \phi_1\mathbb{E}\{X_tX_{t-1}\} + \phi_2\mathbb{E}\{X_tX_{t-2}\} + \mathbb{E}\{X_tW_t\} \\ &= \phi_1\mathbb{E}\{X_tX_{t-1}\} + \phi_2\mathbb{E}\{X_tX_{t-2}\} + \phi_1\mathbb{E}\{W_tX_{t-1}\} \\ &\qquad + \phi_2\mathbb{E}\{W_tX_{t-2}\} + \mathbb{E}\{W_tW_t\}\end{aligned}$$

after re-injecting formula (5.17) into the last expectation of the first equality. First, we notice that $\mathbb{E}\{W_tX_{t-1}\} = 0$ and $\mathbb{E}\{W_tX_{t-2}\} = 0$ because X_{t-1} and X_{t-2} depend only on the past values W_{t-1}, W_{t-2}, $W_{t-3}, \ldots$ which are independent of W_t. Next we rewrite the remaining expectations in terms of the auto-covariance function γ_X. We get:

$$\gamma_X(0) = \phi_1\gamma_X(1) + \phi_2\gamma_X(2) + \sigma^2 \tag{5.18}$$

if we denote by σ^2 the variance of the white noise. Next, multiplying both sides of (5.17) by X_{t-1} and taking expectations we get:

$$\mathbb{E}\{X_tX_{t-1}\} = \phi_1\mathbb{E}\{X_{t-1}X_{t-1}\} + \phi_2\mathbb{E}\{X_{t-2}X_{t-1}\} + \mathbb{E}\{W_tX_{t-1}\}$$

or, in terms of the auto-covariance function γ_X:

$$\gamma_X(1) = \phi_1\gamma_X(0) + \phi_2\gamma_X(1), \tag{5.19}$$

where, as before, we used the fact that $\mathbb{E}\{W_tX_{t-1}\} = 0$ which holds because W_t and X_{t-1} are uncorrelated (remember that X_{t-1} is a function of the W_s for $s \le t-1$). Finally, multiplying both sides of (5.17) by X_{t-2} and taking expectations as before, we get:

$$\mathbb{E}\{X_tX_{t-2}\} = \phi_1\mathbb{E}\{X_{t-1}X_{t-2}\} + \phi_2\mathbb{E}\{X_{t-2}X_{t-2}\} + \mathbb{E}\{W_tX_{t-2}\}$$

or, equivalently:

$$\gamma_X(2) = \phi_1\gamma_X(1) + \phi_2\gamma_X(0) \tag{5.20}$$

since $\mathbb{E}\{W_tX_{t-2}\} = 0$ as well. Summarizing what we just did, assuming that we know the first few values of the auto-covariance function, the numbers $\gamma_X(0)$, $\gamma_X(1)$ and $\gamma_X(2)$ to be specific, we derived a system of three equations, (5.18), (5.19), and (5.20), from which we can solve for ϕ_1, ϕ_2 and σ^2. These equations are called the *Yule-Walker equations* of the model. It is straightforward to solve them to obtain estimates of the parameters of the model from the values of the estimates of the auto-covariance function given by the sample auto-covariance function. This method is not limited to the analysis of the particular example (5.17) of order 2. It is quite general. In fact, this is the standard way to fit an auto-regressive model to sample data.

Finding the Order

In some sense, estimating the order of an AR is a special case of the estimation of the dimension of a linear model. So no one will be surprised if the method of choice for such an estimation is based on a parsimonious balance of fit to the data and the dimension of the proposed model (i.e. number of parameters needed). When fitting general models, as in the case of linear models, information criteria are most commonly used. We shall abide by the rule and use a form of the AIC criterion. However, despite its rather universal character, the AIC criterion is subsumed by more powerful tools in specific cases. As we are about to explain, theoretical properties of the partial auto-correlation function will identify features to look for in order to make sure that an autoregressive model is appropriate, while at the same time, giving us a sharp estimate of the order of the model. Moreover, when we try to fit a moving average model in next subsection, we will see that a vanishing acf will be an indication for a moving average model, and the lag at which it vanishes will provide a sharp estimate of the order.

As we are about to see, the best tool for identifying the order of an AR series is the partial auto-correlation function.

Using the Partial Auto-correlation Function to Find the Order

The concept of auto-regressive process can be used to enlighten the definition of partial auto-correlation function. Indeed, the definition (5.16) of an auto-regressive process and the definition (5.3) of the best linear prediction operator say that the partial auto-correlation coefficient $\phi_{k,k}$ is the last coefficient obtained when one tries to force an AR(k) model on $\{X_t\}_t$ (whether or not $\{X_t\}_t$ is an auto-regressive process). For this reason, if $X \sim AR(p)$, the we should have:

$$\phi_{k,k} = 0$$

whenever $k > p$. This property has very important practical applications, for, if one tries to find out if a time series $x_0, \ldots, x_n$ is a sample from an AR(p), and if one can estimate the partial auto-correlation coefficients $\phi_{k,k}$'s from the data, these estimates should be zero (or essentially zero), whenever k is greater than the order p. This property will be used to suggest auto-regressive models with a specific order. Remember that, according to (5.4), (5.5) and (5.6), γ_X and $\hat{\gamma}_X$ completely determine the partial acf's $\{\phi_{kk}\}_k$ and $\{\hat{\phi}_{kk}\}_k$ respectively.

In `S-Plus`, the partial auto-correlation function is computed by including the option `''partial''` in the command `acf`. This option instructs the program to evaluate formula (5.6).

Prediction

Let us assume that the time series $\{X_t\}_t$ is an AR(p) process for which we know the order p, the coefficients $\phi_1, \ldots, \phi_p$ and the variance σ^2 of the noise. Let us also assume that we have observed the outcomes $X_s = x_s$ of the series up to now, i.e. for $s \leq t$. The issue now is to predict the future values $X_{t+1}, X_{t+2}, \ldots$ of the series. We shall denote by $\widehat{X}_{t+1|t} = E_t(X_{t+1}), \widehat{X}_{t+2|t} = E_t(X_{t+2}), \ldots$ these predictions. Recall the notation $E_t(\,\cdot\,)$ for the prediction operator introduced earlier. As we already mentioned, it could be interpreted as the conditional expectation given the information of all the past observations up to and including time t, or as the orthogonal projection onto the linear space generated by all the random variables known at that time. In any case, rewriting the definition of the AR(p) model as:

$$X_{t+1} = \phi_1 X_t + \cdots + \phi_p X_{t-p+1} + W_{t+1}$$

and applying the prediction operator to both sides of the definition we get:

$$\widehat{X}_{t+1} = \phi_1 X_t + \cdots + \phi_p X_{t-p+1}$$

since $E_t(W_{t+1}) = 0$ by the definition of white noise, and since $E_t(X_s) = X_s$ if $s \leq t$. Similarly, rewriting the definition as:

$$X_{t+2} = \phi_1 X_{t+1} + \phi_2 X_t + \cdots + \phi_p X_{t-p+2} + W_{t+2}$$

and applying once more the prediction operator to both sides, we get:

$$\begin{aligned}
\widehat{X}_{t+2} &= \phi_1 \widehat{X}_{t+1} + \phi_2 X_t + \cdots + \phi_p X_{t-p+2} \\
&= \phi_1(\phi_1 X_t + \cdots + \phi_p X_{t-p+1}) + \phi_2 X_t + \cdots + \phi_p X_{t-p+2} \\
&= (\phi_1^2 + \phi_2) X_t + (\phi_1\phi_2 + \phi_3) X_{t-1} + \cdots + \phi_1\phi_p X_{t-p+1}.
\end{aligned}$$

This procedure can be repeated at will, and we can compute the prediction $\widehat{X}_{t+k}$ for any prediction horizon k. Such a prediction appears as a linear combination of the p most-recently observed values of the series with coefficients computed inductively from the coefficients of the model. Also, because these predictions are given by explicit formulae, one can compute a confidence interval for each of them. Unfortunately, these predictions converge very fast toward the mean of the series, zero in the present situation. So predictions will be uninstructive (i.e. plainly equal to the mean) for far prediction horizons.

The recursive formulae giving the predictions for all the finite horizons are easy to program. They are implemented in `S-Plus` by the function `pred.ar`. We illustrate its use, together with the anti-climatic convergence of the prediction toward the mean of the signal, in Subsection 5.6.3 when we discuss options on the temperature.

Monte Carlo Simulations & Scenarios Generation

Random simulation should not be confused with prediction. For example, it would be unreasonable to add a white noise to a series of predicted values to create a Monte Carlo sample from the series. Proceeding in this way may seem silly, but it is a common error with novices.

For the sake of the present discussion, we assume as before that we know the order and the parameters of the model, as well as the values of X_t, X_{t-1}, ..., X_{t-p+1}, and that we would like to generate N Monte Carlo sample scenarios for the values of X_{t+1}, X_{t+2}, ..., X_{t+M}.

The correct procedure is to generate N samples of a white noise time series of length M, say $\{W_s\}_{s=t+1,\dots,t+M}$, with variance σ^2, and then to use the parameters ϕ_1, ϕ_2, ..., ϕ_p and the definition of the AR(p) model to generate samples of the AR model from these N samples of the white noise. In other words, for each of the N given samples $W_{t+1}, \dots, W_{t+M}$ of the white noise, we generate the corresponding Monte Carlo scenarios of the series (which we denote with a tilde) by computing recursively the values $\widetilde{X}^{(j)}_{t+1}, \dots, \widetilde{X}^{(j)}_{t+M}$ from the formula:

$$\widetilde{X}^{(j)}_{t+k} = \phi_1 \widetilde{X}^{(j)}_{t+k-1} + \phi_2 \widetilde{X}^{(j)}_{t+k-2} + \cdots + \phi_p \widetilde{X}^{(j)}_{t+k-p} + W_{t+j}, \qquad k = 1, 2, \dots, M,$$

for $j = 1, 2, \dots, N$, given the fact that the "tilde"s over the X's (i.e. the simulations) are not needed when the true values are available.

This simulation procedure is very easy to implement. `S-Plus` does it as part of a more general procedure called `arima.sim` which can be used for more general ARIMA models. We give the details of its use later in Subsection 5.5.2.

5.4.4 **Moving Average Time Series**

A time series $X = \{X_t\}_t$ is said to be a moving average time series of order q (with respect to a white noise $W = \{W_t\}_t$) if:

$$X_t = W_t + \theta_1 W_{t-1} + \theta_2 W_{t-2} + \cdots + \theta_q W_{t-q} \tag{5.21}$$

for some real numbers θ_1, θ_2, ..., θ_q. In such a case we use the notation $X \sim MA(q)$.

While the definition formula of an AR process was recursive, the definition formula (5.21) is explicit in terms of the white noise, and as a consequence, some properties of the MA time series can be derived easily from the very form of this formula.
• MA *series are stationary.* This is an immediate consequence of the stationarity of the white noise and the fact that X_t bears to $(W_t, W_{t-1}, \dots, W_{t-q})$ the same relation as X_{t+t_0} to $(W_{t+t_0}, W_{t+t_0-1}, \dots, W_{t+t_0-q})$;
• *The auto-covariance function and the auto-correlation function of an MA(q) series vanish for lags greater than q.* Indeed, the definition formula (5.21) shows that X_{t+s} only depends upon W_{t+s}, W_{t+s-1}, ..., W_{t+s-q}. Consequently, if $t + s - q > t$

(i.e. if $s > q$), this set of white noise terms is disjoint from the set of terms on which X_t depends. This shows that in this case, the random variables X_{t+s} and X_t are independent and $\gamma_X(s) = \text{cov}\{X_{t+s}, X_t\} = 0$.

As we stated earlier, this last property will be instrumental in the identification of the order of the model.

Prediction and Simulation

Simulation of an MA process is trivial. Indeed it is straightforward to follow the definition of the process to generate the values of the desired samples of the MA from the samples of the white noise. The prediction problem is more delicate, and we shall not dwell on it here because of its technical nature. Nevertheless, anticipating a little bit what comes next, we would simply say that, if it were possible to invert an MA model and rewrite it as an AR model (quite likely with different coefficients) then one could use the procedure reviewed above in the case of AR processes. This off-the-wall idea, is not too far-fetched, and it is worth keeping it in mind when one reads the section on invertibility.

A Simulation Example

Even though the auto-correlation function is a powerful tool in itself, it is often very instructive to visualize the serial correlations in a graphical way. We illustrate this fact with the simple example of a simulated MA(2) series. The following `S-Plus` - commands create a sample of size 1024 from a normally distributed moving average time series with coefficients $\theta_1 = \theta_2 = 1$ and unit noise variance $\sigma^2 = 1$. This simulation is a naive implementation of the definition of a moving average process. We shall see later in this chapter, that `S-Plus` offers more powerful simulation methods for general ARIMA processes.

```
> SNOISE <- rnorm(1026)
> MA2 <- SNOISE[1:1024] + SNOISE[2:1025] + SNOISE[3:1026]
> tsplot(SNOISE[1:1024])
> tsplot(MA2)
```

The plots of these two series are reproduced in Figure 5.14. The moving average series appears to be *smoother* than the white noise. As we pointed out in the Simulation part of Subsection 5.4.1, a comparison based on the roughness or smoothness of the series can be very misleading when series are plotted with different time scales. Fortunately, this is not the case here, and this impression is real. The plot of the auto-correlation function of the `MA2` series is given by the `S`-command:

```
> acf(MA2)
```

The result is given in Figure 5.15. It confirms what we already learnt: the auto-correlation function of an MA(2) time series vanishes after lag 2. We now comment

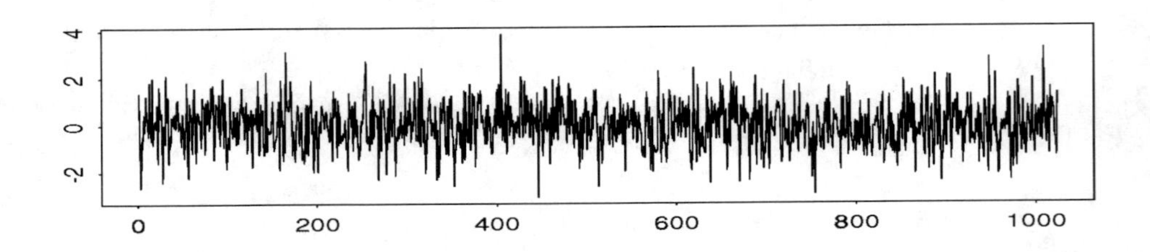

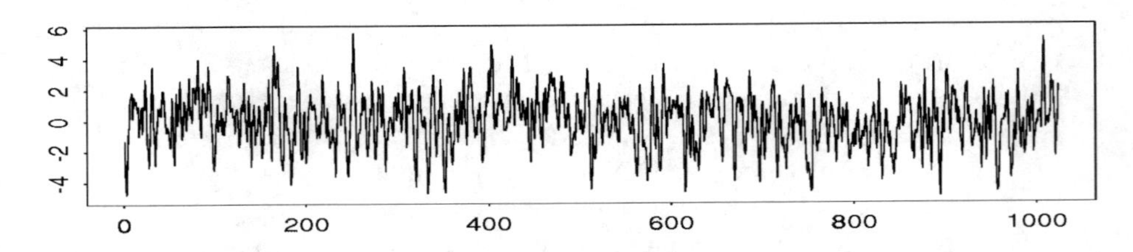

Fig. 5.14. White noise `SNOISE` (top) and a serially correlated MA(2) series constructed from `SNOISE` (bottom).

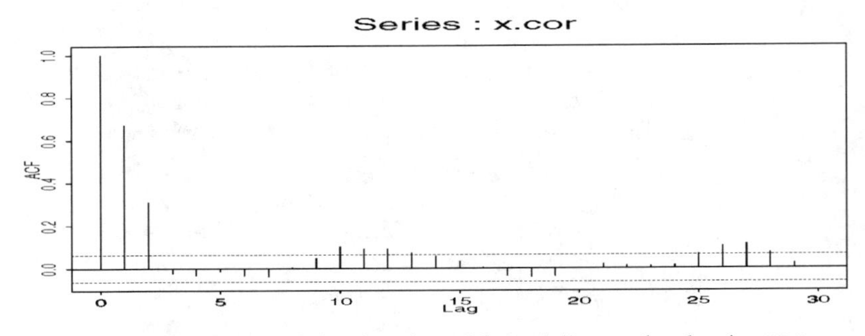

Fig. 5.15. Auto-correlation function of the serially correlated series `MA2`

on Figure 5.16 which shows four scatterplots. It was produced with the command:

```
> lag.plot(MA2,lags=4,layout=c(2,2))}
```

From left to right and top to bottom, they show the scatterplots of all the possible values of the couples (X_t, X_{t-1}), the couples (X_t, X_{t-2}), the couples (X_t, X_{t-3}), and finally the couples (X_t, X_{t-4}) which can be formed from the data. Obviously, the point pattern in the first scatterplot indicates that any two successive entries in the series are correlated. The second scatterplot shows that a significant dependence still exists between entries two time lags apart. However, this dependence does not seem to be as strong. On the other hand, the circular patterns characteristic of the last two scatterplots suggest that there should be no correlation between entries three and four time units apart. This confirms what we learned from the serial correlations in a moving average process.

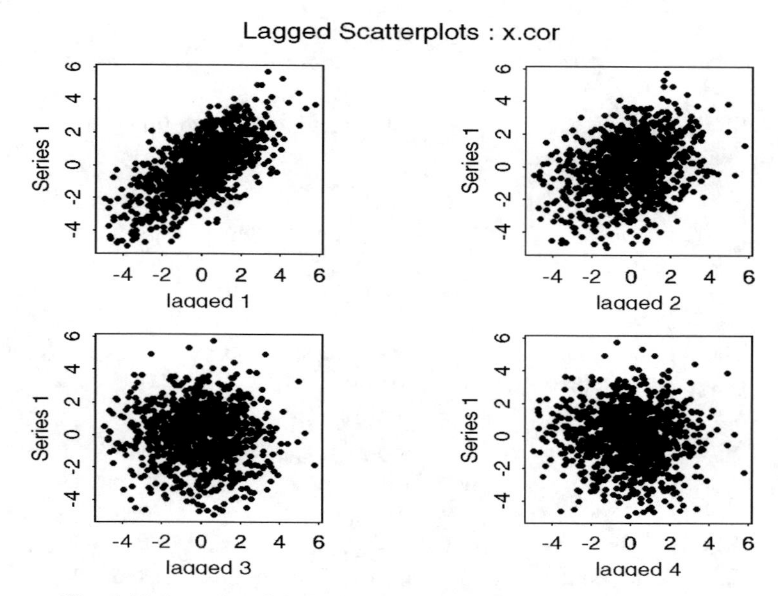

Fig. 5.16. Lag-plot of the series MA2 showing the serial dependence.

5.4.5 Using the Backward Shift Operator B

The definitions given above (as well as the computations to come) provide another justification for the introduction of the backward shift operator B. If X is a time series, then BX is the time series which is equal to X_{t-1} at time t. In other words,

$$BX_t = X_{t-1}.$$

The operator B so defined is called the backward shift operator. For the record we notice that the differentiation operator ∇ can be rewritten in terms of the backward shift operator. Indeed, we have;

$$\nabla = I - B,$$

which follows from the fact that $\nabla X_t = X_t - X_{t-1} = IX_t - BX_t = (I - B)X_t$, if we denote by I the identity operator which leaves the time series unchanged. Applying the operator B twice to the time series X, we get:

$$B^2X_t = BBX_t = BX_{t-1} = X_{t-2}.$$

By induction, we get that $B^kX_t = X_{t-k}$. The definition (5.16) of an auto-regressive time series of order p can be rewritten in the form:

$$\phi(B)X_t = W_t \qquad (5.22)$$

where the function ϕ is the polynomial defined by:

$$\phi(z) = 1 - \phi_1 z - \phi_2 z^2 - \cdots - \phi_p z^p. \tag{5.23}$$

Similarly, the definition (5.21) of a moving average time series of order q can be rewritten in the form:

$$X_t = \theta(B) W_t \tag{5.24}$$

where the function θ is the polynomial defined by:

$$\theta(z) = 1 + \theta_1 z + \theta_2 z^2 + \cdots + \theta_q z^q. \tag{5.25}$$

The forthcoming analysis of auto-regressive and moving average time series relies heavily on the properties of these polynomials.

5.4.6 **Linear Processes**

The definitions of autoregressive and moving average models seem to indicate that these models are very different. This first impression is somewhat misleading, for in fact they are both members of the same family of linear models which we are about to define. Again, we restrict ourselves to the case of mean-zero time series for the sake of simpler presentation. A mean-zero time series $X = \{X_t\}_t$ is said to be a linear process if there exists a white noise $W = \{W_t\}_t$ such that the representation:

$$X_t = \sum_{j=-\infty}^{\infty} \psi_j W_{t-j} \tag{5.26}$$

holds for some (possibly doubly infinite) sequence $\{\psi_j\}_j$ of real numbers satisfying $\sum_j |\psi_j| < \infty$. Restated in a developed form, the definition of a linear process says that:

$$X_t = \cdots\cdots + \psi_{-2} W_{t+2} + \psi_{-1} W_{t+1} + \psi_0 W_t + \psi_1 W_{t-1} + \psi_2 W_{t-2} + \cdots\cdots$$

and the condition imposed on the $|\psi_j|$'s is only here to guarantee that this doubly infinite sum has a meaning. Notice that the fact that this doubly infinite series converges is not a trivial fact, for the sequence of white noise terms W_t is not bounded in general, and consequently, multiplying it by a summable sequence may not be enough to give a meaning to the sum. It is only because of the independence of the terms of the white noise series, that the above definition of X_t is meaningful.

Linear processes are stationary by construction. Indeed, shifting the time indices for X amounts to shifting the time indices for the white noise and, since a time shifted white noise is still a white noise, the shifted version of X bears to the shifted white noise the same relationship as X bears to the original white noise W, so its distribution is the same. A particularly interesting class of linear processes is given by the processes for which $\psi_j = 0$ whenever $j < 0$. These processes are called *causal*. They are characterized by the fact that they are functions of the past of the white noise. Indeed, they are of the form:

$$X_t = \psi_0 W_t + \psi_1 W_{t-1} + \psi_2 W_{t-2} + \cdots\cdots .$$

In particular, we see that any moving average time series is causal since it is of this form by definition, the sum being in fact a finite sum.

Auto-covariance Function

The computation of the auto-covariance function of a linear process goes as follows:

$$\begin{aligned}
\gamma_X(k) = \mathbb{E}\{X_t X_{t-k}\} &= \mathbb{E}\left\{\left(\sum_{i=-\infty}^{+\infty} \psi_i W_{t-i}\right)\left(\sum_{j=-\infty}^{+\infty} \psi_j W_{t+k-j}\right)\right\} \\
&= \sum_{i=-\infty}^{+\infty} \sum_{j=-\infty}^{+\infty} \psi_i \psi_j \mathbb{E}\{W_{t-i} W_{t+k-j}\} \\
&= \sigma^2 \sum_{\substack{-\infty<i,j<+\infty \\ t-i=t+k-j}} \psi_i \psi_j \\
&= \sigma^2 \sum_{i=-\infty}^{+\infty} \psi_i \psi_{k+i} \qquad (5.27)
\end{aligned}$$

Despite its apparent complexity, formula (5.27) can be extremely useful when it comes to actually computing the auto-covariance functions of the AR, MA and ARMA models.

Obviously, one can use formula (5.27) to compute the auto-covariance function of a MA(q) process just by inspection. Indeed one sees immediately that $\gamma_X(k)$ will be zero whenever $|k| > q$. Otherwise the value of $\gamma_X(k)$ is given by formula (5.27) where the summation is only taken from $i = 0$ to $i = q - |k|$.

It is also possible to use formula (5.27) to obtain a closed-form formula for the auto-covariance function of AR(p) and ARMA(p,q) models when p is small.

5.4.7 Causality, Stationarity and Invertibility

Recall that in this section we concern ourselves with time series $X = \{X_t\}_t$ whose definition relies on a white noise series $W = \{W_t\}_t$.

We noticed that MA(q) time series were stationary and causal by definition. What about the AR(p) series? In order to answer this question we consider first the particular case $p = 1$ for the sake of illustration. Iterating the definition we get:

$$\begin{aligned}
X_t &= W_t + \phi_1 X_{t-1} \\
&= W_t + \phi_1 (W_{t-1} + \phi_1 X_{t-2}) \\
&= W_t + \phi_1 W_{t-1} + \phi_1^2 X_{t-2}.
\end{aligned}$$

We can continue the procedure, using the definition to replace X_{t-2}. We get:

$$X_t = W_t + \phi_1 W_{t-1} + \phi_1^2 (W_{t-2} + \phi_1 X_{t-3})$$
$$= W_t + \phi_1 W_{t-1} + \phi_1^2 W_{t-2} + \phi_1^3 X_{t-3}$$

and reiterating the substitution n times we get:

$$X_t = W_t + \phi_1 W_{t-1} + \phi_1^2 W_{t-2} + \cdots + \phi_1^n W_{t-n} + \phi_1^{n+1} X_{t-n-1}.$$

At this stage we would like to take the limit $n \to \infty$, and get a representation of the form:

$$X_t = W_t + \phi_1 W_{t-1} + \phi_1^2 W_{t-2} + \cdots + \phi_1^n W_{t-n} + \cdots \tag{5.28}$$

which would guarantee, if the above infinite sum converges, that the time series X is linear (thus stationary) with a causal representation. The above sum is convergent when $|\phi_1| < 1$, while it diverges in the case $|\phi_1| > 1$. This result is due to the geometric nature of the sequence of coefficients of the shifted white noise. In the case $\phi_1 = 1$, X_t is a random walk, and we know that we cannot have stationarity. The case $\phi_1 = -1$ is the same because the minus sign can be absorbed in the white noise term.

The above result is quite general when restated appropriately. In full generality it says that, an AR(p) time series is stationary and causal when the infinite series (5.28) converges, and this is the case when the (complex) roots of the polynomial $\phi(z) = 0$ lie outside the unit disk of the complex plane (i.e. are of modulii greater than 1).

To check that this is not different from what we just saw in the case $p = 1$, notice that in this case the polynomial $\phi(z)$ is given by $\phi(z) = 1 - \phi_1 z$, and consequently there is only one root, since there is only one root of the equation $1 - \phi_1 z = 0$, namely the number $z = 1/\phi_1$. We saw earlier that the series was stationary and causal if and only if $|\phi_1| < 1$. This is the same thing as $|1/\phi_1| > 1$, which is the condition given in terms of the roots of the polynomial $\phi(z)$.

When it makes sense, formula (5.28) is called the moving average representation (or MA representation) of the autoregressive process X. The existence of this moving average representation can be understood in the following light. If we use the definition of an AR(p) time series in the form $\phi(B)X_t = W_t$, we can formally write:

$$X_t = \frac{1}{\phi(B)} W_t$$

and this would be the desired representation if the rational fraction $1/\phi(z)$ (which is just the inverse of a polynomial) could be written as an infinite power series (i.e. a series in powers of z). Indeed, if we revisit one more time the case $p = 1$, we see that $\phi(z) = 1 - \phi_1 z$ and so:

$$\frac{1}{\phi(z)} = \frac{1}{1 - \phi_1 z} = 1 + \phi_1 z + \phi_1^2 z^2 + \cdots$$

and, even though we shall not try to justify it, the root condition introduced earlier guarantees that the series converges.

The definition of causality states that X_t can be expressed as a function of W_t and its past values $W_{t-1}, W_{t-2}, \ldots$. Trying to give symmetric roles to the two series $X = \{X_t\}_t$ and $W = \{W_t\}_t$, one may wonder when is it the case that W_t can be expressed as a function of X_t and its past values $X_{t-1}, X_{t-2}, \ldots$. If this is the case we say that the time series X is *invertible.*

Recall that, according to its definition, a time series X is an AR(p) if it has the representation:

$$X_t - \phi_1 X_{t-1} - \phi_2 X_{t-2} - \cdots - \phi_p X_{t-p} = W_t.$$

This form of the definition says that an AR(p) series is always invertible in the sense of the above definition. What about moving average series? You have probably already noticed that some sort of duality between auto-regressive and moving average properties. This duality is real as we about to see one more time. The situation of moving average processes with respect to invertibility is the same as the situation of auto-regressive processes with respect to the causality. In order to see that clearly, we consider the simple case of $q = 1$. The definition of an MA(1) series is $X_t = W_t + \theta_1 W_{t-1}$. It can be rewritten as:

$$W_t = X_t - \theta_1 W_{t-1}.$$

Iterating this definition, we replace W_{t-1} and we get:

$$W_t = X_t - \theta_1(X_{t-1} - \theta_1 W_{t-2}) = X_t - \theta_1 X_{t-1} - \theta_1^2 W_{t-2}$$

and if we play the same substitution game over and over we get:

$$W_t = X_t - \theta_1(X_{t-1} - \theta_1 W_{t-2}) = X_t - \theta_1 X_{t-1} - \theta_1^2 X_{t-2} - \cdots \tag{5.29}$$

the sum of this infinite series making sense only when $|\theta_1| < 1$. When this is the case, formula (5.29) is called the auto-regressive representation (or the AR representation) of X. As in the case of moving average representations, this result is quite general. It applies to all the moving average processes as long as the coefficients are such that the above series converges, and this is indeed the case when all the (complex) roots of the characteristic polynomial $\theta(z)$ are outside the unit disk (i.e. are of modulus greater than 1).

Let us show the duality in action one more time. If we rewrite the definition of a MA(q) time series as $X_t = \theta(B)W_t$, then the method used above (in the particular case of a MA(1) model) to show invertibility of the model, can be viewed as a way to rewrite the left hand side of:

$$\frac{1}{\theta(B)} X_t = W_t$$

as an infinite series in positive powers of B. When this infinite series can be derived as a convergent infinite expansion, it provides an AR-representation on which the invertibility condition appears clearly.

	AR(p)	**MA(q)**
Stationarity Causality	$\lvert z\rvert > 1$ if $\phi(z) = 0$	always
Invertibility	always	$\lvert z\rvert > 1$ if $\theta(z) = 0$

Table 5.6. Table of the *invertibility, causality and stationarity* properties illustrating the duality between the auto-regressive and moving average processes.

Remarks

1. The need for the statistical estimates provided by time averages was convenient for justifying the importance of stationarity. At this stage of the analysis, it is practically impossible to justify the fuss about causality and invertibility: why would we care so much about the fact that X_t is a function of the past values of the noise, and why would we need to know that the noise W_t is in turn a function of the past values of the observed series. The importance of these issues will become clear in Chapter 6 when we consider partially observed systems, and filtering issues. For the time being, the reader will have to take our word for it: these matters are of crucial importance in practical applications, if not, we would have crystal balls to predict the market!
2. Notice that adding a constant to a model can have very different effects depending on the model in question. For example, adding a constant term to a regression amounts to adding an intercept. Adding a constant to the equation of a moving average model merely changes the constant mean of the series. On the other hand, adding a constant term to the equation of an auto-regressive model adds a linear drift, i.e. a linear function of time to the series.

Unit-Root Test

Let us assume that the logarithms of a stock price follow either one of the two models

$$X_t = \phi_1 X_{t-1} + W_t$$
$$X_t = \phi_0 + \phi_1 X_{t-1} + W_t$$

for some strong white noise $\{W_t\}$. Testing whether the log-price is a random walk is testing the null hypothesis $H_0 : \quad \phi_1 = 1$ against the alternative $H_1 : \quad \phi_1 < 1$. Notice that when, $\phi_0 = 0$, this alternative guarantees a stationary time series. Such a test is called a unit-root test. Given observations $x_1, \ldots, x_n$ of the log-prices, the least squares estimates of ϕ_1 and of the variance σ of the white noise are given in either model by:

$$\hat{\phi}_1 = \frac{\sum_{i=1}^n x_i x_{i-1}}{\sum_{i=1}^n x_{i-1}^2}, \quad \text{and} \quad \hat{\sigma}^2 = \frac{1}{n-1}\sum_{i=1}^n (x_i - \hat{\phi}_1 x_{i-1})^2 \tag{5.30}$$

where we set $x_0 = 0$ by convention. The quantity:

$$DF = \frac{\hat{\phi}_1 - 1}{\sqrt{\hat{\sigma}^2}} \tag{5.31}$$

is called the Dickey-Fuller test statistics. Its distribution converges when the sample size n grows indefinitely, and the limit can be identified both in the case $\phi_0 = 0$ and in the case $\phi_0 \neq 0$, leading to tests of the hypothesis H_0 against H_1. See Problem 5.9 for the computation of the critical values of the test.

5.4.8 **ARMA Time Series**

A time series $X = \{X_t\}_t$ is said to be an auto-regressive moving average time series of order p and q if there exists a white noise $W = \{W_t\}_t)$ such that:

$$X_t - \phi_1 X_{t-1} - \cdots - \phi_p X_{t-p} = W_t + \theta_1 W_{t-1} + \cdots + \theta_q W_{t-q} \tag{5.32}$$

for some real numbers $\phi_1, \ldots, \phi_p$, and $\theta_1, \ldots, \theta_q$. In such a case we use the notation $X \sim ARMA(p,q)$. Using the shift operator B this definition can be rewritten in the form:

$$\phi(B)X_t = \theta(B)W_t$$

for the polynomials $\phi(z)$ and $\theta(z)$ defined above in formulae (5.23) and (5.25) respectively. Since one can formally write:

$$X_t = \frac{\theta(B)}{\phi(B)} W_t \qquad \text{and} \qquad \frac{\phi(B)}{\theta(B)} X_t = W_t$$

one sees why the stationarity and causality properties only depend upon its auto-regressive part, while its invertibility only depends upon its moving average part. More precisely:

- the ARMA(p,q) series is stationary and causal if all the (complex) roots of $\phi(z)$ have modulii greater than 1;
- the ARMA(p,q) series is invertible if all the (complex) roots of $\theta(z)$ have modulii greater than 1,

since the use of the polynomials $\phi(z)$ and $\theta(z)$ shows that we have the stationarity and the causality of the ARMA(p,q) series if we can prove the stationarity and causality of its AR(p) part, while we have the invertibility of the ARMA(p,q) series if we can prove the invertibility of its MA(q) part.

Remark. By definition, the unit-root test introduced earlier was appropriate for AR models. An extension to ARMA models was proposed. It goes under the name of augmented Dickey-Fuller test. The interested reader is referred to the section Notes & Complements at the end of the chapter for bibliographical references.

5.4.9 **ARIMA Models**

A time series $\{X_t\}_t$ is said to be an ARIMA process if, when differentiated finitely many times, it becomes an ARMA time series. More precisely, one says that $\{X_t\}_t$ is an ARIMA(p,d,q) if its becomes an ARMA(p,q) after d differences. So using the notation introduced earlier:

$$X \sim ARIMA(p, d, q) \Longleftrightarrow \nabla^d X \sim ARMA(p, q).$$

Recall that the operator $\nabla = I - B$ is the first difference operator. Equivalently, using the definition of ARMA processes in terms of polynomials in the shift operator B we see that:

$$X \sim ARIMA(p, d, q) \Longleftrightarrow \phi(B)(I - B)^d X = \theta(B) W$$

for a white noise W and polynomials ϕ and θ.

5.5 FITTING MODELS TO DATA

The purpose of this section is to take advantage of the theoretical properties derived in the first part of the chapter in order to identify models appropriate for given time series data, and to fit these models to the data.

5.5.1 **Practical Steps**

We now summarize the steps recommended for fitting the models we discussed in this chapter.

Searching for Stationarity

The first step of a time series analysis is the identification and the removal of the trend and seasonal components. We should

1. Transform the data into a stationary time series if needed:
   - Remove trends (regression);
   - Remove seasonal components;
   - Differentiate successively;
2. Check if the data form a white noise, in which case the model fitting has to stop.

We shall see in the last chapter that some forms of white noise in the weak sense are still amenable to model fitting, but for the purposes of this chapter, reaching a white noise is our stop criterion.

Fitting an AR

In order to fit an AR model we should follow the steps given below:

1. Computation of the auto-correlation function and checking that it decays fast;
2. Computation of the AIC criterion and the partial auto-correlation function to determine the order of the model;
3. Estimation of the auto-regression coefficients and of the variance of the noise by solving the Yule-Walker equations;
4. Computation of the residuals and testing for white noise to decide if the model fitting is complete or if it needs to be pursued further.

Notice that solving the Yule-Walker equations is not the only way to estimate the coefficients of an AR model. Even though it is used by default in `S-Plus`, there exist other efficient methods. For example, `S-Plus` offers Burg's method as an alternative. It is based on spectral and information theoretic arguments. We shall not discuss them here.

Fitting an MA

In order to fit an MA model we should follow the steps given below:

1. Computation of the auto-correlation function and checking that it vanishes from some point on;
2. Confirmation of the order of the model with the AIC criterion;
3. Estimation of the moving average coefficients by maximum likelihood;
4. As always, computation of the residuals and testing for white noise to decide if the model fitting is complete or if it needs to be pursued further.

If the order of the model has already been determined, `S-Plus` has a method for determining the coefficient of the model. We shall see how to use this function below.

Fitting an ARMA

In order to fit an ARMA model it is recommended to follow the steps:

1. Attempt to fit an AR model to the data and computation of the residuals;
2. Attempt to fit an MA model to the residuals of the AR model fitted first, or to the original data if the AR fit was not deemed satisfactory;
3. Analysis of the residuals and testing for white noise.

Fitting an ARIMA

If the search for stationarity is done by successive differentiations, and if an ARMA model is fitted to the result, then we have fitted an ARIMA model, the order of which is the triplet (p, d, q) of integers, p standing for the order of the AR component, d for the number of differences computed to get to stationarity, and q for the order of the MA component.

So like Mister Jourdain was producing prose without knowing it, we have been fitting ARIMA models by first differentiating a time series until we reached a stationary series, and then fitting an ARMA model to the result.

5.5.2 `S-Plus` Implementation

We illustrate the presciptions given above with the analysis of a couple of simple examples.

Fitting an AR Model

In order to illustrate how one fits an AR model, we revisit the example of the CO_2 concentration data. Recall that `co2d` is the stationary time series we derived after removing the seasonal component and de-trending the original data. We compute the acf and the partial acf of this stationary series with the commands:

```
> acf(co2d)
> acf(co2d,''partial'')
```

The results are reproduced in Figure 5.17. The auto-correlation function seems to decay exponentially, so by computing and plotting the partial auto-correlation function we may be able to confirm our suspicion that an auto-regressive model could be appropriate. The partial auto-correlation function vanishes after lag 2. This reinforces our hunch for an AR model. In that case, the order could be 1 or 2. We fit an AR model with the command `ar` without specifying the parameter `order`.

```
> co2dar <- ar(co2d)
> co2dar$order
[1] 1
> co2dar$ar
              [,1]
[1,]     -0.2608328
> tsplot(co2dar$aic)
```

This prompts `S-Plus` to try a certain number of models with different orders, and choose an order by a criterion which can be set via a parameter in the call of the function `ar`. This criterion is chosen to be the AIC criterion by default. It is recommended to choose the order giving the minimum value to the AIC criterion *minus one*. So in the present situation, Figure 5.18 shows that the order picked by the AIC criterion is 1. As after any other fitting procedure, the next step is to examine the residuals. The plot of their auto-covariance function is given in Figure 5.19. It shows that there is no correlation left, and that pursuing the modeling with an MA part may not be successful. See nevertheless our discussion of the ARCH and GARCH models in Chapter 7.

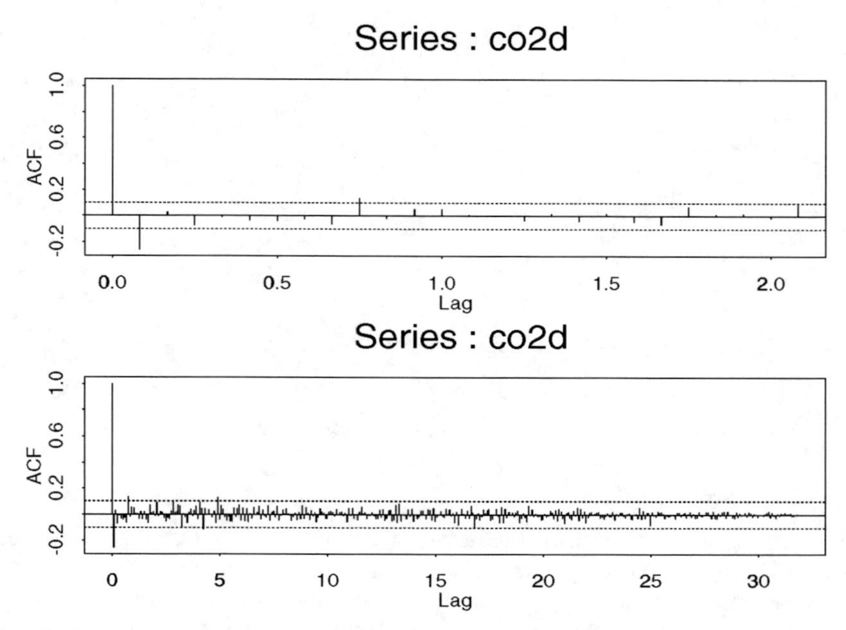

Fig. 5.17. Auto-correlation (top) and partial auto-correlation (bottom) functions of the stationary time series obtained by removing the seasonal component and de-trending the original CO_2 time series.

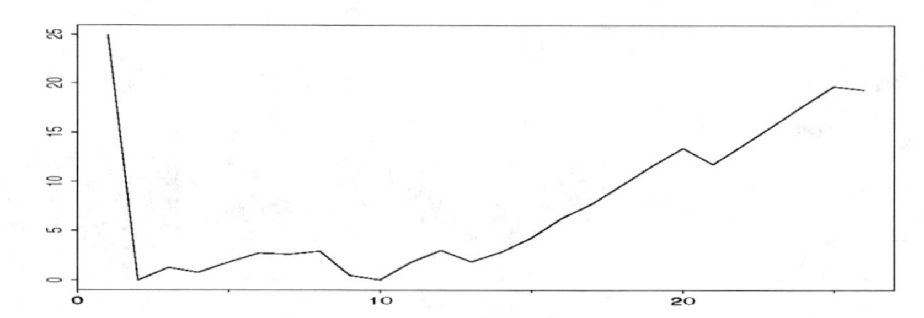

Fig. 5.18. Sequential plot of the AIC (Akaike Information Criterion) for the CO^2 AR model.

Fitting a Moving Average Model

Instead of using a real life time series for which we never know whether or not a model is appropriate, we illustrate the fitting procedure with a simulated time series. In this case, we know in advance what to expect. We could use the series `MA2` constructed earlier, but in order to illustrate the simulation capabilities of `S-Plus` we choose to construct a brand new series. We use the function `arima.sim` to simulate a sample from an MA(4) model and we immediately compute the empirical estimate of the auto-correlation function of the time series generated in this way.

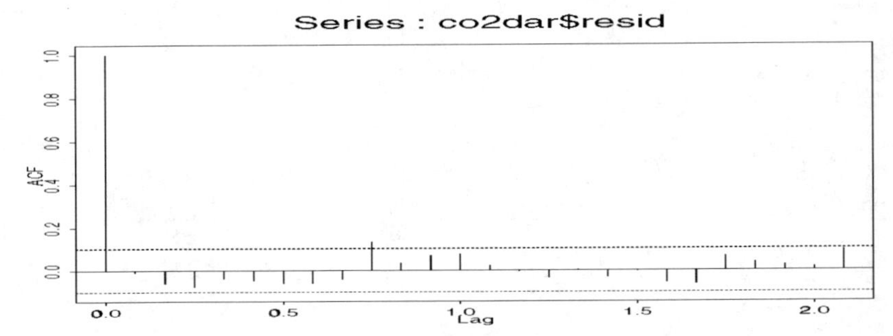

Fig. 5.19. Auto-correlation of the residuals of the CO_2 AR model.

```
> MA4 <- arima.sim(1024,model=list(ma=c(-.5,-.25,.1,.6)))
> tsplot(MA4)
> acf(MA4)
```

In `S-Plus`, ARIMA models are given by specifying a parameter called `model`. The latter is a list. Its `ar` component gives the coefficients ϕ_i of the AR-polynomial $\phi(z)$ while the component `ma` gives the values of the coefficients θ_j of the MA-polynomial $\theta(z)$. We give further details below. The two plots are given in Figure 5.20. The values of the auto-correlation function do not seem to be significantly different from zero after lag 4. This is consistent with a moving average model of order 4. Naturally, we then try to fit an MA(4) model. If we assume that we only know

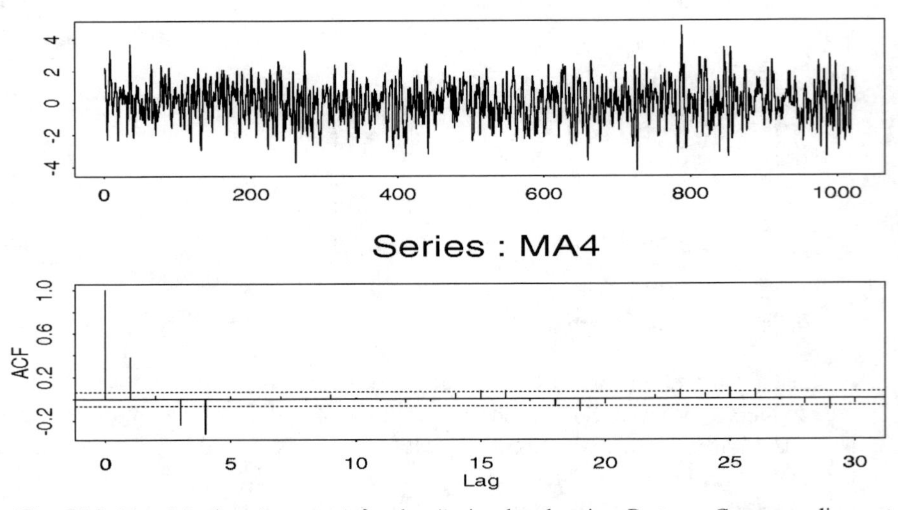

Fig. 5.20. Top: Moving Average (of order 4) simulated series. Bottom: Corresponding auto-correlation function. Notice that it is vanishing for lags larger than 4).

the order of the moving average (and not the actual coefficients), we fit a univariate MA(4) model with the `S-Plus` function `arima.mle` which will be discussed in the next subsection. Printing the results gives the following output:

```
> MA4FIT <- arima.mle(MA4,model=list(order=c(0,0,4)))
> MA4FIT
$model:
$model$order:
[1] 0 0 4
$model$ndiff:
[1] 0
$model$ma:
[1] -0.50235 -0.25128  0.08360  0.56694
$var.coef:
             ma(1)        ma(2)        ma(3)        ma(4)
ma(1)    0.0006627    0.0004443    0.0003845    0.0001965
ma(2)    0.0004442    0.0009023    0.0005881    0.0003845
ma(3)    0.0003845    0.0005881    0.0009022    0.0004442
ma(4)    0.0001965    0.0003845    0.0004443    0.0006627
$method:
[1] "Maximum Likelihood"
$series:
[1] "MA4"
$aic:
[1] 2977.876
$loglik:
[1] 2969.876
$sigma2:
[1] 1.061917
$n.used:
[1] 1024
$n.cond:
[1] 0
$converged:
[1] F
$conv.type: [1] "iteration limit"
```

The object returned by the function `arima.mle` contains the information about the model, together with the value of the AIC criterion, the maximum value of the log-likelihood as computed, the estimates of the coefficients (those at which the maximum of the likelihood was attained) and their variance/covariance matrix in case we need confidence intervals or confidence regions. The likelihood is maximized by an iterative procedure, and the output contains information about the convergence or lack thereof. The experiment is satisfactory in the sense that as expected, the estimated coefficients are reasonably close to the actual coefficients used to generate the MA sample.

Fitting an ARIMA model

If we already know the order of an ARIMA model, in other words, if we know the three integers (d for the order of differentiation, p for the auto-regressive order and q for the moving average order), then it is easy to fit an ARIMA model by maximum likelihood. The `S-Plus` function `arima.mle` will do that for us if we give the argument `order=c(p,d,q)`. The output is an arima model object. Besides the input parameters, it has an attribute called `model`. The latter is a list with `ar` and `ma` components which give the ϕ and θ coefficients of the ARMA part of the model. See the appendix at the end of this chapter for a discussion of the sign conventions used by `S-Plus`.

Remark. Clearly, the function `arima.mle` requires the knowledge of the order of the model. It cannot be used to fit an ARIMA model if one does not already know the order (p, d, q), it can only be used to fit an ARIMA model of a specific order.

Diagnostics

One of the most insidious mischiefs of powerful computer statistical packages is the fact that it is always possible to fit a model to data, even when the model in question is not appropriate. So it is a good practice, once a model is fitted, to run diagnostics in search of warnings. For example, after fitting an ARMA(2,1) model to a time series TS, the resulting model can be diagnosed by the `S-Plus` function `arima.diag`.

```
TS.fit <- arima.mle(TS,model=list(ar=2,ma=1))
TS.diag <- arima.diag(fit)
TS.diag$gof
```

Results of tests of goodness of fit of the model are given in the *goodness of fit* component `TS.diag$gof` which contains, among other things, results of the portmanteau test for lack of correlation in the residuals. This is the test for white noise which we alluded to in Subsection 5.4.1. The reader is invited to consult the help file for details on the meanings of the outputs.

Simulation

`S-Plus` offers a function to effortlessly produce samples of ARIMA time series. This function is called `arima.sim`. The parameters of the model need to be provided each time the function is called. These parameters should be in the form of the output of the function `arima.mle`.

```
> SIM <- arima.sim(model, n=100, innov=NULL, n.start=100)
```

The parameter `n` gives the length of the series to simulate, while `n.start` gives the number of generated values to discard. The use of `n.start` is desirable if we want to make sure that the simulation is representative of an equilibrium situation.

By default, the function `arima.sim` uses a normal random sequence created with the function `rnorm` as a sample for the white noise from which the sample of the ARIMA model is created. When the `innov` parameter is provided, the simulation is not based on such a normal random sequence. Instead, the sequence `innov` is used for that purpose. This is especially useful if we want to repeat simulations with different features while keeping the same source of randomness, but it is also convenient if we want to simulate series with a non-Gaussian white noise with heavy tails. We can use the simulation techniques developed in Chapter 1 to simulate samples from a heavy tail distribution, and use such a sample as the parameter `innov`.

Other parameters can be used. The interested reader is invited to check the help file of the function `arima.sim`. See our analysis of the Charlotte temperature below for an illustration of the use of simulation for Monte Carlo computations of probabilities and expectations.

Prediction

In the general case of an ARIMA model, the predictions of the future values are computed with the function `arima.forecast`. The parameters of the model need to be fed to the function. As in the case of the function `arima.sim`, the model parameters should be in the form of the output of the function `arima.mle` as seen above. Predictions are obtained with a command of the form:

```
> PRED <- arima.forecast(seriesData, model, n)
```

where `seriesData` is the data of the time series whose future values we want to predict, `model` is the ARIMA model in the format of the output of `arima.mle`, and `n` is the number of time periods to be predicted after the end of `seriesData`. The object `PRED` is a list containing a vector `PRED$mean` of length `n` for the actual predictions, and a vector `PRED$sdt.err` for the standard deviations of the estimates. See the analysis of the temperature at Charlotte NC for an example.

5.6 PUTTING A PRICE ON TEMPERATURE

There is a renewal of interest in temperature time series in many economic sectors. We believe that this current wave is mostly due to the great excitement caused by the growing use of weather derivatives for risk management, by the prospect of large speculative profits, and presumably by the mystery surrounding the difficult pricing issues associated with these instruments. A better understanding of the statistics of these series will help dissipate this mystery. This is what we attempt to do in this section.

Our goal is to develop an understanding of the statistics of these time series which would be appropriate for the pricing of meteorological derivatives. Options on heating and cooling degree days are very actively traded in the so-called *over the counter* (OTC for short) market. Moreover, as of September 22nd 1999, the Chicago Mercantile Exchange (CME, for short) started offering futures contracts on monthly indexes computed from the cumulative HDD's and CDD's in three cities, as well as options on these futures contracts. Contrary to our use of non-parametric regression to price options on the S&P 500 index, we shall not attempt to price these options directly. Price discovery is too much of a challenge in these markets, data are too scarce, and illiquidity problems are *muddying the water* too much for us to take a chance in this touchy business. The following two subsections are devoted to the introduction of the terminology and notation needed for the analysis of these new financial instruments.

5.6.1 **Generalities on Degree Days**

Temperature data are readily available, especially on the world wide web. For a very large number of locations, usually US meteorological stations (near or at airports), daily temperatures are provided free of charge. The information is given in the form of a high and a low for each day. We use the standard notation $MinT$ and $MaxT$ for the minimum and the maximum temperatures on a given day. These temperatures are expressed in Farenheit degrees. From these, one usually computes the so-called average temperature $AvgT$ defined as:

$$AvgT = \frac{MinT + MaxT}{2}. \tag{5.33}$$

This definition may not correspond exactly to the intuitive notion of the average temperature on a given day. Nevertheless, it is the quantity which is used to define the important concepts of heating degree day (HDD for short) and cooling degree day (CDD for short). On each given day, say t, the heating degree day HDD_t is defined as the number:

$$HDD_t = (65 - AvgT_t)^+ = \max\{65 - AvgT_t, 0\}. \tag{5.34}$$

The intuitive notion of heating degree day is simple: this number tells us by how many degrees one should heat on day t. The threshold at which one does not need to heat is (arbitrarily) fixed at 65. Apparently, this is the temperature at which the furnaces used to be turned on. But let's face it, the notion of temperature above which one should start the heater is very subjective: residents of Southern California and Minnesota may not agree on the choice of this threshold. In any case, the value of 65 is generally accepted, and it is used in all degree days computations. We shall accept it as is. Formula (5.34) says that $HDD_t = 0$ on days where the average temperature (as given by $AvgT_t$) is greater than 65, and that on the other days, HDD_t is equal to the amount by which the average temperature falls under 65. Similarly, the cooling degree day CDD_t is defined as the number:

$$CDD_t = (AvgT_t - 65)^+ = \max\{AvgT_t - 65, 0\}, \tag{5.35}$$

and its intuitive interpretation is similar. Some daily temperature services provide daily values for $AvgT$, HDD and CDD along with the values of $MinT$ and $MaxT$.

Given a time period P, which can be a month, or a summer, or any other kind of pre-specified time period, the cumulative numbers of degree days for this period are defined as:

$$HDD^{(P)} = \sum_{t \in P} HDD_t \qquad \text{and} \qquad CDD^{(P)} = \sum_{t \in P} CDD_t.$$

As we shall see shortly, these quantities are the *underlying indexes* on which the temperature futures and option contracts are written.

5.6.2 **Temperature Options**

There are many kinds of weather related financial instruments. We already mentioned catastrophic bonds when we discussed the PCS index in Chapter 1. In this section, we concentrate on options on the temperature. We list a few examples for the sake of illustration. Gas retailers are bound to lose money during mild winters because of low demand. Notice also that extremely cold winter can also be the source of losses since, tied by delivery contracts, the gas retailer may have to buy gas at a prohibitive price on the spot market to satisfy an unexpectedly high demand. Similarly, electric power utility companies suffer from cool summers because of lower demand, but they also suffer from too hot a summer when they are forced to purchase electricity on the spot market to satisfy an unexpectedly high demand. The highly speculative and volatile nature of the electricity spot market, especially in the de-regulated markets, has been a source of great concern both for users and producers. So it seems natural to expect that gas producers and retailers will try to protect themselves against warm winters, and electricity producers will try to protect themselves against cool summers. They are the *naturals* for a market which brings together the banking and the insurance sides of the financial industry. The instruments we discuss below have been the most popular since the inception of this market a few years ago. Even though we mentioned only energy providers as naturals for this market, the astute reader will reckon that any business exposed to weather risk (and according to some estimates this may represent up to 70% of the economy) should benefit from the existence of a liquid market for such instruments.

Futures contracts on monthly cumulative degree days have been introduced by the CME , but because of the small volume, we shall restrict our attention to the OTC market. There, the owner of an option will receive the amount $\xi = f(DD)$ at the end of the period P, the pay-off function f being computed on the cumulative index $DD = HDD^{(P)}$ for the HDD seasons or $DD = CDD^{(P)}$ for the CDD seasons. A CDD season typically starts May 15 and ends September 15. It can also start June 15. The HDD season starts November 15 (December 15 in some cases) and ends usually March 15. The remaining months are called the shoulder months, and the lack of underlying instruments during these months is a peculiarity of the OTC weather market.

Typical Examples of Pay-off Functions

Here are some examples of pay-off functions which have been used in the over the counter markets. Figure 5.21 gives the plots of two of the most popular pay-off functions.

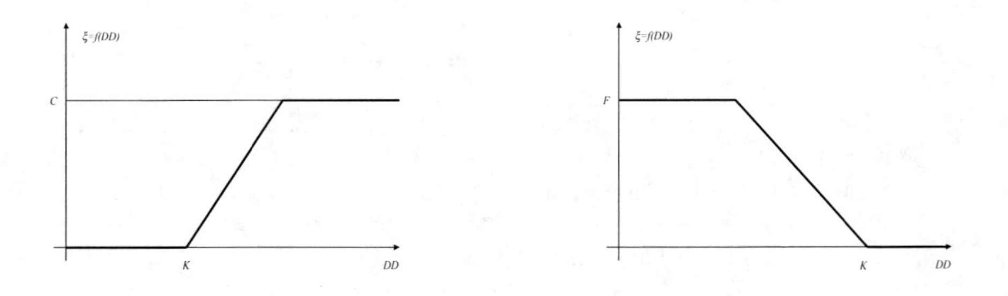

Fig. 5.21. Pay-off functions of a call with a cap (left) and a put with a floor (right).

Call with a Cap For these options, the pay-off $\xi = f(DD)$ is of the form:

$$\xi = \min\{\alpha(DD - K)^+, C\}. \tag{5.36}$$

The graph of the pay-off function f is given on the left pane of Figure 5.21. The buyer of such an option pays the up-front premium at the signature of the contract, and if $DD > K$ at the end of the strike period, the writer (seller) of the option pays the buyer the nominal pay-off rate α times the amount by which DD exceeds the strike, the overall payment being limited by the cap C. The values of US$ 2,500.00 and US $ 5,000.00 have been quite commonly used for the pay-off rate α, while typical values of the cap C are US$ 500,000.00 or US$ 1,000,000.00. See real life examples later in the text and in the problems at the end of the chapter.

The option is said to be *at the money* when the strike K is near the historical average of DD, and *out of the money* when K is far from this average. As far as we know, most of the options written so far have been deep out of the money: buyers are willing to give up some upside for some downside protection.

There is no agreement on what is a reasonable period over which to compute the so-called *historical average.* Where to strike an option? Some recommend to use no more than ten to twelve years in order to capture recent warming trends, while others suggest to use longer periods to increase the chances of having a stable historical average.

Example The marketing department of my favorite cruise line has known for quite some time that the summer sales suffer when the Spring is warm in the North East of the US. People do not suffer from cabin fever which usually follows cold winters, and they tend not to rush to warmer horizons for their summer vacations. So Royal Caribbean will want to buy an out of the money call on Spring CDD's, possibly with a cap since after all, it will sell a minimum number of cruises no matter what.

Put with a Floor For these options, the pay-off $\xi = f(DD)$ is given by:

$$\xi = \min\{\alpha(K - DD)^+, F\}. \tag{5.37}$$

The graph of the pay-off function f is given on the right pane of Figure 5.21. In exchange for the premium, the buyer of such an option receives, at the end of the strike period, if $DD < K$, α times the amount by which the strike exceeds DD, the overall payment being limited by the floor F.

Example In order to hedge the risk of a warm winter a heating oil provider will buy a put on HDD's over the winter season.

Collar Buying a collar contract is essentially equivalent to being long a call with a cap and short a put with a floor. More precisely, the pay-off is given by:

$$\xi = \min\{\alpha(DD - K_c)^+, C\} - \min\{\beta(K_p - DD)^+, F\}. \tag{5.38}$$

The graph of the pay-off function f is given in Figure 5.22. It is possible to adjust all the parameters $\alpha, \beta, C, F, K_c, K_p$ for the premium to be zero. Since there is no up-front cost to get into one of these contracts, this form of collar is quite popular.

Example In order to protect its revenues against the possible losses of a mild Winter, a local Gas Company will bet into a collar contract with no up-front payment. It will pay the counter party if the Winter is cold (which is not a problem since it will enjoy the income from gas sales) and the counter party will pay the Gas Company if the Winter is warm (offsetting the losses due to slow gas sales).

The weather derivatives introduced above can be viewed as European *vanilla* options, if we view them as written on an underlying of the form $HDD^{(P)}$ or $CDD^{(P)}$. But it might be easier to view them as written on the temperature $AvgT_t$ as underlying. In this case, they are more of the Asian type than of the European type since the payoff is computed on the average of the underlying number of daily degree days over a period ending at maturity. For this reason, pricing may be more difficult in this case.

It is clear that a good understanding of the underlying is necessary before we attempt to price some of these derivatives. In any case, the time evolution of the underlying index, as evidenced for example by Figure 5.23, is very different from the time evolution of the indexes and the stock prices considered so far. This is a strong indication that Samuelson's geometric Brownian motion model is not appropriate, and that we need a better grasp of the statistics of the underlying index before we can proceed to the analysis of these derivatives.

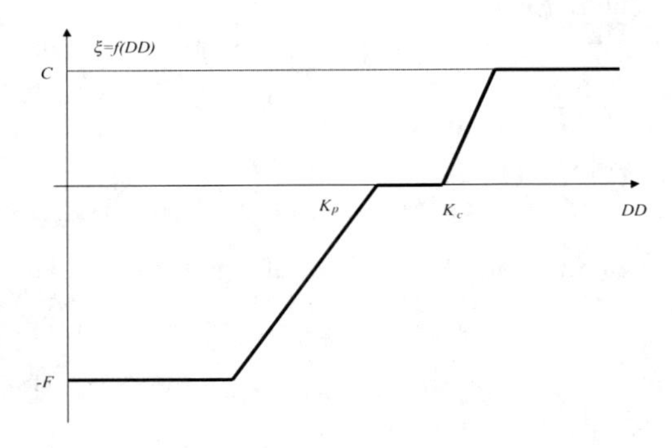

Fig. 5.22. Pay-off function of a collar.

5.6.3 Statistical Analysis of Temperature Historical Data

We propose to illustrate the major issues in this modeling effort by the detailed analysis of the temperature at a given specific meteorological station. We chose the location of Charlotte, NC for the sake of illustration. The temperature information was encapsulated in an `S-Plus` `timeSeries` object called `Charlotte.ts` containing the daily minimum, maximum and average temperatures from January 1, 1961 to March 31, 1999. Its first five entries look like:

```
> Charlotte.ts[1:5,]
 Positions MinT MaxT AvgT
 01/01/1961 35   55   45
 01/02/1961 28   52   40
 01/03/1961 29   43   36
 01/04/1961 22   52   37
 01/05/1961 24   58   41
```

and the plot of the entire time series of average temperatures (i.e. the last column) is given on the top of Figure 5.23, while the bottom part of this figure zooms in on the first ten years. As expected, these plots show the existence of a strong yearly seasonal component.

Identifying and Removing the Seasonal Component

The first order of business is to identify the trends and seasonal components. As in the case of the analysis of the CO_2 concentration data, we use our function `sstl` to identify and extract the seasonal component.

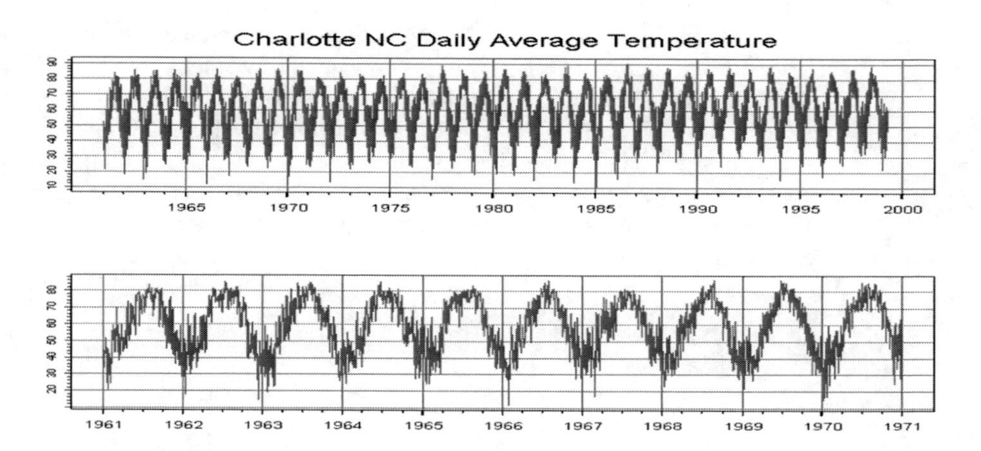

Fig. 5.23. Average daily temperature measured at Charlotte, NC.

```
> Charlotte.stl <- sstl(Charlotte.ts[,3])
> plot(Charlotte.stl$sea)
> plot(Charlotte.stl$rem)
```

The results are reproduced in Figure 5.24. Notice that the seasonal component was

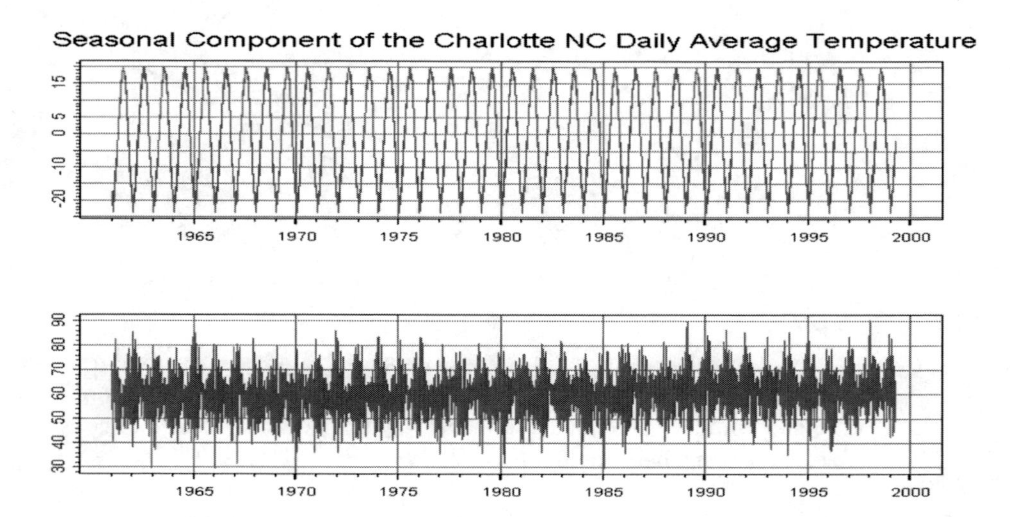

Fig. 5.24. Plot of the seasonal component (top) of the remainder (bottom) of the daily average temperature in Charlotte, NC.

centered around zero, and as a consequence, the remainder term is centered around

60. At this stage, and for the sake of simplicity, we shall assume that this time series looks stationary enough. In other words, we are not trying to look for an upward trend which would indicate the effect of global warming, or we are not trying to detect a break in the baseline level of the remainder term which could have been produced by a change in location of the measuring station, or a change in urbanization around the station, or . . .. We know that such a case has been made for Charlotte (and this is why we chose this example) but we shall ignore it for the sake of the present introductory discussion. In particular, we shall consider that it makes sense to look at its marginal distribution. We use the exploratory data analysis tools introduced in Chapter 1. The

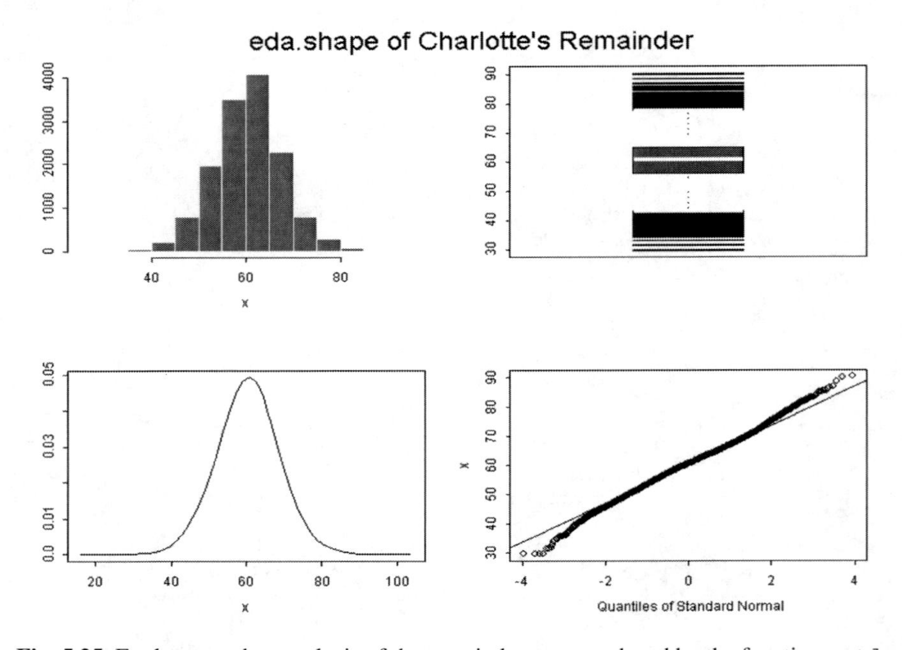

Fig. 5.25. Exploratory data analysis of the remainder term produced by the function `sstl`.

marginal distribution is not exactly Gaussian, because as we can see from the Q-Q plot, the tails are slightly heavier than what we would expect from a normal sample. The goal of this chapter being to model the time dependence of the series, we look for clues on which model to use.

Analysis of the Serial Correlations

As explained in the text above, in order to detect and identify the possible serial correlations present in the data, the first step is to compute and plot the auto-correlation function of the series. The results are reproduced in Figure 5.26. The rapid decay of the correlations for short lags (left part of the plot) suggests that fitting an auto-

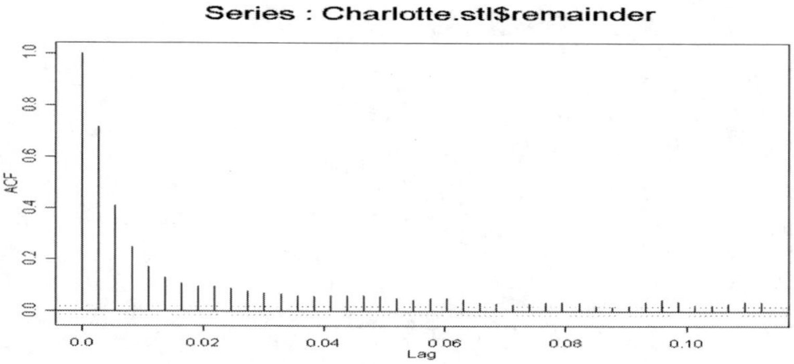

Fig. 5.26. Empirical auto-correlation function of the Charlotte remainder term.

regressive model could be reasonable. Nevertheless, the persistence of significantly different from zero correlations at large lags in the right part of the plot, may be a difficult feature to fit.

Could an AR Model Do the Trick?

We fit an auto-regressive model to the data (the remainder term, to be specific) using the command:

```
> ChAR <- ar(Charlotte.stl$rem)
> ChAR$order
[1] 17
```

Given the relatively fast decay of the auto-correlation function, the order 17 chosen by the function `ar` looks suspiciously large. In order to avoid over-fitting, we double check on the reasons for the choice of such a large order by plotting the AIC and the partial auto-correlation function.

```
> tsplot(ChAR$aic, main="AIC Criterion")
> acf(Charlotte.stl$rem,type="partial"))
```

A careful look at the plot of the AIC (left pane of Figure 5.27) confirms that the minimum is attained for the order 17. This is rather difficult to see on the plot given in Figure 5.27 because of the large scale of the first few values. This plot also shows that the major relative drop occurs before the value 4 of the argument. This suggests that an AR(3) model could be appropriate. In order to confirm this guess, we computed and plotted the partial auto-correlation function in the right part of Figure 5.27. The partial auto-correlation function vanishes for lags greater than or equal to 4, so 3 is the right order, and we proceed to fit an AR(3) model with the command:

```
> MyChAR <- ar(Charlotte.stl$remainder, order =3)
```

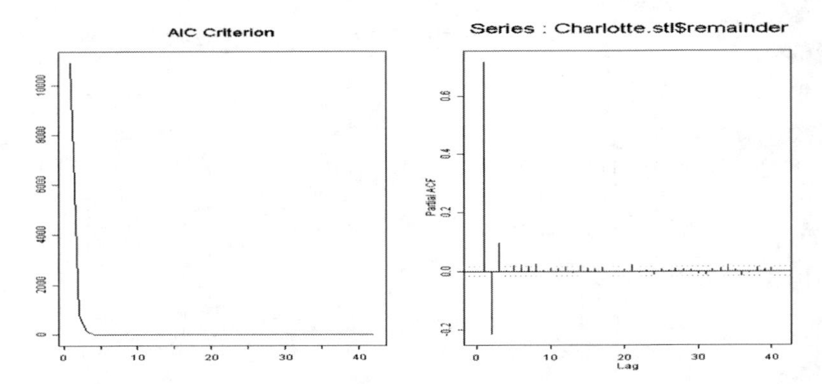

Fig. 5.27. AIC criterion (left) and partial auto-correlation function (right) of the Charlotte remainder term.

We assess the goodness of the fit by looking at the residuals. As seen in Figure 5.28, these residuals look pretty much Gaussian, at least much more than the temperature remainder itself. But most importantly, the plot of the auto-correlation function re-

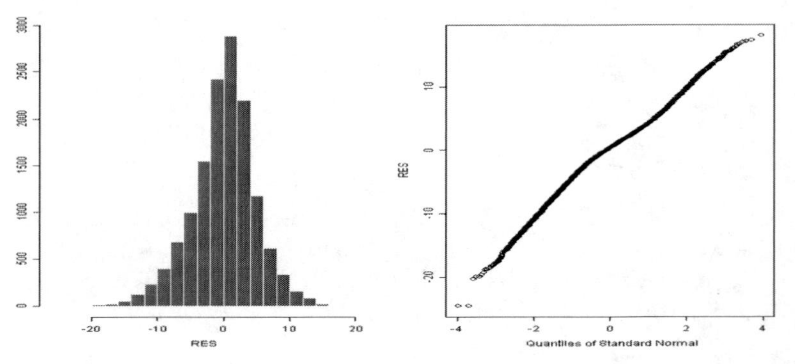

Fig. 5.28. Histogram (left) and normal Q-Q plot (right) of the residuals from the fit of an AR(3) model to the remainder term of the Charlotte average daily temperature.

produced in Figure 5.29 looks much better: it seems that all the serial correlation in the data was captured by the AR(3) model.

Application to Temperature Options

Let us now imagine that in early 1999, we were interested in buying a call option on CDD's for the period covering July and August 1999 with a given strike K. How could we have used the analysis done so far to quantify the risk/reward profile of such a purchase? Our first reaction should be to take the prediction formulae developed for

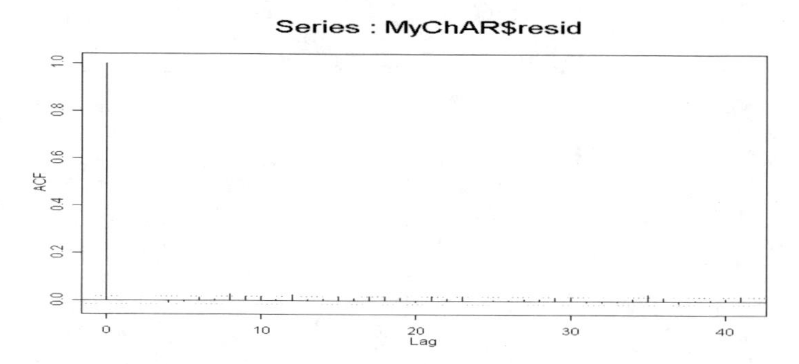

Fig. 5.29. Auto-correlation function of the residuals from the fit of an AR(3) model to the remainder term of the Charlotte average daily temperature.

AR models, and use them to predict the summer values of the remainder term, add the mean level and the seasonal component to obtain a prediction for the actual daily temperatures, from which we could finally compute a prediction for the number of cooling degree days and the pay-off of the option. This strategy has two fatal flaws. First, the prediction horizon is too long for the actual prediction to be significantly different from zero. Indeed, as we saw earlier, the predictions of the future values of an AR model converge exponentially fast toward the mean. But most importantly, we need predictions for nonlinear functions of the daily temperatures, and computing a nonlinear function of a prediction can lead to unexpected results. We illustrate the first of these two flaws in Figure 5.30. Assuming that we are on March 31, 1999, we used the model fitted above to compute the prediction of the daily temperature all the way to August 31. We use the function `pred.ar`, requesting prediction from the model fitted above for a 153 days horizon.

```
> ChPred <- pred.ar(series=Charlotte.stl$rem,
                              ar.est=MyChAR, ahead = 153)
```

The resulting prediction `ChPred` is plotted on the top of Figure 5.30. As expected, the prediction returns very fast to the overall mean of the AR series. So, it cannot be of any interest for long-term predictions. Adding the seasonal component would produce a more realistic graph, but the result would remain, we cannot use this tool for our temperature option problem.

So we decided to use the model fitted above to set up Monte Carlo simulations instead. It is important to understand that there is no black magic here. The model remains the same, and we do not suddenly become immune to the shortcomings mentioned above. Indeed, should we try to compute the prediction of the daily temperature, we would get essentially the same estimate as the one plotted above. The main difference lies in the versatility of the tools. We can compute many expectations and probabilities depending on possibly nonlinear functions of the evolution of the temperature. That's the main difference!

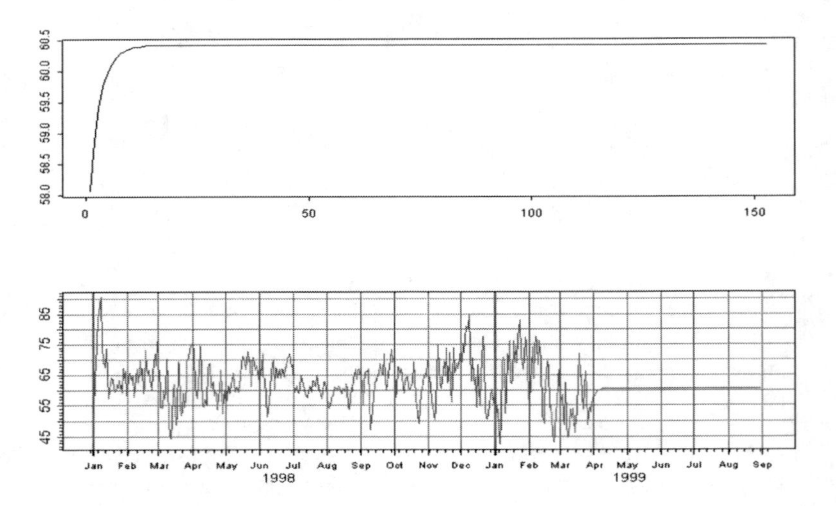

Fig. 5.30. Prediction of the remainder term over a 153 days horizon. Raw prediction (top) and prediction concatenated to the actual remainder up to March 31, 1999.

Let us first estimate the probability that the option will end up being exercised. This is the probability that the number $CDD^{(P)}$ is greater than or equal to the strike K. A Monte Carlo estimate of this probability can be computed in the following way:

- Choose a large number of scenarios, say $N = 10,000$;
- Use the fitted model to generate N samples $x_t^{(j)}$, $x_{t+1}^{(j)}$, …, $x_T^{(j)}$ where t corresponds to March 31st, 1999, and T to August 31st, 1999, and $j = 1, 2, \dots, N$;
- Add the mean temperature and the seasonal component to obtain N scenarios of the temperature in Charlotte starting from March 31st, 1999, and ending August 31st, 1999;
- For each single one of these N scenarios, compute the number of CDD's between July 1st and August 31st;
- Compute the number of times this total number of CDD's is greater that the strike K, and divide by N.

The above procedure is very versatile. Not only can it be used to estimate probabilities, but it can also be used to estimate expected values. For example, if a tick rate is agreed upon (say $\alpha = \$5,000$ per degree day) one can estimate the expected amount the writer of the option will have to pay to the buyer of the option. As before, we can generate a large number of Monte Carlo scenarios for the temperature, compute the pay-off of the option for each of these scenarios, and compute the average (over the scenarios) of the cost to the writer. Similarly, one can compute the expected amount the buyer of the option will receive. In fact, given the choice of utility functions for the buyer and the seller of the options, one can compute the expected terminal utilities for both of them, and these numbers should give a good indication of what the writer would be willing to sell the option for, and of what kind of premium the buyer would be willing to pay in order to get the protection provided by the option.

APPENDIX: MORE S-Plus IDIOSYNCRACIES

This appendix is devoted to two major components of any time series toolbox: the implementation of the shift operator, and the specification of ARIMA models. We provide illustrations of the theoretical concepts discussed in the text and we emphasize some of the pitfalls of the S-Plus implementations.

A1. Unexpected Behavior of the Lag Method

We first consider the very simple example of the following set of S-Plus commands.

```
> X <- c(1,2,3,4,5,6,7,8,9,10)
> Y <- lag(X,k=2)
> Z <- cbind(X,Y)
> tsplot(Z)
```

The result is shown on the left pane of Figure 5.31. Strangely, it seems that nothing

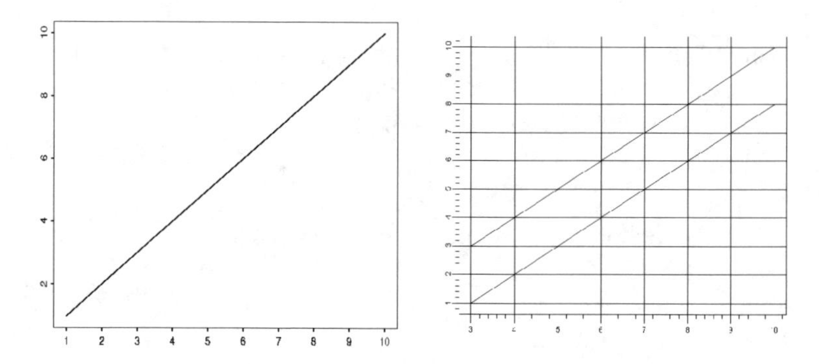

Fig. 5.31. Left: Plot of X and its lagged version: Nothing happened! Right: Plot of the signalSeries object X.ss and its shifted version: the shift was implemented!

happened, X and Y appear to be identical! Indeed, the lag function does not change the values of the entries of the vectors. It merely tries to manipulate the time stamps, but since numerical vectors do not carry time stamps, no change was made.

An instance of the lag method was implemented for time series objects in earlier versions of S-Plus. Because the time series objects of these versions are obsolete, we refrain from giving the effects of the lag method on these objects and instead, we consider the `signalSeries` and `timeSeries` objects which subsumed them in the current version of `S-Plus`. For them, the shift operator B is implemented by the method `shift`.

The Lag Method for `signalSeries` *and* `timeSeries` *Objects*

To illustrate the differences when we are dealing with the new "series" objects of `S-Plus`, we first modify the previous attempts in the following way.

```
> X.ss <- signalSeries(data=x,from=1,by=1)
> Y.ss <- shift(X.ss,k=2)
> Z.ss <- seriesMerge(X.ss,Y.ss)
> plot(Z.ss)
```

The results are reproduced on the right pane of Figure 5.31. This time the expected behavior takes place.

A2. Comments on the `S-Plus` ARIMA Conventions

Here is another set of idiosyncrasies of the `S-Plus` environment, and new conflicts with a widely accepted set of notations. We choose to discuss a specific example to make our point.

Standard ARIMA Conventions

Throughout the book, we tried to follow the notation most frequently used in the literature on time series, and especially in the discussions of the linear theory of ARIMA models. According to a widely accepted practice, in order to write the model:

$$X_t - .5X_{t-1} + .2X_{t-2} = W_t + .3W_{t-1} - .1W_{t-2} \tag{5.39}$$

in parametric form, we should define the coefficients:

$$\phi_1 = .5 \qquad \phi_2 = -.2 \qquad \theta_1 = +.3 \qquad \theta_2 = -.1. \tag{5.40}$$

and write the model (5.39) in the form:

$$X_t - \phi_1 X_{t-1} - \phi_2 X_{t-2} = W_t + \theta_1 W_{t-1} + \theta_2 W_{t-2}. \tag{5.41}$$

`S-Plus` *Convention*

Unfortunately, this is not the convention followed by `S-Plus`. Indeed, to give a more symmetric role to the two sets of coefficients, in order to write the model (5.39) in parametric form `S-Plus` needs the coefficients:

$$\phi_1 = .5 \qquad \phi_2 = -.2 \qquad \theta_1 = -.3 \qquad \theta_2 = +.1. \tag{5.42}$$

so that the model can be:

$$X_t - \phi_1 X_{t-1} - \phi_2 X_{t-2} = W_t - \theta_1 W_{t-1} - \theta_2 W_{t-2}. \tag{5.43}$$

So the reader/S-Plus-user should **be aware** of the fact that the signs of the θ's are different in (5.41) and (5.43). We illustrate this point with specific S-Plus commands involving this model.

```
> MA <- arima.sim(model=list(ar=c(.5,-.2),
                                   ma=c(-.3,.1)),n=1024)
> MA.fit <- arima.mle(MA,model=list(ar=c(.5,-.2),
                                        ma=c(-.3,.1)))
```

The first command generates a sample of length 1024 from the model given by the list specified as argument. Notice that S-Plus will give warnings (and sometimes refuse to perform) if the model is not stationary and/or invertible. The use of the list defining the model is slightly different in the second command. Indeed, S-Plus does not assume that the actual values of the coefficients are known, since this function is trying to estimate these values. By counting the number of coefficients given, it merely infers that the model should be ARIMA(2,0,2) (this information could also have been given by model=list(order=c(2,0,2)) and it then uses the actual values given as arguments as initial values (i.e. starting points) for the optimization procedure seeking the maximum likelihood estimates. S-Plus uses zeroes as initial values when the latter are not provided, as in the present situation. We illustrate these comments by printing the fitted arima object:

```
> MA.fit
Call: arima.mle(x=MA,model=list(ar=c(0.5,-0.2),
                                 ma=c(-0.3,0.1)))
Method:  Maximum Likelihood
Model :  2 0 2

Coefficients:
AR : 0.48102 -0.15653
MA : -0.29322 0.12498
Variance-Covariance Matrix:
         ar(1)       ar(2)      ma(1)      ma(2)
ar(1)  0.12482 -0.016415  0.12493  0.078728
ar(2) -0.01642  0.003910 -0.01626 -0.008957
ma(1)  0.12493 -0.016269  0.12602  0.079251
ma(2)  0.07872 -0.008957  0.07925  0.051733

Optimizer has NOT converged
Due to: iteration limit AIC: 2942.3442
```

Notice that, despite the fact that the optimizer did not converge (because the maximum number of iterations allowed was not large enough) the estimates of the coefficients are reasonably close to the true ones. Moreover the estimate of the variance/covariance matrix of the coefficients can be used to compute approximate confidence intervals/regions for these coefficients.

PROBLEMS

Ⓣ **Problem 5.1** *Let us assume that $\{W_t\}_t$ is a variance-one white noise, and let us consider the time series $\{X_t\}_t$ defined by:*

$$X_t = W_t + (-1)^{t-1}W_{t-1}.$$

Compute the mean and auto-covariance functions of the time series $\{X_t\}_t$. Is it stationary? Say why.

Ⓣ **Problem 5.2** *1. Let W be a normal (Gaussian) random variable with mean 0 and variance 1. Compute the values of the moments:*

$$\mathbb{E}\{W\}, \qquad \mathbb{E}\{W^2\}, \qquad \mathbb{E}\{W^3\}, \qquad \mathbb{E}\{W^4\}, \qquad \mathbb{E}\{W^5\}, \qquad \mathbb{E}\{W^6\}.$$

2. Let us now assume that W is normal with mean 0 and variance σ^2. Determine (as functions of σ) the values of the same moments as before.
3. Let us assume that $\{W_t\}_{t=0,\dots,N}$ is a Gaussian white noise with variance σ^2, and for each $\alpha \in \mathbb{R}$, let us define the time series $\{X_t\}_{t=0,\dots,N}$ by the formula:

$$X_t = \alpha W_t + W_t^3, \qquad t = 0, \dots, N. \tag{5.44}$$

Compute the mean, variance and auto-covariance functions of the time series $\{X_t\}_{t=0,\dots,N}$. Is it stationary?
4. Obviously, X_t strongly depends upon W_t, however, find a value of a for which W_t and X_t are uncorrelated?

Ⓣ **Problem 5.3** *Let us assume that $\theta \in (-1, +1)$ is known, that $\{W_t\}$ is a Gaussian white noise with variance one, and that $\{W'_t\}$ is a Gaussian white noise with variance θ^2. Show that the MA(1) time series $\{X_t\}_t$ defined by:*

$$X_t = W_t + \theta W_{t-1}$$

and the time series $\{Y_t\}_t$ defined by:

$$Y_t = W'_t + \frac{1}{\theta}W'_{t-1}$$

have the same auto-covariance functions. Do they have the same auto-correlation functions?

Ⓣ **Problem 5.4** *1. Find the AR representation of the MA(1) time series*

$$X_t = W_t - .4W_{t-1}$$

where $\{W_t\}_t$ is a white noise $N(0, \sigma^2)$.
2. Find the MA representation of the AR(1) time series

$$X_t - .2X_{t-1} = W_t$$

where, as before, $\{W_t\}_t$ is a white noise $N(0, \sigma^2)$.

Ⓣ **Problem 5.5** *Let us consider the ARMA time series* $\{X_t\}_t$ *defined by:*

$$X_t - .6X_{t-1} = W_t - .9W_{t-1}$$

where $\{W_t\}_t$ *is a white noise with variance one.*
1. Rewrite the model using the shift operator B*.*
2. Is the model stationary? Say why.
3. Is the model invertible? Say why.
4. Express the model in an MA representation if it exists.
5. Express the model in an AR representation if it exists.

Ⓣ **Problem 5.6** *For each of the following ARMA(1,1) models:*
(i) $X_t - X_{t-1} = W_t - 1.5W_{t-1}$
(ii) $X_t - .8X_{t-1} = W_t - .5W_{t-1}$
for which we assume that $\{W_t\}_t$ *is a* $N(0, \sigma^2)$ *white noise,*
1. Rewrite the model using the backward shift operator B *and determine the polynomials* $\phi(B)$ *and* $\theta(B)$*.*
2. Check if the model is stationary and/or invertible and explain your answers.

Ⓣ **Problem 5.7** *Let us assume that* $\{W_t\}_t$ *is a white noise process with mean* 0 *and variance* $\sigma^2 = 1$*. Consider the time series* $\{X_t\}_t$ *defined by:*

$$X_t + 0.4X_{t-1} = W_t - 2.5W_{t-1} + W_{t-2}.$$

1. Rewrite the model using the backward shift operator B*.*
2. Is the model invertible? Say why.
3. Does an auto-regressive representation exist? If yes, give it.
4. Is the model stationary (causal)? Say why.
5. Does a moving average representation exist? If yes, give it.
6. Is the time series $\{Y_t\}_t$ *defined by:*

$$Y_t = (-1)^t X_t$$

stationary? Explain your answer.

Ⓣ **Problem 5.8** *Let us assume that the time series* $\{X_t\}_t$ *defined by:*

$$X_t - 2X_{t-1} + X_{t-2} = W_t - .3W_{t-1} - .5W_{t-2}$$

where $\{W_t\}_t$ *is a* $N(0, \sigma^2)$ *white noise.*
1. Rewrite the model using the shift operator B*.*
2. Is the model stationary? Say why.
3. Is the second difference $D_t = (1 - B)^2 X_t$ *stationary? Say why.*
4. Compute the auto-covariance function of the second difference D_t*.*

ⒺⓈ **Problem 5.9** *The goal of this problem is to compute the critical values of the Dickey-Fuller unit-root test by Monte Carlo simulation, and to use the results to test real log-price data for stationarity. We assume that the time series* $\{X_t\}_t$ *is of the form* $X_t = \phi_1 X_{t-1} + W_t$ *for some strong white noise* $\{W_t\}_t$ *with unknown variance* σ^2*.*
0. Generate a sample of size $N = 5000$ *from the standard normal distribution and use it to compute a sample of the same size for the time series* $\{X_t\}_t$ *with* $\sigma^1 = 1$ *and* $\phi_1 = 1$*.*

1. For each $n = 1, 2, \ldots, N$ compute the values of the estimates $\hat{\phi}_1$ and $\hat{\sigma}^2$, and of the Dickey-Fuller statistic DF (given by formulae (5.30) and (5.31) in the text) from the sample of size n formed by the first n entries of the sample of size N generated in part 0 for $\{X_t\}_t$. Produce a sequential plot of the estimates of ϕ_1 and σ^2 and check that they do converge toward their true values (1 for both of them.)
2. We now fix $N = 1000$ and $NS = 500$. Repeat the steps above NS times to produce NS independent samples of size N. Collect the results in a numerical matrix with N rows and NS columns which you call `XX`. For each $n = 100, 110, 120, \ldots, 1000$ compute the values of the quantiles of the sample of size NS given by the n-th row of the matrix `XX` for the values $q = .01, .02, .03, .04, .05, .1, .25, .5, .75, .9, .95, .96, .97, .98, .99$, and organize the results in a numerical matrix with 101 rows and 15 columns which you call `QQ`.
3. Produce a sequential plot of each of the columns of `QQ` and check that each quantile converges as n grows indefinitely. It is in this sense that the distribution of the Dickey-Fuller statistic converges.

From now on we use the quantiles given by the last row of the matrix `QQ` as a proxy for the critical values of the unit-root test.
4. Compute the Dickey-Fuller statistic DF for the daily log returns `BLRet` and `CLRet`, and give the p-value of the unit-root tests in each case. Is the stationarity assumption rejected at the level 1%?

Ⓔ **Problem 5.10** *1. Create a signal series, say `RW`, with the data in the file `rw.asc`. Use the tools seen in the text to discuss the dependence (or lack thereof) between the successive entries of this time series.*
2. Compute the first difference (i.e. with lag $k = 1$), call it `WN`, and address the same question of the dependence of the successive entries and compare the results with those obtained for the series `RW`. Are your observations consistent with the claim: "the series `RW` was created as the cumulative sum of the entries of a white noise series".
3. Import the file `hs.asc` into `S-Plus`. This should create an object of class `timeSeries` called `HS.ts`. These data give the values of the daily closing values of the Hang Seng Index of Hong Kong Stock Prices from January 1, 1986 to June 25, 2003. Go through the various steps of the analysis of questions 1. and 2. for the series `HS.ts` and compare the results. What does that tell you about the Hang Seng index?

ⒺⓈ **Problem 5.11** *1. Set the `S-Plus` random seed to 14, generate the ensuing realization of length 1024 of a $N(0, 1)$ white noise $\{W_t\}_{t=1,\ldots,1024}$, and generate the corresponding realization of the AR(3) time series $\{X_t\}_{t=1,\ldots,1024}$ defined by:*

$$(1 - .07B - .02B^2 - .3B^3)X_t = W_t.$$

and the convention $X_0 = X_{-1} = X_{-2} = 0$.
2. Fit autoregressive models of orders going up to 9 and produce the corresponding AIC. Choose the best model according to this criterion, determine the coefficients and forecast the next 16 values of the time series. Produce a plot of the predictions together with an approximate 95% confidence interval for these predictions.
3. With the same white noise series as before (which you can regenerate by resetting the seed to 14 if you need) generate a realization of length 1000 of the ARMA(3,4) process X_t defined by:

$$(1 - .07B - .02B^2 - .3B^3)X_t = (1 - .4B - .3B^2 - .2B^3 - .05B^4)W_t.$$

4. As before, fit autoregressive models of orders going up to 9*, and produce the corresponding AIC values. What is the best model suggested by this criterion? Comment. Fit such a model, and as before, forecast the next* 16 *values of the time series and produce a plot of the predictions together with an approximate* 95% *confidence interval for these predictions.*
5. Ignoring the suggestion of AIC, fit an AR(3) model and compute the estimated residuals. Fit moving averages models (of orders up to 5*) to the time series of these estimated residuals and choose the best one. Use the ARMA model so-obtained to forecast the next* 16 *values of the original time series and produce a plot of the predictions together with an approximate* 95% *confidence interval for these predictions. Compare the results with those obtained in part 4.*

Ⓔ **Problem 5.12** *This problem gives an example of identification of a periodic function using nonlinear least squares regression.*

Brain imaging is a new technology which is helping make quantum leaps in our understanding. It is intended to provide in vivo and noninvasively, information which enables scientists and clinicians to discover where in the brain, normal and abnormal cognitive processes take place. Many early experiments in brain function imaging involved taking long sequences of snapshots of the brain while the patient *was asked to perform elementary tasks at regular time intervals. These sequences of images were then analyzed for the identification of regions of the brain actively responding to the stimuli. This experimental design was chosen because of the data analysis used to identify the active regions of the brain. Indeed, since the stimuli were presented at regular time intervals, the parts of the brain responding to these stimuli must vary over time with the same period, and their identification was based on these premises.*

Magnetic Resonance Imaging (MRI for short) is currently the technique of choice for humans because of its 3-D capabilities, but optical imaging is still preferred for animal research because of a better time and space resolution, especially when the response is visible in the outer layer of the brain.

We consider data from an optical imaging experiment conducted on the Irvine campus of the University of California, on the brain of a mouse whose right whisker was bent 11 *times at regular intervals (twice per second). During the experiment* 198 *snapshots of the skull of the mouse were taken with a CCD camera, the skull being (ever so slightly) thinned to enable the light to go through, and so doing, to illuminate the brain and return to the sensors of the camera.*

We extracted the time sequence of the 198 *values of the light intensity at pixel* $(58, 66)$ *found to be in the active region responding to the stimulus and at pixel* $(1, 1)$ *which was not part of the active region responding to the stimulus. We stored them in* `S-Plus` *numeric vectors called* `ACT` *and* `NONACT`*. They can be created by opening and running the (data) script file* `brainpixels.asc`*.*
1. Crete a `timeSeries` *object with the numerical data contained in the vector* `ACT` *and extract a periodic component of frequency* 18 *(recall that the stimulus was presented regularly* 11 *times in the sequence of* 198 *snapshots.)*

Our goal is to represent the periodic component of the reflectance by a function of the form:

$$\varphi(t) = \rho \sin(\omega t + \phi) \tag{5.45}$$

and we plan to use nonlinear regression for that.
2. Assume first that you know that the frequency ω *should be equal to* $2\pi/18$*. So, instead of a search for* 3 *parameters and a regression function of the form (5.45), restrict yourselves to* 2 *parameters and a regression function of the form (5.46) given by:*

$$\varphi(t) = \rho \sin(2\pi t/18 + \phi). \tag{5.46}$$

Estimate these parameters by nonlinear regression and plot on the same graph, the original points of the periodic component together with the graph of the periodic function fitted with the `S-Plus` *function* `nls`*. Compare the results to the actual periodic function extracted by the function* `sstl`*.*
3. Rerun the same experiment (i.e. identification of the periodic component using the `S-Plus` *function* `nls`*) without assuming that the period is known, in other words, using the period as a parameter. Comment on the quality of the results.*

Problems 5.13 and 5.14 use the data contained in the text file `renotemp.asc`. Open this file in `S-Plus` and run it like a script to create an object of class `timeSeries` `RENO.ts`. It contains the daily average temperature in Reno from March 1st 1937 to November 8, 2001.

Ⓔ **Problem 5.13** *The purpose of this problem is to perform a simple regression analysis of the Reno CDD's option discussed in the text.*
0. Compute for each year in the period from 1961 to 2001 inclusive, the yearly cumulative number of CDD's in Reno from June 1st to September 30th each year. Compute the average of these yearly indexes over the last 15 *years of this period, and call this average* K.
From now on we consider a European call option with strike K *computed above (i.e. at the money), with rate* α *equal to US$ 5,000 per degree day, and cap US$ 1,000,000, written on the cumulative number of CDD's in Reno from June 1, 2002 to September 30, 2002.*
1. Use the data at hand and a simple linear regression to explain the dependence of this yearly CDD index upon the year of the computation, and use this model to give an estimate of the value of this yearly CDD index in Reno for the period covered by the option. Give a 95% *confidence interval for this prediction.*
2. Use this model to estimate the probability that the option described in the text is exercised.
3. Estimate the expected loss to the writer of the option (i.e. the party selling the option), and estimate the amount of reserve (in US$) the writer of the option should have in order to cover her losses 95% *of the time.*

Ⓔ **Problem 5.14** *Follow the steps of the analysis of the daily temperature in Charlotte given in Section 5.6 of the text to fit a model to the daily temperature in Reno, use this fitted model and follow the prescriptions given in the text to provide new answers to the questions 2 and 3 of Problem 5.13 above.*

NOTES & COMPLEMENTS

The random walk was introduced at the very beginning of the 20-th century by Bachelier as a model for the stock market. Indeed, a time series plot of a sample from a random walk looks very much like a typical stock chart, and the fact that the increments of the random walks are independent is a good proxy for the expected efficiency of markets. The classical theory of linear processes with the AR, MA, ARIMA techniques presented in this chapter, is usually associated with the names of Box and Jenkins who made the theory and the practice of these models popular through a series of books. See for example [11]. We refer the reader interested in the extensions of this theory to Priestley's book [69] for an introduction to nonlinear time series analysis, and to Rosentblatt's book [75] for an extensive discussion of the identification problems which can be handled with the use of higher order statistics.

Unit root tests and cointegration are important fixtures of econometric theory and practice, and our sketchy presentation does not do justice to their importance. The reader is referred to the financial econometrics books [16] by Campbell, Lo and McKinlay, [41] by Gourieroux and Jasiak, and to the book [89] of Zivot and Wang for examples of `S-Plus` experiments. A form of the Dickey-Fuller test is implemented in the `S+Finmetrics` by the function `unitroot`.

An important component of stationary time series and signal analysis was purposely ignored in this chapter: spectral analysis. The latter is based on the mathematical theory of the Fourier transform, and it offers a dual perspective on the properties of signals and time series. One of its cornerstones is the Shannon sampling theorem for band-limited signals. For details on the sampling theorem and for the spectral analysis of the linear processes studied in this chapter, we refer the interested reader to standard textbooks such as [13] for example. We chose to stay in the time domain, and consequently ignore the frequency domain, not only to avoid the technical dealings with complex analysis, but also to keep the treatment of causality and nonanticipativeness at a simple and intuitive level. `S-Plus` has an extensive set of powerful tools to analyze time series and signal series objects from this point of view. They range from commercial libraries such as the `S+Wavelet` offered by Insightful, to free public domain libraries such as `Swave` available on the web at the URL

```
http:\\www.princeton.edu\~rcarmona\downloads
```

While `S+Wavelet` is mostly concerned with discrete wavelet and other time-frequency transforms, see [15], `Swave` concentrates on implementations of the continuous wavelet and Gabor transforms. as described in the book [20]. Many *charting* time series analysis tools have been developed with trading systems in mind. They usually go under the name of *technical analysis,* and they are based on indicators giving buy and sell signals from the behavior of various moving averages. We hope that the reader realized that we do not want to have anything to do with these pseudo-theories and their uses.

Despite growing recognition of its importance, the weather market is still not a part of mainstream finance. We proposed the analysis of temperature options as the main application of the theory presented in this chapter because we believe that temperature options offer an example of financial engineering at its best. The interplay between statistical modeling, mathematical analysis, financial risk management, and Monte Carlo computations makes it the epitome of case studies. While proofreading the manuscript of this book, I discovered the collective monograph [8], and I highly recommend it to anyone trying to understand the weather market. Also, I noticed at the same time that a chapter on weather derivatives was added to the newest edition [47] of Hull's book. These two publications are a clear indication of the growing respectability of the weather market among the business and academic communities. The reader interested in the statistical analysis for climate research is referred to the book [86] by von Storch and Zwiers.

The idea and the data of Problem 5.12 are borrowed from a study of optical brain imaging [19] [18], by the author, R. Frostig and collaborators, where principal component analysis was used to identify the response of the brain to various stimuli.

6

MULTIVARIATE TIME SERIES, LINEAR SYSTEMS AND KALMAN FILTERING

This chapter is devoted to the analysis of the time evolution of random vectors. The first section presents the generalization to the multivariate case of the univariate time series models studied in the previous chapter. Modern accounts of time series analysis increasingly rely on the formalism and the techniques developed for the analysis of general stochastic systems. Even though financial applications have remained mostly immune to this evolution, because of its increased popularity and its tremendous potential, we decided to include this alternative approach in this chapter. The tone of the chapter will have to change slightly as we discuss concepts and theories which were introduced and developed in engineering fields far remote from financial applications. The practical algorithms were developed mostly for military applications. They led to many civil and technological breakthroughs. Here, we present the main features of the filtering algorithms, and we use financial examples as illustrations, restricting ourselves to linear systems.

6.1 Multivariate Time Series

Multivariate time series need to be introduced and used when significant dependencies between individual time series cannot be ignored. As an illustration, we discuss further the weather derivative market, and some of the natural issues faced by its participants. An active market maker in these markets will want to hold options written on HDD's and/or CDD's in different locations. One good reason for that may be the hope of taking advantage of the possible correlations between the weather (and hence the temperature) in different locations. Also large businesses with several units spread out geographically are likely to want deals structured to fit their diverse weather exposures, and this usually involves dealing with several locations simultaneously. Think for example of a chain of amusement parks, or a large retailer such as Wal-Mart or Home Depot: their weather exposures are tied to the geographic locations of the business units, and an analysis of the aggregate weather exposure cannot be done accurately by considering the locations separately and ignoring the correlations. As we are about to show, the simultaneous tracking of the weather in several locations can be best achieved by considering *multivariate* time series. We discuss a specific example in Subsection 6.1.4 below.

The goal of this section is to review the theory of the most common models of multivariate time series, and to emphasize the practical steps to take in order to fit these models to real data. Most of the Box-Jenkins theory of the ARIMA models presented earlier can be extended to the multivariate setting. However, as we shall see, the practical tools become quite intricate when we venture beyond the auto-regressive models. Indeed, even though the definition of moving average processes can be generalized without change, fitting this class of models becomes even more cumbersome than in the univariate case. As a side remark we mention that `S-Plus` does not provide any tool to fit multivariate moving average models.

In terms of notation, we keep using the convention followed so far in the book: as a general rule, we use bold face characters for multivariate quantities (i.e. vectors for which $d > 1$) and regular fonts for univariate quantities (i.e scalars for which $d = 1$). In particular, throughout the rest of this section, we shall use the notation $\mathbf{X} = \{\mathbf{X}_t\}_t$ to denote a multivariate time series. In other words, for each time stamp t, $\mathbf{X}_t$ is a d-dimensional random vector.

6.1.1 Stationarity and Auto-Covariance Functions

Most of what was said concerning stationarity extends without any change to the multivariate case. This includes the definition and the properties of the shift operator B, and the derivative (time differentiation) operator ∇. Remember that stationarity is crucial for the justification of the estimation of the statistics of the series by time averages. The structure of the auto-covariance/correlation function is slightly different in the case of multivariate time series. Indeed, for each value k of the time lag, the lag-k auto-covariance is given by a matrix $\gamma(k) = [\gamma_{ij}(k)]_{i,j}$ defined by:

$$\gamma_{ij}(k) = \mathbb{E}\{X_t^{(i)} X_{t+k}^{(j)}\} - \mathbb{E}\{X_t^{(i)}\}\mathbb{E}\{X_{t+k}^{(j)}\}.$$

In other words, the quantity $\gamma_{ij}(k)$ gives the lag-k cross-correlation between the i-th and j-th components of $\mathbf{X}$, i.e. the covariance of the scalar random variables $X_t^{(i)}$ and $X_{t+k}^{(j)}$.

6.1.2 Multivariate White Noise

As in the case of univariate series, the white noise series are the building blocks of the time series edifice. A multivariate stationary time series $\mathbf{W} = \{\mathbf{W}_t\}_t$ is said to be a white noise if it is:

- mean-zero (i.e. $\mathbb{E}\{\mathbf{W}_t\} = \mathbf{0}$ for all t);
- serially uncorrelated
  - either in the strong sense, i.e. if all the $\mathbf{W}_t$'s are independent of each other;
  - or in the weak sense, i.e. if $\mathbb{E}\{\mathbf{W}_t \mathbf{W}_s^t\} = 0$ whenever $s \neq t$.

The equality specifying the fact that a white noise needs to be mean zero is an equality between vectors. If d is the dimension of the vectors $\mathbf{W}_t$, then this equality is equivalent to a set of d equalities between numbers, i.e. $\mathbb{E}\{W_t^{(i)}\} = 0$ for $i = 1, \ldots, d$. As usual, we use the notation $W_t^{(i)}$ for the i-th component of the vector $\mathbf{W}_t$. On the other hand, the last equality is an equality between $d \times d$ matrices, and whenever $s \neq t$, it has to be understood as a set of $d \times d$ equalities $\mathbb{E}\{W_t^{(i)} W_s^{(j)}\} = 0$ for $i, j = 1, \ldots, d$.

This is exactly the same definition as in the scalar case in the sense that there is complete de-correlation in the time variable. But since the noise terms are vectors, there is also the possibility of correlation between the various components. In other words,

at each time t, the components of $\mathbf{W}_t$ *can be "correlated".*

Hence, in the case of a white noise, we have

$$\gamma_{i,j}(k) = \mathbb{E}\{W_t^{(i)} W_{t+k}^{(j)}\} = \gamma_{i,j}\delta_0(k)$$

where $\delta_0(k)$ is the usual *delta* function which equals 1 when $k = 0$ and 0 when $k \neq 0$, and $\gamma = [\gamma_{i,j}]_{i,j=1,\ldots,k}$ is a time independent variance/covariance matrix for a d-dimensional random vector. Using the jargon of electrical engineers we would say that the components are *white in time and possibly colored in space.*

6.1.3 **Multivariate AR Models**

A d-dimensional time series $\mathbf{X} = \{\mathbf{X}_t\}_t$ is said to be an auto-regressive series of order p if there exist $d \times d$ matrices $A_1, A_2, \ldots, A_p$ and a d-variate white noise $\{\mathbf{W}_t\}_t$ such that:

$$\mathbf{X}_t = A_1\mathbf{X}_{t-1} + A_2\mathbf{X}_{t-2} + \cdots + A_p\mathbf{X}_{t-p} + \mathbf{W}_t. \tag{6.1}$$

As before we ignored the mean term (which would be a d-dimensional vector in the present situation) by assuming that all the components have already been centered around their respective means. Notice that the number of parameters is now $pd^2 + d(d+1)/2$ since we need d^2 parameters for each of the p matrices A and we need $d(d+1)/2$ parameters for the variance/covariance matrix Σ_W of the white noise (remember that this matrix is symmetric so that we need only $d(d+1)/2$ coefficients instead of d^2!). Except for the fact that the product of matrices is not commutative, AR models can be fitted in the same way as they are fitted in the univariate case, for example by solving the system of Yule-Walker linear equations obtained from the empirical estimates of the auto-covariance function and the consistency relations with the definition (6.1). In fact, fitting auto-regressive models with the `S-Plus` function `ar` can be done with multivariate time series in exactly the same way it is done with univariate series.

Fitting Multivariate AR Models in `S-Plus`

Fitting a multivariate AR model is done in `S-Plus` with the same command as in the univariate case. In order to give an example, let us assume that `TEMPS` is a multivariate timeSeries `S-Plus` object. We shall create such a tri-variate time series of daily temperatures in Subsection 6.1.4 below. The object `TEMPS` could also be a matrix. In any case, we fit a multivariate AR model with the command:

```
> TEMPS.ar <- ar(TEMPS)
```

The order chosen by the fitting algorithm can be printed in the same way. The order chosen for the three city temperature remainders analyzed below is 5. The order was chosen because of the properties of the AIC, but as in the univariate case, the computation of the partial auto-correlation functions should be used to check that the choice of the order is reasonable.

```
> TEMPS.ar$order
[1] 5
> acf.plot(TEMPS.ar,partial=T)
```

Figure 6.1 shows that this value is indeed reasonable.

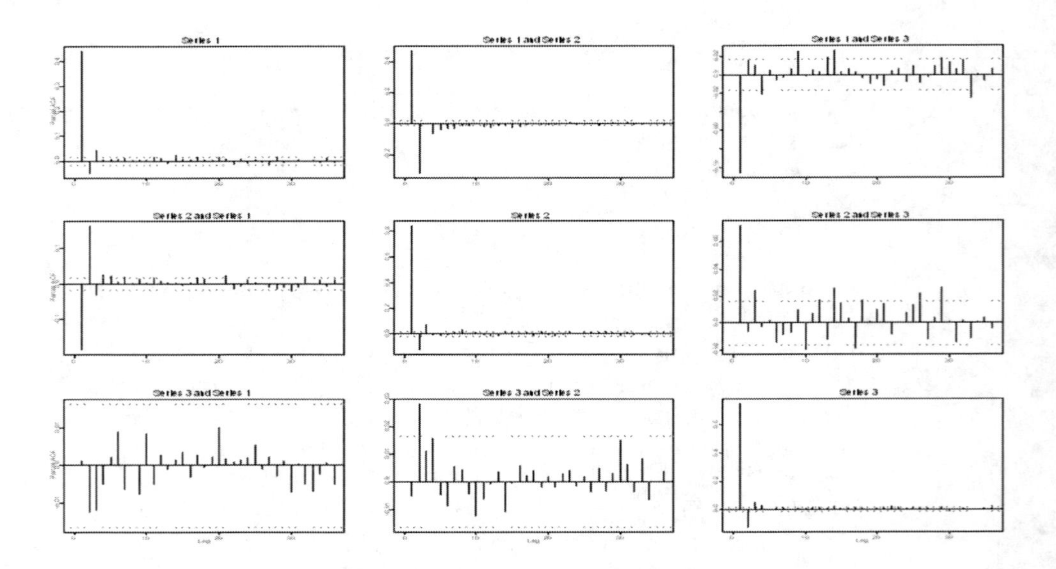

Fig. 6.1. Partial auto-correlation functions of the 3-variate time series of temperature remainders.

Prediction

Except for the technical complexity of the computations which now involve matrices instead of plain numbers, our discussion of the univariate case applies to the multivariate case considered in this section.

For example, if $\hat{A}_1$, $\hat{A}_2$, ... , $\hat{A}_p$ and $\hat{\Sigma}_W$ are the estimates of a d-dimensional AR(p) model (i.e. the parameters of a model fitted to a d-variate data set), then at any instant t (by which we mean at the end of the t-th time interval, when we know the exact values of the outcome $\mathbf{X}_s$ for $s \le t$), the prediction of the next value of the series is given by:

$$\widehat{\mathbf{X}}_{t+1} = \hat{A}_1\mathbf{X}_t + \hat{A}_2\mathbf{X}_{t-1} + \cdots + \hat{A}_p\mathbf{X}_{t-p+1}. \tag{6.2}$$

One can similarly derive the prediction of the value of the series two steps ahead, or for any finite number of time steps in the future. As in the univariate case, when the time series is stable, the longer the horizon of the prediction, the closer it will be to the *long term average* of the series. We shall illustrate this fact on the numerical example we consider below.

When it comes to the implementation in `S-Plus`, the function `pred.ar` can be used in the multivariate case as well. See the discussion of the temperature basket option for an example of how to use this function.

Monte Carlo Simulations & Scenarios Generation

As in the univariate case, random simulation should not be confused with prediction. For the sake of the present discussion, let us assume that we have the values of the vectors $\mathbf{X}_t$, $\mathbf{X}_{t-1}$, ..., $\mathbf{X}_{t-p+1}$ at hand, and that we would like to generate a specific number, say N, of Monte Carlo scenarios for the values of $\mathbf{X}_{t+1}$, $\mathbf{X}_{t+2}$, ..., $\mathbf{X}_{t+M}$.

The correct procedure is to generate N samples of a d-dimensional white noise time series of length M, say $\{\mathbf{W}_s\}_{s=t+1,\dots,t+M}$, with the variance/covariance matrix $\hat{\Sigma}_W$, and then to use the parameters $\hat{A}_1$, $\hat{A}_2$, ..., $\hat{A}_p$ and the recursive formula (6.1) giving the definition of the AR(p) model to generate samples of the AR model from these N samples of the white noise. In other words, for each of the N given samples $\mathbf{W}^{(j)}_{t+1}$, ... , $\mathbf{W}^{(j)}_{t+M}$ of the white noise, we generate the corresponding Monte Carlo scenarios of the series by computing recursively the values $\widehat{\mathbf{X}}^{(j)}_{t+1}$, ... , $\widehat{\mathbf{X}}^{(j)}_{t+M}$ from the formula:

$$\widehat{\mathbf{X}}^{(j)}_{t+k} = \hat{A}_1\widehat{\mathbf{X}}^{(j)}_{t+k-1} + \hat{A}_2\widehat{\mathbf{X}}^{(j)}_{t+k-2} + \cdots + \hat{A}_p\widehat{\mathbf{X}}^{(j)}_{t+k-p}, \qquad k = 1, 2, \dots, M \tag{6.3}$$

for $j = 1, 2, \dots, N$, given the fact that the "hats" over the $\mathbf{X}$'s (i.e. the simulations) are not needed when the true values are available.

	Option #1	Option #2	Option #3
Underlying Series	CDD's in Des Moines	CDD's in Portland	CDD's in Cincinnati
Period	June, July& August	June,July& August	June, July& August
Strike	830	420	980
Rate	US$5,000	US$5,000	US$5,000
Cap	US$1,000,000	US$1,000,000	US$1,000,000

Table 6.7. Characteristics of the three call options offered in lieu of risk mitigation.

6.1.4 **Back to Temperature Options**

We use this subsection to give a practical example of the use of multivariate time series fitting.

Hedging Several Weather Exposures

Let us assume for the sake of illustration, that we are chief risk officer of a company selling thirst quenchers, and that most of the summer revenues come from sales in Des Moines, Portland and Cincinnati. Concerned that an unusually cool summer may hurt the company's revenues, we approach a financial institution to mitigate the possible losses caused by our weather exposure. Let us assume that this institution offers to sell us the three call options whose characteristics are given in Table 6.1.4.

A naive approach would consist of treating the three options separately, performing three times the analysis outlined in our discussion of temperature options at a single location. This is not a good idea, for it would ignore the obvious dependencies between the three temperature series. In order to avoid this shortcoming, we propose to analyze the temperatures in these three different locations *simultaneously*. We read the temperature data from these three cities: Cincinnati (more precisely the meteorological station at Covington airport), Des Moines, and Portland into three time series to which we apply the function `sstl` separately to remove the three seasonal components. Once this is done, and after convincing ourselves that there is no obvious global warming trend to speak of, we bind the three remainder series into a 3-dimensional series which we proceed to model. As in the case of univariate series, we first look for serial correlations by computing and plotting the auto-correlation function.

```
> CVG.stl <- sstl(CVG.ts)
> DM.stl <- sstl(DM.ts)
> PDX.stl <- sstl(PDX.ts)
> TEMPS <- seriesMerge(CVG.stl$rem, DM.stl$rem,PDX.stl$rem)
```

We already explained how to fit a multivariate auto-regressive model to this trivariate time series. We gave the `S-Plus` commands, we computed the order of the fitted model, and we plotted the partial auto-correlation function as a check. We now show how to use this model in order to estimate the probabilities and expectations we need to compute to understand the risks associated with the purcahse of the three options described in Table 6.1.4. As in the univariate case, we use Monte Carlo techniques to estimate, for example, the probability that at least one of the three options will be exercised. We choose this simple problem for the sake of illustration. Indeed, this probability is not of much practical interest. It would be more interesting to associate a cost/penalty/utility to each possible temperature scenario, and to compute the expected utility over all the possible scenarios. In any case, this probability can be computed in the following way:

- Choose a large number of scenarios, say $N = 10,000$;
- Use the fitted model to generate N samples $\mathbf{x}_t^{(j)}$, $\mathbf{x}_{t+1}^{(j)}$, ..., $\mathbf{x}_T^{(j)}$ where t corresponds to March 31st, 1999, and T to August 31st, 1999, $j = 1, 2, \ldots, N$, and each $\mathbf{x}_{t+s}^{(j)}$ is a three dimensional vector representing the temperature remainders in the three cities, on day $t + s$ for the j-th scenario. Remember that according to the explanations given above, this is done by generating the white noise first, and then computing the $\mathbf{x}_{t+s}^{(j)}$ inductively from formula (6.3).
- Add the mean temperatures and the seasonal components to each of the three components to obtain N scenarios of the temperatures in Covington, Des Moines, and Portland starting from March 31st, 1999, and ending August 31st, 1999;
- For each single one of these N tri-variate scenarios, compute the number of CDD's between June 1st and August 31st in each of the three cities;
- Compute the number of times at least one of the three total numbers of CDD's is greater that the corresponding strike, and divide by N.

The Case of a Basket Option

Instead of purchasing three separate options, similar risk mitigation can be achieved by the purchase of a single option written on an index involving the temperatures in the three locations. The simplest instruments available over the counter are written on an underlying index obtained by computing the plain average, or the sum, of the indexes used above for the three cities. In other words, we would compute the total number of CDD's in Des Moines over the summer, the total number of CDD's in Portland over the same period, and finally the total number of CDD's in Cincinnati in the same way, add these three numbers, and compare this compound index to the strike of the option. In other words, instead of being the sum of the three separate pay-offs, the pay-off of the option is a single function of an aggregate underlying index. Such options are called *basket options*. Obviously, the averages could be weighted averages instead of plain averages, the analysis would be the same. Modeling the separate underlying indexes simultaneously is a must in dealing with basket options. There is no way to avoid modeling the dependencies between these indexes. However, the Monte Carlo analysis presented above can be performed in exactly the same way.

The versatility of the Monte Carlo approach can be demonstrated one more time with the analysis of basket options. For each of the N tri-variate scenarios generated as above, one compute the numbers of CDD's in each of the three locations, and add them up together: this is the underlying index. It is then plain to decide if the option is exercised, or to compute the loss/utility associated to such a scenario for the three locations, ... and to compute probabilities of events of interest or expectation of loss/utility functions, by averaging over the scenarios.

6.1.5 **Multivariate MA & ARIMA Models**

This subsection is included for the sake of completeness only. None of the concepts and results presented in this subsection will be used in the sequel (including the problems).

A d-dimensional time series $\mathbf{X} = \{\mathbf{X}_t\}_t$ is said to be a moving average series of order q if there exist $d \times d$ matrices $B_1, B_2, \ldots, B_q$ and a d-variate white noise $\{\mathbf{W}_t\}_t$ such that:

$$\mathbf{X}_t = \mathbf{W}_t + B_1\mathbf{W}_{t-1} + B_2\mathbf{W}_{t-2} + \cdots + B_q\mathbf{W}_{t-q}$$

for all times t. As before we ignored the mean term (which would be a d-dimensional vector in the present situation) by assuming that all the components have already been centered around their respective means. As in the case of a multivariate AR model, the number of parameters can be computed easily. This number of parameters is now $qd^2 + d(d+1)/2$. As in the univariate case, fitting moving average models is more difficult than in the case of auto-regressive processes. It can be done by maximum likelihood, but the computations require sophisticated recursive procedures and they are extremely involved. This is the reason why `S-Plus` does not provide a function (similar to `ar`) to fit multivariate moving average time series.

A d-dimensional time series $\mathbf{X} = \{\mathbf{X}_t\}_t$ is said to be an ARMA(p,q) time series if there exist $d \times d$ matrices $A_1, A_2, \ldots, A_p, B_1, B_2, \ldots, B_q$, and a d-variate white noise $\{\mathbf{W}_t\}_t$ such that:

$$\mathbf{X}_t - A_1\mathbf{X}_{t-1} - A_2\mathbf{X}_{t-2} - \cdots - A_p\mathbf{X}_{t-p} = \mathbf{W}_t + B_1\mathbf{W}_{t-1} + \cdots + B_q\mathbf{W}_{t-q}$$

Finally, as in the univariate case, an ARIMA(p,r,q) time series is defined as a time series which becomes an ARMA(p,q) after r differentiations.

Prediction and Simulation in `S-Plus`

Contrary to the univariate case, `S-Plus` does not provide any function to produce random samples of multivariate ARIMA time series. It does not have a function to compute predictions of future values from a general multivariate ARIMA model either. This is not an issue in the case of AR models. Indeed, who needs such a function when formula 6.2, giving the predictions of futures values of an AR model, is so simple that it can be computed easily without the need for a special function. In fact as we already explained, because of its simplicity, `S-Plus` provides the code of such a function as part of the `ar` help file. This function is called `pred.ar` and we used it both in the univariate case (when we analyze the temperature data of Charlotte) and in the multivariate case (when we modeled simultaneously the daily temperatures at several locations).

6.1.6 **Cointegration**

The concept of cointegration is of crucial importance in the analysis of financial time series. We define it in the simplest possible context: two $I(1)$ time series. See the Notes & Complements section at the end of the chapter for further details and references.

Two unit-root time series are said to be cointegrated if there exists a linear combination of its entries which is stationary. The coefficients of such a linear combination are said to form a cointegration vector. For example, it has been argued that yields of different maturities are cointegrated.

Example

We first discuss an academic example for the purpose of illustration. Let us consider two time series $\{X_n^{(1)}\}_{n=1,\dots,N}$ and $\{X_n^{(2)}\}_{n=1,\dots,N}$ defined by:

$$X_n^{(1)} = S_n + \epsilon_n^{(1)} \quad \text{and} \quad X_n^{(2)} = S_n + \epsilon_n^{(2)}, \qquad n = 1, \dots, N$$

where $S_n = S_0 + W_1 + \cdots + W_n$ is a random walk starting from $S_0 = 0$ constructed form the $N(0,1)$ white noise $\{W_n\}_{n=1,\dots,N}$, and where $\{\epsilon_n^{(1)}\}_{n=1,\dots,N}$ and $\{\epsilon_n^{(2)}\}_{n=1,\dots,N}$ are two independent $N(0,.16)$ white noise sequences which are assumed to be independent of $\{W_n\}_{n=1,\dots,N}$. Since the time series $\{X_n^{(1)}\}_{n=1,\dots,N}$ and $\{X_n^{(2)}\}_{n=1,\dots,N}$ are of the type *random walk plus noise,* they are both integrated of order one, and in particular, non-stationary. However, the linear combination:

$$X_n^{(1)} - X_n^{(2)} = \epsilon_n^{(1)} - \epsilon_n^{(2)}, \qquad n = 1, \dots, N$$

is stationary since it is a white noise with distribution $N(0,.32)$. The random walk $\{S_n\}_n$ is a common trend (though stochastic) for the series $\{X_n^{(1)}\}_n$ and $\{X_n^{(2)}\}_n$, and it disappears when we compute the difference. Cointegration is the state of several time series sharing a common (non-stationary) trend, the remainder terms being stationary.

To emphasize this fact, we consider the case of two models of the same type *random walk plus noise,* but we now assume that the random walks are independent. More precisely, we assume that $\{X_n^{(1)}\}_{n=1,\dots,N}$ and $\{X_n^{(2)}\}_{n=1,\dots,N}$ are defined by:

$$X_n^{(1)} = S_n^{(1)} + \epsilon_n^{(1)} \quad \text{and} \quad X_n^{(2)} = S_n^{(2)} + \epsilon_n^{(2)}.$$

for two random walks $S_n^{(1)} = S_0^{(1)} + W_1^{(1)} + \cdots + W_n^{(1)}$ and $S_n^{(2)} = S_0^{(2)} + W_1^{(2)} + \cdots + W_n^{(2)}$ starting from $S_0^{(1)} = 0$ and $S_0^{(2)} = 0$ respectively, constructed from two independent $N(0,1)$ white noise series $\{W_n^{(1)}\}_{n=1,\ldots,N}$ and $\{W_n^{(2)}\}_{n=1,\ldots,N}$ of size $N = 1024$, and where as before the two independent white noise sequences $\{\epsilon_n^{(1)}\}_{n=1,\ldots,N}$ and $\{\epsilon_n^{(2)}\}_{n=1,\ldots,N}$ have the $N(0,.16)$ distribution, and are independent of $\{W_n^{(1)}\}_{n=1,\ldots,N}$ and $\{W_n^{(2)}\}_{n=1,\ldots,N}$. As before, the time series $\{X_n^{(1)}\}_n$ and $\{X_n^{(2)}\}_n$ are both of the type *random walk plus noise,* but their trends are independent, and no linear combination can cancel them. Indeed it is easy to see that all the linear combinations of $\{X_n^{(1)}\}_n$ and $\{X_n^{(2)}\}_n$ are of the type *random walk plus noise.* Indeed:

$$\begin{aligned} a_1 X_n^{(1)} + a_2 X_n^{(2)} &= a_1 S_n^{(1)} + a_1 \epsilon_n^{(1)} + a_2 S_n^{(2)} + a_2 \epsilon_n^{(2)} \\ &= S_n + \epsilon \end{aligned}$$

where $\{S_n\}_n$ is the random walk starting from $S_0 = a_1 S_0^{(1)} + a_2 S_0^{(2)}$ constructed from the $N(0, a_1^2 + a_2^2)$ white noise $W_n = a_1 W_n^{(1)} + a_2 W_n^{(2)}$ and $\epsilon_n = a_1 \epsilon_n^{(1)} + a_2 \epsilon_n^{(2)}$ is a $N(0, .16a_1^2 + .16a_2^2)$ white noise. So for the *random walk plus noise* series $\{X_n^{(1)}\}_n$ and $\{X_n^{(2)}\}_n$ to be cointegrated, the random walk components have to be the same.

Problem 6.5 is devoted to the illustration of these facts on simulated samples.

6.1.6.1 *Testing for Cointegration*

Two times series $\{X_n^{(1)}\}_n$ and $\{X_n^{(2)}\}_n$, are cointegrated if neither of these series is stationary, and if there exists a vector $\mathbf{a} = (a_1, a_2)$ such that the time series $\{a_1 X_n^{(1)} + a_2 X_n^{(2)}\}_n$ is stationary. As we already mentioned, such a vector is called a cointegration vector.

• Testing for cointegration is easy if the cointegration vector $\mathbf{a} = (a_1, a_2)$ is known in advance. Indeed, this amounts to testing for the stationarity of the appropriate time series $\{a_1 X_n^{(1)} + a_2 X_n^{(2)}\}_n$. This can be done with a unit root test. The `S-Plus` toolbox `S+FinMetrics` provides the function `unitroot` for such a test.

• In general, the cointegration vector cannot be zero, so at least one of its components is non-zero. Assuming that for example that $a_2 \neq 0$ for the sake of definiteness, and dividing by a_2, we see that the series $\{(a_1/a_2) X_n^{(1)} + X_n^{(2)}\}_n$ has to be stationary, or equivalently (if we isolate the mean of this stationary series):

$$X_n^{(2)} = \beta_0 + \beta_1 X_n^{(1)} + \epsilon_n \tag{6.4}$$

for some mean-zero stationary time series $\{\epsilon_n\}_n$. Formula (6.4) says that in order for two time series to be cointegrated, a form of linear model with stationary errors should hold. But as we have seen at the beginning of Chapter 3, linear regression for time series is a *very touchy business* ! Indeed, the residuals are likely to have a strong serial auto-correlation, an obvious sign of lack of indepndence, but also a sign of lack of stationarity. In such a case, one cannot use statistical inference and standard diagnostics to assess the significance of the regression. In the particular example treated in Chapter 3, we overcame these difficulties by considering the log-returns instead of the raw indexes. In general, when the residuals exhibit strong serial correlations, it is recommended to replace the original time series by their first differences, and to perform the regression with these new time series.

This informal discussion is intended to stress the fact that testing for cointegration is not a simple matter. The `S-Plus` toolbox `S+FinMetrics` has several functions to estimate possible cointegration vectors, and test for cointegration using these estimates. The interested reader is encouraged to consult the Notes & Complements at the end of the chapter for references.

6.2 State Space Models: Mathematical Set Up

A state-space model is determined by two equations (or to be more precise, two systems of equations). The first equation:

$$\mathbf{X}_{n+1} = F_n(\mathbf{X}_n, \mathbf{V}_{n+1}) \tag{6.5}$$

describes the time evolution of the state $\mathbf{X}_n$ of a system. Since most of the physical systems of interest are complex, it is to be expected that the state will be described by a vector $\mathbf{X}_n$ whose dimension $d = d_X$ may be large. The integer n labels time and for this reason, we shall use the labels n and t interchangeably. Models are often defined and analyzed in continuous time, and in this case we use the variable t for time. Most practical time series come from sampling of continuous time systems. In these situations we use the notation t_n for these sampling times. Quite often t_n is of the form $t_n = t_0 + n\Delta t$ for a sampling interval Δt. But even when the sample times are not regularly spaced, we shall still use the label n for the quantities measured, estimated, computed, ... at time t_n.

For each $n \geq 0$, F_n is a vector-valued function (i.e. a function taking values in the space $\mathbb{R}^d$ of d-dimensional vectors) which depends upon the (present) time n, the (current) state of the system as given by the state vector $\mathbf{X}_n$, and a system noise $\mathbf{V}_{n+1}$ which is a (possibly multivariate) random quantity. Throughout this chapter we shall assume that the state equation (6.5) is linear in the sense that it is of the form:

$$\mathbf{X}_{n+1} = F_n\mathbf{X}_n + \mathbf{V}_{n+1} \tag{6.6}$$

for some $d \times d$ matrix F_n and a $d \times 1$ random vector $\mathbf{V}_{n+1}$. The system matrix F_n will be independent of n in most applications. The random vectors $\mathbf{V}_n$ are assumed to be mean zero and independent, and we shall denote by Σ_V their common variance/covariance matrix. Moreover, the noise term $\mathbf{V}_{n+1}$ appearing in (6.5) and (6.6) is assumed to be independent of all the $\mathbf{X}_k$ for $k \leq n$. This is emphasized by our use of the index $n+1$ for this noise term.

Notice that, except for a change in notation, such a linear state space system with state matrix constant over time, is nothing but a multivariate AR(1) model! Nothing very new so far. Except for the fact which we are about to emphasize below: such an AR(1) process may not be observed in full. This will dramatically enlarge the realm of applications for which these models can be used, and this will enhance their usefulness.

The second equation (again, to be precise, one should say the second system of equations) is the so-called observation equation. It is of the form:

$$\mathbf{Y}_n = G_n(\mathbf{X}_n, \mathbf{W}_n). \tag{6.7}$$

It describes how, at each time n, the actual observation vector $\mathbf{Y}_n$ depends upon the state $\mathbf{X}_n$ of the system. Notice that in general, we make several simultaneous measurements, and that, as a consequence, the observation should be modeled as a vector $\mathbf{Y}_n$ whose dimension d_Y will typically be smaller than the dimension $d = d_X$ of the state $\mathbf{X}_n$. The observation functions G_n are $\mathbb{R}^{d_Y}$-valued functions which depend upon the (present) time n, the (current) state of the system $\mathbf{X}_n$, and an observation noise $\mathbf{W}_n$ which is a (possibly multivariate) random quantity. Throughout this chapter we shall also assume that the observation equation (6.7) is linear in the sense that it is of the form:

$$\mathbf{Y}_n = G_n\mathbf{X}_n + \mathbf{W}_n \tag{6.8}$$

for some $d_Y \times d_X$ matrix G_n and a $d_Y \times 1$ random vector $\mathbf{W}_n$. As for the system matrix F, in most of the applications considered in this chapter, the observation matrices G_n will not change with n, and when this is the case, we denote by G their common value. The random vectors $\mathbf{W}_n$ modeling the observation noise are assumed to be mean zero, independent (of each other) identically distributed, and also independent of the system noise terms $\mathbf{V}_n$. We shall denote by Σ_W their common variance/covariance matrix.

The challenges of the analysis of state-space models are best stated in the following set of bullet points:

At any given time n, and for any values $\mathbf{y}_1, \dots, \mathbf{y}_n$ of the observations (which we view as realizations of the random vectors $\mathbf{Y}_1, \dots, \mathbf{Y}_n$), find estimates for:

- The state vector $\mathbf{X}_n$ at the same time. This is the so-called *filtering* problem;
- The future states of the system $\mathbf{X}_{n+m}$ for $m > n$. This is the so-called *prediction* problem;
- A past occurrence $\mathbf{X}_m$ of the state vector (for some time $m < n$). This is the so-called *smoothing* problem.

Note that as usual, when we say estimate, we keep in mind that an estimate does not have much value if it does not come with a companion estimate of the associated error, and consequently of the confidence we should have in these estimates.

Remark. Linearization of Nonlinear Systems In many applications of great practical interest, the observation equation is of the form

$$\mathbf{Y}_n = \Phi(\mathbf{X}_n) + \mathbf{W}_n \tag{6.9}$$

for some vector-valued function Φ whose i-th component $\Phi_i(\mathbf{x})$ is a nonlinear function of the components of the state $\mathbf{x}$. This type of observation equation (6.9) is in principle not amenable to the theory presented below because the function Φ is not linear, i.e. it is not given by the product of a matrix with the vector $\mathbf{X}$. The remedy is to replace this nonlinear observation equation by the approximation provided by the first order Taylor expansion of the function Φ:

$$\Phi(\mathbf{X}_n) = \Phi(\mathbf{X}_{n-1}) + \nabla\Phi(\mathbf{X}_{n-1})[\mathbf{X}_n - \mathbf{X}_{n-1}] + HOT's,$$

where we encapsulated all the higher order terms of the expansion into the generic notation $HOT's$ which stands for "Higher Order Terms". If we are willing to ignore the $HOT's$, which is quite reasonable when $\|\mathbf{X}_n - \mathbf{X}_{n-1}\|$ is not too large, or if we include them in the observation noise term, then once in this form, the observation equation can be regarded as a linear equation given by the observation matrix $\nabla\Phi(\mathbf{X}_{n-1})$.

This linearization idea gives the rationale for what is called the extended Kalman filter.

6.3 Factor Models as Hidden Markov Processes

Before we consider the filtering and prediction problems, we pause to explain the important role played by the analysis of state space systems in financial econometrics. This section provides a general discussion somewhat of an abstract nature, and it can safely be skipped by the reader only interested in the mechanics of filtering and prediction.

Factor Models

Attempts at giving a general definition of factor models are usually hiding their intuitive simplicity. So instead of shooting for a formal introduction, we shall consider specific examples found in the financial arena. Let us assume that we are interested in tracking the performance of d financial assets whose returns over the n-th time period we denote by $Y_{1,n}$, $Y_{2,n}$, ..., $Y_{d,n}$. The assumption of a k-factor model is that the dynamics of these returns are driven by k economic factors $X_1, X_2, \ldots, X_k$ which are also changing over time. So we denote by $X_{1,n}, X_{2,n}, \ldots, X_{k,n}$ the values of these factors at time n. The main feature of a linear factor model is to assume that the individual returns $Y_{i,n}$ are related to the factors $X_{j,n}$ by a linear formula:

$$Y_{i,n} = g_{i,1,n}X_{1,n} + g_{i,2,n}X_{2,n} + \cdots + g_{i,k,n}X_{k,n} + w_{i,n} \tag{6.10}$$

for a given set of (deterministic) parameters $g_{i,j,n}$ and random residual terms $w_{i,n}$. Grouping the d returns $Y_{i,n}$ in a d-dimensional return vector $\mathbf{Y}_n$, grouping the k factors $X_{j,n}$ in a k-dimensional factor vector $\mathbf{X}_n$, grouping the noise terms $w_{i,n}$ in a d-dimensional residual vector $\mathbf{W}_n$ and finally. grouping all the parameters $g_{i,j,n}$ in a $d \times k$ matrix G_n, we rewrite the system of equations (6.10) in the vector/matrix form:

$$\mathbf{Y}_n = G_n \mathbf{X}_n + \mathbf{W}_n$$

which is exactly the form of the observation equation (6.8) of a linear state-space model introduced in the previous section. In analogy with the CAPM terminology, the columns of the observation matrix G_n are called the *betas* of the underlying factors.

Assumptions of the Model

Most econometric analyses rely heavily on the notion of information: we need to keep track at each time n of the information available at that time. For the purposes of this section, we shall denote by $\mathcal{I}_n$ the information available at time n. Typically, this information will be determined by the knowledge of the values of the return vectors $\mathbf{Y}_n$, $\mathbf{Y}_{n-1}$, ... and the factor vectors $\mathbf{X}_n$, $\mathbf{X}_{n-1}$, ... available up to and including that time. In order to take advantage of this information, the properties of the models are formulated in terms of conditional probabilities and conditional expectations given this information. To be more specific, we assume that, conditionally on the past information $\mathcal{I}_{n-1}$, the quantities entering the factor model (6.10) satisfy:

- the residual terms $\mathbf{W}_n$ have mean zero, i.e. $\mathbb{E}\{\mathbf{W}_n|\mathcal{I}_{n-1}\} = 0$
- the variance/covariance matrix Σ_W of the residual terms does not change with time, i.e. $\text{var}\{\mathbf{W}_n|\mathcal{I}_{n-1}\} = \Sigma_W$
- the residual terms $\mathbf{W}_n$ and the factors $\mathbf{X}_n$ are uncorrelated:

$$\mathbb{E}\{\mathbf{X}_n \mathbf{W}_n^t|\mathcal{I}_{n-1}\} = 0.$$

It is clear that the residual terms behave like a white noise, and for this reason, it is customary to assume directly that they form a white noise independent of the random sequence of the factors.

Remarks

1. There is no uniqueness in the representation (6.10). In particular, one can change the definition of the factors and still preserve the form of the representation. Indeed, if U is any invertible $k \times k$ matrix, one sees that:

$$\mathbf{Y}_n = G_n \mathbf{X}_n + \mathbf{W}_n = G_n U^{-1} U \mathbf{X}_n + \mathbf{W}_n = \tilde{G}_n \tilde{\mathbf{X}}_n + \mathbf{W}_n$$

provided we set $\tilde{G}_n = G_n U^{-1}$ and $\tilde{\mathbf{X}}_n = U \mathbf{X}_n$. So the returns can be explained by the new factors given by the components of the vector $\tilde{\mathbf{X}}_n$. Not only does this argument show that there is no hope for any kind of uniqueness in the representation (6.10), but it also shows that one can replace the original factors by appropriate linear combinations, and this makes it possible to make sure that the rank of the matrix G_n can be equal to the number k of factors. Indeed, if that is not the case, one can always replace the original factors by a minimal set of linear combinations with the same rank.

2. The aggregation of factors by linear combinations does have advantages from the mathematical point of view, but it also has serious drawbacks. Indeed, some of the factors may be observable (this is, for example, the case if one uses interest rates, or other published economic indicators in the analysis of portfolio returns) and bundling them together with generic factors, which may not be observable, may dilute this desirable (practical) property of some of the factors.

Dynamics of the Factors

The most general way to prescribe the time evolution of the factors is by giving the conditional distribution of the factor vector $\mathbf{X}_n$ given its past values $\mathbf{X}_{n-1}$, $\mathbf{X}_{n-2}$, $\dots$.

A typical example is given by the ARCH prescription which we will analyze in Chapter 7. This model is especially simple in the case of a one-factor model, i.e. when $k = 1$. In this case, the dynamics of the factor is prescribed by the conditional distribution

$$X_n | X_{n-1}, X_{n-2}, \dots \sim N(0, \sigma_n^2) \qquad \text{with} \qquad \sigma_n^2 = \alpha_0 + \sum_{i=1}^{p} \alpha_i X_{n-i}^2. \tag{6.11}$$

In the present chapter, we restrict ourselves to linear factor models for which the dynamics of the factors are given by an equation of the form:

$$\mathbf{X}_n = F_n \mathbf{X}_{n-1} + \mathbf{V}_n \tag{6.12}$$

for some $k \times k$ matrix F_n (possibly changing with n) and a k-variate white noise $\mathbf{V}$ which is assumed to be independent of the returns $\mathbf{Y}$. In other words, the conditional distribution of $\mathbf{X}_n$ depends only upon its last value $\mathbf{X}_{n-1}$, the latter providing the conditional mean $F_n \mathbf{X}_{n-1}$, while the fluctuations around this mean are determined by the distribution of the noise $\mathbf{V}_n$. Given all that, we are now in the framework of the linear state-space models introduced in the previous section.

Remarks

1. As we shall see in Section 6.6 on the state-space representation of time series, it is always possible to accommodate more general dependence involving more of the past values $\mathbf{X}_{n-2}$, $\mathbf{X}_{n-3}$, $\dots$ in equation (6.12). Indeed, increasing the dimension of the factor vector, one can always include these past values in the current factor vector to reduce the dependence of $\mathbf{X}_n$ upon $\mathbf{X}_{n-1}$ only.

2. Factors are called exogenous when their values are dependent upon quantities (indexes, instruments, ...) external to the system formed by the returns included in the state vector. They are called endogenous when they can be expressed as functions

of the individual returns entering the state vector. There is a mathematical result that states that any *linear* factor model can be rewritten in such a way that all the factors become endogenous. This anti-climatic result is limited to the linear case, and because of its counter-intuitive nature, we shall not use it.

6.4 Kalman Filtering of Linear Systems

This section is devoted to the derivation of the recursive equations giving the optimal filter for linear (Gaussian) systems. They were discovered simultaneously and independently by Kalman and Bucy in the late fifties, but strangely enough, they are most of the time referred to as the Kalman filtering equations.

6.4.1 One-Step-Ahead Prediction

To refresh our memory, we restate the definition of the linear prediction operator originally introduced in Chapter 5 when we discussed partial auto-correlation functions. Given observations $\mathbf{Y}_1, \ldots, \mathbf{Y}_n$ up to and including n (which we should think of as the present time), and given a random vector $\mathbf{Z}$ with the same dimension as $\mathbf{Y}$, we denote by $E_n(\mathbf{Z})$ the best prediction of $\mathbf{Z}$ by linear combinations of the $\mathbf{Y}_m$ for $0 \leq m \leq n$. Since "*best prediction*" is implicitly understood in the least squares sense, $E_n(\mathbf{Z})$ is the linear combination $\alpha_1\mathbf{Y}_1 + \cdots + \alpha_n\mathbf{Y}_n$ which minimizes the quadratic error:

$$\mathbb{E}\{\|\mathbf{Z} - (\alpha_1\mathbf{Y}_1 + \cdots + \alpha_n\mathbf{Y}_n)\|^2\}.$$

The best linear prediction $E_n(\mathbf{Z})$ should be interpreted as the orthogonal projection of $\mathbf{Z}$ onto the span of $\mathbf{Y}_1, \ldots, \mathbf{Y}_n$. Unfortunately, in some applications, it may not be natural to restrict the approximation to linear combinations of the $\mathbf{Y}_j$'s. If we lift this restriction and allow all possible functions of the $\mathbf{Y}_j$'s in the approximation, then $E_n(\mathbf{Z})$ happens to be the conditional expectation of $\mathbf{Z}$ given the knowledge of the $\mathbf{Y}_m$ for $1 \leq m \leq n$. Notice that all the derivations below are true in both cases, i.e. whether $E_n(\mathbf{Z})$ is interpreted as the best linear prediction or the conditional expectation. In fact, when the random vectors $\mathbf{Z}$ and $\mathbf{Y}_m$ for $m \leq n$ are jointly Gaussian (or more generally jointly elliptically distributed), then the two definitions of $E_n(\mathbf{Z})$ coincide. For these reasons, we shall not insist on which actual definition of $E_n(\mathbf{Z})$ we use. For the sake of definiteness, we shall assume that the observations and the state vectors are jointly Gaussian and we shall understand the notation E_n as a conditional expectation.

The strength of the theory developed in this section is based on the fact that it is possible to compute the best one-step-ahead predictions recursively. More specifically, if the best prediction $\widehat{\mathbf{X}}_n$ of the unobserved state $\mathbf{X}_n$ (computed on the basis of the observations $\mathbf{Y}_m$ for $1 \leq m \leq n-1$) is known, the computation of the next best guess $\widehat{\mathbf{X}}_{n+1}$ requires only the knowledge of $\widehat{\mathbf{X}}_n$ and the new observation $\mathbf{Y}_n$. Well, this is almost true. The only *little lie* comes from the fact that, not only should we know the current one-step-ahead prediction, but also its prediction quadratic error as defined by the matrix:

$$\Omega_n = \mathbb{E}\{(\mathbf{X}_n - \widehat{\mathbf{X}}_n)(\mathbf{X}_n - \widehat{\mathbf{X}}_n)^t\}.$$

6.4.2 Derivation of the Recursive Filtering Equations

We now proceed to the rigorous derivation of this claim, and to the derivation of the formulae which we shall use in the practical filtering applications that we consider in this text. This derivation is based on the important concept of innovation. The innovation series is given by its entries I_n defined as:

$$I_n = \mathbf{Y}_n - E_{n-1}(\mathbf{Y}_n).$$

The innovation I_n gives the information contained in the latest observation $\mathbf{Y}_n$ which was not already contained in the previous observations. The discussion which follows is rather technical and the reader interested in practical applications more than theoretical derivations can skip the next two pages of computations, and jump directly to the boxed recursive update formulae.

First we remark that the innovations I_n are mean zero and uncorrelated (and consequently independent in the Gaussian case). Indeed:

$$\mathbb{E}\{I_n\} = \mathbb{E}\{\mathbf{Y}_n\} - \mathbb{E}\{E_{n-1}(\mathbf{Y}_n)\} = \mathbb{E}\{\mathbf{Y}_n\} - \mathbb{E}\{\mathbf{Y}_n\} = 0$$

and a geometric argument (based on the properties of orthogonal projections), gives the fact that $I_n = \mathbf{Y}_n - E_{n-1}(\mathbf{Y}_n)$ is orthogonal to all linear combinations of the $\mathbf{Y}_1, \ldots, \mathbf{Y}_{n-1}$, and consequently orthogonal to $I_{n-1}, I_{n-2}, \ldots$ which are particular linear combinations. In other words, except possibly for the fact that they may not have the same variance/covariance matrix, the I_n form an independent (or at least uncorrelated) sequence, and they can be viewed as a white noise of their own.

We now proceed to the computation of the variance/covariance matrices of the innovation vectors I_n. Notice that $E_{n-1}(\mathbf{W}_n) = \mathbf{0}$ since $\mathbf{W}_n$ is independent of the past observations $\mathbf{Y}_1, \ldots, \mathbf{Y}_{n-1}$. Consequently, applying the prediction operator E_{n-1} to both sides of the observation equation we get:

$$E_{n-1}(\mathbf{Y}_n) = E_{n-1}(G\mathbf{X}_n) + E_{n-1}(\mathbf{W}_n) = GE_{n-1}(\mathbf{X}_n) = G\widehat{\mathbf{X}}_n$$

and consequently:

$$\begin{aligned} I_n &= G\mathbf{X}_n + \mathbf{W}_n - G\widehat{\mathbf{X}}_n \\ &= G(\mathbf{X}_n - \widehat{\mathbf{X}}_n) + \mathbf{W}_n. \end{aligned}$$

$\mathbf{W}_n$ is independent of $\mathbf{X}_n$ by definition, moreover, $\mathbf{W}_n$ is also independent of $\widehat{\mathbf{X}}_n$ because the latter is a linear combination of $\mathbf{Y}_1, \ldots, \mathbf{Y}_{n-1}$ which are all independent of $\mathbf{W}_n$. Consequently, the two terms appearing in the above right hand side are independent, and the variance/covariance matrix of I_n is equal to the sum of the variance/covariance matrices of these two terms. Consequently:

$$\begin{aligned}\Sigma_{I_n} &= G\mathbb{E}\{(\mathbf{X}_n - \widehat{\mathbf{X}}_n)(\mathbf{X}_n - \widehat{\mathbf{X}}_n)^t\}G^t + \mathbb{E}\{\mathbf{W}_n\mathbf{W}_n^t\} \\ &= G\Omega_n G^t + \Sigma_W. \end{aligned} \tag{6.13}$$

This very geometric argument also implies that the operator E_n which gives the best linear predictor as a function of $\mathbf{Y}_1, \ldots, \mathbf{Y}_{n-1}$, and $\mathbf{Y}_n$ can also be viewed as the best linear predictor as a function of $\mathbf{Y}_1, \ldots, \mathbf{Y}_{n-1}$ and I_n. In particular this implies that:

$$\begin{aligned}\widehat{\mathbf{X}}_{n+1} &= E_{n-1}(\mathbf{X}_{n+1}) + \mathbb{E}\{\mathbf{X}_{n+1}|I_n\} \\ &= E_{n-1}(F\mathbf{X}_n + \mathbf{V}_{n+1}) + \mathbb{E}\{\mathbf{X}_{n+1}I_n^t\}\Sigma_{I_n}^{-1}I_n \\ &= F\widehat{\mathbf{X}}_n + \mathbb{E}\{\mathbf{X}_{n+1}I_n^t\}\Sigma_{I_n}^{-1}I_n, \end{aligned} \tag{6.14}$$

where we computed the conditional expectation as an orthogonal projection, and where we used the fact that $E_{n-1}(\mathbf{V}_{n+1}) = \mathbf{0}$ since $\mathbf{V}_{n+1}$ is independent of $\mathbf{Y}_1$, $\ldots$, $\mathbf{Y}_{n-1}$, and $\mathbf{Y}_n$. Note that we assume that the variance/covariance matrix of the innovation is invertible. Also, note that:

$$\begin{aligned}\mathbb{E}\{\mathbf{X}_{n+1}I_n^t\} &= \mathbb{E}\{(F\mathbf{X}_n + \mathbf{V}_{n+1})[(\mathbf{X}_n - \widehat{\mathbf{X}}_n)^t G^t + \mathbf{W}_n^t]\} \\ &= \mathbb{E}\{F\mathbf{X}_n(\mathbf{X}_n - \widehat{\mathbf{X}}_n)^t G^t\} \\ &= F\mathbb{E}\{(\mathbf{X}_n - \widehat{\mathbf{X}}_n)(\mathbf{X}_n - \widehat{\mathbf{X}}_n)^t\}G^t \\ &= F\Omega_t G^t, \end{aligned} \tag{6.15}$$

where we used the facts that:

1. $\mathbf{V}_{n+1}$ is independent of $\mathbf{X}_n$, $\widehat{\mathbf{X}}_n$ and $\mathbf{W}_n$;
2. $\mathbf{W}_n$ is independent of $\mathbf{X}_n$; and finally
3. $\mathbb{E}\{\widehat{\mathbf{X}}_n)(\mathbf{X}_n - \widehat{\mathbf{X}}_n)^t\} = 0$.

At this stage, it is useful to introduce the following notation.

$$\begin{cases} \Delta_n = G\Omega_n G^t + \Sigma_W \\ \Theta_n = F\Omega_n G^t \end{cases} \tag{6.16}$$

Notice that (6.13) shows that the matrix Δ_n is just the variance/covariance matrix of the innovation random vector I_n. The matrix $\Theta_n\Delta_n^{-1}$ which appears in several of the important formulae derived below is sometimes called the "Kalman gain" matrix in the technical filtering literature. This terminology finds its origin in the fact that using (6.15) and (6.16), we can rewrite (6.14) as:

$$\widehat{\mathbf{X}}_{n+1} = F\widehat{\mathbf{X}}_n + \Theta_n\Delta_n^{-1}I_n. \tag{6.17}$$

This formula gives a (recursive) update equation for the one-step-ahead predictions, but hidden in the correction term $\Theta_n\Delta_n^{-1}I_t$, is the error matrix Ω_n. So this update equation for the one-step-ahead prediction of the state cannot be implemented without being complemented with a practical procedure to compute the error Ω_n. We do that now.

$$\begin{aligned}\Omega_{n+1} &= \mathbb{E}\{(\mathbf{X}_{n+1} - \widehat{\mathbf{X}}_{n+1})(\mathbf{X}_{n+1} - \widehat{\mathbf{X}}_{n+1})^t\} \\ &= \mathbb{E}\{\mathbf{X}_{n+1}\mathbf{X}_{n+1}^t\} - \mathbb{E}\{\mathbf{X}_{n+1}\widehat{\mathbf{X}}_{n+1}^t\} - \mathbb{E}\{\widehat{\mathbf{X}}_{n+1}\mathbf{X}_{n+1}^t\} + \mathbb{E}\{\widehat{\mathbf{X}}_{n+1}\widehat{\mathbf{X}}_{n+1}^t\} \\ &= \mathbb{E}\{\mathbf{X}_{n+1}\mathbf{X}_{n+1}^t\} - \mathbb{E}\{\widehat{\mathbf{X}}_{n+1}\widehat{\mathbf{X}}_{n+1}^t\}\end{aligned}$$

because:

$$\mathbb{E}\{\mathbf{X}_{n+1}\widehat{\mathbf{X}}_{n+1}^t\} = \mathbb{E}\{\widehat{\mathbf{X}}_{n+1}\mathbf{X}_{n+1}^t\} = \mathbb{E}\{\widehat{\mathbf{X}}_{n+1}\widehat{\mathbf{X}}_{n+1}^t\}.$$

Consequently:

$$\begin{aligned}\Omega_{n+1} &= \mathbb{E}\{(F\mathbf{X}_n + \mathbf{V}_{n+1})(F\mathbf{X}_n + \mathbf{V}_{n+1})^t\} \\ &\qquad - \mathbb{E}\{(F\widehat{\mathbf{X}}_n + \Theta_n\Delta_n^{-1}I_n)(F\widehat{\mathbf{X}}_n + \Theta_n\Delta_n^{-1}I_n)^t\} \\ &= F\mathbb{E}\{\mathbf{X}_n\mathbf{X}_n^t\}F^t + \mathbb{E}\{\mathbf{V}_{n+1}\mathbf{V}_{n+1}^t\} \\ &\qquad - F\mathbb{E}\{\widehat{\mathbf{X}}_n\widehat{\mathbf{X}}_n^t\}F^t + \Theta_n\Delta_n^{-1}\mathbb{E}\{I_nI_n^t\}\Delta_n^{-1}\Theta_n^t \\ &= F(\mathbb{E}\{\mathbf{X}_n\mathbf{X}_n^t\} - \mathbb{E}\{\widehat{\mathbf{X}}_n\widehat{\mathbf{X}}_n^t\}F^t + \Sigma_V + \Theta_n\Delta_n^{-1}\Theta_n^t \\ &= F\Omega_nF^t + \Sigma_V + \Theta_n\Delta_n^{-1}\Theta_n^t,\end{aligned}$$

where we used the facts that:

1. $\mathbf{V}_{n+1}$ is independent of $\mathbf{X}_n$;
2. I_n is independent of (or at least orthogonal to) $\widehat{\mathbf{X}_n}$;
3. the matrix Δ_n is symmetric since it is a variance/covariance matrix.

For later reference, we summarize the recursion formulae in a box:

$$\boxed{\begin{aligned}\widehat{\mathbf{X}}_{n+1} &= F\widehat{\mathbf{X}}_n + \Theta_n\Delta_n^{-1}(\mathbf{Y}_n - G\widehat{\mathbf{X}}_n) \\ \Omega_{n+1} &= F\Omega_nF^t + \Sigma_V - \Theta_n\Delta_n^{-1}\Theta_n^t\end{aligned}}$$

Recall the definitions (6.16) for the meanings of the matrices Δ_t and Θ_t.

6.4.3 Writing an S Function for Kalman Prediction

As we explain in the appendix giving our short introduction to `S-Plus`, it is very easy to add to `S-Plus` the functionality of our favorite homegrown functions: roughly speaking, in order to write your own `S` function, you encapsulate a list of `S` commands delimited by (opening and closing) curly brackets in a text file which you source with the command:

```
> source(``myfunctions.s'')
```

if the name of your text file is myfunction.s. The name of the S object returned by your function should be the last line of the body of the function (i.e. the list of S commands). If your function is of interest for its side effects, and in particular if your function is not intended to return anything, NULL should appear on this last line of text. Moreover, if your S function needs arguments, these arguments should be listed on the first line of the S code. See the following example for details. Our final remark is that the same text file can contain the codes of several functions.

We already gave several examples of homegrown S-Plus functions. The function eda.shape in the appendix and a couple of time manipulation function in the previous chapter. The following code produces a function which computes the one-step-ahead prediction of the Kalman filter of a linear state-space model.

```
kalman <- function(FF,SigV,GG,SigW,Xhat,Omega,Y)
{
  Delta <- GG %*% Omega %*% t(GG) + sigW
   Theta <- FF %*% Omega %*% t(GG)
   X <- FF%*%Xhat + Theta%*%solve(Delta)%*%(Y-GG%*%Xhat)
   Om <- FF%*%Omega%*%t(FF) + SigV
                        - Theta%*%solve(Delta)%*%t(Theta)
   Ret <- list(xpred = X, error=Om)
   Ret
}
```

The following remarks should help with understanding the features of this S-Plus function.

1. As we already mentioned, the transpose of a matrix is obtained by the function t, so that t(A) stands for the transpose of the matrix A.
2. The inverse of a matrix A is given by solve(A) (see the online help for the explanation of this terminology).
3. S-Plus has a certain number of *reserved* symbols which cannot be used as object names if one does not want to mask the actual S objects. This is the case for the symbols t, c, ... which we already encountered, but also for F which means FALSE in S. For this reason we used the notation FF for the system matrix. Similarly we used the notation GG for the observation matrix.
4. A call to the function defined by the S code above must be of the form:

   ```
   > PRED  <- kalman(FF,SigV,GG,SigW,Xhat,Omega,Y)
   ```

   and the S object PRED returned by the function is what is defined on the last line of the body of the function. In the present situation, it is a list with two elements, the one-step-ahead prediction for the next state of the system, and the estimate of the quadratic error. These elements can be extracted from the list by PRED$xpred and PRED$error respectively.
5. The above code was written with pedagogy in mind, not efficiency. It is not optimized. In particular, the inversion of the matrix Delta, and the product of the inverse by Theta are computed twice. This is a waste of computer resources.

This code can easily be streamlined, and made more efficient, but we refrained from doing so for the sake of clarity.

6.4.4 **Filtering**

We now show how to derive a recursive update for the filtering problem by deriving the zero-step-ahead prediction (i.e. the simultaneous estimation of the unobserved state) from the recursive equations giving the one-step-ahead prediction of the unobserved state. From this point on, we use the notation:

$$\widehat{\mathbf{Z}}_{n|m} = E_m(\mathbf{Z}_n)$$

for the $(n-m)$ step(s) ahead prediction of $\mathbf{Z}_n$ given the information contained in $\mathbf{Y}_1$, ..., $\mathbf{Y}_m$. With this notation, the one step ahead prediction analyzed in the previous subsection can be rewritten as:

$$\widehat{\mathbf{X}}_{n+1} = \widehat{\mathbf{X}}_{n+1|n}.$$

For the filtering problem, the quantity of interest is the best prediction:

$$\widehat{\mathbf{X}}_{n|n} = E_n(\mathbf{X}_n)$$

of $\mathbf{X}_n$ by a linear function of the observations $\mathbf{Y}_n$, $\mathbf{Y}_{n-1}$, ..., $\mathbf{Y}_1$. As in the case of the one-step prediction, we cannot find recursive update formulae without involving the update of the error covariance matrix:

$$\Omega_{n|n} = \mathbb{E}\{(\mathbf{X}_n - \widehat{\mathbf{X}}_{n|n})(\mathbf{X}_n - \widehat{\mathbf{X}}_{n|n})^t\}.$$

The innovation argument used earlier gives:

$$\begin{aligned} E_n(\mathbf{X}_n) &= E_{n-1}(\mathbf{X}_n) + \mathbb{E}\{\mathbf{X}_n I_n^t\}\mathbb{E}\{I_n I_n^t\}^{-1} I_n \\ &= \widehat{\mathbf{X}}_n + \mathbb{E}\{\mathbf{X}_n(G(\mathbf{X}_n - \widehat{\mathbf{X}}_n) + \mathbf{W}_n)^t\}\Delta_n^{-1} I_n \\ &= \widehat{\mathbf{X}}_n + \Omega_n G^t \Delta_n^{-1} I_n \end{aligned}$$

and computing Ω_n as a function of $\Omega_{n|n}$ we get:

$$\Omega_n = \Omega_{n|n} + \Omega_n G^t \Delta_n^{-1} G \Omega_n^t.$$

We summarize these update formulae in a box for later references:

$$\begin{aligned} \widehat{\mathbf{X}}_{n|n} &= \widehat{\mathbf{X}}_n + \Omega_n G^t \Delta_n^{-1}(\mathbf{Y}_n - G\widehat{\mathbf{X}}_n) \\ \Omega_{n|n} &= \Omega_n - \Omega_n G \Delta_n^{-1} G \Omega_n^t. \end{aligned}$$

Remark. Later in the text, we will illustrate the use of the filtering recursions on the example of the "time varying beta's" version of the CAPM model for which the observation matrix G changes with n. A careful look at the above derivations shows that the recursive formulae still hold as long as we replace the matrix G by its value at time n. The above `S-Plus` function needs to be modified, either by passing the whole sequence $\{G_n\}_n$ of matrices as parameter to the function, or by adding the code necessary to compute the matrices G_n on the fly whenever possible. This will appear natural in the application that we give later in the chapter.

6.4.5 More Predictions

Using the same arguments as above we can construct all sorts of predictions.

k-Steps-Ahead Prediction of the Unobserved State of the System

Let us first compute the *two steps ahead prediction.*

$$\begin{aligned}\widehat{\mathbf{X}}_{n+2|n} = E_n(\mathbf{X}_{n+2}) &= E_n(F\mathbf{X}_{n+1} + \mathbf{V}_{n+2})\\ &= FE_n(\mathbf{X}_{n+1})\\ &= F\widehat{\mathbf{X}}_{n+1|n}\end{aligned}$$

with the notation of the previous subsection for the one-step-ahead prediction. We used the fact that $\mathbf{V}_{n+2}$ is independent of the observations $\mathbf{Y}_m$ with $m \leq n$. The above result shows how to compute the two-steps-ahead prediction in terms of the one-step-ahead prediction. One computes the *three steps ahead prediction* similarly. Indeed:

$$\begin{aligned}\widehat{\mathbf{X}}_{n+3|n} = E_n(\mathbf{X}_{n+3}) &= E_n(F\mathbf{X}_{n+2} + \mathbf{V}_{n+3})\\ &= FE_n(\mathbf{X}_{n+2})\\ &= F\widehat{\mathbf{X}}_{n+2|n} = F^2\widehat{\mathbf{X}}_{n+1|n}\end{aligned}$$

Obviously, the k-steps-ahead prediction is given by the formula:

$$\widehat{\mathbf{X}}_{n+k|n} = F^{k-1}\widehat{\mathbf{X}}_{n+1|n} \tag{6.18}$$

and the corresponding error

$$\Omega_{n+k|n} = \mathbb{E}\{(\mathbf{X}_{n+k} - \widehat{\mathbf{X}}_{n+k|n})(\mathbf{X}_{n+k} - \widehat{\mathbf{X}}_{n+k|n})^t\}$$

is given by the formula:

$$\Omega_{n+k|n} = F^{k-1}\Omega_{n+1|n}(F^{k-1})^t.$$

Notice that $\Omega_{n+1|n}$ was denoted by Ω_{n+1} earlier in our original derivation of the one-step ahead prediction. We summarize these results in a box for easier reference.

$$
\begin{aligned}
\widehat{\mathbf{X}}_{n+k|n} &= F^{k-1}\widehat{\mathbf{X}}_{n+1|n} \\
\Omega_{n+k|n} &= F^{k-1}\Omega_{n+1|n}(F^{k-1})^t.
\end{aligned}
$$

***k*-Steps-Ahead Predition of the Observations**

Finally, we remark that it is also possible to give formulae for the k-steps-ahead predictions of the future observations. Indeed:

$$
\begin{aligned}
\widehat{\mathbf{Y}}_{n+1|n} = E_n(\mathbf{Y}_{n+1}) &= E_n(G\mathbf{X}_{n+1} + \mathbf{W}_{n+1}) \\
&= GE_n(\mathbf{X}_{n+1}) \\
&= G\widehat{\mathbf{X}}_{n+1|n}
\end{aligned}
$$

because $\mathbf{W}_{n+1}$ is independent of all the $\mathbf{Y}_m$ for $m \leq n$. More generally:

$$
\widehat{\mathbf{Y}}_{n+k|n} = GF^{k-1}\widehat{\mathbf{X}}_{n+1|n} \tag{6.19}
$$

and if we use the notation

$$
\Delta_n^{(k)} = \mathbb{E}\{(Y_{n+k} - \widehat{\mathbf{Y}}_{n+k|n})(Y_{n+k} - \widehat{\mathbf{Y}}_{n+k|n})^t\}
$$

for its prediction quadratic error, it follows that:

$$
\Delta_n^{(k)} = GF^{k-1}\Omega_{n+1|n}(GF^{k-1})^t.
$$

As before we highlight the final formulae in a box.

$$
\begin{aligned}
\widehat{\mathbf{Y}}_{n+k|n} &= GF^{k-1}\widehat{\mathbf{X}}_{n+1|n} \\
\Delta_n^{(k)} &= GF^{k-1}\Omega_{n+1|n}(GF^{k-1})^t.
\end{aligned}
$$

6.4.6 Estimation of the Parameters

The recursive filtering equations derived in this section provide optimal estimators, and as such, they can be regarded as some of the most powerful tools of data analysis. Unfortunately, their implementation requires the knowledge of the parameters of the model. Because of physical considerations, or because we are often in control of how the model is set up, the observation matrix G is very often known, and we shall focus the discussion on the remaining parameters, namely, the state transition matrix F, the variance/covariance matrices Σ_V and Σ_W of the system and observation noises, and the initial estimates of the state and its error matrix. This parameter estimation problem is extremely difficult, and it can be viewed as filtering's Achilles heel.

Several tricks have been proposed to get around this difficulty, including the parameters in the state vector being one of them. The present theory implies that this procedure should work very well when the parameters enter linearly (or almost linearly) in the state and observation equations. However, despite the fact that the theory is not completely developed in the nonlinear case, practitioners are using this trick in many practical implementations, whether or not the system is linear.

For the sake of completeness we describe the main steps of the maximum likelihood approach to parameter estimation in this setup. This approach requires the computation and optimization of the likelihood function. The latter can be derived when all the random quantities of the model are jointly Gaussian. This is the case when the white noise series $\{\mathbf{V}_n\}_n$ and $\{\mathbf{W}_n\}_n$ are Gaussian, and when the initial value of the state is also Gaussian (and independent of both noise series). So instead of looking for exact values of the initial state vector and its error matrix, one assumes that this initial state vector $\mathbf{X}_0$ is Gaussian, and we add its mean vector μ_0 and its variance/covariance matrix Σ_0 to the list of parameters to estimate. The computations of the previous subsection showed that the innovations satisfy:

$$I_n = \mathbf{Y}_n - G\widehat{\mathbf{X}}_n$$

and we also argued the fact that they are independent (recall that we are now restricting ourselves to the Gaussian case), and we derived recursive formulae to compute their variance/covariance matrices Δ_n. The above equation, together with the fact that one can compute the one-step-ahead predictions $\widehat{\mathbf{X}}_n$ recursively from the data, allows us to compute the innovations from the values of the observations, and reformulate the likelihood problem in terms of the innovations only. This simple remark simplifies the computations. Given all this, one can write down the joint density of the I_n's in terms of the observations $\mathbf{Y}_n = \mathbf{y}_n$ and the successive values of the matrices Δ_n and of the one-step-ahead predictions $\widehat{\mathbf{X}}_n$ which one computes inductively from the recursive filtering equations. Because of the special form of the multivariate normal density, the new log-likelihood function is of the form:

$$-2\log L_{I_1,\dots,I_n}(\Theta) = \text{cst} + \sum_{j=1}^{n} \log\det(\Delta_j(\Theta)) + \sum_{j=1}^{n} I_j(\Theta)\Delta_j(\Theta))^{-1} I_j(\Theta)^t$$

where cst stands for a constant of no consequence, and where we emphasized the dependence of the innovations I_j and their variance/covariance matrices Δ_j upon the vector Θ of unknown parameters, $\Theta = (\mu_0, \Sigma_0, F, \Sigma_V, \Sigma_W)$. Notice that the dimension of this parameter vector may be large. Indeed its components are matrices and vectors whose dimensions can be large. In any case, this log-likelihood function is a non-convex function of the multi-dimensional parameter Θ and its maximization is quite difficult. Practical implementations of optimization algorithms have been used by the practitioners in the field: Newton-Raphson, EM algorithm, ... are among those, but no single method seems to be simple enough and reliable enough for us to discuss this issue further at the level of this text.

6.5 Applications to Linear Models

Linear models were introduced in Chapter 3 as a convenient framework for linear regression. Their versatility made it possible to apply their theory to several specific classes of nonlinear regression problems such as polynomial and natural spline regression. We now recast these linear models in the framework of partially observed state space systems, and we take advantage of the filtering tools developed in this chapter to introduce generalized forms of linear models, and to tackle several new problems which could not have been addressed with the tools of Chapter 3.

6.5.1 State Space Representation of Linear Models

Combining the original notation of Section 3.5 with the notation introduced in this chapter for state space models, we give a new interpretation to the multiple linear regression setup:

$$y_n = \mathbf{x}_n\beta + \epsilon_n \tag{6.20}$$

where $\{\epsilon_n\}_n$ is a white noise in the strong sense, where $\mathbf{x}_n = (x_{1,n}, x_{2,n}, \ldots, x_{k,n})$ is a $k = p+1$ dimensional vector of explanatory variables, and $\beta = (\beta_1, \beta_2, \ldots, \beta_k)$ is a k-dimensional vector of unknown parameters. Notice that we are now using the lower case n to number the observations, while we used the lower case i when we introduced the linear models in Section 3.5. The novelty of the present approach is to interpret equation (6.20) as the observation equation of a state space system whose dynamics are very simple since they do not change over time. To be more specific, we consider a state vector $\mathbf{X}_n$ always equal to the vector β of parameters. Hence, the dynamical equation reads:

$$\mathbf{X}_{n+1} = \mathbf{X}_n. \tag{6.21}$$

In other words, the state matrix F_n does not change with n, and is always equal to the $k \times k$ identity matrix, while the state noise is identically zero. Setting $G_n = \mathbf{x}_n$, equation (6.20) coincides with the observation equation (6.8) if we set $\mathbf{Y}_n = y_n$ and $\mathbf{W}_n = \epsilon_n$. This rewriting of a linear model as a state space model can be artificial at times, but it makes it possible to apply the powerful tools of filtering theory to these models. We demonstrate by example some of benefits of this reformulation.

Recursive Estimation

Recasting linear models as state space models is especially useful when we need to recompute the least squares estimates of the parameters after an observation is added. Indeed, if we denote by $\hat\beta_n$ the least squares estimate of β computed from the data $(\mathbf{x}_1, y_1)$, ... , $(\mathbf{x}_n, y_n)$, then the Kalman recursive filtering equations give a very convenient way to compute the least squares estimate $\hat\beta_{n+1}$ as an update of the previous estimate $\hat\beta_n$ using the current observation $(\mathbf{x}_{n+1}, y_{n+1})$. If we use the notation K_n for the Kalman gain matrix introduced earlier, we get:

$$\hat{\beta}_{n+1} = \hat{\beta}_n + K_{n+1}(y_{n+1} - \mathbf{x}_{n+1}\hat{\beta}_n. \tag{6.22}$$

with the corresponding update for the estimated error variance:

$$\Omega_{n+1} = [I - K_{n+1}\mathbf{x}_{n+1}]\Omega_n.$$

Recursive Residuals

Some of the most efficient tests for change in a model are based on the analysis of the residuals $r_n = y_n - \mathbf{x}_n\hat{\beta}_n$. Again, the recursive filtering equations derived in the previous section make it possible to update these residuals recursively without having to recompute them from scratch each time a new observation is made available. The recursive standardized residuals w_n are usually defined by the formula:

$$w_n = \frac{y_n - \mathbf{x}_n\hat{\beta}_{n-1}}{\sqrt{1 + \mathbf{x}_n(\mathbf{X}^t_{n-1}\mathbf{X}_{n-1})^{-1}\mathbf{x}^t_n}}.$$

Unexpectedly, the recursive residuals defined in this manner form a sequence of independent $N(0, \sigma^2)$ random variables. This makes their distribution theory very easy. These remarkable properties were first identified and used by Brown, Durbin and Evans who derived a series of useful tests, called CUSUM tests, for the adaptive detection of changes in a linear model. It appears that `S-Plus` does not have any implementation of the CUSUM test. Its function `cusum` serves another purpose.

6.5.2 Linear Models with Time Varying Coefficients

This subsection discusses a generalization of linear models prompted by the previous discussion on the first application of filtering to linear models. There, the state equation was trivial because the state did not change with time. Since filtering theory deals with state vectors varying with time, it is natural in this context to consider linear models where the parameter vector β can change with time. Indeed, the theory presented in this chapter allows us to consider time-varying parameters β_n satisfying:

$$\beta_{n+1} = F\beta_n + \mathbf{V}_{n+1} \tag{6.23}$$

for some white noise $\{\mathbf{V}_n\}_n$ and some deterministic matrix F. Writing the corresponding linear model one component at a time as before, we get the same observation equation as (6.20):

$$y_n = \mathbf{x}_n\beta_n + \epsilon_n \tag{6.24}$$

Notice that the update equation for $\hat{\beta}_n$ is slightly more involved than (6.22), since we do not have $F = I$ and $\Sigma_V = 0$ as before.

We give a detailed implementation of this idea in the application to the CAPM discussed in the next subsection.

6.5.3 CAPM with Time Varying β's

The Capital Asset Pricing Model (CAPM for short) of Lintner and Sharpe was introduced in Subsection 3.5.4. There we argued that, even when empirical evidence was not significant enough to reject the zero-intercept assumption of the model, instability over time of the betas did not always support the model despite its economic soundedness and its popular appeal. We now try to reconciliate CAPM with empirical data by allowing the betas to vary with time. Recall that according to CAPM, we assume that, at each time t, the excess return $\widetilde{R}_{j,t}$ of the j-th asset over the risk-free rate r of lending and borrowing is given, up to an additive noise term, by a multiple of the excess return $\widetilde{R}_t^{(m)}$ of the market portfolio at the same time t. In other words, the CAPM model states that:

$$\widetilde{R}_{j,t} = \beta_j \widetilde{R}_t^{(m)} + \epsilon_{j,t} \tag{6.25}$$

for some white noise $\{\epsilon_{j,t}\}_t$ specific to the j-th asset.

Time Varying Reformulation

As we mentioned in Chapter 3, there is empirical evidence that the hypotheses underlying CAPM theory do not always hold. And even if one is willing to accept them for their appealing economic rationale, the estimates of the β_j appear to be quite unstable, varying with economic factors. Given that fact, we propose to generalize the CAPM model by allowing the betas to vary with time. For the sake of illustration, we choose the simplest possible model, and we assume for example that our time-varying betas follows a random walk. So switching to the notation n for time instead of t, for each index j, we assume that:

$$\beta_{j,n+1} = \beta_{j,n} + v_{j,n+1} \tag{6.26}$$

for some white noise $\{v_{j,n}\}_n$ with unknown variance σ_j^2. Choosing for state at time n the vector of betas under consideration, i.e. setting $\mathbf{X}_n = \beta_n$, we can rewrite the various equations (6.26) in a vector form:

$$\mathbf{X}_{n+1} = \mathbf{X}_n + \mathbf{V}_{n+1}, \tag{6.27}$$

giving the dynamics of the state. Here the state matrix F is the $d_X \times d_X$ identity matrix where d_X is the number of stocks. For the sake of simplicity, we shall only consider this model one stock at a time, in which case $d_X = 1$, $\mathbf{X}_n = \beta_{j,n}$ and $\mathbf{V}_n = v_{j,n}$ since there is only one value for the index j, and the state matrix F is the number one. Since the excess returns can be observed, we use (6.25) as our observation equation. We can do that provided we set $\mathbf{Y}_n = \widetilde{R}_{j,n}$, in which case $d_Y = 1$, we choose $\mathbf{W}_n = \epsilon_{j,t}$ for the observation noise, and the 1×1 matrix $\widetilde{R}_t^{(m)}$ for the observation matrix G_n. Notice that the observation matrix changes with n. As we pointed out earlier, this is not a major problem. Indeed, even though the derivation of the recursive filtering equations was done with a time independent observation matrix G, the same formulae hold when it varies with n, we just have to use the right matrix at each time step.

So, given an initial estimate for the value of β_0, and given an initial estimate for its prediction error, we can implement the recursive filtering equations derived in this chapter to track the changes over time of the beta of each stock. Notice that this assumes that we know the state and observation variances σ_v^2 and σ_ϵ^2. Further analysis is proposed in Problems 6.11 and 6.12.

It is important to notice that the estimates of the betas are non-anticipative in the sense that the estimate $\hat{\beta}_n$ at time n depends only upon the *past* values of the observed excess returns, i.e. the values of $\widetilde{R}_{j,m}$ for $m \leq n$ and not on the values of $\widetilde{R}_{j,m}$ for $m > n$.

Filtering Experiment

For the sake of illustration, we revisit an example which we already discussed in Subsection 3.5.4, namely the American Electric Power (AEP) weekly excess returns over the period starting 01/01/1995, and ending 01/01/2003. Our analysis of a time-dependent CAPM will shed some light on the effects of the energy crisis, and clearly exhibit facts which are impossible to uncover in the classical time-independent setup.

Figure 6.2 shows the result of the non-anticipative estimation of the time varying betas by the Kalman filter. It shows that the beta remains smaller than one for most of the period, consistent with the fact that the stock was not of the risky type before 2002. But despite the fact that AEP was one of the very few energy companies to weather the crisis with little damage, its beta became greater than one in 2002 implying (according to the commonly admitted interpretation of the size of the beta) that this stock became risky at that time. This phenomenon is even more pronounced for companies which were affected more dramatically by the energy crisis. This is illustrated for example in Problem 6.11.

6.6 State Space Representation of Time Series

As we already noticed, we are only considering state-space models whose dynamics are given by a multivariate AR(1) series. Since some of the components of the state vector may remain unobserved, we can always add new components to the state vector without changing the observations. This feature of the partially observed systems makes it possible to rewrite the equation defining an AR(p) model as an AR(1)-like equation, simply by adding the components $\mathbf{X}_{t-1}, \ldots, \mathbf{X}_{t-p}$ to the value of a (new and extended) state at time t. Adding the past values to the current value of the state enlarges the dimension, and this did not make sense when we were assuming that the entire state vector was observed at any given time. However, now that the observations can be incomplete, this transformation makes sense. This section takes advantage of this remark, but not without creating new problems. Indeed, a given

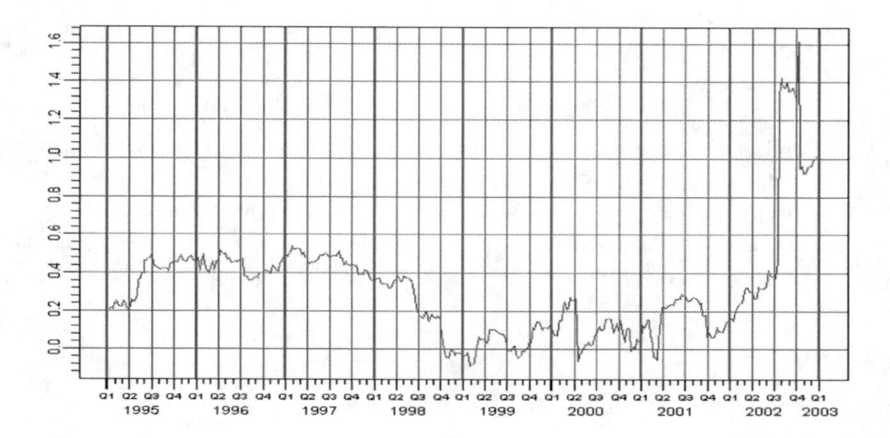

Fig. 6.2. Non-anticipative estimates of the AEP time varying betas given by Kalman filtering over the period starting 01/01/1995 and ending 01/01/2003.

set of observations vectors may correspond to many state space vectors, and consequently, as in the case of the factor models, we should be prepared to face a *lack of uniqueness* in the representation of a given stochastic system as a state space system.

6.6.1 The Case of AR Series

Let us first consider the simple example of an AR(1) model. Let us assume for example that $X \sim AR(1)$, and more precisely that:

$$X_t = .5X_{t-1} + W_t.$$

Switching to the index n, one can write:

$$\begin{cases} X_{n+1} = .5X_n + W_{n+1} \\ \quad Y_n = X_n \end{cases}$$

This means that considering an AR(1) model is equivalent to considering a state space system described by the one dimensional (i.e. $d_X = 1$) state vector $\mathbf{X}_n = X_n$, its dynamics being given by the first of the equations above, the second equation giving the observation equation with $\mathbf{Y}_n = X_n$. Notice that the observations are perfect since there is no noise term in the observation equation.

Since this example is too simple to be indicative of what is really going on, we consider the case of an AR(2) model. Let us assume for example that:

$$X_t = .5X_{t-1} + .2X_{t-2} + W_t \tag{6.28}$$

for some white noise $\{W_t\}$. Switching once more to the notation n for the time stamp, we rewrite this definition in the form:

$$\begin{bmatrix} X_{n+1} \\ X_n \end{bmatrix} = \begin{bmatrix} .5 & .2 \\ 1 & 0 \end{bmatrix} \begin{bmatrix} X_n \\ X_{n-1} \end{bmatrix} + \begin{bmatrix} W_{n+1} \\ 0 \end{bmatrix} \tag{6.29}$$

Indeed, if we look at this equality between two vectors, the equality of the first component of the left hand side with the first component of the right hand side gives back the definition (6.28) of X_n, while the equality between the second components is merely a consistency relation giving no extra information. Now, if for each time index n we define the two-dimensional (i.e. $d_X = 2$) random vector $\mathbf{X}_n$ by:

$$\mathbf{X}_n = \begin{bmatrix} X_n \\ X_{n-1} \end{bmatrix} \tag{6.30}$$

the deterministic matrix F and the random vector $\mathbf{V}_n$ by:

$$F = \begin{bmatrix} .5 & .2 \\ 1 & 0 \end{bmatrix} \qquad \text{and} \qquad \mathbf{V}_n = \begin{bmatrix} W_n \\ 0 \end{bmatrix},$$

then equation (6.29) becomes:

$$\mathbf{X}_{n+1} = F\mathbf{X}_n + \mathbf{V}_{n+1}$$

which can be viewed as the equation giving the dynamics of the state $\mathbf{X}$. Using perfect observation, i.e. setting $G = [1, 0]$ and $\mathbf{W}_n \equiv 0$, we see that we can represent our AR(2) series as a state-space model with perfect observation. We now show how this procedure can be generalized to all the AR models.

Let us assume that $\{X_n\}_n$ is a time series of the AR(p) type given by the standard auto-regressive formula:

$$X_n = \phi_1 X_{n-1} + \cdots + \phi_p X_{n-p} + W_n \tag{6.31}$$

for some set of coefficients ϕ_1, ..., ϕ_p, and a white noise $\{W_n\}_n$ of unknown variance. For each time stamp n we define the p-dimensional column vector $\mathbf{X}_n$ by:

$$\mathbf{X}_n = [X_n, X_{n-1}, \ldots, X_{n-p+1}]^t.$$

There is no special reason to define $\mathbf{X}_n$ from its transpose, we are merely trying to save space and make typesetting easier! Using this state vector, we see that we can derive an observation equation of the desired form by setting $\mathbf{Y}_n = X_n$, $G = [1, 0, \ldots, 0]$, and $\mathbf{W}_n \equiv 0$. This is again a case of perfect observation. Finally we write the dynamics of the state vector $\mathbf{X}$ in the form of definition (6.31). Adding the consistency relations for the remaining components, this equation can be written in the form:

$$\mathbf{X}_{n+1} = \begin{bmatrix} X_{n+1} \\ X_n \\ \vdots \\ X_{n+2-p} \end{bmatrix} = \begin{bmatrix} \phi_1 & \phi_2 & \cdots & \phi_{p-1} & \phi_p \\ 1 & 0 & \cdots & 0 & 0 \\ \vdots & \vdots & \vdots & \vdots & \vdots \\ 0 & 0 & \cdots & 1 & 0 \end{bmatrix} \begin{bmatrix} X_n \\ X_{n-1} \\ \vdots \\ X_{n+1-p} \end{bmatrix} + \begin{bmatrix} W_{n+1} \\ 0 \\ \vdots \\ 0 \end{bmatrix}$$

which can be viewed as the state equation giving the dynamics of the state vector $\mathbf{X}_n$ if we define the state matrix F and the state white noise $\mathbf{V}_n$ by:

$$F = \begin{bmatrix} \phi_1 & \phi_2 & \cdots & \phi_{p-1} & \phi_p \\ 1 & 0 & \cdots & 0 & 0 \\ \vdots & \vdots & \vdots & \vdots & \vdots \\ 0 & 0 & \cdots & 1 & 0 \end{bmatrix} \qquad \text{and} \qquad \mathbf{V}_n = \begin{bmatrix} W_n \\ 0 \\ \vdots \\ 0 \end{bmatrix}.$$

6.6.2 The General Case of ARMA Series

We now assume that the time series $\{X_n\}_n$ satisfies:

$$X_n - \phi_1 X_{n-1} - \cdots - \phi_p X_{n-p} = W_1 + \theta_1 W_{n-1} + \cdots + \theta_q W_{n-q}$$

for some white noise $\{W_n\}_n$ with unknown variance. Adding zero coefficient terms if necessary, we can assume without any loss of generality that $p = q + 1$. Using the notation introduced in the previous chapter this can be written in a condensed manner in the form $\phi(B)X = \Theta(B)W$. Let U be an AR(p) time series with the same coefficients as the AR - part of X, i.e. satisfying $\phi(B)U = W$. If for example the moving average representation exists (but we shall not really need this assumption here) we saw that:

$$\phi(B)U = W \qquad \text{is equivalent to} \qquad U = \frac{1}{\phi(B)}W \tag{6.32}$$

and consequently:

$$X = \frac{\theta(B)}{\phi(B)}W = \theta(B)\frac{1}{\phi(B)}W = \theta(B)U$$

which implies that:

$$\begin{aligned} X_n &= U_n + \theta_1 U_{n-1} + \cdots + \theta_q U_{n-q} \\ &= [1, \theta_1, \ldots, \theta_q] \begin{bmatrix} U_n \\ U_{n-1} \\ \vdots \\ U_{n-q} \end{bmatrix} \end{aligned}$$

which can be viewed as an observation equation if we set:

$$\mathbf{Y}_n = X_n, \qquad G = [1, \theta_1, \ldots, \theta_q], \qquad \text{and} \qquad \mathbf{W}_n \equiv 0$$

and if we define the state vector $\mathbf{X}_n$ as the column vector with $p = q + 1$ row entries U_n, U_{n-1}, ..., U_{n-p+1}. So we have $d_Y = 1$ and $d_X = p = q + 1$. Now that we have the observation equation, we derive the state equation. Because of its definition

(6.32), the time series U is an AR(p) series and $\phi(B)U = W$ can be rewritten in the standard form:

$$U_n - \phi_1 U_{n-1} - \cdots - \phi_p U_{n-p} = W_n$$

which is exactly the form we used in the previous subsection to rewrite the AR series in a state-space form. So using the same procedure (recall that we arranged for $p = q + 1$)

$$\mathbf{X}_{n+1} = \begin{bmatrix} U_{n+1} \\ U_n \\ \vdots \\ U_{n+2-p} \end{bmatrix} = \begin{bmatrix} \phi_1 & \phi_2 & \cdots & \phi_p \\ 1 & 0 & \cdots & 0 \\ \vdots & \vdots & & \vdots \\ 0 & 0 & \cdots & 1 \end{bmatrix} \begin{bmatrix} U_n \\ U_{n-1} \\ \vdots \\ U_{n+1-p} \end{bmatrix} + \begin{bmatrix} W_{n+1} \\ 0 \\ \vdots \\ 0 \end{bmatrix}$$

which can be viewed as the desired state equation giving the dynamics of the state vector $\mathbf{X}_n$, if we define the matrix F as the $p \times p$ matrix appearing in the above equation, and if we define the noise vector $\mathbf{V}_{n+1}$ as the p-dimensional column vector with components W_{n+1}, 0,. . ., 0.

Remarks

1. As we already pointed out, the above representation is not unique.We gave an algorithmic procedure which works in all cases, but one should keep in mind that there are many ways to define a state vector and its dynamics, while keeping the same observations.

2. The above procedure can be implemented (modulo minor technical changes to accommodate the dimensions) in the case of multivariate time series, providing a state space representation of multivariate ARMA series.

6.6.3 Fitting ARMA Models by Maximum Likelihood

When explaining the limitations of `S-Plus` in fitting ARIMA models, we said that the standard way to fit an MA(q) was to use the maximum likelihood method, but that the computation of the likelihood function was difficult, and that the only practical way to do so was to use a recursive computation reminiscent of the filtering paradigm. Moreover, we also said at that time, that we suspected that this was the main reason why fitting MA models was not implemented in `S-Plus` in the multivariate case. Now that we have seen the recursive Kalman filters, and that we know how to rewrite ARMA models in a state space form, it is time to revisit this fitting issue and shed some light on the matter.

The only reason for not implementing the maximum likelihood estimation of an ARMA model is the lack of filtering tools.

Indeed, given the order of an ARMA model, one can

- Use the algorithmic procedure presented in the previous subsection to rewrite the model as a state space model;

- Follow the step outlined in Subsection 6.4.6 to compute the likelihood of any set of observations;
- Run our favorite optimization program to solve for the maximum of this likelihood function.

6.7 Example: Prediction of Quarterly Earnings

This section is devoted to a detailed discussion of a specific application of the recursive filters to the analysis of a financial problem. We show how to use the state space representations and the recursive filtering equations to analyze a scalar time series which appears naturally as the sum of trend, seasonal and irregular components. Contrary to the strategy adopted in the case of temperature data, we do not assume that the trend and seasonal components are deterministic, and we do not start by removing them to reduce the problem to the analysis of a stationary time series. Instead, we include the trend and seasonal components in a structural model which we represent as a partially observed state-space model, which we predict using the recursive filtering equations.

We choose to illustrate this approach with the example of the prediction of the quarterly earnings of a public company. Figure 6.3 gives the plot of the time series of

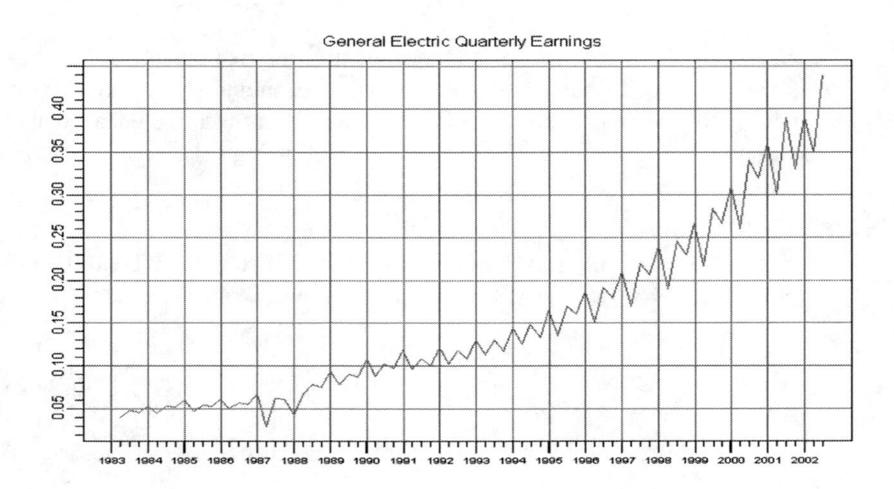

Fig. 6.3. Quarterly earnings per share of General Electric from March 31, 1983 to June 30, 2002.

the quarterly earnings of General Electric starting March 31, 1983 and ending June 30, 2002. Earnings per share used to be published yearly in 1982 and before. This series is obviously non-stationary. In particular, it shows a significant upward trend and a seasonal component that cycles every four quarters, or once per year. Moreover, the scale of this seasonal component seems to fluctuate erratically, possibly getting larger over time. Since square root and logarithmic transformations do not seem to resolve the non-stationarity problem, we express the series directly as the sum of a trend component, a seasonal component and a white noise:

$$Y_n = T_n + S_n + \epsilon_n \tag{6.33}$$

where the series $\{T_n\}_n$ models the trend, $\{S_n\}_n$ models the seasonal component, and $\{\epsilon_n\}_n$ is a white noise. Because the trend could be suspected to be exponential, we choose to model the trend as a time series satisfying an evolution equation of the form:

$$T_{n+1} = \phi T_n + \epsilon^{(T)}_{n+1} \tag{6.34}$$

where $\{\epsilon^{(T)}_n\}_n$ is a white noise independent of $\{\epsilon_n\}_n$, and where the real coefficient ϕ is chosen to satisfy $\phi > 1$. The evolution equation of T_n is the equation of an AR(1), but the assumption we make on the coefficient, namely $\phi > 1$, guarantees that this AR series is not stationary. Instead of assuming that the seasonal component is a deterministic periodic function of period 4, we assume that it is a series satisfying a dynamical equation of the form:

$$S_{n+1} = -S_n - S_{n-1} - S_{n-2} + \epsilon^{(S)}_{n+1} \tag{6.35}$$

where $\{\epsilon^{(S)}_n\}_n$ is a white noise independent of the two previous ones. The idea behind this choice is that at least in expectation, S should sum up to zero over a complete period of four quarters. All these requirements can be encapsulated inside a partially observed state-space model in the following way. First we define the state vector:

$$\mathbf{X}_n = [T_n, S_n, S_{n-1}, S_{n-2}]^t$$

so that $d_X = 4$, and together with the obvious consistency conditions, equations (6.34) and (6.35) can be used to give the dynamics of the state via the equation:

$$\mathbf{X}_{n+1} = \begin{bmatrix} T_{n+1} \\ S_{n+1} \\ S_n \\ S_{n-1} \end{bmatrix} = \begin{bmatrix} \phi & 0 & 0 & 0 \\ 0 & -1 & -1 & -1 \\ 0 & 1 & 0 & 0 \\ 0 & 0 & 1 & 0 \end{bmatrix} \begin{bmatrix} T_n \\ S_n \\ S_{n-1} \\ S_{n-2} \end{bmatrix} + \begin{bmatrix} \epsilon^{(T)}_{n+1} \\ \epsilon^{(S)}_{n+1} \\ 0 \\ 0 \end{bmatrix}$$

which can be viewed as the state equation giving the dynamics of the state vector $\mathbf{X}_n$, if we define the matrix F as the 4×4 matrix appearing in the above equation, and if we define the noise vector $\mathbf{V}_{n+1}$ as the 4-dimensional column vector with components $\epsilon^{(T)}_{n+1}$, $\epsilon^{(S)}_{n+1}$,0,0. Formula (6.33) used to define the model can now be used to give the observation equation for the one-dimensional observation vector $\mathbf{Y}_n = y_n$:

$$\mathbf{Y}_n = [1, 1, 0, 0] = \begin{bmatrix} T_n \\ S_n \\ S_{n-1} \\ S_{n-2} \end{bmatrix} + \epsilon_n$$

which is of the desired form if we set $G = [1, 1, 0, 0]$ and $\mathbf{W}_n = \epsilon_n$. The parameters of the model are the standard deviations σ, $\sigma^{(T)}$ and $\sigma^{(S)}$ of the noise components introduced above, as well as the parameter ϕ. The data is not very noisy, the challenge lies in the fact that the time series is short for the filter to settle down and have a chance to self-correct possible errors in parameters and initializations. For the purposes of the present experiment we chose $\sigma = .0001$, $\sigma^{(T)} = .0001$ and $\sigma^{(S)} = .00001$, and $\phi = 1.03$ corresponding to an annual growth of about 3%. As explained earlier, even more problematic is the choice of initial estimates for the state and its measurement-error matrix. To obtain the results we report below we chose $\widehat{\mathbf{X}_0} = [.05, -.01, .03, -.01]^t$ and .02 times the 4×4 identity matrix for the error matrix. Figure 6.4 shows a couple of years of quarterly earnings, together with the plot of the predicts for a two-years horizon computed at the end of the third quarter of 2000. We chose this date to be able to compare the predictions with the actual data which are superimposed on the same plot. We also give upper and lower 95% confidence bounds for the predictions.

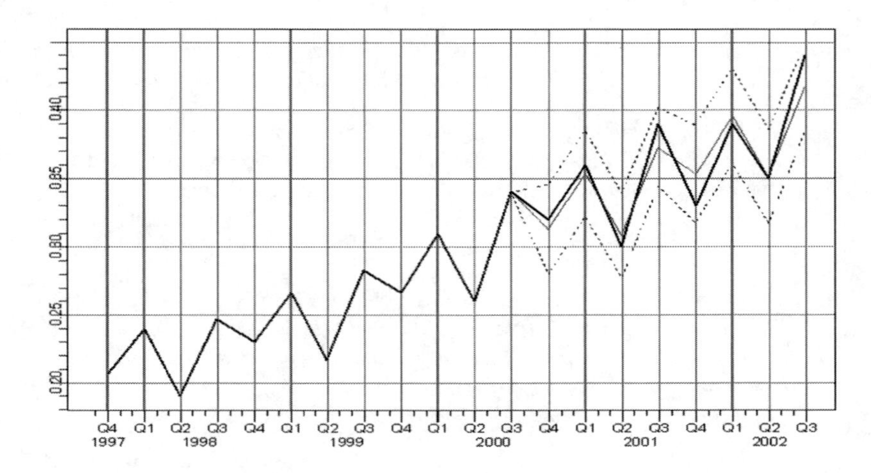

Fig. 6.4. Predictions computed at the end of the third quarter of 2000, of the General Electric quarterly earnings as computed with the Kalman filter based on the state-space model described and parameter values given in the text. The dark solid line gives the actual values which did occur, the light solid line give the predictions, and the dotted lines gives the bounds of a 95% confidence interval.

PROBLEMS

Ⓔ Ⓢ **Problem 6.1** *The purpose of this problem is to examine the various ways the components of a multivariate time series can be dependent. We consider the following AR(1) model for a bivariate time series* $\{\mathbb{X}_t\}_{t=0,1,\cdots}$*:*

$$\mathbf{X}_t = A\mathbf{X}_{t-1} + \mathbf{W}_t, \qquad t = 1, 2, \cdots$$

where:

$$\mathbf{X} = \begin{bmatrix} X_t^{(1)} \\ X_t^{(2)} \end{bmatrix}, \qquad \text{and} \qquad A = \begin{bmatrix} .2 & 0 \\ 0 & .4 \end{bmatrix},$$

and where $\{\mathbf{W}_t\}_t$ *is a bivariate Gaussian white noise with covariance matrix*

$$\Sigma = \begin{bmatrix} .2 & 0 \\ 0 & 1 \end{bmatrix}.$$

1. Give a condition on the initial random variables $X_0^{(1)}$ *and* $X_0^{(2)}$ *which guarantees that all the random variables* $X_t^{(1)}$ *and* $X_s^{(2)}$ *are independent for the various values of* t *and* s*. Explain why. From now on we assume that this condition is satisfied.*
2. Explain how one can change some of the entries of the matrix A *to make sure that the time series* $\{X_t^{(1)}\}_t$ *and* $\{X_t^{(2)}\}_t$ *are dependent.*
3. Setting back the entries of the matrix A *to their original values, explain how one can change some of the entries of the matrix* Σ *to make sure that the time series* $\{X_t^{(1)}\}_t$ *and* $\{X_t^{(2)}\}_t$ *are dependent.*
4. Can you identify the differences in the statistical properties resulting from these two procedures? You can use simulations if you cannot quantify these differences theoretically.

The goal of problems 6.2, 6.3 and 6.4 is to quantify various risk exposures associated with temperature basket options similar to the one analyzed in the text. The data needed for these problems are contained in the text files `temps1.asc` and `temps2.asc`. Open these files in `S-Plus`, run them as `S-Plus` scripts, to get the desired multivariate `timeSeries` objects `TEMPS1.ts` and `TEMPS2.ts`.

Ⓔ **Problem 6.2** *This problem uses the data contained in the* `timeSeries` *object* `TEMPS1.ts`*. For each day of the period* P *starting from 09/06/1948 (the first row) and ending 11/08/2001(the last row) it gives the daily average temperatures recorded at the meteorological stations of the international airports of Reno (first column), Las Vegas (second column) and Sacramento (third column). We consider two time periods.*

- *Period* P *is the period of the measurements reported in* `TEMPS1.ts`
- *Period* P' *is the period of the option for which one needs to predict the number of heating degree days. It starts on 12/1/2001 and ends 2/28/2002 inclusive.*

We consider a basket option written on the average number of heating degree days over these three cities. On each given day one computes the average of the three following numbers: the number of heating degree days in Reno, the number of heating degree days in Las Vegas, and the number of heating degree days in Sacramento. One then computes the sum of these daily averages over the period P'*, and one compares this sum to the strike of the option, using the*

same rules as usual to compute the pay-off of the option. The purpose of this problem is to perform a simple regression analysis of such a basket option.
0. For each year in the period from 1961 to 2000 inclusive, compute the yearly index of the option, compute the average of these yearly indexes over the last 15 *years of this period, and call this average* K.

From now on we consider a European call option at the money (i.e. with strike K *computed above), with tick* α *equal to US$ 5,000 per degree day, and cap US$ 1,000,000, written on the cumulative average number of HDD's in Reno, Las Vegas and Sacramento from December 1, 2001 to February 28, 2002.*
1. Use the linear regression approach introduced in Problem 5.13 to give an estimate of the underlying index on which the basket option is written, i.e. of the average of the three numbers of HDD in the cities of Reno, Las Vegas and Sacramento over the period P', *and give a* 95% *confidence interval for this prediction.*
2. Answer questions 2. and 3. of Problem 5.13 in the present situation.

Ⓔ **Problem 6.3** *Follow the steps of the joint analysis of the daily average temperatures at the Des Moines, Portland and Cincinnati given in Subsection 6.1.4 of the text, and use the data of the period* P *to fit a trivariate model to the daily average temperatures in Reno, Las Vegas and Sacramento of the timeSeries object* `TEMPS1.ts`. *Use this fitted model and follow the prescriptions given in the text to derive an estimate of the value of the average of the three numbers of HDD's in the cities of Reno, Las Vegas and Sacramento over the period* P', *and to provide new answers to the questions of Problem 6.2 above.*

Ⓔ **Problem 6.4** *This problem provides another numerical illustration of the points addressed in Problems 5.13, 6.2 and 6.3. The data needed for this problem are contained in the timeSeries object* `TEMPS2.ts` *giving the daily average temperatures, starting from 1/1/1972 (the first row) ending 12/31/2001 (the last row) at the meteorological stations of Newark, La Guardia and Philadelphia. The data have been cleaned in the sense that all the February 29th's of the leap years of this period have been removed. For the purpose of the questions below, we shall use two time periods.*

- *Period* P *is the period of the time stamps of the timeSeries object.*
- *Period* P' *is the period for which you will need to predict the number of cooling degree days (CDD's for short). It starts on 5/1/2002 and ends 9/30/2002 inclusive.*

0. For the location Newark, and for each year in the period P, *compute the yearly CDD index, i.e. the total number of CDD's between May 1, and September 30 of that year (these two days being included). Bundle these yearly aggregates into a vector which we shall call* `INDEX`.
1. Using only the numerical values contained in the vector `INDEX`

a) Compute a prediction for the value of the yearly cumulative number of CDD's in Newark for the period P'.

b) Give a 95% *confidence interval for this prediction*

c) Let us assume that on January 1, 2002, you buy a call option on the number of CDD's in Newark over the period P', *with strike* K *being equal to the average of the yearly numbers of CDD's over the last* 15 *years, nominal pay-off rate (also called the tick) $ 5000, and cap $ 1000000 for a premium of $ 400000. Use the linear model so fitted to give an estimate of the probability that you will loose money on this deal?*

d) Compute your profit/loss, should the prediction you computed in part a) be perfect in the sense that it is equal to the actual number of CDD's which accumulated over the period P'.

	Newark	La Guardia	Philadelphia
Number of CDD's during P'	1325	1355	1463.5

2. Use the daily data in the period P to fit a model to the average daily temperature in Newark, and use this estimated model to propose new answers to the questions a) through d) above. Compare the results for each of these questions.
3. Give an estimate and a 95% confidence interval for the average of the three numbers of CDD's in the cities of Newark, La Guardia and Philadelphia over the period P',
4. Let us assume that on January 1, 2002, you want to buy from me a "basket" call option at the money, on the average of the numbers of CDD's in Newark, La Guardia and Philadelphia over the period P', with the same tick and the same cap as above. Say how much you would be willing to buy this option for, and explain why.
FYI: *Even though you should not use these numbers in the solutions of the questions above, you should know the actual numbers of CDD's over the period P' in these three cities. They are reproduced in Table 6.4.*

Ⓢ **Problem 6.5** *The purpose of this problem is to illustrate by simulation the cointegration properties stated in the example given at the beginning of Subsection 6.1.6*
1. Construct a sample $\{W_n\}_{n=1,\dots,N}$ of size $N = 1024$ from the normal distribution $N(0,1)$, construct the random walk $S_n = S_0+W_1+\ldots+W_n$ starting from $S_0 = 0$, and construct two independent white noise sequences $\{\epsilon_n^{(1)}\}_{n=1,\dots,N}$ and $\{\epsilon_n^{(2)}\}_{n=1,\dots,N}$ from the distribution $N(0,.16)$ which are independent of $\{W_n\}_{n=1,\dots,N}$. Give on the same figure the plots of the graphs of the time series $\{X_n^{(1)}\}_{n=1,\dots,N}$ and $\{X_n^{(2)}\}_{n=1,\dots,N}$ defined by:

$$X_n^{(1)} = S_n + \epsilon_n^{(1)} \quad \text{and} \quad X_n^{(2)} = S_n + \epsilon_n^{(2)}.$$

2. Give the plot of the linear combination of $\{X_n^{(1)}\}_{n=1,\dots,N}$ and $\{X_n^{(2)}\}_{n=1,\dots,N}$ given by the cointegration vector identified in the text, compare to the previous plot and explain why it confirms cointegration.
3. Construct two independent samples $\{W_n^{(1)}\}_{n=1,\dots,N}$ and $\{W_n^{(2)}\}_{n=1,\dots,N}$ of size $N = 1024$ from the normal distribution $N(0,1)$, construct the random walks $S_n^{(1)} = S_0^{(1)} + W_1^{(1)} + \cdots + W_n^{(1)}$ and $S_n^{(2)} = S_0^{(2)} + W_1^{(2)} + \cdots + W_n^{(2)}$ starting from $S_0^{(1)} = 0$ and $S_0^{(2)} = 0$ respectively, and construct as before two independent white noise sequences $\{\epsilon_n^{(1)}\}_{n=1,\dots,N}$ and $\{\epsilon_n^{(2)}\}_{n=1,\dots,N}$ from the distribution $N(0,.16)$ which are independent of $\{W_n^{(1)}\}_{n=1,\dots,N}$ and $\{W_n^{(2)}\}_{n=1,\dots,N}$. Give on the same figure the plots of the graphs of the time series $\{X_n^{(1)}\}_{n=1,\dots,N}$ and $\{X_n^{(2)}\}_{n=1,\dots,N}$ defined by:

$$X_n^{(1)} = S_n^{(1)} + \epsilon_n^{(1)} \quad \text{and} \quad X_n^{(2)} = S_n^{(2)} + \epsilon_n^{(2)}.$$

4. Give the plot of the same linear combination of $\{X_n^{(1)}\}_{n=1,\dots,N}$ and $\{X_n^{(2)}\}_{n=1,\dots,N}$ as before, compare to the plot obtained in part 2, and explain why there is no cointegration this time.
5. Give the scatterplot of $X_n^{(2)}$ against $X_n^{(1)}$, add the least squares regression line, test if the regression is significant, and give a time series plot of the residuals as well as their acf. Comment.

Ⓢ **Problem 6.6** *We consider the state-space model given by:*

$$\begin{cases} X_{t+1} = FX_t + V_t \\ Y_t = GX_t + W_t \end{cases}$$

where the covariance matrices of the white noises $\{V_t\}_t$ *and* $\{W_t\}_t$ *are the identity matrices and where the other parameters are given by:*

$$F = \begin{bmatrix} .2 & -.1 & 3 \\ 1 & 0 & 0 \\ 0 & 1 & 0 \end{bmatrix}, \qquad \textit{and} \qquad G = \begin{bmatrix} 1 & 1 & 0 \\ 0 & 1 & 1 \end{bmatrix}.$$

We assume that the values of the one step ahead estimates $\hat{X}_{t_0}$ *of the state vector, and* Ω_{t_0} *of its covariance matrix are given by:*

$$\hat{X}_{t_0} = \begin{bmatrix} 1.2 \\ .3 \\ .45 \end{bmatrix}, \qquad \textit{and} \qquad \Omega_{t_0} = \begin{bmatrix} 1.25 & 1 & 1 \\ 1 & 1.25 & 1 \\ 1 & 1 & 1 \end{bmatrix}.$$

We also assume that the next 5 *values of the observation vector* Y *are given by:*

$$Y_{t_0} = \begin{bmatrix} -.1 \\ 1 \end{bmatrix}, \quad Y_{t_0+1} = \begin{bmatrix} .3 \\ .9 \end{bmatrix}, \quad Y_{t_0+2} = \begin{bmatrix} .47 \\ -.8 \end{bmatrix}, \quad Y_{t_0+3} = \begin{bmatrix} .85 \\ -1.0 \end{bmatrix}, \quad Y_{t_0+4} = \begin{bmatrix} .32 \\ .9 \end{bmatrix}.$$

For each time $t = t_0 + 1$, $t = t_0 + 2$, $t = t_0 + 3$, $t = t_0 + 4$ *and* $t = t_0 + 5$, *use Kalman filtering to compute the one step ahead estimates* $\hat{X}_t$, *its prediction quadratic error* Ω_t, *and of the next observation vector* $\hat{Y}_{t|t-1}$, *and its prediction quadratic error* $\Delta^{(1)}_{t-1}$.

Ⓢ **Problem 6.7** *The goal of this problem is to write* `S`*-functions to simulate and visualize the time evolution of the state and the observation of a linear state-space system of the form:*

$$\begin{cases} \mathbf{X}_{t+1} = F\mathbf{X}_t + \mathbf{V}_t \\ \mathbf{Y}_t = G\mathbf{X}_t + \mathbf{W}_t \end{cases} \tag{6.36}$$

where both the state vector $\mathbf{X}_t$ *and the observation vector* $\mathbf{Y}_t$ *are of dimension* 2, *and where* $\{\mathbf{V}_t\}_t$ *and* $\{\mathbf{W}_t\}_t$ *are independent Gaussian white noises with variance/covariance matrices* Σ_V *and* Σ_W *respectively. As usual we assume that* $\mathbf{V}_t$ *and* $\mathbf{W}_t$ *are independent of* $\mathbf{X}_t$, $\mathbf{X}_{t-1}$, $\mathbf{X}_{t-2}$, ... *We shall also visualize the performance of the forecasts derived from the Kalman filter theory.*

1. Write an `S`*-function, say* `ksim`*, with parameters F, G, SigV, SigW, X0, Omega0, N and SEED which:*

- *initializes the seed of the random generator of* `S-Plus` *to SEED*
- *uses the random number generation functions of* `S-Plus` *to create a* $N \times 2$ *array of type numeric containing realizations of the* N *values* $\{\mathbf{V}_t;\ t = 0, 1, \ldots, N-1\}$ *of the white noise and a* $2 \times (N+1)$ *array of type numeric containing realizations of the* $N+1$ *values* $\{\mathbf{X}_t;\ t = 0, 1, \ldots, N\}$ *of the state vector* $\mathbf{X}_t$ *of the linear system given by equation (6.36) starting with* $\mathbf{X}_0 = \texttt{X0}$
- *Create an* $(N+1) \times 2$ *array of type numeric containing realizations of the* $N+1$ *values* $\{\mathbf{W}_t;\ t = 0, 1, \ldots, N\}$ *of the observation white noise and uses the values of the state vector created above to produce an* $(N+1) \times 2$ *array of type numeric containing realizations of the* $N+1$ *values* $\{\mathbf{Y}_t;\ t = 0, 1, \ldots, N\}$ *of the observations* $\mathbf{Y}_t$ *satisfying the observation equation*

- *assume that X_0 is perfectly known at time $t = 0$ (this is obviously an unrealistic assumption but please bear with me, this is merely homework stuff) and compute for $t = 1, \dots, N$ the value of $\hat{\mathbf{X}}_{t|t} = \mathbb{E}\{\mathbf{X}_t|\mathbf{Y}_{\le t}\}$, $\hat{\mathbf{X}}_t = \mathbb{E}\{\mathbf{X}_{t+1}|\mathbf{Y}_{\le t}\}$ and $\hat{\mathbf{Y}}_t = \mathbb{E}\{\mathbf{Y}_{t+1}|\mathbf{Y}_{\le t}\}$ and return the list of the five Nx2 arrays of type numeric say `$xt`, `$yt`, `$hxtt`, `$hxt` and `$hyt` containing the values of $\mathbf{X}_t$, $\mathbf{Y}_t$, $\hat{\mathbf{X}}_{t|t}$, $\hat{\mathbf{X}}_t$ and $\hat{\mathbf{Y}}_t$ for $t = 1, \dots, N$*

We use the notation $\mathbf{Y}_{\le t} = \{Y_t, Y_{t-1}, \dots, Y_0\}$ for the set of all the observations prior to time t. Recall that the notation $E_t(\mathbf{X})$ was used for what is now denoted by $\mathbf{X}_{t|t}$.
2. Run the `S` *command*

```
> KTST <- ksim(F, G, SigV, SigW, X0, Omega0, N, SEED)
```

with:

$$\mathtt{F} = \begin{bmatrix} .2 & -.1 \\ 1 & .3 \end{bmatrix}, \qquad \text{and} \qquad \mathtt{G} = \begin{bmatrix} 1 & 1 \\ 0 & 1 \end{bmatrix},$$

$$\mathtt{SigV} = \begin{bmatrix} .2 & 0 \\ 0 & 1 \end{bmatrix}, \quad \mathtt{SigW} = \begin{bmatrix} 1 & 0 \\ 0 & .4 \end{bmatrix}, \quad \mathtt{X0} = \begin{bmatrix} 1.2 \\ .45 \end{bmatrix} \quad \text{and} \quad \mathtt{Omega0} = \begin{bmatrix} 1.25 & 0 \\ 0 & 1.25 \end{bmatrix},$$

and finally with $N = 125$ and $SEED = 14$.
3. Write an `S` *function, say* `kanim` *which, with the appropriate parameters taken from the list* `KTST` *obtained in the previous question, will produce the following three animations.*

- *plots on the same graphic window of the successive values of* `KTST$xt` *and* `KTST$hxtt`
- *plots on the same graphic window of the successive values of* `KTST$xt` *and* `KTST$hxt`
- *plots on the same graphic window of the successive values of* `KTST$yt` *and* `KTST$hyt`

(T) **Problem 6.8** *Find the state-space representation for the following time series models:*
1. $(1 - \phi_1 B)Y_t = (1 - \theta_1 B)\epsilon_t$
2. $Y_t = \phi_1 Y_{t-1} + \epsilon_t - \theta_1 \epsilon_{t-1} - \theta_2 \epsilon_{t-2}$
where in both cases, $\{\epsilon_t\}_t$ denotes a white noise.

(T) **Problem 6.9** *Let $\{X_t\}_t$ be the AR(1) time series:*

$$X_t = \phi X_{t-1} + V_t$$

and let Y_t be the noisy observation:

$$Y_t = X_t + W_t$$

where we assume that $\{V_t\}_t$ and $\{W_t\}_t$ are independent Gaussian white noises with variances σ_V^2 and σ_W^2 respectively. Give the best predictions (in the sense of the expected squared error) for X_{t+1}, Y_{t+1} and their variances given the past observations Y_t, Y_{t-1}, ...

(T) **Problem 6.10** *Derive a state space representation for the ARIMA(1,1,1) model:*

$$(1 - \phi B)(1 - B)X_t = (1 - \theta B)W_t$$

when $\{W_t\}_t$ is a $N(0, \sigma^2)$ white noise, and identify all the parameters of the model in terms of ϕ, θ, and σ^2. Recall that $(1 - B)$ merely stands for the operation of differentiation.

(E) **Problem 6.11** *For each of the energy stocks DUKE, PCG, SO and TXU used in Subsection 3.5.4, reproduce all the steps of the filtering analysis of the time varying extension of the CAPM theory performed in the text for AEP. Compare the four time series of beta estimates so obtained and comment.*

Ⓔ **Problem 6.12** *This problem is devoted to the estimation of the time-varying betas for the daily stock data contained in the files* ibm.asc *and* merck.asc. *As in the text we use the daily* $S\&P500$ *as a proxy for the market portfolio.*
1. For each of these two stocks, propose an estimate for σ_v^2 *and* σ_ϵ^2, *run the Kalman filter to produce estimates of the time varying betas and plot the resulting time series of betas.*
2. Do you think that the excess returns of IBM and Merck are correlated? How would you modify the procedure followed in the previous question to account for the dependence between these excess returns?

Ⓔ **Problem 6.13** *The data needed for this problem is contained in the text files* pepsiqeps.asc *and* ibmqeps.asc. *Open them in* S-Plus *and run them as a scripts. You should get two timeSeries objects. The first one gives the quarterly earnings per share of Pepsi Co. and IBM.*
1. Use the Pepsi data and reproduce the filtering analysis done in the text for the General Electric quarterly earnings.
2. The goal of this second question is to perform the same analysis with the IBM data contained in the file ibmqeps.asc.
2.1. Fit a state space model and use it to predict the last two years of quarterly earnings using only the data up to the last quarter of 2000.
2.2. Fit a state space model to the data of the quarterly earnings over the period starting with the first quarter of 1996, and ending with the third quarter of 2000. Use this model and Kalman filtering as above to predict the last two years of quarterly earnings using only the data used to fit the model.
2.3. Fit a model to the data up to the last quarter of 1990, and use it to predict the quarterly earnings of the period 1991-1995 using only the data up to the last quarter of 1990.
2.4. Comment on the performance of the prediction in each case.

NOTES & COMPLEMENTS

Despite serious technical complications, the linear theory of time series presented in the previous chapter in the univariate case can be extended to multivariate time series models. A thorough account can be found in Hamilton's exhaustive exposé in his book [43]. Cointegration is a concept of great theoretical significance, and it is now an integral part of most econometric theory textbooks. It was introduced in the seminal work [34] of Engle and Granger. This fundamental contribution was cited as the main reason to grant the 2003 Nobel prize to these authors. The interested reader can also consult Johansen's book [49] for the general statistical theory of cointegrated time series. Cointegration appears in economic theories as a way to imply equilibrium relationships between economic time series. These equilibria are the result of relaxation of the dynamics toward *steady states* and identifying them usually requires long spans of low frequency data, for some form of ergodicity to take effect. In financial applications, cointegration appears as a way to identify arbitrage opportunities. See for example Section 8.6 entitled *Threshold cointegration and arbitrage* of Tsay's book [83]. A complete exposé of cointegration theory would be beyond the scope of this book. Nevertheless, we mention that its importance is now recognized by the financial engineering community as well as the econometric community, and new research programs have been started to understand and control better pricing and risk management issues in the presence of cointegrated price series. Spread options in the energy markets, and basket options in the equity markets are two of the many examples of instruments naturally written on cointegrated price series. Our interest in these financial products drove our attempt to introduce the concept of cointegration, even if we knew that we could make justice to its importance. The recent textbooks of Chan [24], Tsay [83], and especially Zivot and Wang [89] can be consulted for examples, error correction forms, and statistical tests.

Even after so many years, the best way to learn about the classical theory of CUSUM tests is still from the original paper of Brown, Durbin and Evans [14].

The use of linear state-space models in the analysis of time series is now part of the folklore. Most of the modern textbooks on time series do include at least one chapter on the state-space models and the powerful results of the filtering theory of partially observed systems. See for example [13], [43] or [24]. Their use in financial applications has experienced a similar growth. See for example [52]. A discussion in the delicate estimation of the parameters of the model, including an application of the EM algorithm, can also be found in the book of Shumway and Stoffer [78].

The shortcomings of the CAPM model and its failure to pass the empirical tests of its validity (recall Section 3.5.4 of Chapter 3) have been studied extensively. The discussion of this chapter, especially the use of filtering theory to track the values of a potentially time varying beta, are borrowed from the book of Gençay, Selçuk and Whitcher, where references to the extensive literature devoted to the *fixing* of the CAPM model can be found. The idea of the application of recursive filtering equations to the prediction of the quarterly earnings of a company was borrowed from Shumway and Stoffer [78]

7

NONLINEAR TIME SERIES: MODELS AND SIMULATION

Most financial time series data exhibit nonlinear features which cannot be captured by the linear models seen in the previous two chapters. In this last chapter, we present the elements of a theory of nonlinear time series adapted to financial applications. We review a set of standard econometric models which were first introduced in the discrete time setting. They include the famous, ARCH, GARCH, ... models, but we also discuss stochastic volatility models and we emphasize the differences between these concepts which are too often confused. However, because of the growing influence of the theoretical developments of continuous time finance in the everyday practice, we spend quite a significant part of the chapter analyzing the time series models derived from the discretization of continuous time stochastic differential equations. This new point of view can bring a fresh perspective. Indeed, the classical calculus based on differential equations can be used as a framework for time evolution modeling. Its stochastic extension is adapted to the requirements of modeling of uncertainty, and powerful intuition from centuries of analyses of physical and mechanical systems can be brought to bear. We examine its implications at the level of simulation. The last part of the chapter is devoted to a new set of algorithms for the filtering of nonlinear state space systems. We depart from the time honored tradition of the extended Kalman filter, and we work instead with discrete approximations called particle filters. This modern approach is consistent with our strong bias in favor of Monte Carlo simulations. We illustrate the versatility of these filtering algorithms with the example of price volatility tracking.

7.1 FIRST NONLINEAR TIME SERIES MODELS

This introductory section builds on some of the ideas of the linear theory presented in Chapter 5. It is devoted to the discussion of a couple of nonlinear time series models based on natural generalizations of classical linear time series models introduced in that chapter. The first of these two models was inspired by the desire to develop a time series analog of the so-called fractional Brownian motion whose modeling potential for financial data was popularized by Mandelbrot. The second model is a straightforward generalization of the notion of auto-regressive model.

7.1.1 **Fractional Time Series**

The process of fractional Brownian motion has a certain number of desirable properties which are present quite often in financial data. Even though self-similarity has limited use outside of continuous time models, long range dependence is a feature which most of the linear models analyzed in Chapter 5 do not share. The time series model introduced here is an attempt to capture this feature.

Definition 2. *If p and q are integers and $d \in (0, 1)$, we say that the time series $\{X_t\}_t$ is an ARIMA(p,d,q) series if $(I - B)^d X_t$ is a stationary ARMA(p,q) time series where the fractional difference operator $(I - B)^d$ is defined by the infinite sum:*

$$(I - B)^d = I + \sum_{j=1}^{\infty} \frac{d(d-1)\cdots(d-j+1)}{j!}(-1)^j B^j. \tag{7.1}$$

As usual we use the notation B for the backward shift operator. The time series of the type defined above are called fractional processes or fractional time series. This definition calls for a few remarks.

• The infinite series in the right hand side of formula (7.1) mimics the Taylor expansion of the fractional powers:

$$(1 - z)^d = 1 + \sum_{j=1}^{\infty} \frac{d(d-1)\cdots(d-j+1)}{j!}(-1)^j z^j$$

which converges for $|z| < 1$, hence the terminology of fractional differentiation.

• The cases $d = 0$ and $d = 1$ appear as limiting cases of the above definition. The case $d = 0$ corresponds to the usual stationary ARMA(p,q) model, while the case $d = 1$ corresponds to the classical ARIMA(p,1,q) model. Fractional processes provide a continuum of models interpolating between these extreme cases.

• Fractional time series are asymptotically stationary when $d < 1/2$. This means that even though they are not technically speaking stationary, they behave as if they were in the regime $t \to \infty$. In fact, it is possible to prove that in this case (i.e. when $d < 1/2$) the auto-correlation function $\rho_X(h)$ converges toward zero like a power for large lags (i.e. when $h \to \infty$). More precisely:

$$\rho_X(h) \sim h^{2d-1} \qquad \text{as} \quad h \to \infty. \tag{7.2}$$

Notice that $2d - 1 < 0$ since $d < 1/2$. This slow polynomial decay of the acf is in contrast with what we found in the case of the classical ARMA models for which the acf either vanishes after some lag (like in the case of the MA series) or decays at an exponential rate (like in the case of the AR series). The long range dependence (also called long range memory) is the reason why these models were introduced. They exhibit persistence while most linear models don't. The slow decay of the empirical estimate of the acf is a strong indication that a fractional model may be relevant.

• Formula (7.2) can be turned into a method of estimation of the exponent d. Indeed, at least asymptotically, one should have:

$$\log \rho_X(h) \sim \beta_0 + (2d-1)\log h$$

for some constant β_0. Consequently, after estimating the sample acf $\hat{\rho}_X(h)$ in the usual way, a simple linear regression of $\log \hat{\rho}_X(h)$ against the logarithm of the lag $\log h$ should give a slope equal to $2d-1$, from which one can derive an estimate of the value of d. Unfortunately, this estimate is very poor in the case of non-Gaussian time series.

`S-Plus` does not have a special function for the estimation and simulation of the fractional time series defined above. See nevertheless the discussion below of the fractionally integrated GARCH models.

7.1.2 Nonlinear Auto-Regressive Series

The first examples of genuinely nonlinear time series models are provided by the nonlinear auto-regressive models. Like their linear counterparts, they generalize the simple model:

$$X_t = \mu + \sigma W_t$$

to the case where the mean is a function of the past values of the series itself. But instead of assuming that this function is linear as in the case of the classical AR(p) models, we shall now assume that it can be any nonlinear function of X_{t-1}, X_{t-2}, ..., X_{t-p}. In other words, we shall assume the existence of a deterministic function $\mu : \mathbb{R}^p \hookrightarrow \mathbb{R}$ such that:

$$X_t = \mu(X_{t-1}, X_{t-2}, \dots, X_{t-p}) + \sigma W_t.$$

However, we shall not limit the nonlinear dependence on the past lags of the series to the mean term. We shall also allow it in the variance of the series. More precisely:

Definition 3. *If p is an integer, we say that the time series $\{X_t\}_t$ is a nonlinear AR(p) series if there exist a white noise $\{W_t\}_t$ and two deterministic functions:*

$$\mu : \mathbb{R}^p \ni (x_1, \dots, x_p) \hookrightarrow \mu(x_1, \dots, x_p) \in \mathbb{R}$$

and

$$\sigma : \mathbb{R}^p \ni (x_1, \dots, x_p) \hookrightarrow \sigma(x_1, \dots, x_p) \in \mathbb{R}_+$$

such that:

$$X_t = \mu(X_{t-1}, X_{t-2}, \dots, X_{t-p}) + \sigma(X_{t-1}, X_{t-2}, \dots, X_{t-p}) W_t. \tag{7.3}$$

Let us consider the simplest case of a nonlinear AR(1) for the sake of illustration. Such a series is of the form:

$$X_t = \mu(X_{t-1}) + \sigma(X_{t-1}) W_t$$

Moreover, if we further assume that the white noise $\{W_t\}_t$ is an $N(0,1)$ i.i.d. sequence, then we see that conditioned on its past, X_t is still normally distributed. More precisely:

$$X_t|X_{t-1} = N(\mu(X_{t-1}), \sigma(X_{t-1})^2)$$

which shows that the marginal distribution of X_t is a mixture of normal distributions, and as such, it has heavy tails and excess kurtosis as proved in Section 7.3 below and Problem 7.1.

7.1.3 Statistical Estimation

As before we limit the scope of our discussion to the particular case $p = 1$. The general case can be treated exactly in the same way, but the notation become so cumbersome that we refrain from discussing the general case all together. There are two ways to approach the statistical estimation of a nonlinear AR model. Either by parametrization of the unknown functions $\mu(x)$ and $\sigma(x)$ or by appealing directly to nonparametric techniques.

• ***Parametric Approach*** Let us assume that the functions μ and σ are known up to a parameter θ. This parameter θ can be estimated by the maximum likelihood method when the white noise is Gaussian. Indeed, in this case, the likelihood function is the product of normal densities, and its maximization is not more difficult than in the case of the estimation of the mean and the variance of a normal sample. The maximum likelihood estimate (MLE for short) is much more difficult to find in the general case. A reasonable approximation can be obtained by *acting as if* the white noise was still Gaussian. The estimate so obtained is usually called the quasi-MLE or the pseudo-MLE. So given a sample $x_1, x_2, \ldots, x_{T-1}, x_T$ of size T, the maximum likelihood estimator $\hat{\theta}_T$ is given by:

$$\begin{aligned}\hat{\theta}_T &= \arg\min_\theta \log L(x_1, \ldots, x_T|\theta) \\ &= \arg\min_\theta \sum_{t=1}^{T} \log f(x_t|x_{t-1}, \theta)\end{aligned}$$

if we use the notation $f(x|x', \theta)$ for the conditional density of X_t given that $X_{t-1} = x'$. Since in general we cannot compute it, we compute instead the quasi-maximum likelihood estimator $\hat{\theta}_T$ is given by:

$$\hat{\theta}_T = \arg\min_\theta \sum_{t=1}^{T} \left(-\frac{1}{2}\log 2\pi - \frac{1}{2}\log \sigma^2(x_{t-1}, \theta) - \frac{[x_t - \mu(x_{t-1}, \theta)]^2}{2\sigma^2(x_{t-1}, \theta)} \right) \quad (7.4)$$

which is obtained by assuming that the white noise is Gaussian. How difficult this optimization problem is depends upon the explicit form of the functions $\mu(\,\cdot\,, \theta)$ and $\sigma(\,\cdot\,, \theta)$. But what is remarkable is that, asymptotically, the resulting estimator has essentially the same desirable properties as in the case of Gaussian white noise. Indeed, it can be proven that, if $\{W_t\}_t$ is an i.i.d. sequence of mean-zero variance-one random variables, then the quasi-MLE $\hat{\theta}_T$ is consistent in the sense that whatever the true value θ of the parameter is, we have:

$$\lim_{T \to \infty} \hat{\theta}_T = \theta,$$

and it is asymptotically normal in the sense that the distribution of $\sqrt{T}(\hat{\theta}_T - \theta)$ converges toward a normal distribution with mean zero and a variance given by the Fisher information matrices which we shall not make explicit here. This last property makes it possible to derive (asymptotically correct) tests of significance and confidence intervals for the parameter.

• ***Nonparametric Approach*** If we do not know enough about the functions $\mu(x)$ and $\sigma(x)$, and if we cannot reduce their estimation to estimating one or a small number of scalar parameters, there is always the possibility of appealing to nonparametric estimation techniques. In particular, building on the expertise we developed in Chapter 4, we can solve the estimation problem by choosing a kernel function K, a bandwidth $b_T > 0$, and for each value of x, computing the estimates:

$$\hat{\mu}_T(x) = \frac{\sum_{t=2}^{T} x_t K\left(\frac{x - x_{t-1}}{b_T}\right)}{\sum_{t=2}^{T} K\left(\frac{x - x_{t-1}}{b_T}\right)} \tag{7.5}$$

and

$$\hat{\sigma}_T^2(x) = \frac{\sum_{t=2}^{T} x_t^2 K\left(\frac{x - x_{t-1}}{b_T}\right)}{\sum_{t=2}^{T} K\left(\frac{x - x_{t-1}}{b_T}\right)} - \hat{\mu}_T(x)^2. \tag{7.6}$$

Notice that we applied the procedure learned in Chapter 4 to the couples (x_{t-1}, x_T) with x_{t-1} playing the role of the explanatory variable and x_t playing the role of the response variable. Because we are well versed in the theory of kernel estimation, we know that the properties of these estimators depend upon the value of the bandwidth b_T. Hence, we should not be surprised to learn that the consistency and the asymptotic normality of these estimates will only hold for specific forms of this dependence. To be specific, the estimates are consistent, i.e.

$$\lim_{T \to \infty} \hat{\mu}_T(x) = \mu(x) \qquad \text{and} \qquad \lim_{T \to \infty} \hat{\sigma}_T(x) = \sigma(x),$$

whenever $T \to \infty$ in such a way that $\lim_{T \to \infty} T b_T = \infty$. So as the sample size grows (i.e. as $T \to \infty$) we can take smaller and smaller a bandwidth b_T (which is desirable) but the latter cannot go to zero too fast. Under the same condition the bivariate distribution of the vector

$$\sqrt{T b_T}\left(\begin{bmatrix} \hat{\mu}_T(x) \\ \hat{\sigma}_T^2(x) \end{bmatrix} - \begin{bmatrix} \mu(x) \\ \sigma^2(x) \end{bmatrix}\right)$$

converges toward a bivariate normal distribution with zero mean. This result shows that even though the kernel estimate converges toward the true value, the rate of convergence is smaller since it is given by $\sqrt{T b_T}$ instead of the usual $\sqrt{T}$. This is the price to pay for not having to make assumptions on the particular forms of the functions $\mu(x)$ and $\sigma(x)$. In simple terms, this illustrates a statement we made several times already: the kernel method can be used even when we do not know much about the functions to estimate, but in order to get the same precision as parametric methods, it needs more data.

7.2 More Nonlinear Models: ARCH, GARCH & All That

It is now time to investigate the most popular of the nonlinear time series models: the famous ARCH and GARCH models.

7.2.1 **Motivation**

As a matter of illustration, we revisit the example of the daily log-returns of the Brazilian coffee which we considered earlier in Chapter 2. However, the reader should be aware of the fact that most of what we are about to do or say applies to most financial time series as well. The plot of the auto-correlation function of these log-returns is given in the left pane of Figure 7.1. Obviously, this acf looks pretty much like the auto-correlation function of a white noise. If, ignoring this warning,

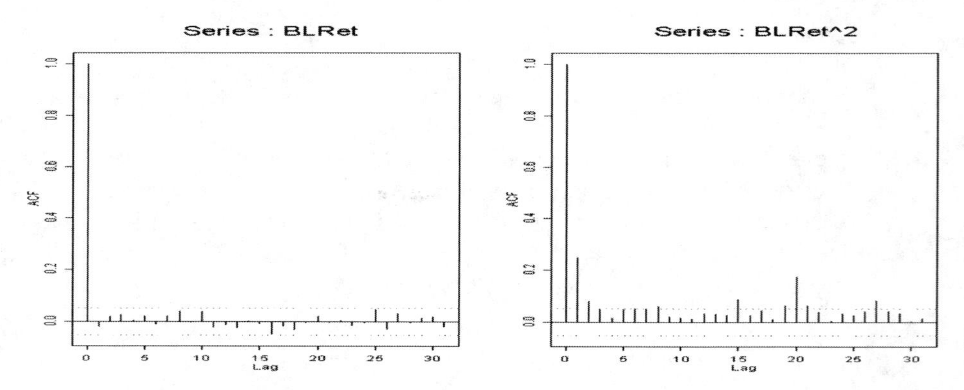

Fig. 7.1. Plot of the auto-correlation function of the Brazilian coffee daily log-returns (left) and of their squares (right).

we try to fit an AR model anyway, we get:

```
> BLRet.ar <- ar(BLRet)
> BLRet.ar$order
[1] 0
```

which confirms our fear that the series may not carry more information than a white noise would. We saw in Chapter 2 that the marginal distribution of these daily log-returns was not Gaussian, and that it had heavy tails. Recall the normal Q-Q plot comparing the distribution of these daily log-returns to the normal distribution in the left pane of Figure 2.7. This remark is crucial at this stage: indeed it is only for Gaussian time series that the absence of correlation implies independence. So, even though we have an uncorrelated sequence, it is still possible that significant forms of dependence between the successive terms of the series exist. Indeed, if the terms of a time series are not jointly Gaussian, the lack of correlation does not imply independence. The existence of dependencies is confirmed by the plot of the auto-correlation function of the square of the time series which we reproduce in the right pane of Figure 7.1. Indeed, this plot does not look like the acf of a white noise. Since squares of random variables are independent whenever the original random variables are independent, this proves that the original series of daily log-returns was not an independent series. So, there is still some hope for us to be able to untangle these dependencies.

7.2.2 **ARCH Models**

The next step is to model the volatility (i.e. the instantaneous standard deviation) as a random process of its own. There are two main reasons for that. The first one is that time series plots of these log-returns show that the variance seems to change over time. This goes under the name of heteroskedasticity. The second is that, as we already saw, one of the easiest ways to create distributions with heavy tails, is to mix together Gaussian distributions with different variances. See Problem 7.1 at the end of this chapter for example. By making sure that the instantaneous standard deviation is random, we may force the distribution of the log-returns to be a mixture of Gaussian distributions, and hence to have heavy tails. These two basic ideas are at the core of the approach taken in the next several sections. First, we introduce ARCH(p) models. Not surprisingly, ARCH stands for:

Auto-**R**egressive **C**onditional **H**eteroskedasticity

A formal definition is as follows.

Definition 4. *A time series $\{X_t\}_t$ is said to be of type ARCH(p) if:*

$$X_t = \sigma_t W_t$$

where $\{W_t\}_t$ is a strong Gaussian white noise (i.e. an i.i.d. sequence of $N(0,1)$ random variables), and where σ_t is a (positive) function of X_{t-1}, X_{t-2}, ... determined by:

$$\sigma_t^2 = \alpha_0 + \sum_{j=1}^{p} \alpha_j X_{t-j}^2$$

with $\alpha_0 > 0$ and $\alpha_j \geq 0$ for $j = 1, 2, \ldots, p$.

In most of the textbook definitions of the ARCH(p) models, it is not required that the white noise $\{W_t\}_t$ is Gaussian. However, for the sake of simplicity, we do make this restrictive assumption. As a warm up, we first consider the simple case of an ARCH(1) model. In this case we have:

$$X_t = \sigma_t W_t \qquad \text{and} \qquad \sigma_t^2 = \alpha_0 + \alpha_1 X_{t-1}^2$$

which gives inductively:

$$\begin{aligned} X_t^2 &= \sigma_t^2 W_t^2 \\ &= \alpha_0 W_t^2 + \alpha_1 X_{t-1}^2 W_t^2 \\ &= \alpha_0 W_t^2 + \alpha_1 \alpha_0 W_{t-1}^2 W_t^2 + \alpha_1 X_{t-2}^2 W_t^2 W_{t-1}^2 \\ &= \alpha_0 W_t^2 + \alpha_0 \alpha_1 W_{t-1}^2 W_t^2 + \alpha_1 X_{t-2}^2 W_t^2 W_{t-1}^2 + \cdots + \alpha_0 \alpha_1^n W_t^2 W_{t-1}^2 \cdots W_{t-n}^2 \\ &\qquad \alpha_1^{n+1} W_t^2 W_{t-1}^2 \cdots W_{t-n}^2 X_{t-n-1}^2 \end{aligned}$$

by successive substitutions using the definition formula. This series converges if

- $\{X_t\}_t$ is causal (so X_{t-n-1} depends on W_s for $s < t - n$)
- $\{X_t\}_t$ is stationary (so that $\mathbb{E}\{X_{t-n-1}\}$ is a constant and hence bounded)
- $|\alpha_1| < 1$

In this case:

$$X_t^2 = \alpha_0 \sum_{j=0}^{\infty} \alpha_1^j W_t^2 W_{t-1}^2 W_{t-2}^2 \cdots W_{t-j}^2$$

from which we get:

$$\mathbb{E}\{X_t\} = 0 \qquad\qquad \mathbb{E}\{X_t^2\} = \frac{\alpha_0}{1 - \alpha_1}$$

and the final formula:

$$X_t = \sigma_t W_t \quad \text{with} \quad \sigma_t = \sqrt{\alpha_0 \left(\sum_{j=0}^{\infty} \alpha_1^j W_{t-1}^2 W_{t-2}^2 \cdots W_{t-j}^2 \right)}$$

It is important to notice that:

$$\mathbb{E}\{X_{t+h} X_t\} = \mathbb{E}\{\mathbb{E}\{X_{t+h} X_t | W_{t+h-1}, W_{t+h-2}, \ldots\}\} = 0$$

so that, in the sense of "weak white noise" introduced in Chapter 5, the time series $\{X_t\}_t$ is a white noise, not a white noise in the strong sense of i.i.d. sequences, but STILL A WHITE NOISE !!!

7.2.3 GARCH Models

The above discussion of ARCH models can be regarded as a warm up for the introduction of models with a broader appeal in the financial econometric community.

General Definition of a GARCH Model

We will say that the time series $\{X_t\}_t$ is GARCH(p,q) if:

$$X_t = \mu_t + \sigma_t W_t$$

where as before, we assume that the noise sequence $\{W_t\}_t$ is i.i.d. $N(0,1)$, so that the conditional distribution of $\tilde{X}_t = X_t - \mu_t$ given $\tilde{X}_{t-1}$, $\tilde{X}_{t-2}$, ... is $N(0, \sigma_t^2)$ with:

$$\sigma_t^2 = \sigma^2 + \sum_{j=1}^{p} \phi_j \sigma_{t-j}^2 + \sum_{j=1}^{q} \theta_j \tilde{X}_{t-j}^2. \tag{7.7}$$

Obviously this definition generalizes the notion of ARCH(p) models which can be recovered by setting $q = 0$. The term μ_t should be understood as a mean. It could be zero (as assumed in the previous subsection) or it could be the result of the fit of an ARMA model, in which case $\tilde{X}_t$ should be viewed as the residual time series after the fit of such a linear time series model. It is sometimes called the regression term.

The case $p = q = 1$ of GARCH(1,1) models is by far the most frequently used model. In such a case, we try to model the time series $\tilde{X}_t$ as:

$$\tilde{X}_t = \sigma_t W_t$$

in such a way that the conditional distribution of $\tilde{X}_t$ given $\tilde{X}_{t-1}$, $\tilde{X}_{t-2}$, ... is $N(0, \sigma_t^2)$ with:

$$\sigma_t^2 = \sigma^2 + \phi_1 \sigma_{t-1}^2 + \theta_1 \tilde{X}_{t-1}^2$$

> This formula for the instantaneous variance is screaming for an interpretation in terms of ARMA(1,1) models !

We address this question next.

Summary

Even though we stated the definitions of ARCH and GARCH models with variance one strong white noise series, it is easy to see that the main properties of the models remain true if we only assume that the innovations are only white noise in the weak sense. In this subsection, we shall only consider weak sense white noise series, and instead of working in the general case of ARCH(p) and GARCH(p,q) models, we state results only for the case $p = 1$ and $q = 1$.

The formal way to identify a ARCH(1) time series is to write it down first in terms of its conditional standard deviation, i.e. in the form $X_t = \sigma_t W_t$ where $\{W_t\}_t$

is a (stationary) weak white noise with variance one. Then the ARCH(1) prescription becomes:

$$\sigma_t^2 = \alpha_0 + \alpha_1 X_{t-1}^2.$$

which can be rewritten in the form:

$$X_t^2 = \alpha_0 + \alpha_1 X_{t-1}^2 + \tilde{W}_t$$

for some new weak white noise series $\{\tilde{W}_t\}_t$. This shows that (provided we generalize the definitions to accept weak white noise series)

$$\{X_t\}_t \sim \text{ARCH(1)} \quad \Longleftrightarrow \{X_t^2\}_t \sim \text{AR(1)}.$$

We now consider the case of GARCH(1,1) series. In this case the determining condition on the conditional variance reads:

$$\sigma_t^2 = \alpha_0 + \alpha_1 X_{t-1}^2 + \beta_1 \sigma_{t-1}^2.$$

which can be rewritten in the form:

$$X_t^2 = \alpha_0 + (\alpha_1 + \beta_1) X_{t-1}^2 + \tilde{W}_t - \beta_1 \tilde{W}_{t-1}$$

for some new weak white noise series $\{\tilde{W}_t\}_t$. Under the same proviso as before, this shows that

$$\{X_t\}_t \sim \text{GARCH(1,1)} \quad \Longleftrightarrow \{X_t^2\}_t \sim \text{ARMA(1,1)}.$$

7.2.4 `S-Plus` **Commands**

```
> BLRet.g <- garch(BLRet~1,~garch(1,1))
R-Square = 8.598804e-05 is less than tolerance = 0.0001
Convergence reached
```

It is now time to consider a first practical example to see why, when and how one fits a ARCH and/or a GARCH model to data.

```
> BLRet.g
Call: garch(formula.mean=BLRet~1, formula.var=~garch(1,1))
Mean Equation: BLRet ~ 1
Conditional Variance Equation:  ~ garch(1, 1)
Coefficients:
   C 4.510e-04
   A 3.404e-05
   ARCH(1) 1.222e-01
   GARCH(1) 8.638e-01
```

The correspondence between the coefficients `C,A, ARCH(1)` and `GARCH(1)` appearing in the `S-Plus` output and the parameters appearing in the definition formula (7.7) is given by:

◇ C stands for the mean μ
◇ A stands for the mean variance σ^2
◇ ARCH(j) stands for θ_j
◇ GARCH(j) stands for ϕ_j

The `S-Plus garch` objects can be summarized in still another way. One can use the command `summary(BLRet.g)` which produces on the top of the information given above, an entire set of tests and p-values which we shall not reproduce here.

7.2.5 **Fitting a GARCH Model to Real Data**

Enough of theoretical discussions: it is time to see these ARCH and GARCH models in action. We use the example of the stock price of Enron to demonstrate how the concepts introduced so far can be implemented on real data. At the risk of being accused of nostalgia for the exciting period of the seemingly endless expansion in the energy trading boom, we chose to restrict ourselves to the period pre-bankruptcy. The data we used is plotted in Figure 7.2

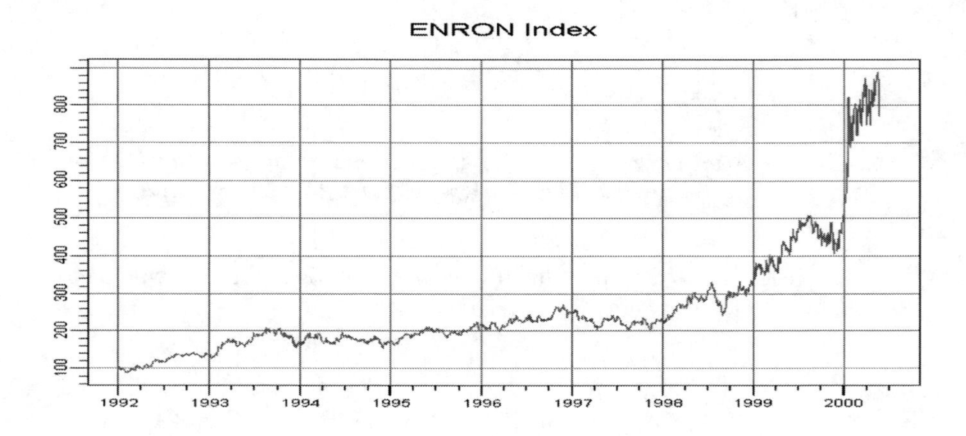

Fig. 7.2. Time series plot of the index created from the values of Enron's stock

Indexes are normalized aggregates of stock prices. Even when the index is computed from a single stock, it differs from the actual stock price in many ways. However, we should view them as providing essentially the same returns as the stock price would since they are merely normalized to have a specific value (say 100) on a given date. We give the `S-Plus` commands used to produce these `timeSeries` objects in an appendix at the end of the chapter for the sake of completeness.

We use the following `S-Plus` commands to compute the log-returns from the Enron index, to compute and plot their acf, to fit an AR model, and to plot the AIC criterion which determines the order of the AR model chosen by the program, 4 in this case. The graphical results are reproduced in Figure 7.3.

```
> LRET <- diff(log(ENRON.ts[,3]))
> acf(LRET)
> LRET.ar <- ar(LRET)
> LRET.ar$order
[1] 4
> tsplot(LRET.ar$aic)
> title("AIC criterion for ENRON LRET")
```

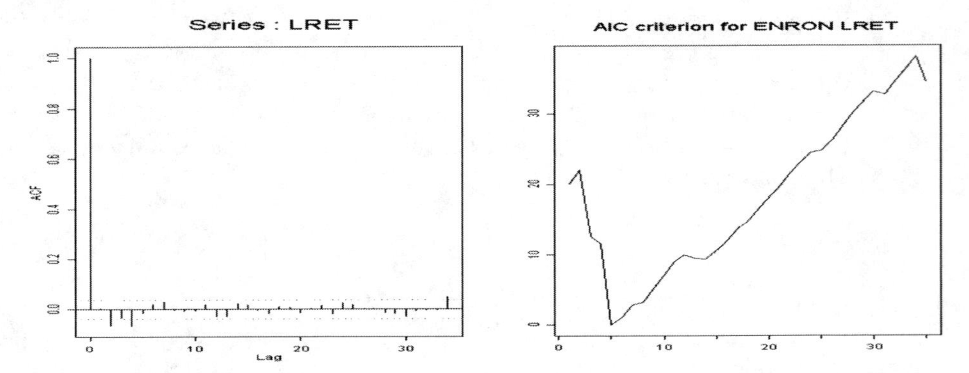

Fig. 7.3. Auto-correlation function of the log returns computed from the Enron index (left) and AIC criterion from the fit of an AR model to these log returns (right).

The time series plot of the residuals of this AR fit together with their Q-Q norm plot are obtained with the commands:

```
> plot(LRET.ar$resid)
> title("AR Residuals for ENRON LRET")
> VRES <- as.matrix.data.frame(seriesData(LRET.ar$resid))
> qqnorm(VRES)
> qqline(VRES)
> title("Normal Q-Q plot of the AR Residuals")
```

Notice that we had to deal with the annoying data typing of `S-Plus` which would not let us produce the desired Q-Q plot with the data from the residuals. Indeed, in the `Windows` version of `S-Plus`, the object `seriesData(LRET.ar$resid)` happens to be a data frame to which the function `qqnorm` cannot be applied. So we need to coerce it to a matrix/vector with the command `as.matrix.data.frame` before we can apply the function `qqnorm`. The results are shown in Figure 7.4. If lack of serial correlation is a possibility in light of what we see in the left pane of

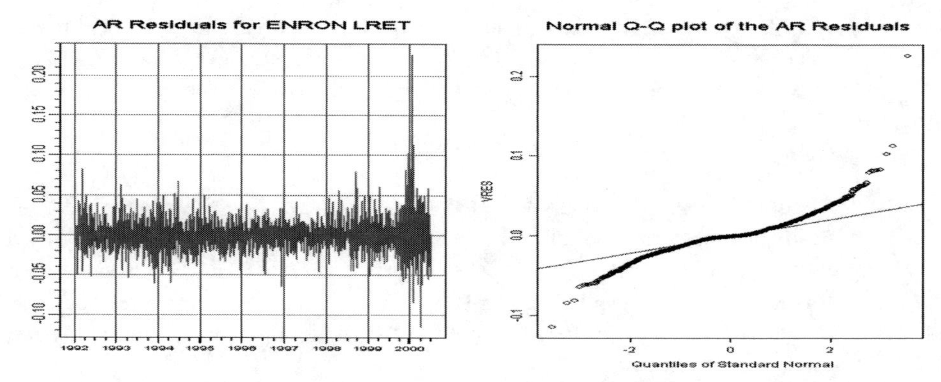

Fig. 7.4. Time series plot of the residuals of the AR model as fitted to the log returns computed from the Enron index (left) and their Q-Q normal plot (right).

Figure 7.4, it is clear that too many measurements end up several standard deviation away from the mean. The marginal distribution of these residuals is presumably not normal. This is confirmed by the normal Q-Q plot of these residuals reproduced in the right pane of this figure. This plot shows that the distribution of the residuals has heavy tails. So even if the AR model was able to capture the serial correlation contained in the log returns, we cannot be sure that the residuals are independent, and that there is no more serial structure left. We settle this question by computing the acf of the squares of these residuals.

```
> acf(LRET.ar$resid)
> acf(LRET.ar$resid^2)
```

The results are shown in Figure 7.5. We are now in a rather delicate situation. Indeed,

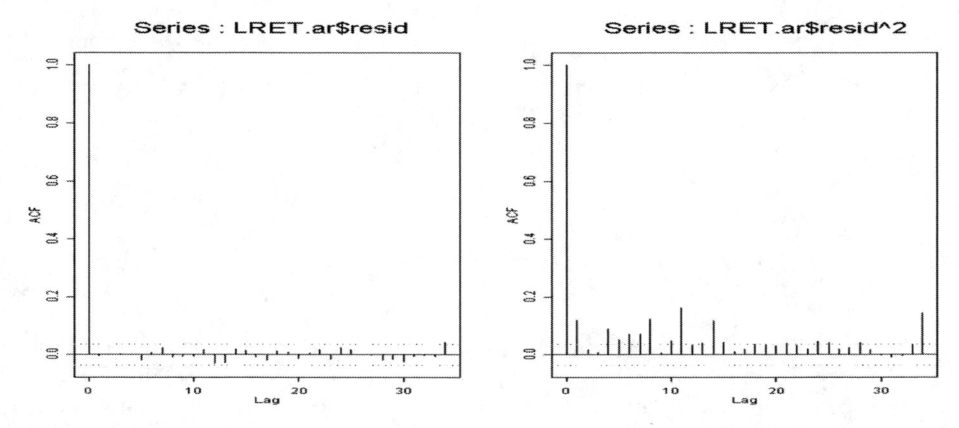

Fig. 7.5. Auto-correlation function of the raw residuals from the fit of an AR model to the Enron log-returns (left) and acf of the squares of these residuals (right).

after transforming the data to stationarity (by first taking their logarithms and then the difference of the latter), and fitting a time series model (say and AR(4)), we end up with a residual time series with an `acf` of the white noise type, i.e. of the form:

$$1, 0, 0, \cdots .$$

Should we consider ourselves done? as we did in Chapter 5 when we were restricted to linear time series models? or should we try to further analyze the data by digging into the residual series? Our discussion of the ARCH/GARCH models indicates that this is a reasonable option at this stage. For this reason, we proceed with the fitting of a GARCH(1,1) to the residuals of the AR model fitted earlier to the log-returns of the Enron Index data.

Fitting a GARCH(1,1) Model to ENRON Index Log Return

Because we fitted an AR(4) model to the log-returns, the first four entries of the fitted model could not be computed (indeed, their computation involves entries with negative time stamps !) and the first four entries of the time series of residuals are NA's. Since many `S-Plus` functions do not handle properly NA's, we remove them, and we fit a GARCH(1,1) model to the remaining series. We use the following commands:

```
> LRES <- LRET.ar$resid[5:length(VRES)]
> LRES.g <- garch(LRES ~ 1,  ~ garch(1, 1))
Iteration   0  Step Size =  1.00000  Likelihood =  2.72059
Iteration   0  Step Size =  2.00000  Likelihood =  2.71671
.........   .. ..........   ....... ............  ........
Iteration   4  Step Size =  1.00000  Likelihood =  2.73205
Iteration   4  Step Size =  2.00000  Likelihood =  2.73202
 Convergence R-Square = 0.0000480889 is less than tolerance
 = 0.0001 Convergence reached.

Call: garch(formula.mean=LRES~1, formula.var=~garch(1, 1))
 Mean Equation: LRES ~ 1
 Conditional Variance Equation:  ~ garch(1, 1)
 Coefficients:
       C -1.656e-004
       A  2.266e-006
 ARCH(1)  4.084e-002
GARCH(1)  9.519e-001
```

Recall that the meaning of the parameters C, A, ARCH(j) and GARCH(j) was given in Subsection 7.2.4. Trying to find if the serial correlation present in the AR residuals has been removed by the GARCH model is a reasonable step to take at this stage. This can be done with the commands:

```
> acf(LRES.g$resid)
> acf(LRES.g$resid^2)
```

The results are reproduced in Figure 7.6. These plots of the auto-correlation functions of the residuals and squared residuals of the fitted GARCH(1,1) model show that the serial correlation has not been completely removed. We shall see examples of GARCH fits with a better performance. In dealing with ARCH and GARCH models,

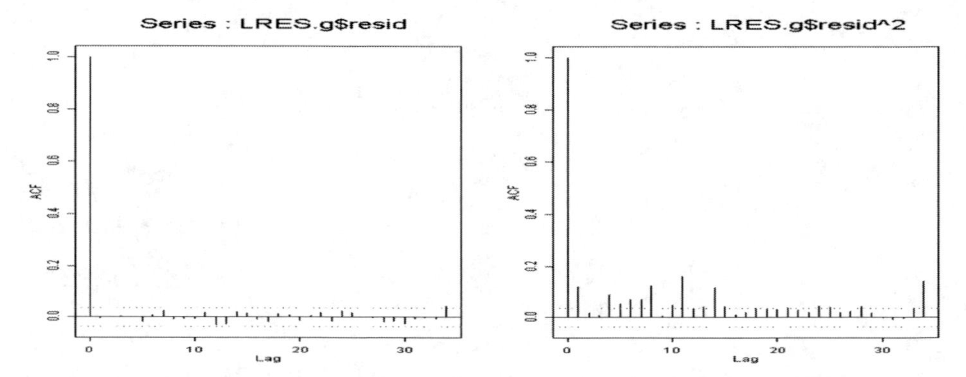

Fig. 7.6. Plots of the acf of the residuals of the GARCH(1,1) model fitted to the AR residuals from the Enron log-returns (left), and of the acf of their squares (right).

part of the fitting procedure involves the estimation of the instantaneous conditional variance, and it is always instructive to take a look at such an estimate. In S-Plus, this estimate can be retrieved with the extension $\cdots$ $sigma.t at the end of the name of the ARCH or GARCH object produced by the fitting procedure. We illustrate its usefulness by plotting the estimated variance in the present situation:

```
> plot(LRES.g$sigma.t,main="Conditional Variance of the
                                    Fitted GARCH(1,1)")
```

The result is reproduced in the top pane of Figure 7.7. This estimator of the conditional variance appears to be quite realistic and quite accurate. We see a large peak indicative of the huge growth and the subsequent crisis of 2000, and regularly spaced smaller peaks reminiscent of the seasonal nature of the business. So the estimation of the conditional standard deviation seems to be a powerful tool. However, just to make sure that we would not be fooled by this capability of S-Plus, we decided to compare this estimate to the result of a *pedestrian* approach: on each day, we consider the entries of the series which occured in the window of the last 60 days, and we compute the empirical variance the series in this window. Notice that this computation is non-anticipative because the sliding window ends at the time of the computation, i.e. at any given time, we use only past values to compute the variance estimate. We wrote a small function localvar to do that. Its use is illustrated by the commands:

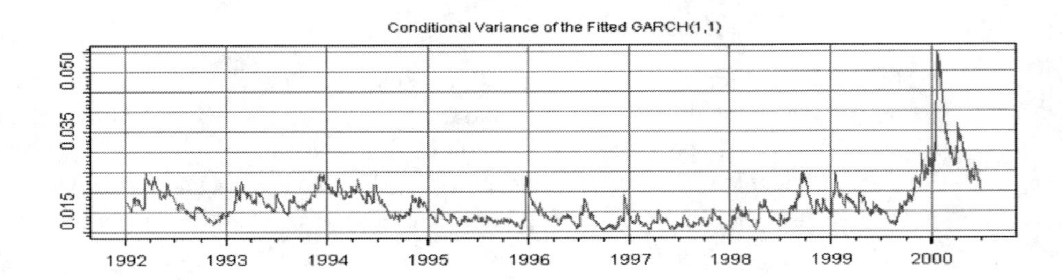

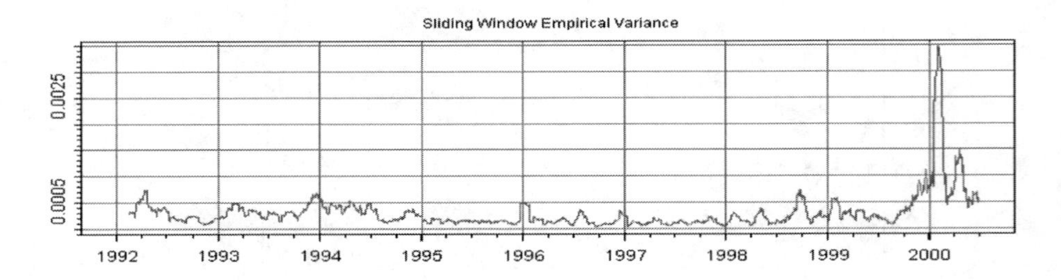

Fig. 7.7. Conditional variance as estimated in the fitting of a GARCH(1,1) model to the AR residuals from the Enron log-returns (top) and instantaneous empirical variance computed in a non-anticipative sliding window (bottom).

```
> RESLV <- localvar(LRES)
> plot(RESLV,main="Sliding Window Empirical Variance")
```

whose result is reproduced in the lower part of Figure 7.7. Notice that the results are very similar to the plot of `LRES.g$sigma.t` obtained by fitting the GARCH model. Given the complexity of the GARCH fitting procedure, this remark is rather anti-climatic. It shows that GARCH should not be fitted for the mere purpose of computing an instantaneous conditional standard deviation.

Monte Carlo Simulations

`S-Plus` provides a function to generate Monte Carlo samples from a GARCH model. Here is an example of its use.

```
> LRES.sim <- simulate.garch(LRES.g,n=1024,n.start=1000)
> tsplot(LRES.sim$et,main="Simulated GARCH(1,1) with same
                                  parameters as in LRES.g")
> tsplot(LRES.sim$sigma.t,main="Conditional Variance of
                                 the Simulated GARCH(1,1)")
```

The results are reproduced in Figure 7.8 Several remarks are in order at this point. As in most simulations of time series models, the `start` parameter is very important.

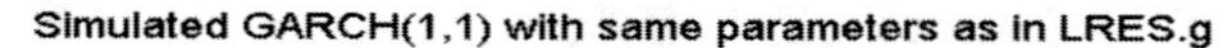

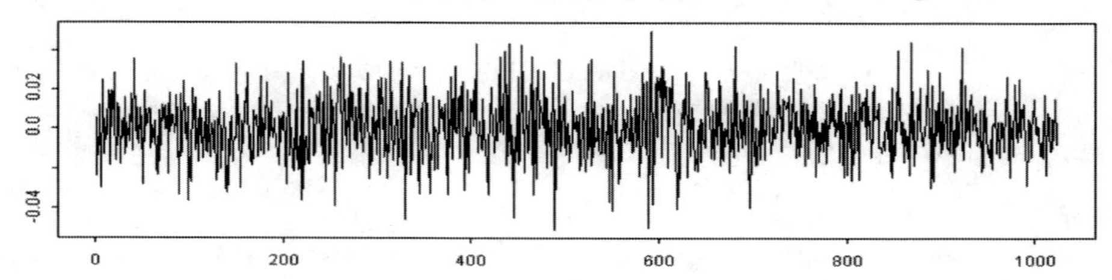

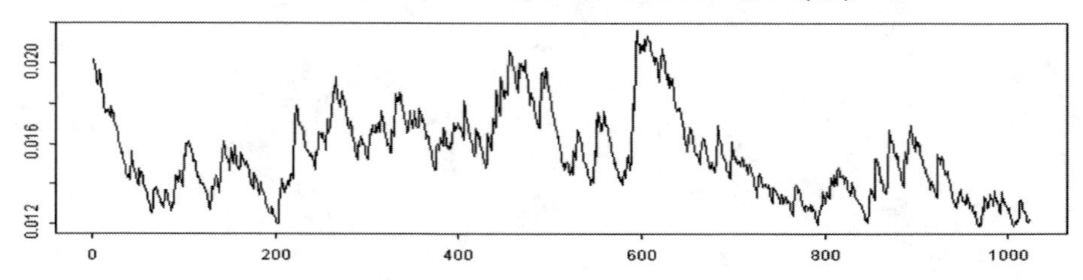

Fig. 7.8. Simulated stretch of length 1024 of the GARCH(1,1) model fitted to the AR residuals from the Enron log-returns (top) and conditional variance (bottom).

It represents the number of samples which are disregarded. They are generated in order for the series to reach a steady state regime in which the dynamics are stable. One should experiment and try several values of this parameter to get a feeling for its effect. In particular, too small a value of this parameter will produce a sample from a non-stationary time series. This is usually detected because of significant differences between the first part of the simulation which exhibits non-stationary behavior, and the later part where the stationary regime is reached, and the effects of the initial values have been wiped out of the statistics.

It is always rather difficult to assess the quality of Monte Carlo simulations. In particular, there is no reason to believe that the simulation appearing in the top pane of Figure 7.8 is not appropriate. On the other hand, the simulation of the conditional standard deviation found in the bottom pane is presumably very poor. Indeed, it does not have any of the deterministic features (in particular it does not have a significant seasonal component) which we identified in the data. Next, we expand on this problem.

Simulation Can Be a Touchy Business

A wrong value for the `start` parameter is not the only possible misuse of the Monte Carlo simulation function `simulate.garch`. Indeed, quite unrealistic simulated samples can be produced when a deterministic seasonal pattern exists, and the GARCH model simply believes that it is part of the typical fluctuations of the random variance. We illustrate this phenomenon in the case of intra-day data, and we refer the reader to Problem 7.4 for a similar example in the case of temperature data.

We use high-frequency data of all the quotes on the stock of IBM throughout the month of June 1999. We used the `S-Plus` function `align` to produce regularly spaced quotes averaging to reduce the bid-ask spread so that we have only one single price every 30 seconds. There were 22 trading days on that month, so the data ends up being a 22×779 matrix `IBM`. For the sake of the present illustration, we shall only use the first row (i.e. the first trading day). We fit a GARCH(1,1) model to the log-returns and we plot the estimate of the conditional standard deviation in the way described above.

```
> I <- 1
> tsplot(IBM[I,],main=paste("IBM High Frequency (30 s).
                on First Day of Trading in June 1999"),sep=""")
> IBMLR <- diff(log(IBM[I,]))
> IBM.g <- garch(IBMLR~1,~garch(1,1))
> tsplot(IBM.g$sigma.t,main=c("Conditional
                                   Volatility on that Day"))
```

Figure 7.9 gives the plots produced by these commands. The instantaneous volatility

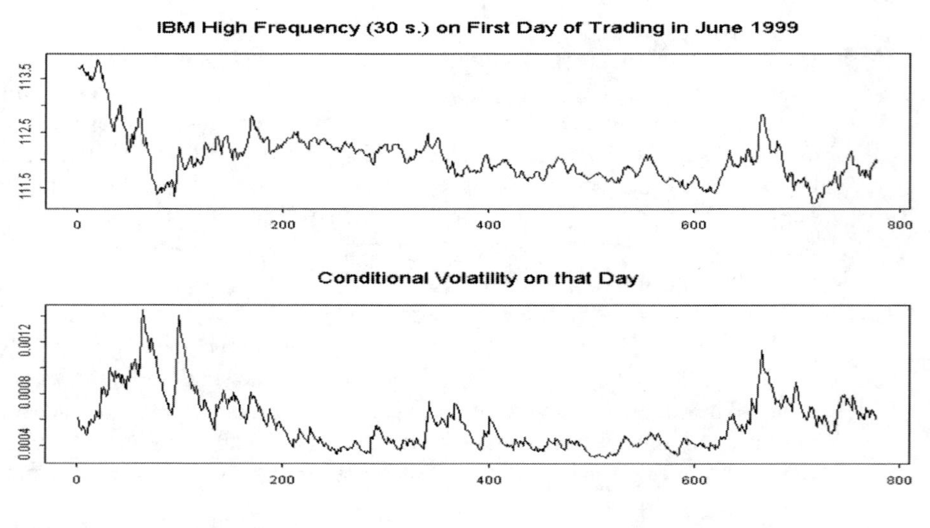

Fig. 7.9. IBM Quotes Regularly Spaced by 30 sec. on the first trading day of June 1999 (top) and estimateof the conditional volatility (bottom) given by a GARCH(1,1) model fitted to the log-returns.

shows an interesting pattern: high levels in the morning (around 9:30 or 10:00 am, and in the later part of the trading day, with a smaller surge before lunch time. This pattern is pretty typical, and it can be identified most days, so we would like to see it as a deterministic seasonal component. Whether or not it is a deterministic component, the GARCH fitting procedure, picked it up. So this looks pretty nice, and it is indeed. The problems arise when we try to use the model fitted in this way, in order to simulate Monte Carlo samples to be used as possible scenarios of a trading day. As before, we can use commands of the following type to produce such scenarios:

```
> IBM.sim <- simulate.garch(IBM.g,n=779,n.start=1000)
> tsplot(IBM.sim$et,main="Simulated GARCH(1,1)
                     with same parameters as in IBM.g")
> tsplot(IBM.sim$sigma.t,main="Conditional Variance of
                               the Simulated GARCH(1,1)")
```

The results are reporduced in Figure 7.10.
There one cannot find the special pattern of high activity levels at specific times of the day. The simulation algorithm uses the coefficients estimated for the GARCH(1,1) model, and using them, it produces a sample which is as stationary as possible. In particular, the simulation algorithm makes sure that the relative frequencies of the periods with high and low levels of volatility come with their expected frequencies, say two or three periods of high volatility per day example. However, there is absolutely no reason for the algorithm to try to set these periods of high volatility in the early morning and/or in the later part of the trading day. Because of stationarity, they appear any time throughout the day. These deterministic features of the term structure of volatility are not part of the model which is simulated. This may not be a problem if we are interested in statistics involving the whole day, but if we care about statistics depending on the time of the day, the scenarios produced by this particular use of the function `simulate.garch` become misleading, and possibly useless. Worse than that, they may end up being very dangerous since they may lead to gross errors.

7.2.6 **Generalizations**

ARCH and GARCH models have been generalized in many different directions, and the menagerie of models available to the data analyst is amazing. This great variety can be intimidating to the non-specialist. Moreover, because of the anxiety resulting from this lack of "one size fits all", and of the difficulties encountered in interpreting and using the results of the fits, many potential users have shied away from the GARCH methodology in favor of more robust and stable alternatives.

More Univariate GARCH Models

We shall refrain from discussing here the various forms of GARCH which have been proposed to accommodate the features missed by the standard ARCH and GARCH

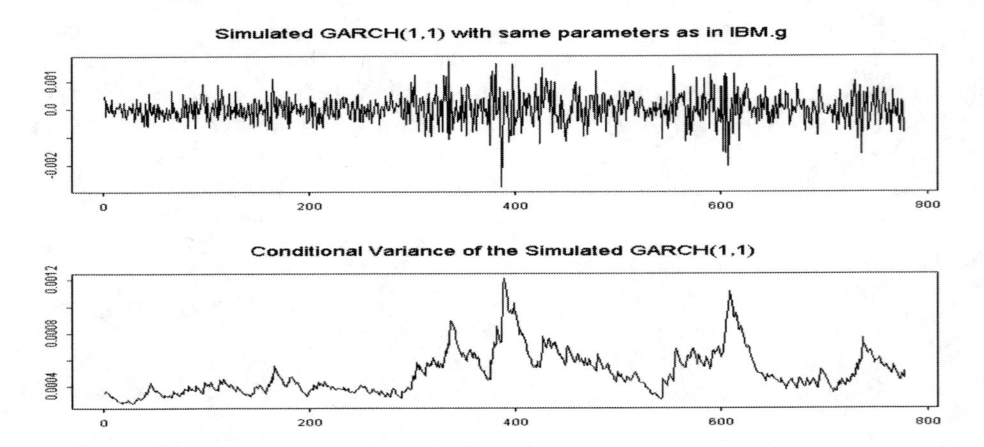

Fig. 7.10. Simulation of one trading day for IBM high-frequency data (top), and corresponding instantaneous volatility (bottom).

models introduced in this chapter. The main reason for that is the high level of complexity of the technicalities of these generalizations. We merely refer the interested (or desperate) reader to the Notes & Complements at the end of the chapter for precise references of texts in which these generalizations are described. However, we list the models which can be fitted using the `S+FinMetrics` module of `S-Plus`.

- **LGARCH** Leverage GARCH
- **PGARCH** Power GARCH
- **EGARCH** Exponential GARCH
- **TGARCH** Threshold GARCH
- **CGARCH** Component GARCH
- **GARCH-M** GARCH in the Mean
- **FIGARCH** Fractionally Integrated GARCH

Multivariate Models

In the same way AR models were effortlessly generalized to the multivariate setting, GARCH models can be extended as well, and the `S-Plus` fitting methods keep the same structure. A multivariate time series $\{\mathbf{X}_t\}_t$ is said to be a GARCH series if it is of the form:

$$\mathbf{X}_t = \mu + \mathbf{W}_t$$

where

- μ and $\mathbf{W}_t$ are $d \times 1$ vectors
- the distribution of $\mathbf{W}_t$ conditioned on $\mathbf{W}_{t-1}$, $\mathbf{W}_{t-2}$, ..., can be:
  - d-variate normal;
  - d-variate Student;

- with $d \times d$ variance/covariance matrix $\mathbf{V}_t$ satisfying

$$\mathbf{V}_t = A + \sum_{k=1}^{p} A_k * [\mathbf{W}_{t-k}\mathbf{W}_{t-k}^t] + \sum_{h=1}^{q} \mathbf{B}_h * \mathbf{V}_{t-k}$$

where
- $\mathbf{A}$, $\mathbf{A}_k$ and $\mathbf{B}_h$ are symmetric $d \times d$ matrices
- $*$ stands for the Hadamard product (entry by entry)
- as usual ^{t} stands for the transpose of a matrix/vector

Finally, note that one can replace μ by the result of a regression on exogenous variables.

7.3 Stochastic Volatility Models

A stochastic volatility model (SV model for short) is based on a couple of independent variance one white noise series $\{W_t\}_t$ and $\{u_t\}_t$ and the defining dynamical equations:

$$\begin{cases} X_t = \sigma_t W_t \\ \log \sigma_t = \phi_0 + \phi_1 \log \sigma_{t-1} + \gamma u_t \end{cases} \tag{7.8}$$

where ϕ and η are constants. We shall assume that $|\phi| < 1$ so that the second equation gives a stationary AR(1) process for the logarithm of the conditional variance σ_t^2. This implies that the time series $\{X_t\}_t$ is stationary since both series $\{W_t\}_t$ and $\{\sigma_t\}_t$ are. Notice that $\{X_t\}_t$ is a (weak) white noise, and that σ_t^2 is its conditional variance. So this SV model has a lot in common with the ARCH and GARCH models discussed in this chapter. But there is a fundamental difference: *the model is driven by two independent sources of randomness.* This innocent looking difference has dramatic consequences, and this model will behave quite differently from the ARCH and GARCH models.

Information Structure

The first problem to consider is related to the presence of two driving white noise terms, and for that reason, it is specific to the SV models: which is the right notion of past information. At any given time t, should the past information be the information contained in the past values of the series as encapsulated in $\mathbf{X}_{\leq t-1} = \{X_{t-1}, X_{t-2}, \dots, X_1\}$, or in the past values of the series together with the past values of the volatility (or equivalently the noise driving it) as captured by:

$$(\mathbf{X}_{\leq t-1}, \mathbf{u}_{\leq t-1}) = \{(X_{t-1}, u_{t-1}), (X_{t-2}, u_{t-2}), \dots, (X_1, u_1)\}.$$

This question is not purely academic, it has very practical consequences. For example, if one uses the first notion of past information, the corresponding notion of conditional variance should be:

$$h_t = \text{var}\{X_t|\mathbf{X}_{t-1}\} = \text{var}\{X_t|X_{t-1}, X_{t-2}, \dots, X_1\} \tag{7.9}$$

like in the case of the ARCH and GARCH models. On the other hand, if we use the second notion of past information, then the natural conditional variance to use is given by:

$$\begin{aligned}\sigma_t^2 &= \text{var}\{X_t|(\mathbf{X}_{\le t-1}, \mathbf{u}_{\le t-1})\} \\ &= \text{var}\{X_t|(X_{t-1}, u_{t-1}), (X_{t-2}, u_{t-2}), \dots, (X_1, u_1)\}.\end{aligned}$$

Notice that, because of the tower property of conditional expectations we have:

$$h_t = \mathbb{E}\{\sigma_t^2|\mathbf{X}_{\le t-1}\}.$$

State Space Formulation

Mostly because of the fact that they are driven by two different white noise series, the stochastic volatility systems fit very naturally in the framework of state space systems. Indeed the second equation in (7.8) can be written in the form:

$$x_t = \phi_0 + \phi_1 x_{t-1} + v_t$$

if we set $x_t = \log \sigma_t^2$ and $v_t = 2\eta u_t$, which is exactly the form of a state equation (with linear dynamics in the present situation). Moreover, the first equation in (7.8) can be written in the form:

$$\log|X_t| = \log \sigma_t + \log|W_t|$$

which can be rewritten in the form:

$$y_t = a_0 + a_1 x_t + w_t$$

which is a linear observation equation of the state x_t, provided we set $y_t = \log|Y_t|$, $a_0 = \mathbb{E}\{\log|W_t|\}$, $a_1 = 1/2$ and $w_t = \log|W_t| - \mathbb{E}\{\log|W_t|\}$. So the stochastic volatility model (7.8) rewrites as a linear state space model, and all the estimation/filtering techniques studied in Chapter 6 can be applied. We shall come back to this important remark later in this chapter when we implement nonlinear filtering techniques to the continuous time analog of the stochastic volatility models of this section.

The analysis of AR(1) models done in Chapter 5 implies that:

$$\mu_x = \mathbb{E}\{x_t\} = \frac{\phi_0}{1-\phi_1} \qquad \text{and} \qquad \sigma_x^2 = \text{var}\{x_t\} = \frac{\sigma^2}{1-\phi_1^2}$$

where σ^2 is the variance of the white noise of the AR(1) dynamics of x_t, i.e. $\sigma^2 = 4\gamma^2$ if we come back to the parameters of the original dynamical equation (7.8).

Excess Kurtosis

In this subsection we assume that both noise terms are Gaussian. Notice that, because we assume that the two white noise series are independent, we have:

$$\mathbb{E}\{X_t^k\} = \mathbb{E}\{\sigma_t^k\}\mathbb{E}\{W_t^k\}$$

for all the integers k. This k-th moment will be zero for all odd power k because the distribution of the noise W_t is symmetric. When $k = 2h$ is even, using the expressions for the Laplace transform (moment generating function) and the moments of the Gaussian distributions, we find that:

$$\mathbb{E}\{X_t^{2h}\} = \mathbb{E}\{e^{hx_t}\}\mathbb{E}\{W_t^{2h}\} = e^{h\mu_x + h^2\sigma_x^2/2}\frac{(2h)!}{h!2^h}$$

In particular, the kurtosis of X_t is given by:

$$k\{X_t\} = \frac{\mathbb{E}\{X_t^4\}}{\mathbb{E}\{X_t^2\}^2} = \frac{e^{2\mu_x + 2\sigma_x^2}\frac{4!}{2!2^2}}{e^{2\mu_x + \sigma_x^2}\frac{2!}{1!2}} = 3e^{\sigma_x^2}$$

which is strictly greater than 3, proving that SV models exhibit excess kurtosis and heavy tail distributions.

Leverage Effect

Both GARCH and SV models have been shown to give an account of excess kurtosis and persistence in financial data. There is still another stylized fact which needs to be checked against these models: the so-called leverage effect. The latter is attributed to the fact that large up-moves in the value of a stock are usually accompanied by decreases in volatility while at the contrary down-moves in the value of a stock are usually accompanied by surge in volatility. This effect appears as a form of negative correlation between the changes in prices and the changes in volatility. This is the way we shall detect the leverage effect in the mathematical models.

◇ *The Case of ARCH & GARCH Models*

Let us assume for example that the time series $\{X_t\}_t$ is ARCH(1), and let us try to compute the sign of the correlation coefficient between the changes in X_t and the changes in its conditional variance. At any given time t, we have the information of the past values of the series, so the probabilities, expectations, variances, correlations, ... are computed conditionally on the knowledge of $\mathbf{X}_{\leq t-1} = \{X_{t-1}, X_{t-2}, \ldots, X_1\}$. We shall emphasize that conditioning by adding a subscript $t-1$ to all the expectations, variances and covariances which we compute. Our ARCH(1) assumption can be written in the form

$$X_t = \sqrt{\alpha_0 + \alpha_1 X_{t-1}^2}W_t$$

for some positive coefficients α_0 and α_1, and we have:

$$\mathrm{cov}_{t-1}\{X_t - X_{t-1}, \sigma_{t+1}^2 - \sigma_t^2\} = \mathrm{cov}_{t-1}\{X_t, \sigma_{t+1}^2\}$$

because X_{t-1} and σ_t^2 are known at time $t-1$. Consequently:

$$\begin{aligned}
\mathrm{cov}_{t-1}&\{X_t - X_{t-1}, \sigma_{t+1}^2 - \sigma_t^2\} \\
&= \mathrm{cov}_{t-1}\{X_t, \alpha_0 + \alpha_1 X_t^2\} \\
&= \alpha_1 \mathrm{cov}_{t-1}\{X_t, X_t^2\} \\
&= \alpha_1 \mathrm{cov}_{t-1}\{\sqrt{\alpha_0 + \alpha_1 X_{t-1}^2} W_t, (\alpha_0 + \alpha_1 X_{t-1}^2) W_t^2\} \\
&= \alpha_1 (\alpha_0 + \alpha_1 X_{t-1}^2)^{3/2} \mathrm{cov}_{t-1}\{W_t, W_t^2\} \\
&= \alpha_1 (\alpha_0 + \alpha_1 X_{t-1}^2)^{3/2} \mathbb{E}\{W_t^3\}
\end{aligned}$$

The conclusion is that the leverage effect is present in the model only when the noise distribution is skewed to the left, i.e. when $\mathbb{E}\{W_t^3\} < 0$.

◇ *The Case of the SV Models*

The leverage effect is more difficult to pinpoint in the case of the stochastic volatility models. As explained earlier, the main difficulty lies in the notion of information available at time t, and this difficulty is rooted in the presence of several sources of randomness. Should we assume that we know only the outcomes of all the past values of the series, i.e.

$$\mathbf{X}_{\leq t-1} = \{X_{t-1}, X_{t-2}, \dots, X_1\},$$

or should we assume that we know the past outcomes of both the series and the noise driving the volatility, i.e.

$$(\mathbf{X}_{\leq t-1}, \mathbf{u}_{\leq t-1}) = \{(X_{t-1}, u_{t-1}), (X_{t-2}, u_{t-2}), \dots, (X_1, u_1)\}.$$

Notice that, because we assume that the two noise terms are independent, we have:

$$\begin{aligned}
\mathrm{cov}\{X_t - X_{t-1}, \sigma_t^2 - \sigma_{t-1}^2 | \mathbf{X}_{\leq t-1}, \mathbf{u}_{\leq t-1}\} &= 0 \\
\mathrm{cov}\{X_t - X_{t-1}, \sigma_{t+1}^2 - \sigma_t^2 | \mathbf{X}_{\leq t-1}, \mathbf{u}_{\leq t-1}\} &= 0.
\end{aligned}$$

On the other hand, the conditional covariance:

$$\mathrm{cov}\{X_t - X_{t-1}, \sigma_{t+1}^2 - \sigma_t^2 | \mathbf{X}_{\leq t-1}, \mathbf{u}_{\leq t-1}\}$$

can be different from zero. Here we use the notation $h_t = \mathrm{var}\{X_t | \mathbf{X}_{\leq t-1}\}$ already defined in (7.9) for the conditional variance of X_t given its own past values (excluding the knowledge of the past values of the noise driving the volatility). The use of h_t as a conditional variance is very natural since we observe the values of X, while we have no way to guess the values of the noise u in general. Unfortunately, it is very difficult to handle this quantity mathematically, and proving rigorously the presence of a negative correlation in SV models is usually very difficult.

Comparison with ARCH and GARCH Models

A SV time series $\{X_t\}_t$ is a martingale difference in the sense that:

$$\mathbb{E}\{X_t | \mathbf{X}_{t-1}, \mathbf{u}_{t-1}\} = 0.$$

It is a weak white noise (recall that we assume $|\phi_1| < 1$ to guarantee stationarity). Concentrating on the series of squares, and computing their auto-covariance function we get:

$$\begin{aligned}\text{cov}\{X_t^2, X_{t-s}^2\} &= \mathbb{E}\{e^{x_t + x_{t-s}} W_t^2 W_{t-s}^2\} - \mathbb{E}\{X_t^2\}^2 \\ &= e^{2\mu_x + (1+\phi_1^s)\sigma_x^2} - \mathbb{E}\{e^{x_t}\}^2 \\ &= e^{2\mu_x + \sigma_x^2}(e^{\phi_1^s \sigma_x^2} - 1)\end{aligned}$$

and the auto-correlation function is given by:

$$\rho_{X^2}(s) = \frac{\text{cov}\{X_t^2, X_{t-s}^2\}}{\text{var}\{X_t^2\}} = \frac{e^{\phi_1^s \sigma_x^2} - 1}{3e^{\sigma_x^2} - 1} \sim \frac{e^{\sigma_x^2} - 1}{3e^{\sigma_x^2} - 1}\phi_1^s$$

which prompts the following remarks:

Remarks

1. This acf changes signs when $\phi_1 < 0$, which is not the case for ARCH(1) models.

2. This acf looks very much like the acf of an ARMA(1,1) model. For this reason, one should think that a SV model is closer to a GARCH(1,1) model than to a ARCH(1) model!

The Smile Effect

Going down the list of stylized facts about financial prices, we find the smile effect. We already discuss its importance during our discussion of the nonparametric approach to option pricing. The empirical evidence of smiles had a tremendous influence on the practice of option pricing methodology. But before we define it rigorously, we need to introduce the notion of implied parameter.. For this we need the notion of price, and more precisely, we need to have a pricing formula (a pricing algorithm would do as well) which provides a one-to-one correspondence between the price of a financial instrument and certain parameters (short interest rate, volatility, current value of the underlying index on which a derivative is written, time to maturity, etc.). This one-to-one correspondence makes it possible to infer the value of a parameter which needs to be fed to such a formula or algorithm in order to recover the value of a price actually quoted on the market. Let us consider for example the case of the implied volatility, and let us assume that we are using the Black-Scholes formula to price options. For each option price quoted on the market, one can take note of the strike price, the time to maturity, . . ., and we can find which value of the volatility parameter σ we have to use in the Black-Scholes formula to get the price actually quoted. This value is called the *implied volatility*. As we already explained in our earlier discussion, it is not a statistical estimate of the parameter σ, it is a value implied by a transaction (or a set of transactions).

If on a given day and for a given time to maturity one can observe the prices of several European call options for different strike prices, the Black-Scholes theory tells us that the corresponding implied volatilities should be equal to each other. But there is strong empirical evidence to the contrary. The implied volatilities of these options are different. Quite often, if one plots them against the strike price, these implied volatilities form a convex curve having a minimum when the strike price is equal to the current price of the underlying stock or index. This curve is called the volatility smile. In this case we say that the option is at the money. The implied volatility is higher when the strike price is above the current price (in this case we say that the option is out of the money) and when it is below the current price (in this case we say that the option is in the money).

This empirical fact has prompted analysts and traders to revise the simultaneous use of a pricing model and a pricing formula (the Samuelson's log-normal model and the Black-Scholes formula in the discussion above) when they lead to a contradiction of the type we just described. One of the great successes of the SV models was the discovery that they can explain the volatility smile. Such a derivation would be far beyond the scope of this book, but we could not resist giving one of the main reasons for the popularity of these models. The interested reader should consult the Notes & Complements section at the end of the chapter for references.

7.4 Discretization of Stochastic Differential Equations

Continuous time finance has seen a tremendous growth in the last 30 years. Among the many abstract concepts involved in the theory developed for the analysis of these financial models, martingales and Ito's stochastic calculus are the most obscure to the non-mathematical side of the financial arena. Any attempt to present Ito's theory of stochastic integration, and stochastic differential equations would take us beyond the scope of this book. We invite the interested reader to browse the Notes & Complements at the end of the chapter for references to standard texts on the subject. In this section, we consider discrete time analog of the dynamics they model.

We already encountered one example of Ito stochastic differential equation (SDE, for short). It is the most famous of all the SDE's used by the financial community:

$$dS_t = S_t[\mu dt + \sigma dW_t]. \tag{7.10}$$

It describes the stochastic dynamics of the so-called geometric Brownian motion introduced by Samuelson as a model for the time evolution of stock prices. More generally, stochastic differential equations appear in the form:

$$dX_t = \mu(t, X_t)dt + \sigma(t, X_t)dW_t \tag{7.11}$$

where $(t, x) \hookrightarrow \mu(t, x)$ and $(t, x) \hookrightarrow \sigma(t, x)$ are functions of the time variable t, and the state variable x. These functions will be deterministic and real valued in the applications discussed below. But they could very well be vector valued, or matrix valued, or even random, essentially the same theory would apply. For the sake of the present discussion, one should think of X_t as describing the state at time t of a set of economic factors. The $''dt''$ - term has the usual interpretation of an infinitesimal change in the time variable t. The $''dW_t''$ - term tries to play an analogous role for an infinitesimal random change. Its rigorous definition is very delicate, and far beyond the scope of this book. The intuitive interpretation of dW_t should be that of an infinitesimal random shock of the white noise type, but since the notion of continuous time white noise is very intricate, this term has to be understood as the (stochastic) differential of its antiderivative. Indeed, the latter can be defined more easily, as a stochastic process with independent increments. This process $\{W_t\}_{t\geq 0}$ is usually called a Wiener process, or a process of Brownian motion since one of its early use was to model the motion of particles in suspension, investigated by Brown. Even though Einstein is usually credited for the first development of the theory of Brownian motion, a growing part of the scientific community is now making a case that Bachelier should be getting this credit because of his earlier work on the theory of speculation.

Equation (7.11) is a concise way to describe the dynamics (time evolution) of the (possibly random) quantity X. The coefficient $\mu(t, X_t)$ gives the instantaneous mean of the infinitesimal increment dX_t. The coefficient $\sigma(t, X_t)$ represents its instantaneous standard deviation, also called volatility. So, equation (7.11) is not random when $\sigma \equiv 0$. In this case, it reduces to an ordinary differential equation, and it can be analyzed using classical calculus. Things are much more complicated when σ is not identically zero. However, we shall not need to get involved in the meanders of Ito's theory of stochastic calculus, we shall limit ourselves to discretized versions of these equations, and this will give us a chance to bridge continuous time finance with the time series analysis of financial econometrics.

7.4.1 **Discretization Schemes**

Instead of working directly with a continuous time model, we assume that snapshots of the system are taken at discrete time intervals. We assume that measurements take place at regular times $t_j = t_0 + j\Delta t$, and we use the notation $X_j^{(\Delta t)}$ for the value X_{t_j} of the (random) variable X at time t_j. We sometimes drop the superscript (Δt) specifying the length of the sampling interval from the notation for the sake of easier typesetting. This abuse of notation should not create confusion. The sampling frequency is usually defined as the inverse of the length Δt of the time interval separating two successive measurements.

The Euler Scheme

A natural question is now to identify a discrete time dynamical equation for $X_j^{(\Delta t)}$ which would be consistent with the continuous time dynamics given by equation (7.11). This is usually done in the following manner. We consider the evolution given by the recursive equation:

$$X_j - X_{j-1} = \mu(t_{j-1}, X_{j-1})\Delta t + \sigma(t_{j-1}, X_{j-1})\sqrt{\Delta t}\epsilon_j \qquad (7.12)$$

where $\{\epsilon_j\}_{j\geq 1}$ is an $N(0,1)$ white noise, and where the initial condition X_0 is assumed to be given. A few important remarks are granted at this stage.

Remarks

1. Notice that the (stochastic) differential dX_t appearing in the left hand side of (7.11) was discretized as $X_j - X_{j-1}$, while the X_t's appearing in the right hand side were discretized as X_{j-1}. This is specific to the Ito's stochastic integration theory. The stochastic increments dW_t should be taken ahead of the instant of the discretization. This non-anticipative idiosyncratic feature is crucial in financial applications (unless you own a crystal ball, in which case it should not apply to you!)

2. The stochastic differential dW_t appearing in (7.11) was discretized as $\sqrt{\Delta t}\epsilon_j$. This is a consequence of the fact that the increments $W_t - W_s$ of a Wiener process are mean-zero Gaussian random variables with standard deviations $\sqrt{t-s}$, and because they are independent of each other when computed over non-overlapping intervals.

3. If σ is constant, i.e. if $\sigma(t,x) \equiv \sigma$, and μ is linear, say $\mu(t,x) = \phi_0 + \phi_1 x$, then equation (7.11) is called an equation of the Ornstein-Uhlenbeck type, and it is plain to see that its discretized form (7.12) defines an AR(1) process. In general equation (7.12) defines a (possibly nonlinear) auto regressive process of order 1 as defined in Subsection 7.1.2. So we are still in known territory.

4. It should be understood that, even if they both start from the same initial values, say X_0, the solution X_t of the continuous time stochastic differential equation (7.11), and the solution $X_j^{(\Delta t)}$ of the recursive equation (7.12) have no reason to coincide, *not even at the sampling times* $t_j = t_0 + j\Delta t$. In other words, we should not expect that $X_{t_j} = X_j^{(\Delta t)}$. But there is a justice, and whenever the coefficients μ and σ are smooth, and whenever the discretization step Δt tends to zero, then the difference between these values tends to zero in a controlled manner. We shall not state a precise mathematical theorem, but obviously, this result is a clear justification of the extensive use of discrete models to simulate and approximate continuous time models. See the Notes & Complements at the end of the chapter for references to results showing that the continuous time stochastic volatility models can also appear as diffusion limits of appropriately set up ARCH and GARCH models.

5. Finally, we notice that the recursive form of equation (7.12) is perfectly suited for Monte Carlo simulations. Indeed, it is very easy to simulate a white noise, and from white noise samples, it is plain to implement formula (7.12) to generate samples of X_j.

The discretization given by equation (7.12) is known as the Euler scheme. It not the only way to derive a discrete time approximation to a continuous time evolution. Other procedures have been introduced to speed up the convergence toward the true dynamics. But it is nevertheless the simplest one, and we shall use it for that reason.

7.4.2 **Monte Carlo Simulations: A First Example**

In this subsection, we give examples of direct random simulations based on explicit formulae for the solutions of SDE's. In particular, these simulations do not use the Euler's scheme described earlier. However, due to the fact that they depend upon the existence of exact formulae, their realm of application will be limited.

As we already pointed out in our discussion of nonparametric option pricing of Section 4.7, the form of equation (7.10) is so simple that an explicit formula can be found for the solution. It is given by:

$$S_t = S_0 e^{(\mu-\sigma^2/2)t+\sigma W_t} \tag{7.13}$$

or more generally as:

$$S_t = S_s e^{[\mu-\sigma^2/2](t-s)+\sigma[W_t-W_s]} \tag{7.14}$$

if we know the value S_s of the solution at time $s < t$ instead of 0. As earlier, we emphasize the presence of the unexpected term $-\sigma^2/2$ which should not have appeared according to the rules of classical calculus. Its presence is forced on us by the special rules of the calculus developed by Ito to handle integrals and differentials with respect to the Wiener process $\{W_t\}_t$.

Simulation of a Random Variable

In many instances, one is interested in the the price (or the value of the index) at a given time T in the future. This is for example the case if one is interested in a contingent claim with European exercise and maturity T. In such a case, formula (7.14) can be used with $s = 0$ and $t = T$. It shows that the random variable S_T is log normal with mean $\log S_0 + T(\mu - \sigma^2/2)$ and variance $\sigma^2 T$, and simulation is plain: generating samples from this distribution can be done with simple commands. For example, the S-Plus commands:

```
> N <- 1024; S0 <- 845, MU <- .05; SIG <- .2, TT <- .9
> MUT <- log(S0) + TT*(MU-SIG^2/2)
> SIGT <- SIG*sqrt(TT)
> SAMPLE <- rlnorm(N, meanlog=MUT,sdlog=SIGT)
```

produce a sample of size $N = 1024$ for S_T with $T = .9$, when $S_0 = 845$, $\mu = .05$, and $\sigma = 20\%$. The risk manager will be able to use such a sample to compute means, probabilities, quantiles (such as VaR's), conditional expectations (such as expected shortfall) $\cdots$ involving $S(T)$, as we did in the first chapters of the book.

Simulation of a Time Series

In other circumstances, simulation of the entire series may be needed. In such a case, instead of relying directly on the Euler scheme, one may still use the exact formula (7.14) before calling on random number generators. This is best achieved at the level of the logarithms of prices rather than the level of the prices themselves. Indeed, it is more convenient to simulate first a sample sequence for the log-prices, and then to take the exponentials of the numbers so obtained. This is based on the fact that:

$$\log S_t = \log S_s + [\mu - \frac{\sigma^2}{2}](t-s) + \sigma[W_t - W_s] \tag{7.15}$$

the process $\{\log S_t\}_t$ is a Brownian motion with drift which can easily be simulated. If we want to generate samples $X_0, X_1, \ldots, X_N$ of log prices $X_t = \log S_t$ separated in time by the time interval Δt, we set $X_0 = 0$, we generate an $N(0,1)$ white noise $\epsilon_1, \ldots, \epsilon_N$, and we use the recursive formula:

$$X_{j+1} = X_j + [\mu - \frac{\sigma^2}{2}]\Delta t + \sigma\sqrt{\Delta t}\epsilon_{j+1}.$$

This is a plain consequence of the formula (7.15) if we set $s = j\Delta t$ and $t = (j+1)\Delta t$, and if we use the fact that the increments of $\{W_t\}_t$ over disjoint intervals are independent Gaussian random variables, and that $W_t - W_s \sim N(0, t-s)$. The following `S-Plus` code was used to produce the plot in Figure 7.11.

```
> N <- 1024; DELTAT <- 1/365; MU <- .05; SIG <- .2
> DELTAX <- rnorm(1024, mean=DELTAT*(MU-SIG^2/2),
                                     sd=SIG*sqrt(DELTAT))
> GBMDATA <- c(1,exp(cumsum(DELTAX)))
> GBM.ts <- timeSeries(pos=timeSeq(from=
  timeDate("08/24/1971"),by="days",length=N+1),data=GBMDATA)
> plot(GBM.ts,main="Simulation of a geometric Brownian")
```

After setting the length of the simulated time series and the values of the parameters Δt, μ and σ, we generate the sequence of the increments $X_{j+1} - X_j$ in a vector (sequence) which we called `DELTAX`. This uses the fact that these increments are independent and normally distributed with mean $(\mu - \sigma^2/2)\Delta t$ and with variance $\Delta t\sigma^2$. Then, we use the `S-Plus` function `cumsum` to sum these increments in order to recover the X_j from their increments, and finally, we compute the exponentials giving the desired prices $S_{j\Delta t}$. For plotting purposes, we chose to turn the geometric Brownian motion numeric vector `GBMDATA` into a daily time series starting from August 24, 1971.

We shall revisit the problem of the simulation of geometric Brownian motion in Subsection 7.5.1. There we use an alternative approach based on Euler's discretization scheme.

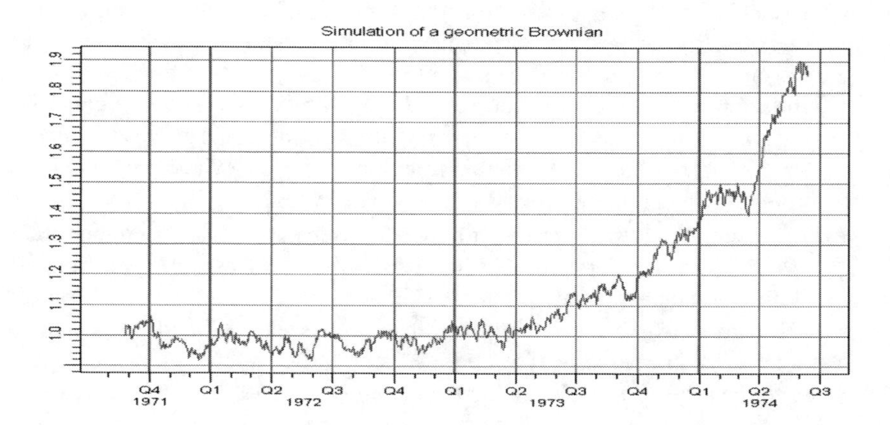

Fig. 7.11. Result of the simulation of a geometric Brownian motion. The similarity with the actual data (see Figure 7.12 below for example) is striking.

7.5 Random Simulation and Scenario Generation

In this section, we use our newly acquired expertise in discretization of continuous time stochastic differential equations to prepare a set of scenarios for the purpose of portfolio risk management.

We assume that an investment portfolio is based on three economic factors which are monitored on a monthly basis. These factors are the cost of short term borrowing, the cost of long term borrowing and a stock indicator. The data which we use for this experiment are monthly quotes (between May 1986 and November 1999) of the one year Treasury Bill yield (which we shall subsequently call the short interest rate, the 30 years US Government Bond yield (which we shall call the long interest rate) and the value of the S&P 500 composite index. Note that the 30 years Treasury Bonds have been retired since we first performed this experiment. These data are further analyzed in Problems 7.5 and 7.6.

One has to keep in mind the fact that we do not intend to address the delicate problem of portfolio risk management, we simply prepare tools for that purpose. Because of the pervasive use of *scenarios* in the industry, we illustrate how Monte-Carlo scenarios are generated, but we leave risk assessment and decision making under uncertainty untouched.

7.5.1 A Simple Model for the S&P 500 Index

As we explained in our discussion of the nonparametric pricing of options in Section 4.7, geometric Brownian motion is the time-honored model for stock price and financial index dynamics. We follow this tradition in this section. In other words, denoting by S_t the value of the index at time t, we assume that its time evolution is given by the geometric Brownian motion model proposed by Samuelson, i.e. by the solution of the stochastic differential equation (7.10) where the constant μ has the interpretation of a mean rate of growth, while the constant $\sigma > 0$ plays the role of the volatility of the index. In this equation, $W(t)$ is a Wiener process, (as introduced to model the physical process of Brownian motion).

Contrary to the approach followed in Subsection 7.4.2, we ignore the fact that we have an explicit form for the distribution of S_t at any given time, and we use Euler's scheme to produce Monte Carlo samples of the time evolution of the index. This gives us the opportunity to illustrate the use of this discretization scheme in a setup where we already used direct simulation, allowing for a comparison of the two methods. So we discretize the (stochastic) differential equation (7.10) directly, and we try to construct solutions of the discretized forms of this equation. Equation (7.10) rewrites:

$$S_{t+\Delta t} - S_t = S_t[\mu \Delta t + \sigma\sqrt{\Delta t}\epsilon_{t+1}] \tag{7.16}$$

where the $\{\epsilon_t\}_t$ is an $N(0,1)$ i.i.d. white noise. Once more, we see one of the main curiosities associated with the process of Brownian motion. The increment $dW(t)$ is essentially proportional to $\sqrt{\Delta t}$ instead of being proportional to Δt! Dividing both sides of equation (7.16) by S_t, we get an expression for the raw return RR_t over the period $[t, t+\Delta t]$:

$$1 + RR_t = \frac{S_{t+\Delta t}}{S_t} = (1 + \mu\Delta t) + \sigma\sqrt{\Delta t}\epsilon_{t+1} \tag{7.17}$$

which shows that, according to this discretization procedure, the raw return RR_t over one time period $[t, t+\Delta t]$ is a normal random variable with mean $\mu\Delta t$ and standard deviation $\sigma\sqrt{\Delta t}$.

The plot of the monthly values of the S&P 500 index is given in the left pane of Figure 7.12. We use the notation $\{S_t\}_t$ for this time series, even though we used earlier the same notation t for the time variable when the latter was assumed to be continuous. This time series is obviously non-stationary. As explained earlier, it is integrated of order one, i.e. if the type $I(1)$. So it will be more convenient to work with the stationary time series of raw returns $\{RR_t = S_{t+1}/S_t\}_t$, or with the stationary time series $\{\log(1 + RR_t)\}_t$ of log-returns. The plot of the log-returns is given in the right pane of Figure 7.12. This series looks definitely more stationary, and as such, it is more amenable to statistical inference.

We already argued in Chapter 1, and in our discussion of non-parametric option pricing of Section 4.7 that the log-normal model is inconsistent with some of the empirical statistics computed from daily stock returns. Monthly data do not behave much differently. Indeed, the time evolutions given by equations (7.10) and (7.17) imply that the distribution of the monthly log-return is normal, but unfortunately, this fact cannot be confirmed by even the simplest of the exploratory data analysis tools such as the histogram. Figure 7.13 shows that the distribution of the log-returns

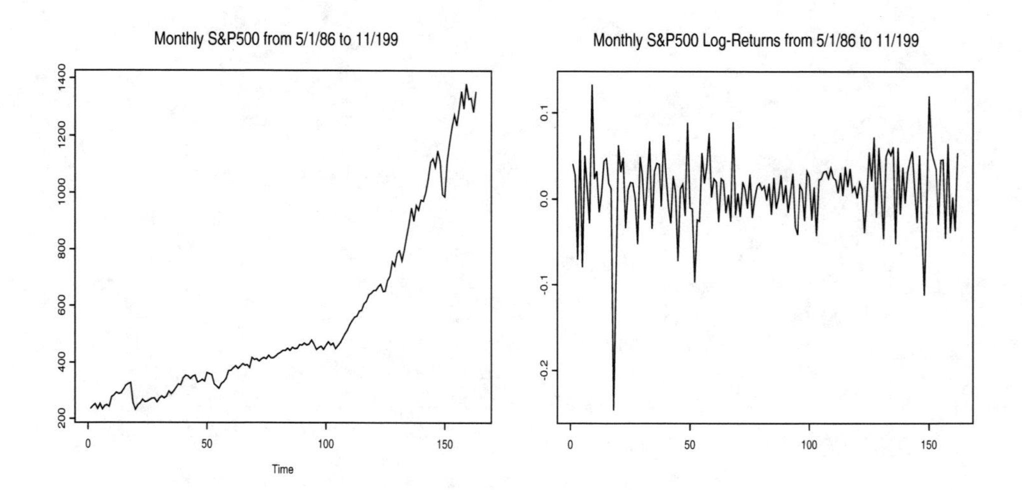

Fig. 7.12. Sequential plot of the monthly values of the S&P 500 composite index (left), and of of the corresponding log-returns for the same period (right).

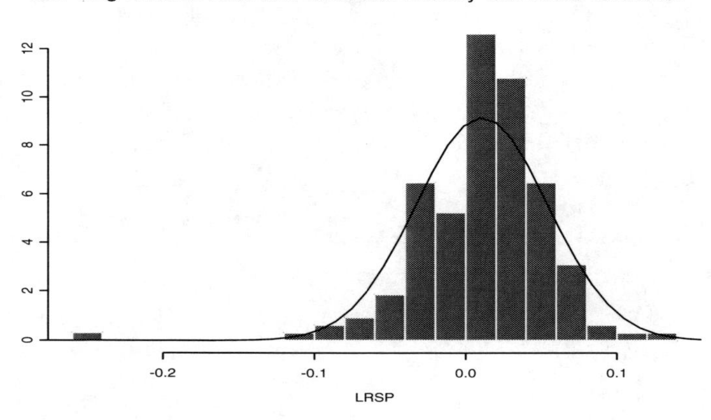

Fig. 7.13. Histogram of the log-return values and density of the normal distribution having the same mean and the same variance.

cannot be Gaussian. This graphical evidence could be complemented by tests of goodness of fit, but we shall not delve on that at this stage, especially since our intention is to keep the log-normal model given by equations (7.10) for the purposes of the simulation analysis tackled in this section. Figure 7.12 was produced with the following `S-Plus`-commands:

```
> tsplot(SP500)
> title("Monthly S&P500 from 5/1/86 to 11/199")
```

```
> LSP <- log(SP500)
> LRSP <- diff(LSP)
> tsplot(LRSP)
> title("Monthly S&P500 Log-Returns from 5/1/86 to 11/199")
```

while Figure 7.13 was produced with the following commands:

```
> mean(LRSP)
[1]  0.01080645
> sqrt(var(LRSP))
[1]  0.04348982
> hist(LRSP,nclass=12,probability=T)
> lines(LX,dnorm(LX,mean=.010806,sd=0.04349))
> title("S&P Log-Returns Hist. and Gaussian Density
                                   with same Moments")
```

7.5.2 Modeling the Short Interest Rate

One of the advantages of modeling the stock index via its logarithm is that, the resulting S_t is always positive, and obviously, this is a desirable feature for a stock index. Unfortunately, the short interest rate r_t cannot be modeled in the same way with the same success. Instead, we shall model its (stochastic) time evolution by a mean reverting Ornstein Uhlenbeck process. This model was introduced in the remark following the definition of the Euler discretization scheme. In the financial community, this model for the dynamics of the short interest rate is known as the Vasicek model. It is given by an equation of the form:

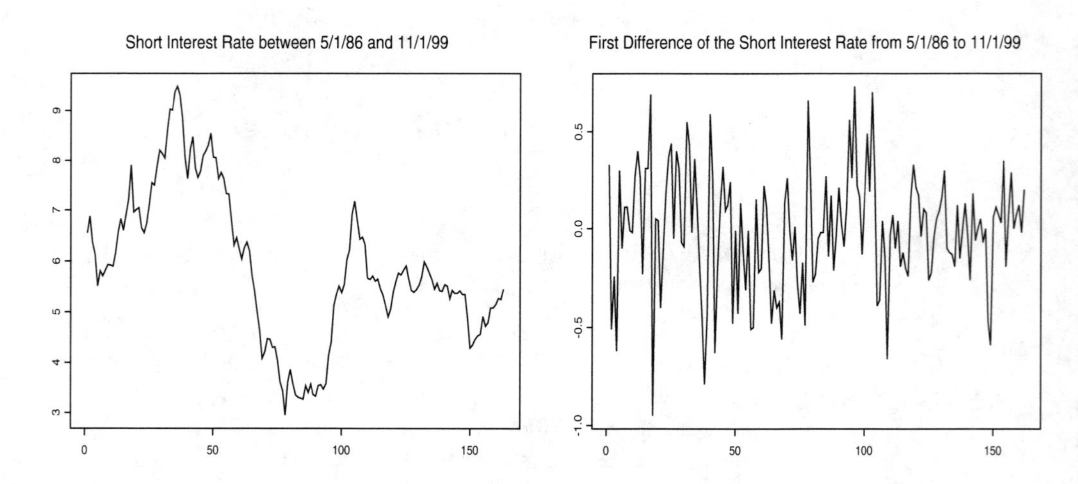

Fig. 7.14. Time series plot of the monthly values of the short interest rate (left) and of its first difference for the same period (right).

$$dr_t = -\lambda_r(r_t - \overline{r})dt + \sigma_r dW^{(r)}(t) \tag{7.18}$$

where λ_r, $\overline{r}$ and σ_r are positive constants and where $W^{(r)}(t)$ is a process of Brownian motion. If it weren't for the presence of this Brownian motion, the equation would read:

$$dr_t = -\lambda_r(r_t - \overline{r})dt,$$

and its solution would be given by an exponential function converging toward $\overline{r}$ when $t \to \infty$. This is the relaxation property given by the Hooke's law in physics. In the present situation, it is perturbed by the (random) kicks $\sigma_r dW^{(r)}(t)$, but the mean reverting tendency remains.

Equation (7.18) has three parameters, λ_s, $\overline{s}$, and σ_s, that have to be estimated from the data. As before, we use Euler's scheme, leading to the following finite difference equation:

$$r_{t+\Delta t} = r_t - \lambda_r(r_t - \overline{r})\Delta t + \sigma_r\sqrt{\Delta t}\,\epsilon^{(r)}_{t+1} \tag{7.19}$$

for an i.i.d. N(0,1) strong white noise $\{\epsilon^{(r)}_t\}_{t=1,2,\dots}$ which we assume independent of the white noise $\{\epsilon_t\}_{t=1,2,\dots}$ driving the stochastic time evolution in our model for the S&P 500 index.

Remark. The mathematical model for the short term interest rate given by equation (7.18) is known in the financial literature as the Vasicek model. According to this model, the short interest rates at different times are jointly Gaussian random variables. This is very convenient because it leads to closed form formulae for the prices of many fixed income derivatives, including the forward and yield curve manipulated in Problem 3.6. This fact is the main reason for the extreme popularity of the model. Nevertheless, this model has annoying shortcomings. One of the most frequently voiced complains is that, since a normal random variable can take values ranging from $-\infty$ to $+\infty$, it is quite possible that $r_{t+\Delta t}$ will become negative in any time period. Most practitioners regard this possibility as heretic. However, some have argued that r_t should be viewed as a *real* interest rate as defined by the difference between the short rate and the inflation rate, and as such, it could become negative. But more realistically, the reason why the Vasicek model is reasonable is the following: normal random variables can take arbitrarily large values indeed, but with overwhelming probability they remain within three standard deviations from their means. So if the constants λ_r giving the rate of recall to the long term relaxation level, and this level $\overline{r}$ are large enough compared to σ_r, then the random quantity given by equation (7.18) will essentially never be negative. Figure 7.14 was produced with the following S-commands:

```
> tsplot(Short)
> title("Short Interest Rate from 5/1/86 to 11/199")
> DS <- diff(Short)
> tsplot(DS)
> tsplot(DS)
> title("First Difference of the Short Interest Rate
                                   from 5/1/86 to 11/199")
```

7.5.3 Modeling the Spread

The long interest rate could be modeled in the same way as we did for the short rate. Indeed, Figure 7.15 shows that most of the features of the short rate r_t are shared by the long rate ℓ_t. The plots of Figure 7.15 were produced with the following `S-Plus`-

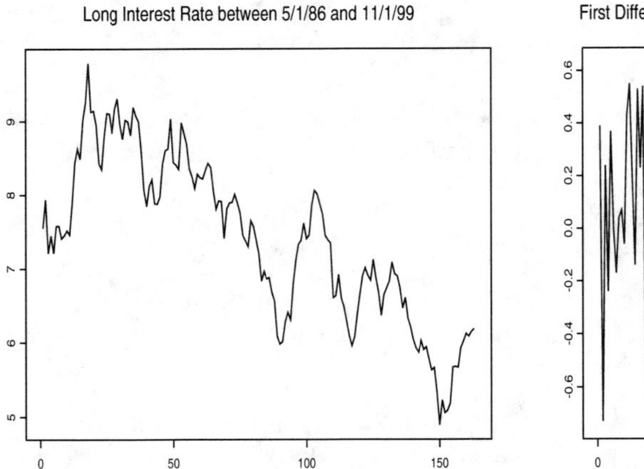

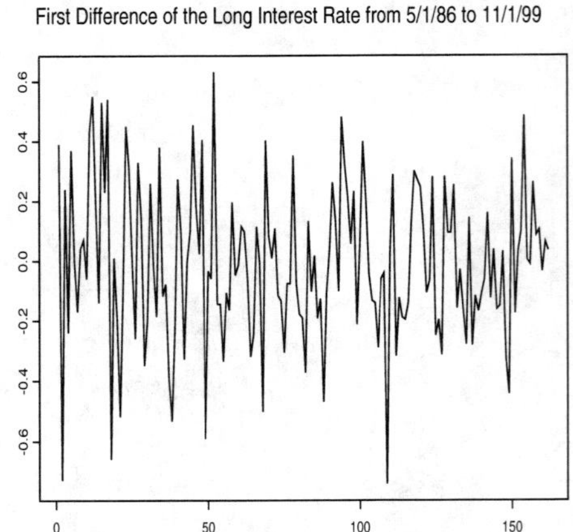

Fig. 7.15. Time series plot of the monthly values of the long interest rate (left) and of its first difference for the same period (right).

commands:

```
> tsplot(Long)
> title("Long Term Interest Rate from 5/1/86 to 11/199")
> DL <- diff(Long)
> tsplot(DL)
> title("First Difference of the Long Term Interest Rate
                                    from 5/1/86 to 11/199")
```

By looking at the simultaneous plots of the long and the short interest rates given m in Figure 7.16, we see that they cross rarely, and that the long interest rate is essentially always higher than the short interest rate. In other words, their difference (which is usually called the spread in interest rate and which quantifies the time value of money) is almost always positive.

At this stage of our discussion, it is important to recall the discussion of Subsection 6.1.6. There, we argued that the time evolutions of the short and long interest rates r_t and ℓ_t, could be modeled by I(1) time series, i.e. by non-stationary time series integrated of order one. However, we insisted that their difference was stationary, i.e. integrated of order zero, and we used that example as an illustration for the notion of cointegration. In this section, we still assume that the difference between the long and short interest rates is stationary. Nevertheless, to make our life easier, we also assume that both the short and long interest rates are stationary, i.e. I(0) by assuming that $\lambda_r > 0$. This is a slight departure from the assumptions of Subsection 6.1.6.

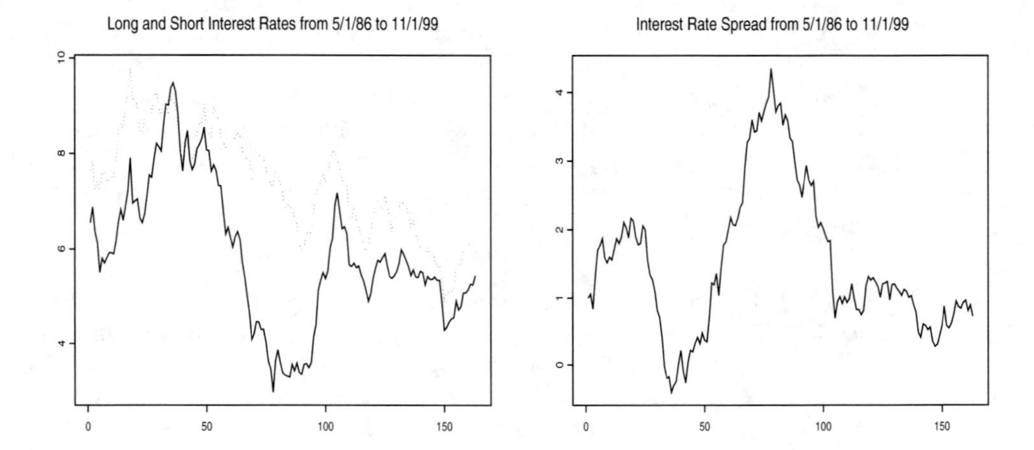

Fig. 7.16. Simultaneous plots of the long and short interest rates (left) and time series plot of the spread (right) for the same period. Notice that the scales of the vertical axes are different.

For reasons to be discussed later, we decide to model the spread $s_t = \ell_t - r_t$ instead of the long interest rate ℓ_t. As before, we model the (stochastic) time evolution by a mean reverting Ornstein Uhlenbeck process.

$$ds_t = -\lambda_s(s_t - s_0)dt + \sigma_s dW_s(t) \tag{7.20}$$

where λ_s is a positive number measuring the speed with which s_t reverts to its long term average $\overline{s}$. As above, this equation has three parameters, λ_s, $\overline{s}$ and σ_s, which will have to be estimated from the data. And again, as before, we shall discretize this equation in the following form:

$$s_{t+\Delta t} = s_t - \lambda_s(s_t - s_0)\Delta t + \sigma_s\sqrt{\Delta t}\,\epsilon_t^{(s)} \tag{7.21}$$

for an i.i.d. N(0,1) random white noise $\{\epsilon_t^{(s)}\}_{t=0,\Delta t,2\Delta t,\dots}$ which we shall assume independent of the other white noises $\{\epsilon_t\}_{t=0,\Delta t,2\Delta t,\dots}$ and $\{\epsilon_t^{(r)}\}_{t=0,\Delta t,2\Delta t,\dots}$.

7.5.4 Putting Everything Together

According to the models presented above, the three economic factors r_t, s_t and S_t are independent of each other. Indeed the three equations (7.10), (7.18), and (7.20) are un-coupled and driven by independent white noise terms. Obviously, this statement applies as well to their time-discretizations (7.19), (7.21), and (7.10).

Because of our discussion in Subsection 6.1.6, using a model without a strong dependence between the short and long interest rate dynamics would not have been acceptable. Notice nevertheless that cointegration does not necessarily preclude the independence of r_t and s_t as we have it so far in our model. In any case, the lack of dependence between the three time evolutions is a serious shortcoming of our model as it stands.

Minor changes in the equations can remedy this lack of correlation. In order to model the fact that the spread has a tendency to increase when the short interest rate decreases to unusually low levels, and that it has a tendency to decrease when the short interest rate is high, one could for example replace (7.20) by an equation of the form:

$$ds_t = [-\lambda_s(s_t - s_0) + \mu_s(r_t - r_0)]dt + \sigma_s dW^{(s)}(t) \qquad (7.22)$$

and (7.21) by the corresponding discretization of (7.22). Using this equation introduces a coupling between the dynamics of r_t and of s_t, and consequently, between the dynamics of r_t and ℓ_t. Unfortunately, this modification of the dynamics of s_t creates a couple of problems. First, it introduces a new parameter, μ_s, which may be difficult to estimate from the data. Second, this monotonic dependence between s_t and r_t may not hold all the time, and as a consequence, it may lead to erroneous forecasts in regimes in which this specific relationship between r_t and s_t does not hold. Thus, we refrain from implementing it. Problem 7.5 is concerned with still another way to build dependencies between the three stochastic differential equations giving the dynamics of r_t, s_t and ℓ_t.

For each time t we define the (random) vector $\mathbf{X}_t$ by:

$$\mathbf{X}_t = \begin{bmatrix} RR_t \\ r_t \\ s_t \end{bmatrix}$$

With this notation, the stochastic dynamics given above can be rewritten as:

$$\mathbf{X}_{t+\Delta t} = F\mathbf{X}_t + B + \Sigma \mathbf{W}_t$$

where the constant matrix F is given by:

$$F = \begin{bmatrix} 0 & 0 & 0 \\ 0 & 1 - \lambda_r \Delta t & 0 \\ 0 & 0 & 1 - \lambda_s \Delta t \end{bmatrix},$$

the constant vector B is given by:

$$B = \begin{bmatrix} \mu \Delta t \\ \lambda_r \overline{r} \Delta t \\ \lambda_s \overline{s} \Delta t \end{bmatrix},$$

and the matrix Σ is given by:

$$\Sigma = \begin{bmatrix} \sigma\sqrt{\Delta t} & 0 & 0 \\ 0 & \sigma_r\sqrt{\Delta t} & 0 \\ 0 & 0 & \sigma_s\sqrt{\Delta t} \end{bmatrix},$$

and finally, where the random noise vector $\mathbf{W}_t$ is defined as:

$$\mathbf{W}_t = \begin{bmatrix} \epsilon_t \\ \epsilon_t^{(r)} \\ \epsilon_t^{(s)} \end{bmatrix},$$

The fact that the matrices F and Σ are diagonal, is the reason why the three equations are not coupled. So if the three components of the noise term $\mathbf{W}$ are independent, the time evolutions of the three components are statistically independent. As explained earlier, including non-zero off diagonal terms in the matrices F and/or Σ couples the equations and creates dependencies between the time evolutions. Obviously, another possibility is to assume that the components of the noise are dependent. See Problem 7.5 for details.

7.6 Filtering of Nonlinear Systems

The state space models analyzed in the previous chapter are linear. Our goal is now to extend the recursive filtering procedures to nonlinear models. With this in mind, we introduce the terminology of hidden Markov models (HMM for short). The latter form the most popular class of models for partially observed stochastic systems, and being able to filter them efficiently is of great practical importance. Rather than attempting to develop the mathematical theory, which would take us far beyond the scope of this book, we restrict ourselves to the discussion of general features, and to a thorough presentation of a recently discovered implementation of great computational usefulness. Like the Kalman filter of the linear models, it is based on a recursive scheme. But since the nonlinearities take us out of the domain of Gaussian distributions, it is not possible to restrict ourselves to tracking conditional means and conditional variances. Tracking the full conditional distribution of the partially observed state is done via particle approximations requiring Monte Carlo simulations of the time evolution of the state, and clever resampling.

7.6.1 Hidden Markov Models

Let us first recall the notation (6.5) and (6.7) introduced earlier for the analysis of general state space models. We assume that at each time n, the state of a system is described by a state vector $\mathbf{X}_n$ in a state space $\mathcal{X}$. In most applications, the state space is a subset of a Euclidean space $\mathbb{R}^{d_X}$, and there is little loss of generality in assuming $\mathcal{X} = \mathbb{R}^{d_X}$. It is assumed that the dynamics of the state vector are given by a Markov chain. Rather than being bugged down by the intricacies of the general theory of Markov chains, we shall assume the following convenient form for the dynamics. We assume that the transition from state $\mathbf{X}_n$ to $\mathbf{X}_{n+1}$ is given by an explicit equation:

$$\mathbf{X}_{n+1} = F_n(\mathbf{X}_n, \mathbf{V}_{n+1}) \tag{7.23}$$

where $\mathbf{V} = \{\mathbf{V}_n\}_{n\geq 1}$ is a white noise i.i.d. sequence in $\mathbb{R}^{d'}$ where the dimension d' could be different from the dimension d_X of the state vector. The functions F_n : $\mathbb{R}^{d_X} \times \mathbb{R}^{d'} \hookrightarrow \mathbb{R}^{d_X}$ will be independent of n in all the applications considered in this chapter. However, it is important to emphasize that we do not assume that they are linear, as we did in the previous chapter. Nevertheless, we assume that these functions are known, and that we have a way to compute $F_n(\mathbf{x}, \mathbf{v})$ for any couple $(\mathbf{x}, \mathbf{v}) \in \mathbb{R}^{d_X} \times \mathbb{R}^{d'}$. The states $\mathbf{X}_n$ of the system are not directly observable, and the data with which we have to work, are given in the form of partial observations $\mathbf{Y}_n$ derived from the states $\mathbf{X}_n$ via a formula of the form:

$$\mathbf{Y}_n = G_n(\mathbf{X}_n, \mathbf{W}_n). \tag{7.24}$$

Even though the functions G_n are linear (as in (6.8) of the previous chapter) in many applications, we shall only assume that the noise is additive. In other words, we assume that the observations are of the form:

$$\mathbf{Y}_n = G_n(\mathbf{X}_n) + \mathbf{W}_n \tag{7.25}$$

for some (possibly nonlinear) function $G_n : \mathbb{R}^{d_X} \hookrightarrow \mathbb{R}^{d_Y}$ and a sequence $\mathbf{W} = \{\mathbf{W}_n\}_{n\geq 1}$ of mean zero independent random vectors in $\mathbb{R}^{d_Y}$ with densities ψ which we assume to be known.

The challenge is framed in the same way as in the linear case: given an integer $m \geq 0$, and observations $\mathbf{Y}_1, \ldots, \mathbf{Y}_n$, we want to *guess* $\mathbf{X}_m$. For the sake of convenience we shall use the notation $\mathbf{Y}_{\leq n}$ for the set of n d_Y-dimensional vectors $\mathbf{Y}_1, \ldots, \mathbf{Y}_n$.

7.6.2 General Filtering Approach

As explained in the previous chapter, the three main problems of state space models are 1) data smoothing ($m < n$), 2) filtering ($m = n$), and 3) prediction ($m > n$), and as before, we concentrate on the filtering and prediction problems.

All the information about the state $\mathbf{X}_n$ which one can hope to derive from the knowledge of the observations $\mathbf{Y}_{\leq n}$ up to (and including) time n, is contained in the conditional distribution of $\mathbf{X}_n$ given $\mathbf{Y}_{\leq n}$ which we denote by:

$$\pi_n(\,\cdot\,, \mathbf{Y}_{\leq n}) = \mathbb{P}\{\mathbf{X}_n \in \cdot\,|\mathbf{Y}_{\leq n}\}. \tag{7.26}$$

This conditional distribution obviously depends upon the past observations $\mathbf{Y}_{\leq n}$. We assume that a given sequence $\mathbf{Y}_1 = \mathbf{y}_1, \ldots, \mathbf{Y}_n = \mathbf{y}_n$ of observations has been acquired, and we drop the dependence of π_n upon $\mathbf{Y}_{\leq n}$ in the notation. This abuse of notation should not hinder the understanding of the arguments.

The main result of the theory of stochastic filtering is that given a sequence of observations $\mathbf{y}_1, \ldots \mathbf{y}_n, \ldots$, the conditional distributions π_n can be computed recursively. In other words, there exists an algorithm:

$$\pi_{n+1} = \rho_n(\pi_n, \mathbf{y}_{n+1}) \tag{7.27}$$

which gives π_{n+1} in terms of π_n and the *new observation* $\mathbf{y}_{n+1}$. This says that the dependence of π_{n+1} upon the past observations $\mathbf{Y}_1 = \mathbf{y}_1, \ldots, \mathbf{Y}_n = \mathbf{y}_n$ is already contained in the conditional distribution π_n, so that, in order to compute π_{n+1}, it is enough to *update* the previous conditional distribution π_n using the algorithm (7.27) and the new observation $\mathbf{y}_{n+1}$.

The proof of this beautiful theoretical result is the conclusion of thirty years of intensive research following the original works of Kalman and Bucy on linear systems. In the case of these linear systems (studied in the previous chapter), if the state white noise $\mathbf{W}$ and the observation white noise $\mathbf{V}$ are Gaussian, then the random vectors $\mathbf{X}_n$ and $\mathbf{Y}_{\leq n}$ are jointly Gaussian, and the conditional distribution of $\mathbf{X}_n$ given $\mathbf{Y}_{\leq n}$ is also Gaussian. π_n being Gaussian, it is entirely determined by its mean vector, say $\hat{\mathbf{X}}_n$, and its variance/covariance matrix, say Ω_n. In this case, the algorithm (7.27) can be rewritten as an update algorithm for $\hat{\mathbf{X}}_n$ and Ω_n:

$$\begin{bmatrix} \hat{\mathbf{X}}_{n+1} \\ \Omega_{n+1} \end{bmatrix} = \rho_n \left(\begin{bmatrix} \hat{\mathbf{X}}_n \\ \Omega_n \end{bmatrix}, \mathbf{y}_{n+1} \right). \tag{7.28}$$

Working out the details of this update algorithm, we would recover the update formulae derived in the previous chapter. So what could be viewed as a miracle, can now be explained. It is only because a Gaussian distribution is completely determined by its mean vector and its variance/covariance matrix, that the optimal filter could be given by a recursive update of the mean and the variance. In general, the conditional distributions π_n are not Gaussian, and the update algorithm cannot be reduced to such a simple form as (7.28).

7.6.3 Particle Filter Approximations

As suggested by the fact that the update algorithm involves the entire conditional distribution, any practical implementation of the solution to the nonlinear filtering problem has to be based on an approximation. The idea of the particle approximation to nonlinear filtering is simple and natural. The basic principle of the particle method is to approximate the desired distribution π_n by the empirical distribution of a sample of randomly chosen states. There would not be anything special to this idea if it weren't for the fact that these approximations can be updated in a very natural way without affecting the quality of the approximation.

Nothing can be so simple without being deep and powerful ! What is remarkable is the fact that it can be implemented in a recursive manner, as in the case of the Kalman filter for linear systems.

At each time n we consider a set $\{x_n^j\}_{j=1,\dots,m}$ of m elements of the state space in which $\mathbf{X}_n$ takes its values, and we consider the approximation to the optimal filter:

$$\pi_n(dx) = \mathbb{P}\{\mathbf{X}_n \in dx | \mathbf{Y}_{\leq n}\} \approx \hat{\pi}_n^{(m)}(dx) = \frac{1}{m}\sum_{j=}^{m} \delta_{x_n^j}(dx)$$

given by the empirical distribution of the sample $\{x_n^j\}_{j=1,\dots,m}$. Choosing m large enough, and choosing the x_n^j appropriately, one can make sure that the approximation is as good as desired. But the main question remains: do we have to recompute the entire approximation each time we have a new observation, or could it be possible to update the approximation to π_n in an efficient way, and get a reasonable approximation of π_{n+1}. In other words, can we:

- Compute $\hat{\pi}_{n+1}^{(m)}$ as $\phi_n(\hat{\pi}_n^{(m)}, \mathbf{y}_{n+1})$?
- and still have: $\lim_{m\to\infty} \hat{\pi}_n^{(m)} = \pi_n$ for each n.

The algorithm which we present now, does just that. To implement it we need two kinds of particles:

- those used to simulate $\mathbb{P}\{\mathbf{X}_n | \mathbf{Y}_{\leq n}\}$
$$x_n^1, .., x_n^m$$
- those used to simulate $\mathbb{P}\{\mathbf{X}_{n+1} | \mathbf{Y}_{\leq n}\}$
$$p_{n+1}^1, \dots, p_{n+1}^m$$
for the one-step-ahead distribution. At this stage it is important to remember the fundamental role played by the one-step-ahead prediction in the linear case.

One Step Ahead Prediction

To describe and justify the algorithm we assume temporarily that

$$x_n^1, \dots, x_n^m$$

form a random sample from the distribution $\pi_n = \mathbb{P}\{X_n | \mathbf{Y}_{\leq n}\}$. If we assume that $v_n^1, .., v_n^m$ are m independent realizations of the noise $\mathbf{V}_n$ (remember that we assume that we know the distribution of the state equation noise terms), and if we define the new particles $p_{n+1}^1, .., p_{n+1}^m$ by:

$$p_{n+1}^j = F_n(x_n^j, v_n^j),$$

in other words by simulating the dynamics given by the state equation, then it is clear that we have a random sample

$$p_{n+1}^1, \dots, p_{n+1}^m$$

from the conditional distribution $\mathbb{P}\{\mathbf{X}_{n+1} | \mathbf{Y}_{\leq n}\}$.

***Filtering, or* Updating**

The second step of the algorithm is less obvious. Assuming that we have a sample

$$p^1_{n+1}, \dots, p^m_{n+1}$$

from the conditional distribution $\mathbb{P}\{\mathbf{X}_{n+1}|\mathbf{Y}_{\le n}\}$, and assuming that a new observation $\mathbf{y}_{n+1}$ is made available, we compute the likelihoods

$$\alpha^j_{n+1} = \mathbb{P}\{\mathbf{Y}_{n+1} = \mathbf{y}_{n+1}|p^j_n\}$$

of the particles p^j_{n+1} given this new observation. These likelihoods can be computed because we assume that we know the distribution of the observation noise, and we also assume that we can invert the observation equation. More precisely, since $\mathbf{Y}_{n+1} = G_{n+1}(\mathbf{X}_{n+1}, \mathbf{V}_{n+1})$, we must have:

$$\alpha^j_{n+1} = \psi(H_{n+1}(y_{n+1}, p^j_{n+1}))|\frac{\partial H_{n+1}}{\partial y}(H_{n+1}(y_{n+1}, p^j_{n+1}))|$$

provided we denote by ψ the density of the observation white noise, and by H_{n+1} the inverse of G_{n+1}. The second step of the algorithm is justified by the following important computation.

$$\begin{aligned}
\mathbb{P}\{\mathbf{X}_{n+1} = p^i_{n+1}|\mathbf{Y}_{\le n+1}\} &= \frac{\mathbb{P}\{\mathbf{X}_{n+1} = p^i_{n+1}, \mathbf{Y}_{n+1}|\mathbf{Y}_{\le n})}{\mathbb{P}\{\mathbf{Y}_{n+1}|\mathbf{Y}_{\le n})} \\
&= \frac{\mathbb{P}\{Y_{n+1}|p^i_{n+1})\mathbb{P}\{\mathbf{X}_{n+1} = p^i_{n+1}|\mathbf{Y}_{\le n})}{\sum_{j=1}^m \mathbb{P}\{\mathbf{Y}_{n+1}|p^j_{n+1}\}\mathbb{P}\{\mathbf{X}_{n+1} = p^j_{n+1}|\mathbf{Y}_{\le n})} \\
&= \frac{\alpha^i_{n+1}\cdot\frac{1}{m}}{\frac{1}{m}\sum_{j=1}^m \alpha^j_{n+1}} \\
&= \frac{\alpha^i_{n+1}}{\sum_{j=1}^m \alpha^j_{n+1}}.
\end{aligned}$$

Indeed, this computation suggests that we define x^j_{n+1} by:

$$x^j_{n+1} = \begin{cases} p^1_{n+1} \text{ with probability} \frac{\alpha^1_{n+1}}{\alpha^1_{n+1}+\dots+\alpha^m_{n+1}} \\ \vdots \\ p^m_{n+1} \text{ with probability} \frac{\alpha^m_{n+1}}{\alpha^1_{n+1}+\dots+\alpha^m_{n+1}} \end{cases}$$

because in doing so, we make sure that $x^1_{n+1}, \dots, x^m_{n+1}$ appear as particles which form a random sample from the conditional distribution $\pi_{n+1} = \mathbb{P}\{\mathbf{X}_{n+1}|\mathbf{Y}_{\le n+1}\}$, putting us at time $n+1$, in the situation we started from at time n. This second step completes at the level of the random samples, the implementation of the update algorithm (7.27) for the conditional distributions.

Algorithm Summary

We can summarize the algorithm as follows:

1. Initialization: generate an initial random sample of m particles $x_1^0, \dots, x_0^m$
2. For each time step n, we repeat the following process:
   - Generate independent particles v_n^j from the distribution of the state noise;
   - Generate the particles p_{n+1}^j using the formula $p_{n+1}^j = F(x_n^j, v_n^j)$;
   - Given a new observation, compute the likelihood α_{n+1}^j;
   - Resample the p_{n+1}^j to produce the x_{n+1}^j's.

The remainder of this chapter is devoted to the discussion of several implementations of this particle filtering algorithm. The applications that we discuss share the following features.

- Nonlinear dynamics for the unobserved state;
- Nonlinear observation equation;
- Parameters as components of the state vector;
- Regime switching can be incorporated in the Markov dynamics of the partially observed state.

7.6.4 Filtering in Finance? Statistical Issues

Some of the financial applications of filtering can be motivated by a quick look at the plots in Figure 7.17 where we show the time evolution of the stock prices of a few energy companies before the crash which followed the California energy crisis and Enron's bankruptcy. One of the time series (obviously Enron) clearly stands out. Is it because of a change in the mean rate of return? Is it because of a sudden change in the volatility? Could it be because of the issuance of debt? A change in rating of some of those debts? A sudden change in other economic factors? Tracking these important parameters is a natural task that most econometricians, economists, analysts, and traders do on a day-to-day basis. We propose to illustrate the use of nonlinear filtering techniques to do just that.

The tracking motivation outlined above fits perfectly in the statistical theory of so-called change point problems. Nevertheless, several important differences need to be pointed out.

- Real time computations are generally not an issue for economic/econometric applications for which computations can be performed *after the fact* (overnight for example). Only highly speculative operations (such as program trading or energy spot price tracking) require real time computations.
- The filtering approach is very natural when it comes to the non-anticipative estimation of some of the parameters needed for scenario generation and risk management. Nevertheless, some insight into these issues can easily be gained without having to introduce the heavy machinery of filtering theory. As we pointed out in Subsection 7.2.5, standard estimation can be performed in trailing sliding

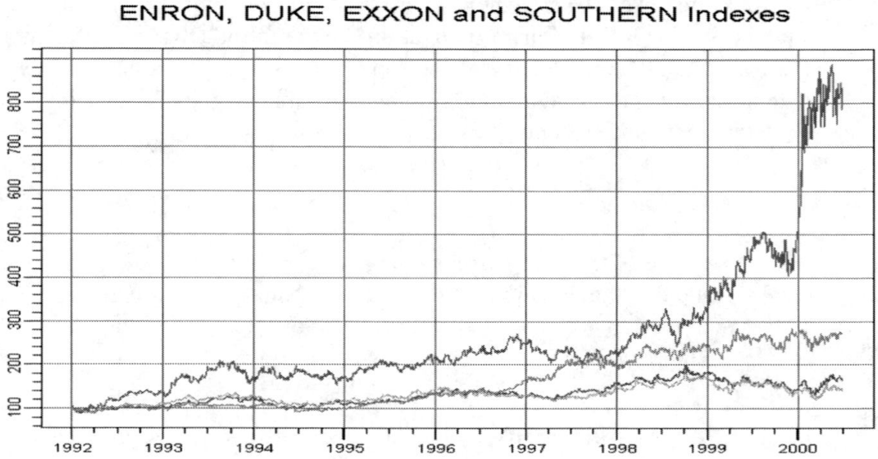

Fig. 7.17. Time series plots of the Enron, Exxon, Duke and Mirant (formally SouthWest) indexes.

windows, and classical multivariate analysis tests can be run inside these windows to test for the equality of rates of mean return or volatility. See the Notes & Complements at the end of the chapter for a discussion of an application to the subindexes of the Dow Jones Total Market Index.

- Even when the problem is naturally framed as a problem of parameter estimation or detection of change in parameters (as opposed to a filtering problem), particle approximations are still of great value. Indeed, parameter estimation and hypotheses testing are very often based on the computation of the likelihood function of the model. This task is unfortunately too difficult in many practical applications. But according to our discussion of Subsection 6.4.6, it is possible to express the likelihood function in a recursive fashion, in terms of the optimal filter. It is not a surprise to learn that many active researchers have based their computation of the likelihood (and consequently their estimation and test procedures) on the particle approximation of the optimal filter. This recursive computation of the likelihood approximation is currently investigated as a leading candidate for efficient estimation procedures and test statistic computations.

7.6.5 **Application: Tracking Volatility**

Stochastic volatility models of continuous time finance are factor models, and most of them are two-factor models, one of these factors being the volatility in question. The dynamics of the two factors are given by stochastic differential equations of the Ito's type driven by two different Wiener processes. In this section, we consider the discrete versions of these models obtained by Euler's scheme.

As already explained when we discussed discrete time stochastic volatility models in Section 7.3, the main motivation for the introduction of these models is the smile effect. The popularity of the stochastic volatility models is deeply rooted in their intuitive rationale, and the fact that they are capable of producing "smiles" like the ones encountered in our discussion of Section 4.7.

The Models

As in the case of the Samuelson's framework leading to the Black-Scholes formula, we assume that the dynamics of the underlying asset price are given by a geometric Brownian motion derived from the stochastic differential equation:

$$dS_t = S_t(\mu dt + \sigma_t dW_t),$$

but instead of assuming that σ_t is a positive constant, we assume that σ_t is another random quantity which changes with time, and the model is based on the choice of another stochastic differential equation for the dynamics of this stochastic quantity. For the sake of definiteness we shall assume that σ_t satisfies:

$$d\sigma_t = -\lambda(\sigma_t - \sigma_\infty)dt + \gamma d\tilde{W}_t \tag{7.29}$$

for some positive constants λ, σ_∞ and γ, and for another Wiener process $\tilde{W}_t$ providing the source of the random kicks driving the time evolution of the volatility. These parameters have clear interpretations. σ_∞ is a mean level of volatility toward which σ_t tries to revert. In fact σ_∞ would be the deterministic limit of σ_t for large values of the time t if it were not for the random kicks given by the Wiener process $\tilde{W}_t$. The constant λ is the rate of mean reversion toward σ_∞, while γ is the volatility of the volatility σ_t (volvol in the jargon of the street). So the resulting force $d\sigma_t$ acting on the volatility is an aggregate of a restoring force $-\lambda(\sigma_t - \sigma_\infty)dt$ and a random kick $\gamma d\tilde{W}_t$. The balance between these two terms determines the actual dynamics of the stochastic volatility model chosen here.

Discretizations

In order to make the above model amenable to implementation and analysis by the filtering techniques introduced in this chapter, we discretize the above equations in order to set up a state space system in the form of a hidden Markov model.

We first consider the equation for the asset price. We choose a sampling interval Δt, and we write the dynamical equation given by Euler's scheme. Mimicking the procedure followed in Subsection 7.5.1, we introduce the variable RR for the raw return over one time period. The Euler discretization scheme gives:

$$RR_t = \frac{S_{t+\Delta t}}{S_t} - 1 = \mu\Delta t + \sigma_t\sqrt{\Delta t}\epsilon_{t+\Delta t}. \tag{7.30}$$

Next, we consider equation (7.29). Using Euler's scheme again, we get:

$$\sigma_{t+\Delta t} = \lambda\sigma_\infty \Delta t + (1 - \lambda\Delta t)\sigma_t + \gamma\sqrt{\Delta t}\tilde{\epsilon}_{t+\Delta t}. \tag{7.31}$$

Here $\{\epsilon_t\}_t$ and $\{\tilde{\epsilon}_t\}_t$ are $N(0,1)$ strong white noise time series independent of each other.

Remarks

1. Equation (7.31) is not the only way to discretize the dynamical equation (7.29). Indeed the latter can be solved *explicitely* and, as we explained in Subsection 7.4.2, it is then preferable to discretize the exact solution (as opposed to solving the discretized equation). If we do that in the present situation, we end up with the equation:

$$\sigma_{t+\Delta t} = \sigma_\infty + e^{-\lambda\Delta t}(\sigma_t - \sigma_\infty) + \sqrt{\frac{\gamma^2}{2\lambda}(1 - e^{-2\lambda\Delta t})}\epsilon^{(\sigma)}_{t+\Delta t}. \tag{7.32}$$

2. The independence of the white noise is not a realistic assumption. Indeed, the leverage effect discussed earlier would suggest that the two white noise time series should be negatively correlated. It is not difficult to force this condition and still developed the filtering apparatus presented below. We refrain from doing it because this requires adding noise terms to the dynamical equations, and as a consequence, rather cumbersome notation, and more involved formulae.

Setting Up Hidden Markov Models

In the present set up, we can assume that the asset prices are observed, hence so are the raw returns. However, it is clear that the volatility σ_t cannot be observed directly.

1. Using the experience gained in rewriting ARMA models as state space models, one is tempted to choose the observation equation

$$Y_t = \begin{bmatrix} 1 & 0 \end{bmatrix} \begin{bmatrix} RR_{t-\Delta t} \\ \sigma_t \end{bmatrix} \tag{7.33}$$

which would lead us to choose the state vector $\mathbf{X}_t$ and the state noise vector $\mathbf{V}_{t+\Delta t}$ as:

$$\mathbf{X}_t = \begin{bmatrix} RR_{t-\Delta t} \\ \sigma_t \end{bmatrix}, \quad \text{and} \quad \mathbf{V}_{t+\Delta t} = \begin{bmatrix} \epsilon_{t+\Delta t} \\ \tilde{\epsilon}_{t+\Delta t} \end{bmatrix}, \tag{7.34}$$

with state equation:

$$\mathbf{X}_{t+\Delta t} = F(\mathbf{X}_t, \mathbf{V}_{t+\Delta t}) \tag{7.35}$$

where the function F is defined by:

$$F(\mathbf{x}, \mathbf{v}) = \begin{bmatrix} \mu\Delta t + x_2\sqrt{\Delta t}v_1 \\ \lambda\sigma_\infty\Delta t + (1 - \lambda\Delta t)x_2 + \gamma\sqrt{\Delta t}v_2 \end{bmatrix}, \quad \text{if} \quad \mathbf{x} = \begin{bmatrix} x_1 \\ x_2 \end{bmatrix}, \quad \mathbf{v} = \begin{bmatrix} v_1 \\ v_2 \end{bmatrix}.$$

Clearly, the observation is partial in the sense that not all the components of the state vector can be observed. However, the observation is perfect in the sense that there is no noise coming to perturb the accuracy of the observation. Strangely enough, this can be a serious handicap, and filtering methods have a harder time handling perfect observations. As a result, the practitioner adds an artificial (small) noise to the right hand side of the observation equation (7.33), just to regularize the situation. We shall not follow this approach here.

2. Another possibility is to choose $X_t = \sigma_t$ for the state of the system and $Y_t = RR_{t-\Delta t}$ for the observation. In this case, the dynamics of the state are given by

$$X_{t+\Delta t} = F(X_t, V_{t+\Delta t}) \tag{7.36}$$

with

$$F(x, v) = \lambda\sigma_\infty \Delta t + (1 - \lambda\Delta t)x + \gamma\sqrt{\Delta t}v$$

if we choose the system noise $V_{t+\Delta t}$ to be $\tilde{\epsilon}_{t+\Delta t}$. Accordingly, the observation equation reads:

$$Y_t = G(X_t, W_t)$$

with

$$G(x, w) = \mu\Delta t + x\sqrt{\Delta t}v$$

provided the observation noise W_t is chosen to be ϵ_t. Notice that the noise is not additive, i.e. the observation function $G(x, w)$ is not of the form $G(x) + w$. However, the function G is invertible, and as we saw when we derived the particle filtering algorithm, that's all we need !

3. The above formulation is quite appropriate for the implementation of the particle filtering algorithm described in the previous section, as long as we have estimates of the parameters μ, σ_∞, λ and $c = \gamma^2/2\lambda$ of the problem. Rather than trying to estimate them first, and then run the filtering algorithm, we include them in the state of the system, letting the filter estimate their values dynamically. This idea is very appealing, and this practice is pretty common in some circles, despite the fact that it still lacks rigorous mathematical justification. For the sake of simplicity, we only include the parameters λ and c in the state, estimating directly the parameters μ and σ_∞ off line. So the final form of the state space model which we use to track (stochastic) volatility is given by the following state dynamical equation:

$$\begin{bmatrix} \sigma_{t+\Delta t} \\ \lambda_{t+\Delta t} \\ c_{t+\Delta t} \end{bmatrix} = \begin{bmatrix} \sigma_0 + e^{-\lambda_t \Delta t}(\sigma_t - \sigma_0) + \sqrt{c_t(1 - e^{\lambda_t \Delta t})}\epsilon^{(\sigma)}_{t+\Delta t} \\ \lambda_t \\ c_t \end{bmatrix}$$

coupled with the observation equation:

$$R_{t-\Delta t} = \mu\Delta t + \sigma_t\sqrt{\Delta t}\epsilon_t.$$

Notice that the dynamics of the parameters are trivial since their true values do not change over time. Only their estimations are changing as we update their conditional distributions when we compute our new best guess for the state of the system given a new set of observations. The convergence of these estimates toward their true values is not guaranteed, but it seems to be happening in many applications, giving an empirical justification for this practice. See also the Notes & Complements at the end of the chapter for references on this approach to adaptive parameter estimation.

Remarks

1. Positivity of σ_t is not guaranteed in the above model. This prompts us to search for other stochastic differential equations for the dynamics of σ_t. We could use some of the models proposed for the short interest rate because of their positivity. This is for example the case of the square root model also called CIR model. Unfortunately, the direct approach (analogous to (7.32)) cannot be used because we do not have random number generators from direct control of the conditional probability density of $\sigma_{t+\Delta t}$ given σ_t, and the Euler's scheme looses the positivity of the solution.

2. Because of these positivity problems, geometric Ornstein-Uhlenbeck processes (i.e. exponentials of Ornstein-Uhlenbeck processes) are often used as models for stochastic volatility.

Empirical Results

We implemented the particle filtering algorithm on the third state space systems described above. The results of some Monte Carlo experiments are reproduced graphically in Figures 7.19 and 7.20. Figure 7.18 gives the plot of the price series we generated to run the filter. We used the values $\Delta t = 0.004$, $\mu = 0.006$, $\lambda = 2$, $c = 0.5$, $\sigma_0 = 0.1$ & $\sigma_\infty = 0.2$ for the parameters. Figure 7.18 gives the plot of

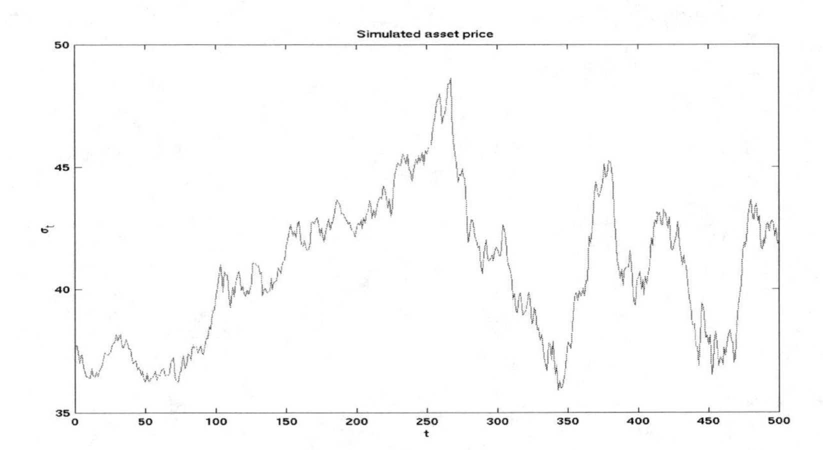

Fig. 7.18. Simulated asset price.

the volatility which was actually used to generate the price shown in Figure 7.18 together with two estimates of this instantaneous volatility. The first one is obtained by the particle filter described above. It tracks the true volatility reasonably well. The other estimate is the result of the empirical estimation of the standard deviation in a sliding non-anticipative window of size 30. As we can see, this naive estimate is too slow to react to changes in the volatility.

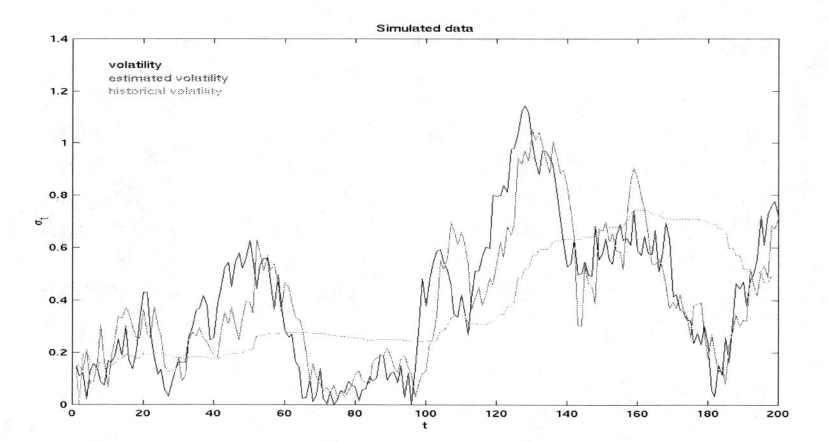

Fig. 7.19. Another example of volatility tracking by particle filtering.

Finally, Figure 7.20 gives another instance of the superiority of the particle filter over the sliding window naive estimate. In fact the particle filter estimate is always better for simulated data. It is not clear how to compare the two on real data. Indeed, we do not know the exact value of the *true* volatility which drove the price dynamics.

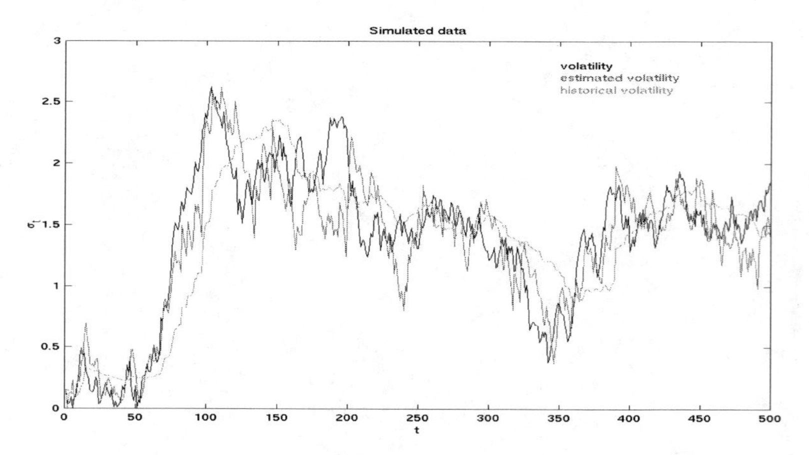

Fig. 7.20. Still another example of volatility tracking by particle filtering.

From the many experiments which we ran on this specific application, we found that the particle filter approximation to the (true) optimal filter

- Provides exceptionally good tracking
  - If initial parameters (mean reverting rates, vol of vol, etc.) in right range
- Lacks robustness (poor estimation)
  - If initial parameters too far
- Reacts and relaxes faster (in all cases) than historical volatility

Possible Extensions

The application considered above can be made more realistic by extending the model to include more features of real data. As a first possible extension, we should mention that the particle filters can be used without much change to the case of a regime switching model in which the volatility is a function of a factor changing value over time in a limited and restricted way, though Markovian. Another possible generalization is to adjust the filtering model to volatility processes with several time scales. These multiscale volatility models are very fashionable, and as long as filtering is concerned, they can be handled without much significant change to what we presented above.

APPENDIX: PREPARING INDEX DATA

Dow Jones graciously provided us with some of the data sets used in this chapter. In typical applications, the raw data is processed in order to facilitate the computation of the many indexes published by Dow Jones. This short appendix shows how one can use simple `S-Plus` commands to get from the raw data to `timeSeries` objects amenable to the kind of analysis we report in the text.

We start from the data contained in the file `enron.tab`. This file is typical of the files used to compute indexes. As shown below, this file contains five columns. The first one gives the date, the second one gives the number of stocks involved in the computation of the index (only one in the present situation), the third one gives the capitalization, the fourth one the divisor (which is capitalization related, but which we will not explain here) and the fifth one gives the value of the index.

```
> ENRON <- scan("Data/enron.tab")
> dim(ENRON) <- c(5,length(ENRON)/5)
> ENRON <- t(ENRON)
> ENRON[1:5,]
          [,1] [,2]       [,3]     [,4]      [,5]
 [1,] 19920101    1 3516527280 35165273 100.00000
 [2,] 19920102    1 3604942823 35165273 102.51429
 [3,] 19920103    1 3655178927 35165273 103.94286
 [4,] 19920106    1 3579824771 35165273 101.80000
```

We shall need a few transformations to be able to use these data efficiently. We first write a small function `makedate` to read the date in an appropriate format.

```
makedate <- function(REAL)
{
# This function takes a real like 19920106.00 and
# returns a date of the form 10/6/1992
    YEAR <- floor(as.integer(REAL)/10000)
    TMP <- as.integer(REAL) - YEAR*10000
    MONTH <- floor(TMP/100)
    DAY <- TMP - MONTH * 100
    TMP.julian <- julian(MONTH,DAY,YEAR)
    dates(TMP.julian)
}
```

With this function at hand, we can easily read the Enron data into a `timeSeries` object. We shall use the function `readindex` whose code is as follows.

```
readindex <- function(FRAME)
{
# Use the data frame FRAME resulting from the import
# of one of the Energy indexes created from the DOW JONES
# data base, and create a tri-variate timeSeries with
# components cap (for capitalization), divisor and index.
    names(FRAME) <- c("date","nbstock","cap","divisor","index")
    timeSeries(pos=makedate(FRAME[,1]), data=FRAME[3:5])
}
```

We can now create the `timeSeries` object `ENRON.ts` which we used in the text. The plot of this time series was given in Figure 7.2

```
> ENRON <- data.frame(ENRON)
> ENRON.ts <- readindex(ENRON)
> plot(ENRON.ts[,3])
> title("ENRON Index")
```

PROBLEMS

Ⓣ **Problem 7.1** *This problem shows once more that a conditionally Gaussian random variable has excess kurtosis.*
Let us assume that X and σ^2 are two random variables and that $X|\sigma^2 \sim N(0, \sigma^2)$, i.e. that conditioned on the value of σ^2, X is mean-zero normal with variance σ^2. Prove that:

$$\frac{\mathbb{E}\{X^4\}}{var\{X\}^2} = 3\left[1 + \frac{var\{\sigma^2\}}{\mathbb{E}\{\sigma^2\}^2}\right]$$

proving the claim of excess kurtosis when σ^2 is not deterministic.

Ⓣ **Problem 7.2** *This problem shows that (linear) AR(p) time series can lead to (nonlinear) ARCH models when they have random coefficients.*

Let $\{\epsilon_t\}_t$ *be a strong univariate white noise with* $\epsilon_t \sim N(0,1)$*, and let* $\{\phi_t\}_t$ *be a p-variate time series independent of* $\{\epsilon_t\}_t$*, and such that all the* ϕ_t *are independent of each other, and for each time t, the vector* $\phi_t = (\phi_{t,1}, \phi_{t,2}, \ldots, \phi_{t,p})$ *is a vector of jointly Gaussian random variables with mean zero and variance/covariance matrix* Σ*. We study the time series* $\{Y_t\}_t$ *defined by:*

$$Y_t = \phi_{t,1}Y_{t-1} + \phi_{t,2}Y_{t-2} + \cdots + \phi_{t,p}Y_{t-p} + \epsilon_t.$$

1. Determine the conditional distribution of Y_t *given* $\underline{Y}_{t-1}$ *by integrating out the* ϕ*-random variables.*
2. Assume that the components $\phi_{t,1}, \phi_{t,2}, \ldots, \phi_{t,p}$ *of* ϕ_t *are independent, and show that at least in this case,* $\{Y_t\}_t$ *has an ARCH representation.*

Ⓣ **Problem 7.3** This problem addresses the very important issue of the stability of models under change of sampling frequency. Here is a typical example: assuming that a GARCH(1,1) model was found appropriate for some daily financial time series, and assuming that we need now to look at the end of the week data entries only, could we assume that this time series will also follow a GARCH(1,1) model, or should we start the model fitting process from scratch?
We assume that a series $\{Y_t\}_t$ *of log-returns has the GARCH(1,1) representation of the form:*

$$Y_t = \sigma_t\tilde{\epsilon}_t, \qquad \sigma_t^2 = c + b\sigma_{t-1}^2 + aY_{t-1}^2$$

where we assume that $\{\tilde{\epsilon}_t\}_t$ *is a strong* $N(0,1)$ *white noise, and where we assume that the coefficients a, b and c are such that* σ_t^2 *is stationary.*
1. Show that the squared log-returns have an ARMA(1,1) representation in the sense that:

$$Y_t^2 = c + (b+a)Y_{t-1}^2 + \epsilon_t - b\epsilon_{t-1}$$

for some weak white noise $\{\epsilon_t\}$ *which you should identify. Show that the number* $\alpha = \mathbb{E}\{\epsilon_t^2\}$ *is independent of t and that:*

$$Y_t^2 = c(1+b+a) + (b+a)^2Y_{t-2}^2 + u_t$$

with $u_t = \epsilon_t + a\epsilon_{t-1} - b(a+b)\epsilon_{t-2}$*.*
2. Compute the number $\beta = \mathbb{E}\{u_t^2\}$ *in terms of the numbers a, b and c, and check that it is independent of t. Show that* $\mathbb{E}\{u_t u_{t-2k}\} = 0$ *for all integers* $k > 1$*, and that:*

$$\frac{\mathbb{E}\{u_t u_{t-2}\}}{\mathbb{E}\{u_t^2\}} = -\frac{b(a+b)}{1+a^2+b^2(a+b)^2}.$$

3. We now use the process $\{v_t\}_t$ *defined by:*

$$v_t = \frac{1}{1-\lambda B^2}u_t$$

for some $\lambda \in (0,1)$ *where B stands for the backward shift operator. Show that:*

$$\mathbb{E}\{v_{2t}v_{2t-2s}\} = \frac{\lambda^{s-1}}{1-\lambda^2}[\lambda\mathbb{E}\{u_t^2\} + (1+\lambda^2)\mathbb{E}\{u_t u_{t-2}\}], \qquad s \geq 1.$$

4. Show that if λ *is chosen as the root of the equation:*

$$\frac{\lambda}{1+\lambda^2} = \frac{b(a+b)}{1+a^2+b^2(a+b)^2}$$

which belongs to the interval $(0,1)$, *then the* v_t *with even indices are uncorrelated, i.e.*

$$\mathbb{E}\{v_{2t}v_{2t+2s}\} = 0, \qquad s \geq 1$$

and that:

$$\mathbb{E}\{v_{2t}^2\} = \frac{\beta}{1+\lambda^2}.$$

5. Show that the square $\overline{Y}_t^2$ *of the series* $\{\overline{Y}_t\}_t$ *defined by:*

$$\overline{Y}_t = Y_{2t}$$

has an ARMA(1,1) representation, and conclude that the series $\{\overline{Y}_t\}_t$ *admits a GARCH(1,1) representation similar to the original GARCH(1,1) representation of* $\{Y_t\}_t$ *which we started from.*

Ⓔ **Problem 7.4** The goal of this problem is to give another example of the misuse of the simulation of GARCH models fitted to data with a deterministic seasonal pattern in the instantaneous variance.

1. We first use the daily average temperatures in La Guardia as encapsulated in the `S-Plus` *time series* `LG.ts`. *Identify the seasonal component of* `LG.ts`, *fit a GARCH(1,1) model to the remainder, and plot the estimate instantaneous conditional variance.*
2. In each CDD season, identify the periods of high volatility and the periods of low volatility. Same question for the HDD seasons.
3. Using the fitted model, simulate a Monte Carlo sample for 5 *years worth of temperature in La Guardia. Do you find the periods of high and low volatility identified in question 2. above? Explain.*
4. Redo questions 1, 2, and 3 for the temperature in Las Vegas contained in the time series `LV.ts`.

Ⓢ **Problem 7.5** *This problem is based on the data contained in the text file* `monthly.asc`. *The first column represents the values of the short interest rate as given by the one year T-bill between May 1986 and November 1999, the second column gives the long interest rate as given by the 30 years US Government Bonds, and the third column gives the values of the S&P 500 index during the same period.*
1. Use the equation:

$$S_{t+\Delta t} - S_t = S_t[\mu\Delta t + \sigma\sqrt{\Delta t}\epsilon_t]$$

used in the text as equation (7.21) to model the monthly S&P 500 values, to estimate the parameters μ *and* σ *from the data. Remember,* $\Delta t = 1/12$ *stands for one month. Explain your work.*
2. Similarly, use the data and the equation:

$$r_{t+\Delta t} = r_t - \lambda_r(r_t - \overline{r})\Delta t + \sigma_r\sqrt{\Delta t}\ \epsilon_t^{(r)}$$

used in the text as equation (7.19) to model the monthly values of the short interest rate, to estimate the parameters λ_r, $\overline{r}$, *and* σ_r. *Be imaginative and explain your work.*
3. Finally, use the data and the equation:

$$s_{t+\Delta t} = s_t - \lambda_s(s_t - \overline{s})\Delta t + \sigma_s\sqrt{\Delta t}\ \epsilon_t^{(s)}$$

used in the text as equation (7.21) to model the monthly values of the interest rate spread, to estimate the parameters λ_s, $\overline{s}$, *and* σ_s. *Again, explain your work.*

4. Assuming that the three white noise sequences $\{\epsilon_t\}_t$, $\{\epsilon_t^{(r)}\}_t$, and $\{\epsilon_t^{(s)}\}_t$ are independent, use the above equations and parameter estimates to generate a $100 \times 120 \times 3$ array containing 100 scenarios of 120 possible monthly values of the short interest rate, the long interest rate and the S&P 500 index, starting from the last entries of the data in the file `monthly.asc`*. Save these three sets of 100 scenarios in text files named* `IndepShort`, `IndepLong` *and* `IndepSP500` *respectively.*

5. From the data, compute the variance/covariance matrix of the log-return of the S&P 500, the first difference of the short interest rate, and the first difference of the spread. Assuming now that the variance covariance matrix of the three white noise sequences $\{\epsilon_t\}_t$, $\{\epsilon_t^{(r)}\}_t$, and $\{\epsilon_t^{(s)}\}_t$ is the same as the empirical variance/cavariance matrix you just computed, generate a new set of $100 \times 120 \times 3$ array containing 100 scenarios of 120 possible monthly values of the short interest rate, the long interest rate and the S&P 500 index. Save the 3 sets of scenarios in text files named `CorrelShort`, `CorrelLong` *and* `CorrelSP500` *this time.*

Ⓣ **Problem 7.6** *In this problem, we assume that the dynamics of a stock are given by Samuelson's geometric Brownian motion model, and we assume that the rate of growth μ and the volatility σ are known.*

1. Use formula (7.15) to derive the distribution of the raw returns RR_t over a one-month period Δt.

2. Compare with the distribution derived from (7.17) given by a blind application of Euler's scheme. Comment.

Ⓢ **Problem 7.7** *The goal of this problem is to implement a volatility tracking algorithm based on the particle filtering of a stochastic volatility model where the dynamics of the volatility are given by a geometric Ornstein Uhlenbeck process.*

1. For the purpose of this question we consider the price of an asset evolving with time according to a geometric Brownian motion with a stochastic volatility satisfying equation (7.29) used in the text. Such a process is usually called an Ornstein-Uhlenbeck process. Explain why, at any given time t, the volatility σ_t has a positive probability of being negative. In doing so, you can choose to work directly with the continuous time equation (7.29) if you feel comfortable, or with its discretized version if you find it more intuitive.

2. Explain why this shortcoming is not shared by the geometric Ornstein-Uhlenbeck model in which the volatility evolves according to the equation:

$$d\sigma_t = \sigma_t[-\lambda(\log \sigma_t - \mu)dt + \gamma d\tilde{W}_t] \tag{7.37}$$

where λ, μ and γ are strictly positive constants. Again, feel free to choose to work with the continuous time equation (7.37) or its discretized form.

Ⓔ **Problem 7.8** *You will need to use the data contained in the files* `enelr.txt` *and* `dynlr.txt`*. Each file contains the daily log return of a large energy company.*

1. Fit an AR model to the part of the time series contained in `dynlr.txt` *corresponding to the period of the 500 trading days ending January 1, 2001. Use this model to predict the values of the log return for all the remaining days after January 1, 2001, and compute the sum of square errors.*

2. Same question for the time series contained in the file `enelr.txt`

3. Concatenate the two time series in a bivariate time series, fit an AR model to the bivariate series so obtained for the same period as before, and use this model to predict the values of the two series following January 1, 2001. Compute the sum of the squares errors and compare to the previous results. Comment your results.

NOTES & COMPLEMENTS

Mandelbrot was presumably the first one to warn the scientific community of the effects of the presence of heavy tails and long range dependence in financial time series. A good account of his work on the subject can be found in [58], [59] and [60]. Nonlinear time series models are studied with great care and lucidity in the recent graduate textbook [36] by Fan and Yao, where the interested reader will find many examples of analyzes relying on nonparametric regression techniques.

Financial time series differ from typical time series in many different ways, but most noticeably, in the properties of the conditional standard deviation, i.e the so-called volatility. Persistence (tendency of large changes to be followed by more large changes), are at the root of the introduction of the ARCH and GARCH models discussed in this chapter. The first ARCH models were introduced in their simplest form by Engle in 1982, see [32]. This introduction had far-reaching implications on subsequent developments in financial economics, hence the award to Engle of the 2003 Nobel prize in economics. ARCH models were later generalized by Engle and Bollerslev in [33], and Bollerslev [9] who introduced the GARCH models. The review article [10] is a comprehensive introduction to the early developments of the subject. Since then, the literature on the subject exploded, and the analysis of these models is now an integral part of financial econometrics. For more recent developments on these models we refer the interested reader to the more modern account given in the monograph [40].

One of the major difficulties with the stochastic volatility models is their statistical estimation. Many methods have been proposed and tested: generalized method of moments (GMM for short), quasi-likelihood method based on Monte Carlo Markov Chain (MCMC for short) computations and importance sampling, and even the simulated EM (expectation maximization algorithm). We chose to recast the SV models in the framework of linear state space models, and refer to the statistical estimation methods used for these models.

In the second half of the nineties, many books on the applications of Ito's stochastic calculus to finance were published. One of the earliest ones, and presumably the best one, was the short book of Lamberton and Lapeyre [56]. Many followed but they were often technically difficult and rarely self-contained. We shall only refer to the introductory texts of Michael Steele [81]. Finally, and at the risk of being accused of blunt self-promotion, I will refer these readers to the companion volume [22]. Discretization schemes for stochastic differential equations have been studied in details. It would have taken us far afield to attempt to review this literature. We limited ourselves to the Euler scheme because of its simplicity and its intuitive appeal. Further properties of these schemes and their relations to the ARCH and GARCH and SV models discussed in the text can be found in the recent text by C. Gourieroux and J. Jasiak [41].

The continuous time stochastic volatility models have been studied extensively by means of the properties of the Ito's stochastic calculus. The names of Heston, and Hull and White are often attached to these models. See for example the account given in the book [47]. The analysis of the volatility smile for these models was done rigorously in Touzi and Renault [72]. For a complete exposé of the facts of continuous time stochastic volatility models from the point of view of Ito's stochastic calculus and asymptotic expansions we refer the interested reader to the excellent monograph [38] by Fouque, Papanicolaou and Sircar.

The abstract form of the equation giving the optimal filter update in the form of a dynamical system is due to Stettner. The update algorithm (7.27) can be viewed as a discrete time dynamical system in the space of probability distributions. Unfortunately, this space is infinite dimensional and the analysis of such a dynamical system is very difficult. This dynamical system can also be viewed as a discrete time version of the stochastic partial differential equation derived for the optimal filter of partially observed systems driven by Ito's stochastic differential equations. This stochastic partial differential equation is known under the name of Kushner equation, or Zakai equation, depending on our looking at the update of the (normalized) density of the optimal filter, or the density of the unnormalized filter. Both are regarded as natural generalizations of the Ricatti equation giving the update of the Kalman filter in the linear case. Indeed, even when we look at conditional density functions for the conditional distributions, the update algorithm appears (because of the infinite dimensionality of the functional space) as a partial differential equation. Because the coefficients of this partial differential equation depend upon the observations, they can be viewed as random. These equations were some of the first stochastic partial differential equations studied, and solving the general stochastic filtering problem was one of the main impetus in the development of the field of stochastic partial differential equations.

Unfortunately, the numerical solution of these equations is extremely computer intensive and, as of today, there is essentially no example of practical problem whose solution can be computed in real time. The particle method is the first method which stands a real chance at cracking this nagging problem.

The particle approximation algorithm presented in the text is due to Kitagawa [53]. The mathematical theory of this approximation (and several variations on this approximation) was subsequently developed by several probabilists including Del Moral, Guillonnet, Lyons, Crisan, Le Gland, , Cérou, . . . and many others. Among other things, they proved that at each fixed time n, the particle approximation $\hat{\pi}_n^{(m)}$ converges toward π_n as $m \to \infty$.

It is interesting to track the volatility of indexes and sub-indexes when the latter are computed by sectors, and even by subdividing the sectors according to the credit ratings of the companies comprising the sectors in question. In his 2001 Princeton PhD, M. Sales considered amon other things the sector "Materials / Mid-Cap's" of DOW JONES TMI (Total Market Index) subindexes. He noticed a general increase in volatility in 1999. But if after breaking the sector into two sub-sectors, say the companies whose bonds are rated "investment grade", and the companies whose bonds are rated "non-investment grade" (using for example Moody's ratings), then as expected, he noticed an increase in volatility in the non-investment grade subsector, NOT the investment grade. However, it should be emphasized that the reverse though less natural, DOES occur as well. `S-Plus` can be used to obtain simultaneous Likelihood Ratio Tests for the significance of the differences in mean returns. Unfortunately, similar tests for the comparison of the variances of correlated samples are not available, and part of Sales' PhD was devoted to the development of such tests.

The application of the particle filter to stochastic volatility tracking was done by A. Papavasiliou's as part of her Princeton PhD. The classical (linear) Kalman filter has been applied to the construction of commodity forward curves. The estimation of these curves is of crucial importance in energy risk management. The reader interested in energy derivatives may consult the book of Pilipovic [67], or our favorite, the recent chapter book of Clewlow and Strickland [27]. The author recently used the particle filtering algorithm to estimate the convenience yield of crude oil and natural gas: as far as we know, these applications are original and they will be published in the near future. This work was motivated by an enlightening discussion in the recent paper of Miltersen and Schwartz [63].

APPENDIX: AN INTRODUCTION TO S AND S-Plus

This appendix gives the script for a possible introductory session intended for first time users of S-Plus. Imagine that you are trying to learn enough of S to use it in a Data Analysis course based on this book, and let us imagine that the instructor of the course decided to help the students by devoting one of her first lectures to a first introduction to S-Plus. This appendix should be understood as a summary of all the commands and examples used in this hypothetical introductory lecture.

But before we start, a little bit of terminology might be helpful to explain why we often use S and S-Plus interchangeably. The reader needs to keep in mind that S is the programming environment/language developed at AT&T, and S-Plus is the commercial package implementing this language. Consult the online manuals, and especially the *Guide to Statistics* for more information on S and S-Plus.

STARTING S-PLUS (UNDER Windows AND UNDER UNIX)

The instructions given below are restricted to versions 6.0 and higher of S-Plus.

Under Windows

After starting the program, you will be faced with a dialog box asking you to choose the S-Plus Chapter in which you want to work. If you are using the program for the first time, there is no existing chapter, and you will need to browse your hard disk and choose a directory (folder) – or possibly create one – in which you will work on this project. It is recommended that you choose a special chapter which you will use exclusively for the purpose of running the examples and doing the homework problems of this book. The notion of S-Plus chapter replaces the the notion of project or workspace of the previous Windows versions of S-Plus (for example in S-Plus 2000).

The first thing need to do is to go to the menu item **File** ▷ **Chapters** and choose **File** ▷ **Chapters** ▷ **Attach/Create New Chapter** if you want to create a new chapter (which you will want to do the first time you run S-Plus), or if you want to attach an existing chapter to be able to access its functions and data. You should choose **File ▷ Chapters ▷ New Working Chapter ...** if you want to switch to an existing chapter different from the default chapter started with S-Plus.

Under Unix

Let us now imagine that the introductory session was intended for Unix or Linux users (we will use the notation $ to denote the Unix prompt), and let us imagine that the name of the class is ORF405.

The first thing we want to do is to create a subdirectory ORF405 in your home directory. The plan is to store all the files relative to the course in this directory, and in particular, to do our S-Plus based homework in this directory. Next we move to this directory, and run the command CHAPTER

```
$ mkdir ORF405
$ cd ORF405
$ S-Plus CHAPTER
Creating data directory for chapter .
S-PLUS chapter ORF405 initialized.
```

The purpose of this command is to customize the directory for future use. Among other things, it creates a subdirectory .Data, and all the objects created by S-Plus when you run it in the directory ORF405 will be saved (automatically) in this subdirectory .Data and all the objects present in this directory will be accessible each time you run S-Plus from ORF405. You can now start the program S-Plus.

```
$ S-Plus
S-PLUS : Copyright (c) 1988, 2002 Insightful Corp.
S : Copyright Lucent Technologies, Inc.
Version 6.1.2 Release 2 for Linux 2.2.12 : 2002
Working data will be in .Data
>
```

The four lines of text reproduced above are a sign that S-Plus started without problem. The next thing we do is to open a graphic window in which we shall display the graphic objects produced by S-Plus.

```
> trellis.device(motif)
```

At this stage a motif window should appear on the desktop. If this is not the case, especially if you get an error message stating that a display cannot be open, one reason could be that you are running S-Plus remotely, and you did not set up the environment correctly. In such a case, you need to type the following commands needed for the X window system to behave appropriately:

```
$ setenv DISPLAY ''network name of the terminal''
```

on the remote server on which you are running S-Plus to make sure that the server knows where to send the X graphic objects and

```
xhost +
```

to make sure that your terminal will accept the X commands to display the graphics.

Important Remark

Most every work with S can be described independently of the platform if one limits ourselves to the list of S-commands used to perform whatever tasks one is interested in. We will adhere to this practice. This is not much of an issue for Unix or Linux users who are traditionally used to doing most of the work through typing. But it is clear that doing so, we are missing on most of the *goodies* of the Graphic User Interface, also called GUI (and pronounced "gooohee"). In any case, I remain convinced that the users preferring the mouse to the keyboard will find easily, and take full advantage of, the GUI equivalents offered by S-Plus.

Despite the merits of this discourse on the virtues of platform independence, I will presumably err on the side of Windows users since after all, most of the participants in this hypothetical introductory class are working with a laptop running a version of the Windows operating system.

CREATING S OBJECTS

Typing the command

```
> X <- 1:16
```

in the Command Window, or at the S-Plus prompt under Unix or Linux, creates a vector of length 16 containing the first 16 integers in increasing order. Notice the arrow "< −" which is typed by using first the "less than" key "<" and then the minus sign "−". This combination reads "X gets . . . to avoid the possible confusion with left arrow key. It is very important in S-Plus, since it provides an assignment command: whatever is on the left of this "gets" is an S object created by S-Plus with the result of the command on the right of the "gets"sign. If an object with this name already exists, it is automatically replaced. In fact, any symbol appearing on the left of the "gets" sign, will be used as the name of an object which S-Plus creates and saves in the local .Data directory. All these saved objects can be found in the Object Explorer under Windows.

Note that the same result would be obtained with the command:

```
> X <- seq(from=1,to=16,by=1)
```

To understand what the function `seq` does, we can use the online help by typing:

```
> help(seq)
```

Now that we know what the function `seq` can do for us, we can type:

```
> help
```

and notice that, instead of getting a help file of some kind, we get something which looks more like code. This is a general rule: if one types the name of an S method or function without parentheses, the system returns the S-code of the function in question. Only if we add the parentheses, will we see the command be executed (or an error message returned telling us that we did not included the parameters expected by the function). We can experiment further by typing:

```
> ls()
```

which lists all the S objects contained in the `_Data` subdirectory, and

```
> ls
```

which merely gives the code of the `ls` method. The command:

```
> dim(X) <- c(4,4)
```

reshapes the one dimensional vector X into a 4 by 4 matrix. The command:

```
> Y <- X*X
```

creates a new S-object. It is also a 4 by 4 matrix. Its entries are obtained by multiplying X by itself *entry by entry*. The symbol * does not give the usual matrix product. The latter is obtained by sandwiching the * symbol in between % 's as in:

```
> Z <- X%*%X
```

Z is now the 4 by 4 matrix equal to the usual matrix product of X by itself. To illustrate further the difference between these products we can do:

```
> dim(X) <- c(8,2)
> Y <- X*X
> Z <- X%*%X
```

and even though Y is computed correctly as the *entry by entry* product of X by itself, the computation of Z is not performed and we get an error message. After all, one cannot multiply (in the usual sense of the product of matrices) a 8 by 2 matrix by a 8 by 2 matrix. But notice that the command:

```
>  Z <- X%*% t(X)
```

does not trigger an error message. This is because the S command t(X) computes the transpose of the matrix X, and that consequently, the dimensions of the matrices involved in this matrix product match the usual requirements of the product of matrices.

Exiting S-Plus can be done at any time with the menu item **File** ▷ **Exit**, or by typing the command:

```
> q()
```

in the command window, or even by closing the S-Plus window. There is not need to save your work if you are working under Unix or Linux. Indeed S-Plus automatically saves all the objects you created during the session. You can check that these objects have been saved by listing the objects in your .Data directory if you are working under Unix. As a consequence of this automatic save, the Unix users of S-Plus need to clean up house regularly in order to avoid using too much disk space with all the S objects saved silently by the program. Things are slightly different in the Windows environment. When you quit the program, a dialog box prompts you to decide which, among the objects created during the session, do you want to save to the disk.

Random Generation and White Noise

The commands

```
> WN <- rnorm(1024)
> tsplot(WN)
```

create a vector WN of length 1024, and produce a sequential plot of the entries of this vector. The entries of WN are realizations of independent normal random variables with the same distribution $N(0,1)$. We specified the length of the random vector but we did not specify the values of the parameters of the distribution. S-Plus uses the default values 0 and 1 for the mean and the standard deviation of the normal distribution. The function tsplot gives a sequential plot of the vector WN, by plotting its values against the variable which takes the values 1, 2, ..., 1024.

NB: A command WN <- rnorm(1024,1.2,4.0) would have created a sample of the same size 1024 of realizations of i.i.d. variates from the normal distribution with mean 1.2 and standard deviation 4, in other words the distribution $N(1.2, 16)$ if we use the standard notation used in the text. The commands:

```
> par(mfrow=c(2,1))
> tsplot(WN)
> tsplot(WN[1:64])
> par(mfrow=c(2,1))
```

divide the graphics window into two subplot areas, one on the top of the other, give a time series plot of the full WN vector on top, a plot of the time series of the first 64 entries of the vector WN, and reset the graphic window to a single plotting area. The results are shown in Figure 7.21. The top plot shows what we should expect from a

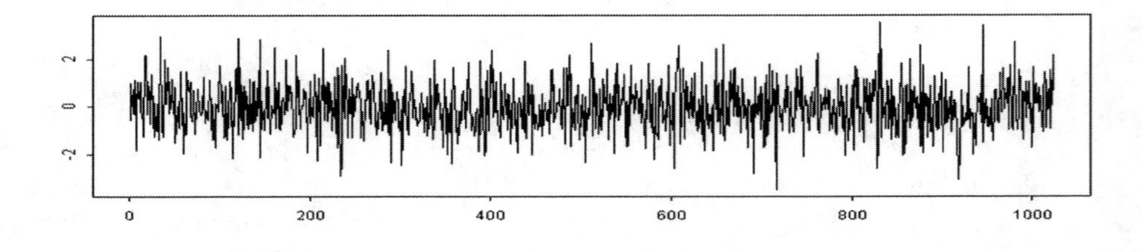

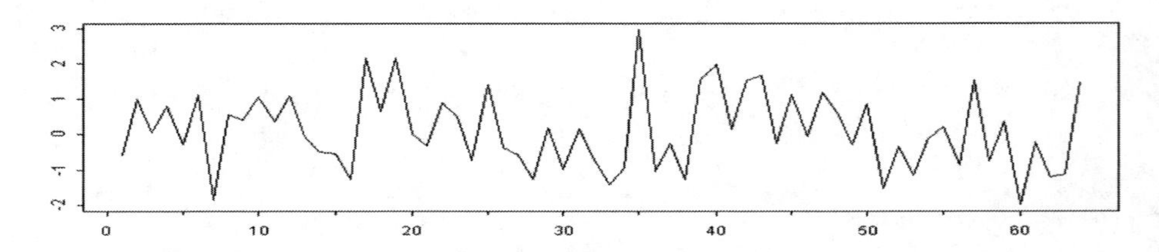

Fig. 7.21. Time series plot of the full white noise series WN (top) and of the series of its first 64 entries (bottom).

white noise: very erratic because of the lack of correlation of the successive entries of the series. Identifying a white noise is of crucial importance to a statistician. Indeed, most model fitting efforts are devoted to the identification and the extraction from the data of organized structures. This is usually done up until the remainder terms (which we call the residuals) form a white noise. From that point on, it is unreasonable to try to go any further, and statistical inference should take place at that time, and definitely before we start trying to fit the noise by mistake.

Strangely enough, the bottom plot of Figure 7.21 looks smoother. The reason is simple: it is viewed at a different scale. We plotted one sixteenth of the data on a plot with the same width! The *white noise look* which was identified earlier has practically disappeared. It is important to realize that, looking at the same data at different scales will leave different visual impressions. Don't be fooled by this artifact of the graphic settings used.

In the above commands, we did subscript the vector WN to consider only the set of its first 64 entries. To do so, we specified the range of the indices we wanted to keep. Subscripting is also conveniently used to extract sub-matrices of matrices. For example, the command X[1:3,1:3] will produce the 3 by 3 matrix in the left most corner in the top of X (recall that X was defined earlier) while X[,2] will produce a vector equal to the second column of X.

MORE FUNCTIONS AND for LOOPS

The function diff should be viewed as a discrete analog of the operation of differentiation for functions of continuous variables. We learned in calculus that the inverse operation is integration. The discrete analog is given by the function cumsum. If we apply to the white noise vector WN:

```
> WN <- rnorm(1024)
> RW <- cumsum(WN)
```

we get a numeric vector RW which represents a sample of length 1024 from a random walk.. The sequential plot of RW (i.e. the plot of the values of RW against the succesive integers) is given in the left pane of Figure 7.22. Let us now try to create an object according to the Samuelson's model for stocks. According to this model, the time evolution of a stock is represented by the exponential of a random walk (with a given volatility) with drift given by the rate of appreciation of the stock. More precisely, we define the objects SIG, MU, TIME and STOCK by:

```
> DELTAT <- 1/252
> SIG <- .2*sqrt(DELTAT)
> MU <- .15*DELTAT
> TIME <- (1:1024)/252
```

for the volatility, the rate of return, the time (expressed in years) and an initial zero vector for the stock. With these objects at hand, one should be able to fill in the entries of the vector STOCK with the commands:

```
> for (I in 1:1024) STOCK[I] <- exp(SIG*RW[I]+MU*TIME[I])
```

This command is typical of the use of the for loops in S-Plus. Unfortunately, in S-Plus like in all other interpreted languages, computations with loops are very slow, and they should be avoided whenever possible. There is a very simple way to avoid the above loop. Indeed, most numerical functions in S-Plus when applied to a vector (resp. matrix, array, ...) will create a vector (resp. matrix, array, ...) with entries given by the computations of the function on the entries of the vector (resp. matrix, array, ...) passed as argument to the function. So in the above example, the command:

```
> STOCK <- exp(SIG*RW+MU*TIME)
```

will give the same result, and the computations will be much faster. The plot of the entries of this vector is given in the right pane of Figure 7.22. The latter was produced with the commands:

```
> par(mfrow=c(2,1))
> tsplot(RW)
> title("Sample of size 1024 from a random walk")
> plot(TIME,STOCK,type="l")
> title("Corresponding geometric random walk with drift")
> par(mfrow=c(1,1))
```

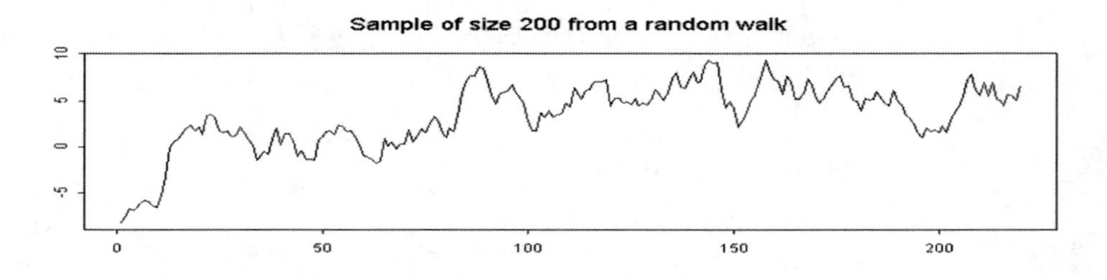

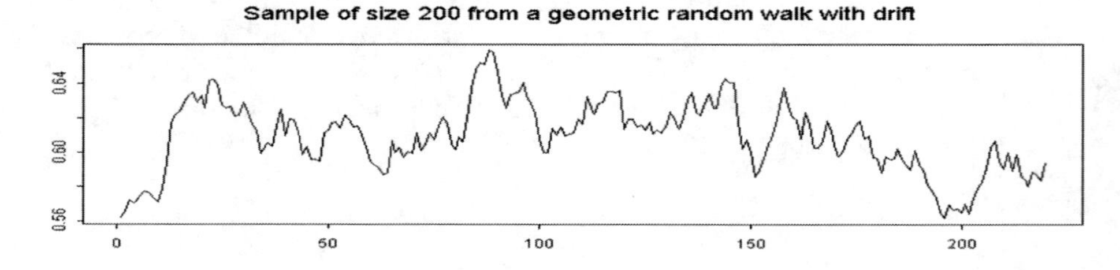

Fig. 7.22. Sequential plot of random walk sample RW (left) and of the corresponding geometric random walk with drift (right).

Notice that we use the command plot instead of tsplot. We need to pass two vectors to the command plot. The first one gives the values of the first coordinates of the points to plot while the second vector gives the second coordinates of these points. We use the option type="l" to joint the points by straight line segments to get the visual impression of a continuous curve instead of isolated points. The plots look very similar, until we notice the difference in scale on the vertical axes.

IMPORTING DATA

In most data analysis applications, the data are not created within the statistical package used for the analysis: they have to be imported. We give several examples as illustrations. These examples can be used as templates for further applications and to work out the problems (homework assignments) given in the text.

Using Data Already in S-Plus

In some instances in the text, we use data sets contained in the `S-Plus` distribution package. Let us consider for example, the data of the record times of the Scottish races. The data set is in the form of a `data frame` called `hills`, and it is part of the library `mass` which we need to link to our `S-Plus` session. See below for the way to link the library. We can choose to manipulate the variables by using their names `hills$dist`, `hills$climb` and `hills$time`, or we can decide to avoid having to type over and over the data frame name `hills`, in which case we need to "attach" the data frame to the `S-Plus` session. This is done in the following way.

```
> library(mass)
> attach(hills)
```

Doing so makes it possible to use all the variables in this frame by simply referring to their names. For example, the following commands produce the results shown in Figure 7.23.

```
> plot(climb,time)
> plot(dist,time)
```

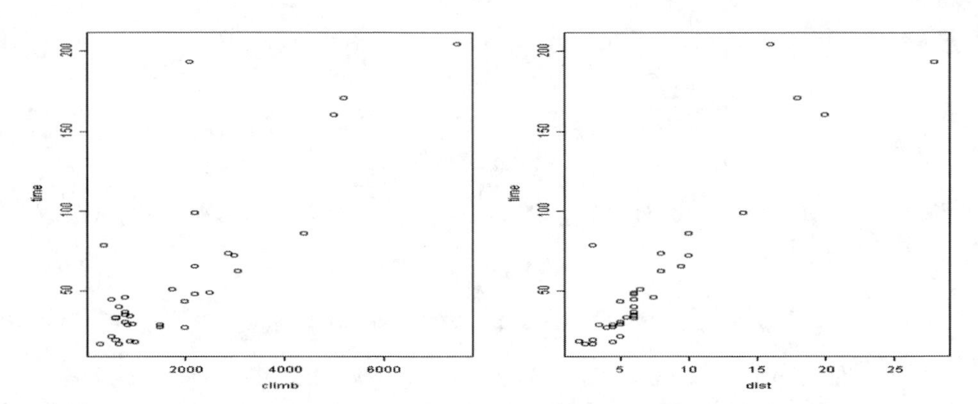

Fig. 7.23. Scatterplot of the record `time` versus `climb` (left) from the `hills` data frame, and of the `time` versus the distance variable `dist` (right).

Importing Text Files in S-Plus

As an example, we show how to import the text (ASCII) file `HOWAREYOU` which is part of the directory `C:\My Documents\ORF405\book\data` containing all the data sets used in the book. We can open the file with a text editor to check its contents. We show the first seven rows of the file for the sake of illustration.

```
0.0359678408504965
0.0635196843758394
0.0750752080020106
0.0628095887777292
0.031177878417794
-0.00520235015344319
-0.0294474884975495
```

Under Windows, we import this file into S-Plus with the dialog created by the **File** ▷ **Import Data** ▷ **From File**, and clicking on the button BROWSE to navigate to the location where the file is, and selecting it. The full pathname of the file HOWAREYOU appears, its being an ASCII file also shows, and OK'ing all that allows S-Plus to create an S object called HOWAREYOU in the current data base. The same result could be obtained with the command:

```
> HOWAREYOU <- scan("C:\\My Documents\\ORF405\\book
                                        \\data\\HOWAREYOU")
```

which could also be used under Unix if the double backslashes \\ were to be replaced by single ones \. As the result of the next command shows, it is a vector of length 5151 and its first 8 elements are given by:

```
> length(HOWAREYOU)
[1] 5151
> HOWAREYOU[1:10]
[1]  0.035968  0.063520  0.075075  0.062810
[5]  0.031178 -0.005202 -0.029447 -0.031487 -0.013135
```

We use this vector in Chapter 5 to illustrate the construction of signal series objects in S-Plus

Importing S-Plus Dumps

The previous example was concerned with plain text files. Next, we consider the case of text files produced by S-Plus, and containing information about S-objects they were issued from. It is very easy to port files from one version of S-Plus to another. For example, files created and manipulated under UNIX or LINUX can be easily read and manipulated under Windows. S objects can be dumped to a text file with the dump command (see Help file for details), and the format of these text files is such that they can be read by any other version of S-Plus irrespective of the platform in question. A text file resulting from dumping S objects should be open under Windows from the dialog created by **File** ▷ **Open**, or by double-clicking the open folder icon in the task bar. S-Plus creates a script window, and displays the contents of the text file in this window. Hitting the **F10** key, or running the script using the menu item **Script** ▷ **Run F10**, is what is needed to create the desired object. These dumps are text files and as such, they can be opened and edited with a text editor. We shall use the extension `.asc` to distinguish them from regular text files for which we try to use the extension `.txt`. Most of the data sets provided for illustration purposes or because they are needed for the problem sets, will be in the form of script files with the extension `.asc`.

Getting Data from the Web

We now show how to import data from the web into `S-Plus`. In the process, we show how starting from scratch, we can perform the analysis of the daily S&P 500 log-returns which we presented in Chapter 1.

We first retrieve the data from the internet. Let us start our favorite web browser and go to

`http://www.yahoo.com`.

Once there, we double click on the Finance special icon/button at the top of the page, or on the link Finance from the list of links right underneath the search box in the middle of the page. Once on the `Finance` page, we double click on the link US Indices under the search box. Next we double click on the Chart link of the Standard and Poor's 500 Index section. Finally, we double click on the link Daily Historical Quotes at the bottom of the page. We choose the start date January 1, 1960 and end date June 6, 2003 (now you know when this part of the book was prepared !) and we download the resulting table in spreadsheet format. We choose a location where to save the file. For the sake of definiteness, I call `filename` the full pathname of the file.

Warning. Web sites change frequently. Yahoo is no exception, so some of the detailed instructions given above may be obsolete by the time the book appear in print, or the reader tries to follow them: improvisation may be the only way out then !

The following instructions are for `Windows` only. See for example the next subsection below for instructions appropriate for the Unix/Linux platforms.

Next we switch to `S-Plus` and we open the dialog box under **File** ▷ **Import Data** ▷ **From File** and use the button **Browse** to get the pathname of the file to appear in the dialog box. Once `ASCII file - comma delimited (csv)` and `DSP500` appear in the `File format` and `Data set` spaces we click `OK`, and *voila* ! the data frame `SP500` appears on the desktop. A sequential plot of the daily close prices can be obtained with the command:

```
> tsplot(DSP500$Close)
```

Note that the dollar sign is used to extract a column from a data frame using the name of the column. The graphic window appears on the desktop and shows the graph reproduced in the left pane of Figure 7.24. We shall revisit this procedure when we study time series and we introduce the `timeSeries` objects used by `S-Plus` for

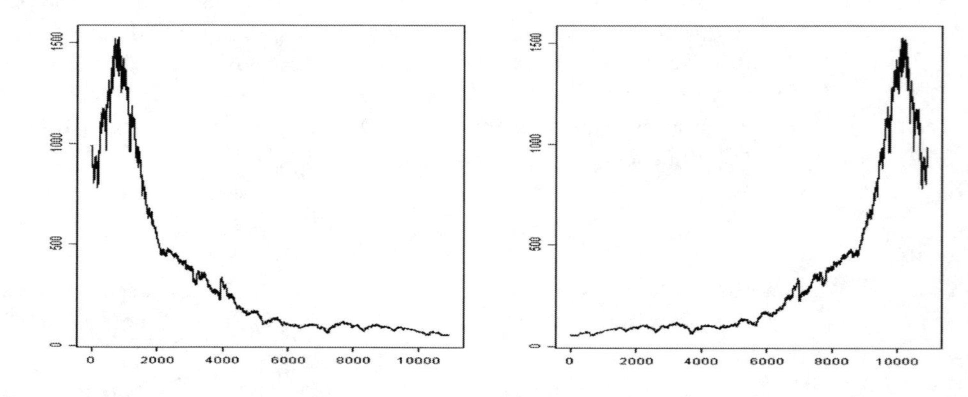

Fig. 7.24. Daily S&P 500 closing prices as imported from the web (left) and same data after we revert to the natural order (right).

their analyses. Obviously, something is wrong. The data were read in the reverse order of what would be natural. So we revert to the natural order using the function `rev` provided by `S-Plus`. We create a new `S` object of the numerical vector class:

```
> DSP <- rev(SP500$Close)
> tsplot(DSP)
```

and we compute and plot the daily log returns with the command: The result is reproduced in the right pane of Figure 7.24. Things look right now.

```
> DSPLR <- diff(log(DSP))
> tsplot(DSPLR)
> title("S&P 500 daily log-returns")
```

The logarithm of the daily close is computed with the function `log`, and the difference of the resulting vector is computed with the function `diff`. Check the help file for details on the function `diff`. Notice that we ignore the weekends (i.e. Monday's close follows the preceding Friday's close) and holidays, and more generally the days the market was close like the week of September 12, 2001 following the terrorist attack on the World Trade Center in New York.

Using the functions `mean` and `var` we compute the mean and the standard deviation of the daily log return. We use the function `sqrt` to compute the square root of the variance.

```
> mean(DSPLR)
[1] 0.0002717871
> sqrt(var(DSPLR))
[1] 0.009175474
```

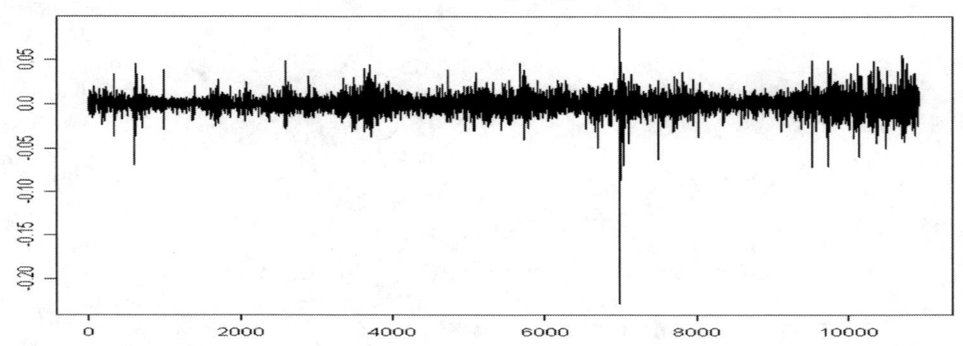

Fig. 7.25. Sequential plot of the daily S&P 500 log-returns.

Looking at the sequential plot of the daily log return (as reproduced in Figure 7.25) we notice we notice a few very large negative values. Looking at the largest of these down moves we see that:

```
> min(DSPLR)
[1]  -0.2289972
> (min(DSPLR)-mean(DSPLR))/sqrt(var(DSPLR))
[1]  -24.98716
```

which shows that this down move was essentially 25 standard deviations away from the mean daily move. So much for the normal distribution as a model for the daily moves of this index. This single day of October 1987, cannot be accounted for by a normal model for the daily log returns !

Still Another Example

Let us illustrate one more of the typical features of data sets which we have to face before we can actually begin the statistical analysis. We show how to handle the presence of headers. The text file `BondPrices61300.txt` contains information downloaded from `Data Stream` about the prices of the treasuries traded on June 6, 2000. Opening this file with a text editor shows the existence of a 12 lines header which we would like to ignore when importing the data into `S-Plus`. As illustrated before, the file can be imported using the menu or a command which can also be used under `Unix`. One can use **File** ▷ **Import Data** ▷ **from File** from the `S-Plus` menu if we are working under `Windows`, then browse the disk to locate the file as before, but this time, we should state after clicking the `Options` tab, that we want to start on row 13 only. Equivalently, one can use the command:

```
> BP61300 <- read.table('BondPrices61300.txt',skip=12,
```

```
              col.names=c('NAME','YEAR','EXPIRY','PRICE',
       'YIELD',RED.YIELD''ACCR.INT.','LIFE','VARIATION'))
```

which can also be used if we are working under `Unix`. This command creates a data frame, whose first five rows can be printed in the command window. We only reproduce the first six columns.

```
> BP61300[1:5,]
            NAME YEAR EXPIRY  PRICE YIELD RED.YIELD
 1 US.TRSY.BONDS 1997   2027  99.50  6.16     6.161
 2 US.TRSY.BONDS 1975   2005 100.03  8.25     8.241
 3 US.TRSY.BONDS 1977   2007 101.56  7.51     7.317
 4 US.TRSY.BONDS 1977   2007 102.78  7.66     7.375
 5 US.TRSY.BONDS 1978   2008 106.43  8.22     7.691
```

Notice that `S-Plus` labels the rows by the successive integers (in increasing order) by default, since we did not provide the `read.table` command with a value for the parameter `row.names`. The `skip=12` part of the command line forced `S-Plus` to ignore the first 12 lines of the ASCII file `w6.txt`. We can get the same result under `Windows` by setting appropriately the relevant options appearing in the **Options** tab of the dialog box. In particular, one would have to decide which row should the reading start from, one should have to give the name of the column variables, ... If we want to plot the yield against the remaining life of these fixed income instruments, we can use the command:

```
> plot(BondPrices61300$LIFE,BondPrices61300$YIELD,type="l")
```

The result is shown in Figure 7.26. The extension `$LIFE` added to the name of the data.frame `BondPrices61300` identifies the variable `LIFE` from this data frame. Similarly `BondPrices61300$YIELD` identify the `YIELD` variable. We used the option `type="l"` to join the points of the scatterplot by line segments, producing in this way a continuous curve, akin the graph of the yield against the time remaining to maturity. If we want to work for a while with these data without having to type the reference `BondPrices61300` to the data frame over and over, we can attach the data frame with the command:

```
> attach(BondPrices61300)
```

and work with the variables in the columns of the data frame by referring directly to their names. For example the plot of Figure 7.26 can be equivalently produced by the command:

```
> plot(LIFE,YIELD)
```

NB: It is important to keep in mind the fact that, contrary to most `Windows` programs, `S-Plus` is case sensitive. This feature is presumably inherited from the `Unix` version of `S-Plus`. Consequently, `BondPrices61300` is not the same thing as `bondprices61300` !

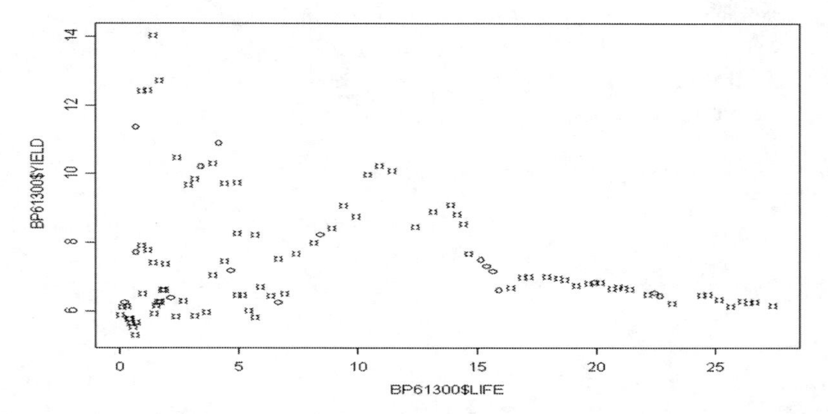

Fig. 7.26. Plot of the yield against the remaining life of the fixed income instruments contained in the text file BondPrices61300.

PROGRAMMING IN S-Plus: A FIRST FUNCTION

When we suspect that a set of S-Plus commands will be needed frequently, it is advantageous to encapsulate the individual commands in a script which we can save as a script, and which we will be able to re-use at our convenience. We can re-run individual commands or entire sets of commands by selecting them (highlighting them) and hitting the key **F10** or using the **Script** ▷ **Run F10** menu.

We can also structure the set of commands to be able to take parameters as arguments and save it as an S object which will be called a function whenever it acts on an argument. We did just that to create functions qexp and myqqnorm which we used earlier in the first chapter. We now give the details of an example which we borrow from the S-Plus manual. , we define function eda.shape defined by typing the following in the command window.

```
> eda.shape <- function(x)
+ {
+ par(mfrow = c(2,2))
+ hist(x)
+ boxplot(x)
+ iqd <- summary(x)[5] -summary(x)[2]
+ plot(density(x,width=2*iqd),xlab="x",ylab="",type="l")
+ qqnorm(x)
+ qqline(x)
+ par(mfrow=c(1,1))
+ }
```

The argument of this function needs to be a numeric vector x containing the numeric data which we want to visualize. Except for the boxplots which are hardly used in the text, all of these commands are used extensively throughout the book. We illustrate the use of this function with a sample of size 1024 which we generate from the standard normal distribution.

```
> eda.shape(rnorm(1024))
```

The result is reproduced in Figure 7.27. As expected, the histogram tries to reproduce

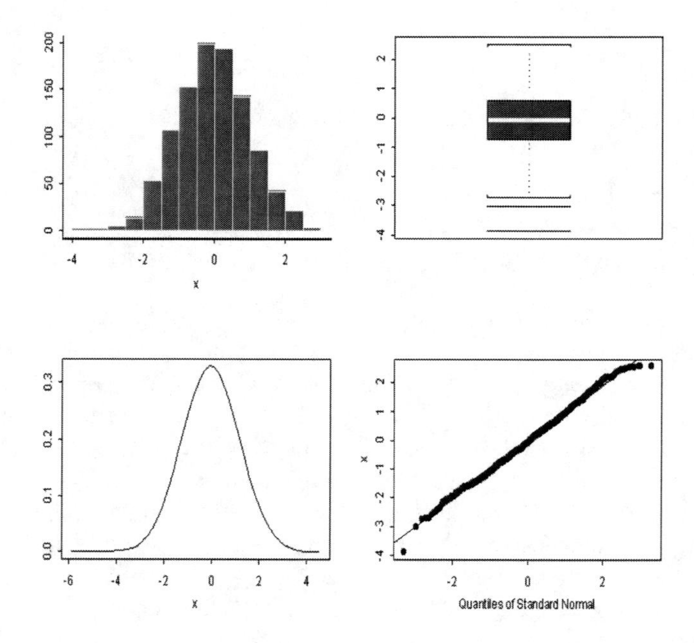

Fig. 7.27. Exploratory data analysis of a sample of size 1024 from the standard normal distribution.

the bell shape normal density, the kernel density estimate gives a faithful representation of the normal density function, and the points of the Q-Q plot are tightly packed around the diagonal. This shows that the random number generator used to create this sample did a reasonable job. At this stage, it is important to emphasize that the function eda.shape is useful for samples whose distributions are nearly normal. This is especially true for the histogram and density estimation parts of the exploratory data analysis. We will use it to check graphically, if the distribution of a sample is normal or does not depart too much from a normal distribution. Indeed, it can give strange results for distributions with drastically different features.

Notes & Complements

When I first started to use S-Plus in teaching, I used to recommend the small book of Phil Spector to newcomers to S and S-Plus, especially if they were to work under Unix. More recently, I changed my recommendation to the thicker book by Krause and Olson [55]. It is best for an introduction to the language S, and to the functionalities (especially of the Windows version) of S-Plus. Advanced users can consult the book entitled S Programming by Venables and Ripley, however the *bible* remains the green book Programming with Data by John Chambers.[23]

References

1. Y. Ait Sahalia and A.W. Lo: Nonparametric Estimation of State Price Densities Implicit in Financial Asset Prices. *The Journal of Finance* **53** (1998) 499-547.
2. S. Amari: *Differential Geometrical Methods in Statistics.* Lect. Notes in Statistics **# 28** (1985) Springer Verlag, New York, NY.
3. N. Anderson, F. Breedon, M. Deacon, A. Derry, and G. Murphy: *Estimating and Interpreting the Yield Curve.* (1996) Wiley, Chichester.
4. A. Antoniadis, J. Berruyer and R. Carmona: *Régression Non-linéaire et Applications.* (1992) Economica.
5. S.P. d'Arcy and V.G. France: Catastrophe Futures: A Better Hedge for Insurers. *J. of Risk and Insurance,* **59** (1992)
6. L. Bachelier: *Théorie de la spéculation.* Thesis, (1900) Paris.
7. Bank for International Settlements: Zero-coupon Yield Curves. Technical Report March (1999).
8. E. Banks ed.: *Weather Risk Management.* (2002) Palgrave, New York, NY.
9. T. Bollerslev: Generalized Auto Regressive Heteroskedasticity. *J. of Econometrics* **31** (1986) 307-327.
10. T. Bollerslev, R.F Engle and J.M. Wooldridge: ARCH Models. In *Handbook of Econometrics,* **IV** (1994) 2959-3038.
11. G.E.P. Box and G.M. Jenkins: *Time Series Analysis: Forecasting and Control.* Revised Edition, (1976) Holden Day, San Francisco.
12. L. Breiman, J.H. Friedman, R.A. Olshen and C.I. Stone: *Classification and Regression Trees.* Wadsworth and Brooks/Cole. (1984) Monterey, CA.
13. P.J. Brockwell and R.A. Davis (1996): *Introduction to Time Series and Forecasting.* Springer Texts in Statistics, Springer Verlag, New York, NY.
14. R.L. Brown, J. Durbin and J.M. Evans: Techniques for testing the constancy of regression relationship over time (with comments). *J. Royal Statistical Society, ser. B* **37** (1975) 149-192.
15. A. Bruce and H.Y. Gao: *Applied Wavelet Analysis with S-plus.* Springer Verlag (1996) New York, N.Y.
16. J.Y. Campbell, A.W. Lo, and A.C. MacKinlay: *The Econometrics of Financial Markets.* Princeton University Press (1997) Princeton, NJ.
17. R. Carmona and J. Morrisson (2000): EVANESCE, an `S-Plus` Library for Heavy Tail Distributions and Copulas. *Dept. of Operations Research & Financial Engineering, Princeton University.* `http://www.princeton.edu/ rcarmona`.

18. R. Carmona, C. Chen, R.D. Frostig and S. Zhong: Brain Function Imaging: a Comparative Study. *Tech. Rep.* (1996) Princeton Univ.
19. R. Carmona, W. Hwang and R.D. Frostig: Wavelet Analysis for Brain Function Imaging. *IEEE Trans. on Medical Imaging* **14** (1995) 556-564.
20. R. Carmona, W. Hwang and B. Torresanni: *Time Frequency Analysis: Continuous Wavelet and Gabor Transform, with an implementation in* `Splus` (1998) Academic Press, New York, N.Y.
21. R. Carmona: *Interest Rates Mathematical Models: from Parametric Statistics to Infinite Dimensional Stochastic Analysis.*SIAM Monographs (2003) (to appear) SIAM, Philadelphia, PA
22. R. Carmona: Stochastic Analysis for Financial Engineering. Lecture Notes, Princeton University (2001) Princeton NJ.
23. J.M. Chambers: *Programming with Data: A Guide to the S Language.* MathSoft (1998) Seattle WA
24. N.H. Chan: *Time Series: Applications to Finance.* John Wiley & Sons, (2002) New York, N.Y.
25. Chicago Board of Trade: *A User Guide: PCS Catastrophe Insurance Options.* (1995)
26. W.S. Cleveland: *The Elements of Graphing Data.* Hobart Press, (1994) Summit, N.J.
27. L. Clewlow and C. Strickland: *Energy Derivatives, Pricing and Risk management.* Lacima Publications (2000) London.
28. C. de Boor: *A Practical Guide to Splines.* Springer Verlag, (1978) New York, NY.
29. D. Drouet Mari and S. Kotz: *Correlation and Dependence.* Imperial College Press (2001) London
30. B. Dupire: Pricing with a Smile. *Risk Magazine,***7** (1994).
31. P. Embrechts, C. Kluppelberg and T. Mikosch: *Modelling Extremal Events for Insurance and Finance.* Springer Verlag (1997) New York, N.Y.
32. R.F. Engle: Autoregressive conditional heteroskedasticity with estimates of the variance of the United Kingdom inflation. *Econometrica,* **50** (1982) 987-1006.
33. R.F. Engle and T. Bollerslev: Modeling the Persistence of Conditional Variances. *Econometric Reviews,* **5** (1986) 1-50.
34. R.F. Engle and C. Granger: Cointegration and Error-Correction: Representation, Estimation and Testing. *Econometrica,* **55** (1987) 251-276.
35. E.F. Fama and R.R. Biss: The Information in Long-Maturity Forward Rates. *American Economic Rev.* **77** (1987) 680-692.
36. J. Fan and Q. Yao: *Nonlinear Time Series: Nonparametric and Parametric Methods.* Springer Verlag (2003) New York, N.Y.
37. H. Föllmer and A. Schied: *Stochastic Finance: An Introduction in Discrete Time.* (2002) De Gruyter Studies in Mathematics. Berlin.
38. J.P Fouque, G. Papanicolaou and R. Sircar: *Derivatives in Financial Markets with Stochastic Volatility.* (2000) Cambridge University Press, London.
39. R. Gençay,F. Selçuk and B. Whitcher: *An Introduction to Wavelets and Other Filtering Methods in Finance and Economics.* Academic Press (2002) New York, NY.
40. C. Gouriéroux: *ARCH Models and Financial Applications.* Springer Verlag (1997) New York, N.Y.
41. C. Gouriéroux and J. Jasiak: *Financial Econometrics.* Princeton University Press (2002) Princeton, N.J.
42. W. Haerdle: *Non-parametric Regression.* Springer Verlag (1994) New York, N.Y.
43. J.D. Hamilton: *Time Series Analysis.* Princeton University Press (1994) Princeton, NJ.

44. Hastie, Tibshirani, J. Friedman: *The Elements of Statistical Learning, Data mining, Inference and Prediction.* Springer Verlag (2001) New York, N.Y.
45. Haussdorff: Dimension und äusseres Mass. *Math. Annalen* **79** (1919) 157-179.
46. P.J. Huber: *Robust Statistics.* J. Wiley & Sons (1981) New York, NY.
47. J.C. Hull: *Options, Futures, and Other Derivatives.* Prentice Hall, 5th edition (2002) New York, NY
48. J.M. Hutchinson, A.W. Lo and T. Poggio: A Nonparametric Approach to the Pricing and Hedging of Derivative Securities via Learning Networks. *Journal of Finance,* **49** (1994)
49. S. Johansen: *Likelihood-Based Inference in Cointegrated Vector Autoregressive Models.* Oxford University Press (1995) Oxford
50. P. Jorion: *Value at Risk: The New Benchmark for Managing Financial Risk.* 2nd ed. McGraw Hill, (2000) New York, NY
51. S. Karlin and H.W. Taylor: *A First Course in Stochastic Processes.* 2nd ed. Academic Press (1975) New York, N.Y.
52. C.J. Kim and C.R. Nelson: *State-Space Models with Regime Switching.* MIT Press, (1999) Cambridge MA.
53. G. Kitagawa: Monte Carlo Filter and Smoother for Non-Gaussian Nonlinear State Space Models. *Journal of Computational and Graphical Statistics,***5**, (1996) 1-25.
54. T. Kohonen: *Self-organizing Maps.* Springer Verlag (1995) New York, NY
55. A. Krause and M. Olson: *The Basics of S and S-PLUS.* 2nd Edition. Springer Verlag (2000) New York, N.Y.
56. D. Lamberton and B. Lapeyre: *Introduction to Stochastic Calculus Applied to Finance.* (1996) CRC Press.
57. R. Litterman and J. Scheinkman: Common factors affecting bond returns. *J. of Fixed Income,* **1**, 49-53.
58. B.B. Mandelbroit: New Methods in Statistical Economics. *J. of Political Economy,* **71** (1963) 421-440.
59. B.B. Mandelbroit: The Variation of Certain Speculative Prices. *J. of Business,* **36** (1963) 394-419.
60. B.B. Mandelbroit: *Fractal, Form Dimension and Chance.* W.H. Freeman & Co. (1977) San Francisco, CA.
61. K.V. Mardia, J.T. Kent and J.M. Bibby: *Multivariate Analysis.* Academic Press, (1979) New York, NY
62. T.C. Mills: *The Econometric Modelling of Financial Time Series.* 2nd Ed. Cambridge University Press, (1999) London
63. K.R. Miltersen and E.S. Schwartz: Pricing of Options on Commidity Futures with Stochastic Term Structure of Convenience Yields and Interest Rates. *J. of Financial and Quantitative Analysis,* (2000)
64. D.C. Montgomery and E.A. Peck: *Introduction to Linear Regression Analysis.* 2nd ed. John Wiley & Sons (1992) New York, NY
65. R.B. Nelsen: *An Introduction to Copulas.* Lect. Notes in Statistics **139** (1999) Springer Verlag, New York, NY.
66. C.R. Nelson and A.F. Siegel: Parsimonious Modeling of Yield Curves. *J. of Business* **60** (1987) 473-489.
67. D. Pilipovic: *Energy Risk: Valuing and Managing Energy Derivatives.* Mc Graw Hill (1997) New York, NY
68. S.R. Pliska: *Introduction to Mathematical Finance. Discrete Time Models.* Blackwell (1997) Oxford, UK.

69. M.B. Priestley: *Nonlinear and Non-stationary Time Series Analysis.* Academic Press, (1988) New York, NY.
70. R. Rebonato: *Interest-Rate Option Models: Understanding, Analyzing and Using Models for Exotic Interest-Rate Options.* Wiley (1996) New York, NY
71. P.A. Ruud: *An Introduction to Classical Econometric Theory.* Oxford University Press, (2000) New York, NY.
72. E. Renault and N. Touzi (1996): Option Hedging and Implied Volatility in a Stochastic Volatility Model. *Mathematical Finance* **6** 215-236.
73. J.Rice: *Mathematical Statistics and Data Analysis.* 2nd ed. Duxbury Press (1995) Belmont CA
74. R. Roll: A Critique of the Asset Pricint Theory's Test, Part One: Past and Potential Testability of the Theory. *Journal of Financial Economics* **4** (1977) 129-176.
75. M. Rosenblatt: *Gaussian and Non-Gaussian Linear Time Series and Random Fields.* Springer Verlag (2000), New York N.Y.
76. P.J. Rousseeuw and A.M. Leroy: *Robust Regression and Outlier Detection.* J. Wiley & Sons (1984) New York, NY.
77. L.O. Scott: Option princing when the variance changes randomly: Theory, estimation and application.*Journal of Financial and Quantitative Analysis* **22** (1987) 419-438.
78. R.H. Shumway and D.S. Stoffer: *Time Series Analysis and Its Applications.* Springer Verlag (2000) Newy York, NY.
79. B.W. Silverman: *Density Estimation for Statistics and Data Analysis.* Chapman & Hall (1986) London.
80. P. Spector: *An Introduction to S and S-Plus.* Duxbury Press (1994) Belmont, CA.
81. J.M. Steele: *Stochastic Calculus and Financial Applications.* Springer Verlag (2000) New York NY
82. L.E.O. Svensson: Estimating and Interpreting Forward Interest Rates: Sweden 1992-94. *NBER working Paper Series #* **4871** (1994)
83. R.S. Tsay: *Analysis of Financial Time Series.* John Wiley & Sons, (2002) New York, N.Y.
84. O. Vasicek and G. Fong: Term structure estimation using exponential splines.*J. of Finance,* **38** (1982) 339-348.
85. W.N. Venables and B.D. Ripley: *Modern Applied Statistics with S-PLUS.* 2nd Edition. Springer Verlag (1997) New York, N.Y.
86. H. von Storch and F.W. Zwiers: *Statistical Analysis in Climate Research.* Cambridge University Press (1999) New York, NY.
87. M.V. Wickerhaüser: *Adapted Wavelet Analysis: from Theory to Software.* A.K.Peters LTd (1994) Wellesley, MA
88. I.H. Witten and E. Frank: *Data Mining.* Academic Press (2000) San Diego CA.
89. E. Zivot and J. Wang: *Modeling Financial Time Series with S-PLUS.* Springer Verlag (2003) New York, N.Y.

Notation Index

Data Set Index

S-Plus Index

P

Q

R

S

T

U

V

Author Index

Subject Index

D

E

Springer Texts in Statistics *(continued from page ii)*

Madansky: Prescriptions for Working Statisticians
McPherson: Applying and Interpreting Statistics: A Comprehensive Guide, Second Edition
Mueller: Basic Principles of Structural Equation Modeling: An Introduction to LISREL and EQS
Nguyen and Rogers: Fundamentals of Mathematical Statistics: Volume I: Probability for Statistics
Nguyen and Rogers: Fundamentals of Mathematical Statistics: Volume II: Statistical Inference
Noether: Introduction to Statistics: The Nonparametric Way
Nolan and Speed: Stat Labs: Mathematical Statistics Through Applications
Peters: Counting for Something: Statistical Principles and Personalities
Pfeiffer: Probability for Applications
Pitman: Probability
Rawlings, Pantula and Dickey: Applied Regression Analysis
Robert: The Bayesian Choice: From Decision-Theoretic Foundations to Computational Implementation, Second Edition
Robert and Casella: Monte Carlo Statistical Methods
Rose and Smith: Mathematical Statistics with *Mathematica*
Santner and Duffy: The Statistical Analysis of Discrete Data
Saville and Wood: Statistical Methods: The Geometric Approach
Sen and Srivastava: Regression Analysis: Theory, Methods, and Applications
Shao: Mathematical Statistics, Second Edition
Shorack: Probability for Statisticians
Shumway and Stoffer: Time Series Analysis and Its Applications
Simonoff: Analyzing Categorical Data
Terrell: Mathematical Statistics: A Unified Introduction
Timm: Applied Multivariate Analysis
Toutenburg: Statistical Analysis of Designed Experiments, Second Edition
Wasserman: All of Statistics: A Concise Course in Statistical Inference
Whittle: Probability via Expectation, Fourth Edition
Zacks: Introduction to Reliability Analysis: Probability Models and Statistical Methods

01/12/04

ALSO AVAILABLE FROM SPRINGER!

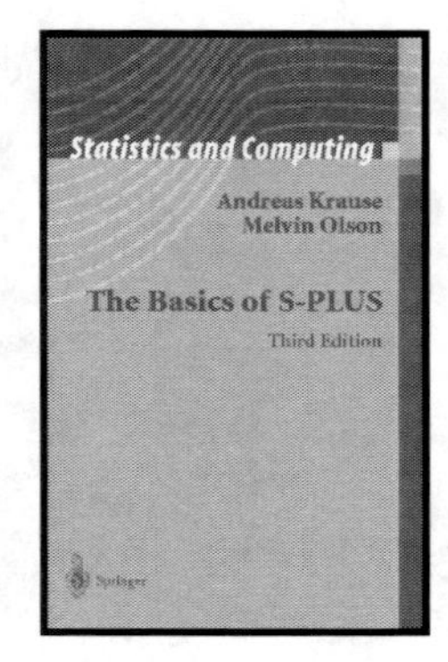

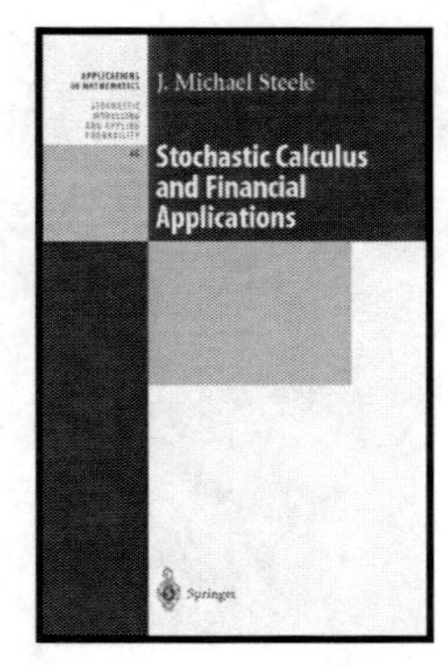

MODERN APPLIED STATISTICS WITH S

Fourth Edition

W.N. VENABLES and **B.D. RIPLEY**

This fourth edition is intended for users of S-PLUS 6.0 or R 1.5.0 or later. A substantial change from the third edition is updating for the current versions of S-PLUS and adding coverage of R. The introductory material has been rewritten to emphasis the import, export and manipulation of data. Increased computational power allows even more computer-intensive methods to be used, and methods such as GLMMs, MARS, SOM and support vector machines are considered. The authors have written several software libraries that enhance S-PLUS and R; these and all the datasets used are supplied with Windows versions of S-PLUS and all versions of R.

2002/520 PAGES./HARDCOVER/ISBN 0-387-95457-0
STATISTICS AND COMPUTING

THE BASICS OF S-PLUS

Third Edition

ANDREAS KRAUSE and **MELVIN OLSON**

This book explains the basics of S-PLUS in a clear style at a level suitable for people with little computing or statistical knowledge. The third edition is based on S-PLUS Version 6 for Windows and Unix and has been completely updated. The book serves equally well as an introduction to the R system, and concludes with a comparison of S-PLUS and R.

2002/450 PAGES/SOFTCOVER/ ISBN 0-387-954562
STATISTICS AND COMPUTING

STOCHASTIC CALCULUS AND FINANCIAL APPLICATIONS

J. MICHAEL STEELE

The Wharton School course on which the book is based begins with simple random walk and the analysis of gambling games. This material is used to motivate the theory of martingales, and, after reaching a decent level of confidence with discrete processes, the course takes up the more demanding development of continuous time stochastic process, especially Brownian motion. Stochastic processes of importance in finance and economics are developed in concert with the tools of stochastic calculus that are needed in order to solve problems of practical importance.

2000/344 PAGES/HARDCOVER/ISBN 0-387-95016-8
APPLICATIONS OF MATHEMATICS, VOL. 45

To Order or for Information:

In the Americas: **CALL:** 1-800-SPRINGER or **FAX:** (201) 348-4505 • **WRITE:** Springer-Verlag New York, Inc., Dept. S5640, PO Box 2485, Secaucus, NJ 07096-2485 • **VISIT:** Your local technical bookstore • **E-MAIL:** orders@springer-ny.com

Outside the Americas: **CALL:** +49/30/8/27 87-3 73 • +49/30/8 27 87-0 • **FAX:** +49/30 8 27 87 301 • **WRITE:** Springer-Verlag, P.O. Box 140201, D-14302 Berlin, Germany • **E-MAIL:** orders@springer.de

PROMOTION: S5640

Printed by Publishers' Graphics LLC